Lecture Notes in Networks and Systems 1740

The series "Lecture Notes in Networks and Systems" publishes the latest developments in Networks and Systems—quickly, informally and with high quality. Original research reported in proceedings and post-proceedings represents the core of LNNS.

Volumes published in LNNS embrace all aspects and subfields of, as well as new challenges in, Networks and Systems.

The series contains proceedings and edited volumes in systems and networks, spanning the areas of Cyber-Physical Systems, Autonomous Systems, Sensor Networks, Control Systems, Energy Systems, Automotive Systems, Biological Systems, Vehicular Networking and Connected Vehicles, Aerospace Systems, Automation, Manufacturing, Smart Grids, Nonlinear Systems, Power Systems, Robotics, Social Systems, Economic Systems and other. Of particular value to both the contributors and the readership are the short publication timeframe and the world-wide distribution and exposure which enable both a wide and rapid dissemination of research output.

The series covers the theory, applications, and perspectives on the state of the art and future developments relevant to systems and networks, decision making, control, complex processes and related areas, as embedded in the fields of interdisciplinary and applied sciences, engineering, computer science, physics, economics, social, and life sciences, as well as the paradigms and methodologies behind them.

Indexed by SCOPUS, EI Compendex, INSPEC, WTI Frankfurt eG, zbMATH, SCImago.

All books published in the series are submitted for consideration in Web of Science.

For proposals from Asia please contact Aninda Bose (aninda.bose@springer.com).

Mustapha Hatti
Editor

Intelligent Urban Systems: IoT-Driven Energy Management and Smart City Optimization

Volume 1

Springer

Editor
Mustapha Hatti
Solar Equipment Development Unit
Renewable Energy Development Center
Bou Ismail, Algeria

ISSN 2367-3370 ISSN 2367-3389 (electronic)
Lecture Notes in Networks and Systems
ISBN 978-3-032-12787-7 ISBN 978-3-032-12788-4 (eBook)
https://doi.org/10.1007/978-3-032-12788-4

This Springer imprint is published by the registered company Springer Nature Switzerland AG
The registered company address is: Gewerbestrasse 11, 6330 Cham, Switzerland

Contents

Editor Biographie

Dr. Mustapha Hatti

was born in El-Asnam (Chlef), Algeria. He studied at El Khaldounia school, then at El Wancharissi high school, obtained his electronics engineering diplomat from USTHB Algiers, and his post-graduation studies at USTO –Oran (Master's degree and doctorate). Worked as research engineer, at CDSE, Ain oussera, Djelfa, CRD, Sonatrach, Hassi messaoud, CRNB, Birine, Djelfa, and senior research scientist at UDES / EPST-CDER, Bou Ismail, Tipasa, ''Habilité à diriger des recherches'' HDR from Saad Dahlab University of Blida, Algeria. Actually, he is Research Director in renewable energy. Since 2013, he is an IEEE senior member, he is the author of several scientific papers, and chapter books, and his areas of interest are smart sustainable energy systems, innovative system, electrical vehicle, fuel cell, photovoltaic, optimization, intelligent embedded systems. President of the Tipasa Smart City association. An eBook editor at International Springer Publishers, guest editor and member of the editorial board of the journal Computers & Electrical Engineering, he has supervised, examined and reviewed several doctoral theses and supervised master's degrees. He organizes the conference entitled Artificial Intelligence in Renewable Energetic Systems

Deep Learning and Detection

State of Health Estimation of Lithium-Ion Batteries in Electric Vehicles

Imane Yahiaoui(✉), Meriem Touati, Mouloud Denaï, Abdellatif Seghiour, and Fouad Kaddour

École Supérieure en Génie Électrique et Énergétique d'Oran (ESGEE), Oran, Algeria
{yahiaoui_imane,touati_meriem}@edu.esgee-oran.dz,
{denai_mouloud,seghiour_abdellatif}@esgee-oran.dz

Abstract. Electric vehicles (EVs) are a key technology for sustainable transportation, relying heavily on lithium-ion batteries for energy storage. To operate these batteries efficiently, it is essential to estimate both their state of health (SOH) and state of charge (SOC). To ensure adequate safety and performance, this research focuses on estimating the SOH using data-driven techniques such as Support Vector Machine (SVM), Gaussian Process Regression (GPR), Neural Networks (NN), and Regression Trees (RT), utilizing a real dataset from Sandia National Laboratories (SNL) with various input configurations. These models were explored and compared based on their performance in SOH prediction. The results show that these data-driven techniques are effective and accurate tools for predicting the battery's state of health

Keywords: Lithium-ion battery · State of health · Electric vehicles · Machine learning · Data driven methods

1 Introduction

Lithium-ion batteries are among the most common types of batteries used in various electronic equipment and appliances, including battery electric vehicles. They are known for their reliability, extended lifespan, and faster charging compared to conventional batteries. However, like all batteries, lithium-ion batteries degrade over time, leading to a decline in performance. Factors such as temperature, charge-discharge cycling, depth of discharge (DoD), and overall operating conditions contribute to this degradation [1, 2]. Critical components like electrodes and separators deteriorate at varying rates, resulting in reduced capacity and power output [3]. For safe and efficient battery management, a battery's State of Health (SOH) must be accurately determined [4]. SOH calculates how much a battery's primary performance metrics, including capacity, internal resistance, and power delivery, have deteriorated from their initial levels [5]. It is commonly represented as a percentage, with 100% being a brand-new battery, and gives a clear indication of the battery's overall quality and remaining useful life.

In recent years, a lot of attention has been focused on the diagnosis of lithium-ion batteries state of health. As a result, researchers have explored various estimation

M. Hatti (Ed.): IC-AIRES 2025, LNNS 1740, pp. 3–13, 2026.
https://doi.org/10.1007/978-3-032-12788-4_1

methods, such as model-based approaches that simulate battery behavior [6], and data-driven methods that use machine learning to detect degradation patterns from large datasets. There are also hybrid methods that combine statistical learning with physical models [7].

In this study, the necessary data for SOH estimation was obtained using a data-driven approach, which served as the foundation for model training. Multiple machine learning and regression techniques were applied in MATLAB to perform SOH estimation, allowing for a comprehensive and comparative analysis of different methodologies.

2 Data Modeling Methodology

Cylindrical Panasonic NCR lithium-ion rechargeable cells, based on lithium nickel cobalt aluminum oxide (NCA) chemistry, are tested in this work. These cells, specifically the NCR18650B model with a nominal capacity of 3.2 Ah, were selected for their widespread use in electric vehicles and portable electronics. The cells were subjected to charge-discharge cycling between 0% and 100% State of Charge (SOC) at a controlled temperature of 25 °C, using a constant current charging rate of 0.5C and a discharging rate of 1C. Experimental data from four individual cells labeled Data a, b, c, and d was collected, providing both high-resolution in-cycle measurements and per-cycle summary metrics.

Figure 1 presents the per-cycle capacity data for four different cells, following the removal of data artifacts, The data is pre-processed for making it suitable according the considered approach and min-max normalisation is used to scale the data to the range of [0,1]. Data for three cells is used to train the models in case of neural network-based models and data for one cell was used for testing the trained models for their ability to predict the SOH at the given cycle.

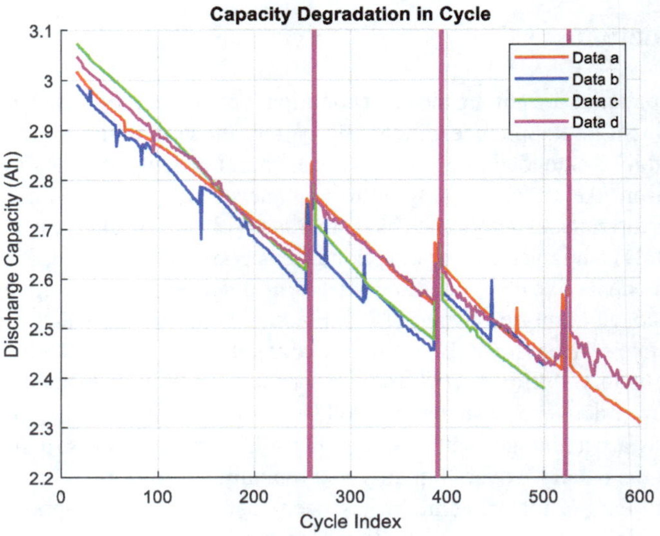

Fig. 1. Capacity degradation in cycle.

The chart compares the per-cycle discharge capacity degradation of four lithium-ion battery cells, labeled as Data a, Data b, Data c, and Data d, over a span of up to 600 cycles. The discharge capacity, measured in Ampere-hours (Ah) on the y-axis, shows a gradual downward trend across all cells from a maximum of around 3.0 to 3.1 Ah down to approximately 2.3 to 2.4 Ah. Data a (red) and Data d (purple) extend to 600 cycles, while Data b (blue) and Data c (green) end near 500 cycles. The curves follow a similar overall rate of decline but feature distinct localized variations, minor downward fluctuations, and prominent vertical capacity recovery jumps at periodic intervals most notably around cycles 260, 390, and 520, which are highlighted by vertical lines. This visualization displays the experimental baseline data before pre-processing and min-max normalization are applied for model training.

3 Results and Discussion

3.1 Regression Tree results

The SOH prediction Regression Tree models (coarse, medium and fine) was performed across three input configurations: single channel (V), multi-channel (V, I, T), and extended multi-channel (V, I, T, SOC, T_amb).

Figure 2 shows the SOH prediction over cycles using three regression tree models. The coarse tree captures the general degradation trend but misses sharp changes. The medium tree provides finer steps, better aligning with the actual SOH. The fine tree follows the real curve closely but introduces more fluctuations.

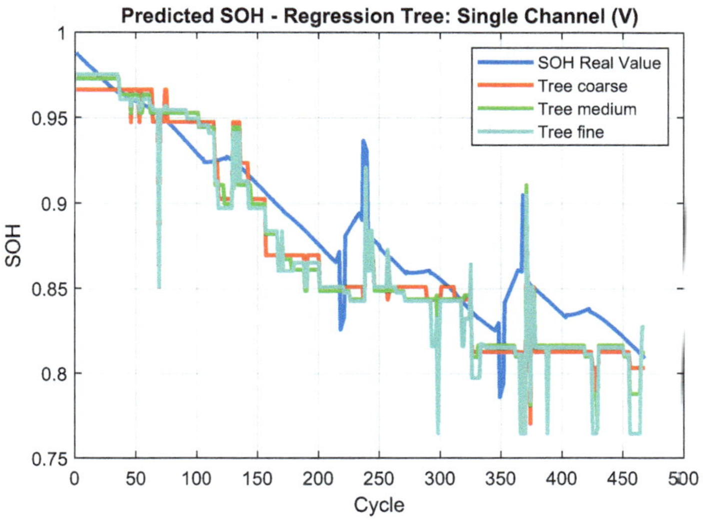

Fig. 2. Predicted SOH using regression trees with single channel (V) input.

Figure 3 presents the SOH prediction using regression tree models with multi-channel input. The fine tree closely fits the training data, though with a risk of overfitting. The medium tree generalizes the SOH trend effectively, while the coarse tree best captures the overall degradation pattern.

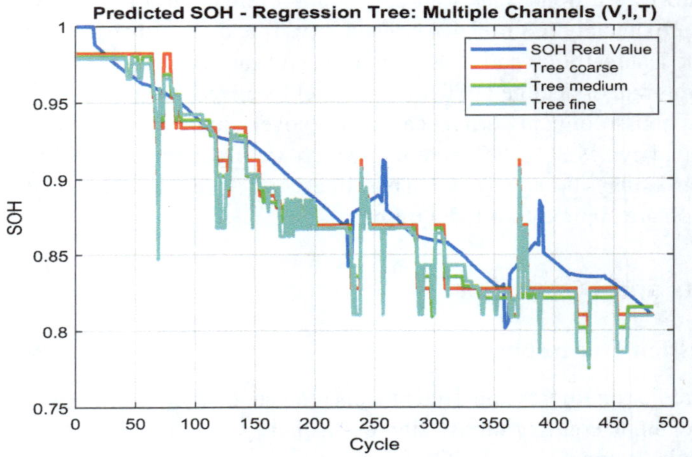

Fig. 3. Predicted SOH using regression trees with multiple channel (V, I, T) input.

Figure 4 illustrates the SOH prediction across cycles using the three regression tree variants. The coarse tree produces step-wise estimations that follow the overall trend. The medium tree aligns well with the real SOH values during early cycles but shows deviations in later stages. The fine tree presents a more detailed prediction with visible fluctuations in the second half of the cycle life.

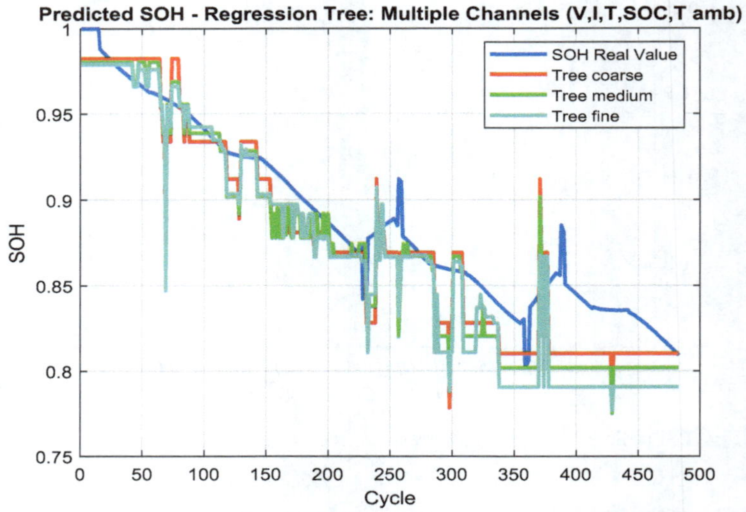

Fig. 4. Predicted SOH using regression trees with multiple channel (V, I, T, SOC, Tamb) input.

Table 1 presents the performance metrics (RMSE, MAE, MAPE) for each RT variant and input configuration. Among all models, the combination of coarse tree with three input channels (V, I, T) achieves the lowest error values across the metrics.

Table 1. Performance metrics of regression tree models for SOH estimation using different input configurations

Input Features	Tree Variant	RMSE	MAE	MAPE
Voltage (V)	Coarse	0.059392	0.043382	1.6198
	Medium	0.062249	0.044064	1.644
	Fine	0.070126	0.048865	1.8316
Voltage, Current, Temp	Coarse	0.052328	0.038660	1.4500
	Medium	0.050903	0.039590	1.4917
	Fine	0.060088	0.041646	1.5686
V, I, T, SOC, T_amb	Coarse	0.053500	0.046705	1.7685
	Medium	0.067004	0.053772	2.0532
	Fine	0.078561	0.060208	2.3030

3.2 Neural Networks Results

The SOH prediction using different neural network models (FNN, CNN, and LSTM) was performed across three input configurations: single channel (V), multi-channel (V, I, T), and extended multi-channel (V, I, T, SOC, T_amb).

In the single-channel case, Fig. 5 presents SOH predictions over the number of cycles using five NN variants. FNN models exhibit significant fluctuations, while CNN2 and LSTM models closely follow the actual SOH curve, showing improved prediction accuracy.

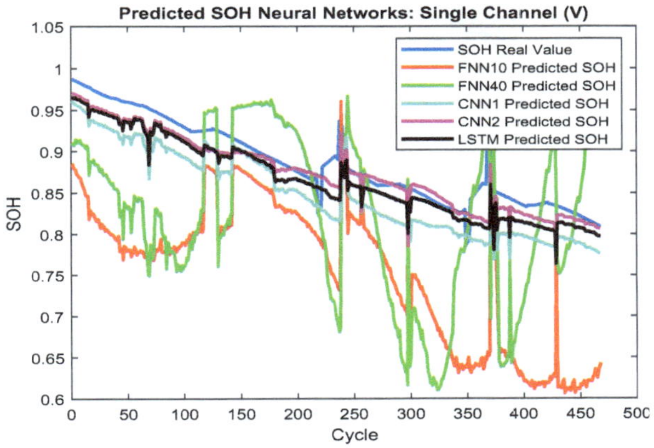

Fig. 5. Predicted SOH using neural networks with single channel (V) input.

With multi-channel (V, I, T) input, performance improved for most models. FNN outputs exhibit fewer fluctuations than in the single-channel case. CNN and LSTM models display stable SOH trends across the cycles (Fig. 6).

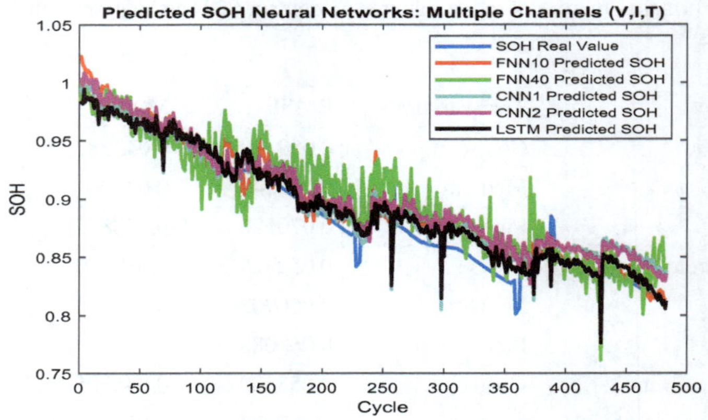

Fig. 6. Predicted SOH using neural networks with multi-channel (V, I, T) input.

In the extended input case (V, I, T, SOC, T_amb). FNN variants show lower performance compared to the previous input channels. CNN and LSTM models exhibit increased fluctuations but LSTM continued to provide the lowest RMSE, MAE, and MAPE values (Fig. 7).

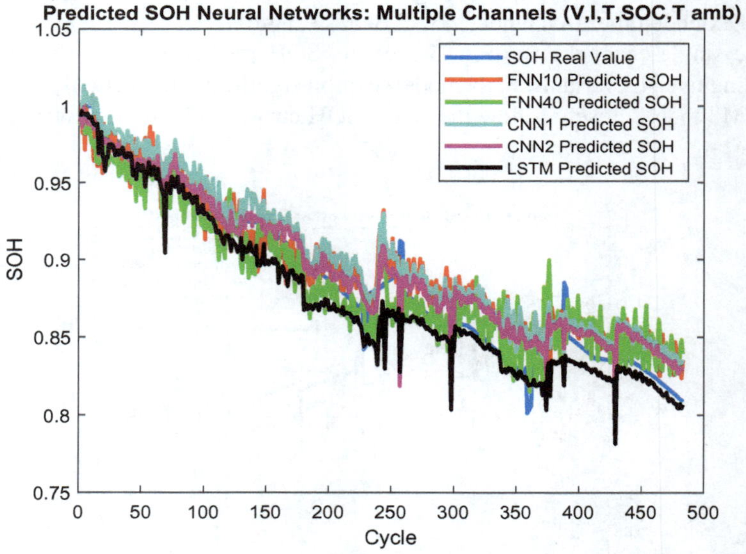

Fig. 7. Predicted SOH using neural networks with multi-channel (V, I, T, SOC, T_amb) input

Table 2 presents the SOH performance metrics for all neural network variants across the three input configurations.

Table 2. SOH performance metrics for neural network variants.

Input Type	Model	RMSE	MAE	MAPE
Single Channel (V)	FNN10	0.40332	0.35614	13.575
	FNN40	0.32864	0.27995	10.591
	CNN1	0.081855	0.068157	2.583
	CNN2	0.050818	0.036327	1.4011
	LSTM	0.04591	0.028864	1.1008
Multi-Channel (V, I, T)	FNN10	0.060883	0.052299	1.9847
	FNN40	0.087949	0.072993	2.7813
	CNN1	0.063101	0.053492	2.0711
	CNN2	0.070823	0.063317	2.431
	LSTM	0.048677	0.040506	1.5485
Multi-Channel (V, I, T, SOC, T_amb)	FNN10	0.065149	0.055373	2.1337
	FNN40	0.056273	0.043592	1.6789
	CNN1	0.071109	0.064757	2.4756
	CNN2	0.054875	0.045511	1.7635
	LSTM	0.035007	0.022901	0.86225

4 Discussion

This section analyzes the performance of the different prediction models Regression Trees (RT) and Neural Networks (NN) in estimating the State of Health (SOH) of lithium-ion batteries across three input configurations such as single channel (V), multi-channel (V, I, T) and extended multi-channel (V, I, T, SOC, Tamb). The discussion is supported by both the visual trends of the SOH plots and the quantitative performance metrics.

4.1 Regression Tree

For the single-channel input (Voltage), the Coarse Tree achieved the lowest RMSE (0.0594) and MAPE (1.6198), outperforming both the Medium and Fine trees. This indicates that simpler tree structures may generalize better when limited input features are available. In contrast, the Fine Tree, despite its complexity, yielded higher errors (RMSE = 0.0701, MAPE = 1.8316), suggesting a tendency toward overfitting and sensitivity to noise.

With the inclusion of multiple input features (Voltage, Current, Temperature), both the Coarse and Medium Trees exhibited improved performance. The Medium Tree recorded the lowest RMSE (0.0509), while the Coarse Tree slightly outperformed it in terms of MAPE (1.4500 vs. 1.4917). These results suggest that the addition of temperature and current data enhances model accuracy without requiring increased complexity. The Fine Tree continued to perform less favorably (RMSE = 0.0601, MAPE = 1.5686), reinforcing the limited benefit of fine granularity in this context.

For the extended input configuration (V, I, T, SOC, T_amb), a slight degradation in performance was observed across all tree variants. While the Coarse Tree remained the most robust (RMSE = 0.0535), all models experienced increased MAPE values. Notably, the Fine Tree exhibited the highest error (MAPE = 2.3030), likely due to the increased variability introduced by SOC and ambient temperature, which may require more advanced feature processing.

In summary, the Coarse Tree consistently provided the most stable and accurate predictions, especially with limited or moderately enriched inputs. The Medium Tree demonstrated improved performance in multi-channel scenarios. Meanwhile, the Fine Tree was generally prone to overfitting, particularly in the presence of noisy or high-dimensional input data (Fig. 8).

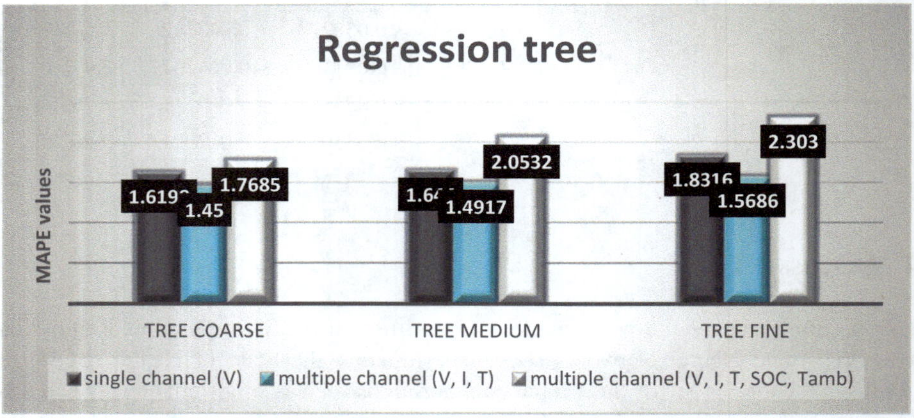

Fig. 8. Summary of performance metric MAPE of regression tree for capacity predictions.

4.2 Neural Networks

For the single channel (Voltage) input, the LSTM outperformed all other models with the lowest RMSE (0.04591) and MAPE (1.1008), confirming its suitability for time-series data like SOH. CNN2 followed closely with RMSE of (0.0508) and MAPE of (1.4011), while both FNN variants (FNN10 and FNN40) exhibited significantly higher error values (MAPE above 10%), indicating their limited ability to model temporal dependencies in low-dimensional input settings.

When using multi-channel input (Voltage, Current, Temperature), the LSTM again yielded the best performance (MAPE = 1.5485), suggesting that including current and temperature provided more meaningful patterns for temporal modeling. Interestingly, FNN10 showed a large improvement (MAPE dropped to 1.98 from 13.57), showing that even basic feedforward architectures benefit from richer input features. However, CNN-based models (CNN1 and CNN2) showed slightly increased MAPE values compared to the single-channel case, indicating possible overfitting or sensitivity to input complexity.

In the case of the extended multi-channel input (Voltage, Current, Temperature, SOC, Ambient Temperature), the LSTM model once again demonstrated superior robustness and generalization with the lowest RMSE (0.0350) and MAPE (0.8622). This reflects its capacity to utilize high-dimensional temporal data. FNN40 and CNN2 also showed improved performance over the previous input setups, with MAPE of (1.6789) and (1.7635) respectively. The improvement in FNNs compared to simpler input configurations suggests that additional features help compensate for their architectural limitations. CNN1, however, did not show significant gains and remained less stable in terms of MAPE.

In summary, the LSTM model consistently provided the most accurate SOH predictions across all input types, with the best results achieved using the extended input set. FNN models significantly benefited from added features but remained less reliable than temporal models. CNN models were moderately successful but less consistent across input types (Fig. 9).

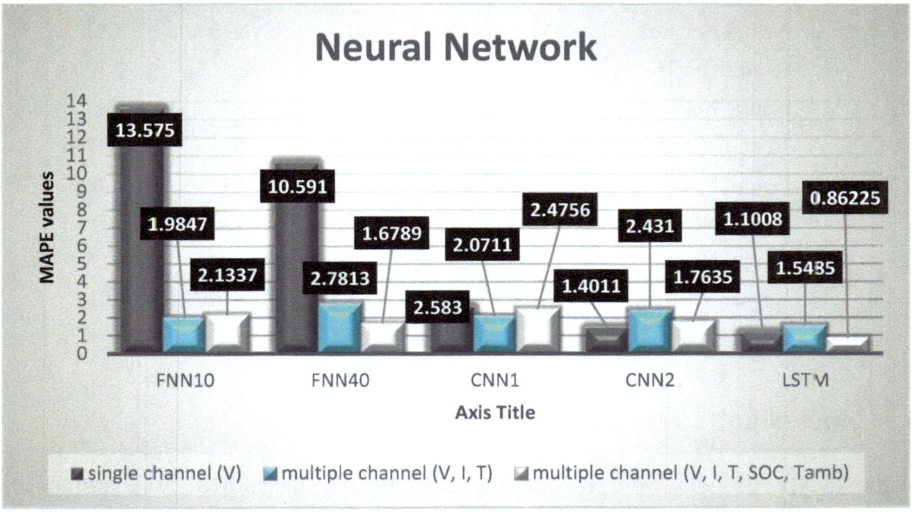

Fig. 9. Summary of performance metric MAPE of neural networks for capacity predictions.

All the models with different algorithm variants and input approaches were ranked based on their MAPE value and summarized in Table 3.

The LSTM model for multiple channel input (V, I, T, SOC, T_amb) performs the best.

Table 3. Comparative performance metrics for neural network variants based on MAPE.

Rank	Model Type	Input Features	Variant	MAPE
1	**Neural Network**	V, I, T, SOC, T_amb	**LSTM**	**0.8623**
2	Neural Network	Single Channel (V)	LSTM	1.1008
3	Neural Network	Single Channel (V)	CNN2	1.4011
4	Regression Tree	V, I, T	Coarse	1.4500
5	Regression Tree	V, I, T	Medium	1.4917
6	Neural Network	V, I, T	LSTM	1.5485
7	Regression Tree	V, I, T	Fine	1.5686
8	Regression Tree	Voltage only	Coarse	1.6198
9	Regression Tree	Voltage only	Medium	1.6440
10	Neural Network	V, I, T, SOC, T_amb	FNN40	1.6789
11	Neural Network	V, I, T, SOC, T_amb	CNN2	1.7635
12	Regression Tree	V, I, T, SOC, T_amb	Coarse	1.7685
13	Regression Tree	Voltage only	Fine	1.8316
14	Neural Network	V, I, T	FNN10	1.9847
15	Regression Tree	V, I, T, SOC, T_amb	Medium	2.0532
16	Neural Network	V, I, T	CNN1	2.0711
17	Neural Network	V, I, T, SOC, T_amb	FNN10	2.1337
18	Regression Tree	V, I, T, SOC, T_amb	Fine	2.3030
19	Neural Network	V, I, T	CNN2	2.4310
20	Neural Network	V, I, T, SOC, T_amb	CNN1	2.4756
22	Neural Network	Single Channel (V)	CNN1	2.5830
21	Neural Network	V, I, T	FNN40	2.7813
23	Neural Network	Single Channel (V)	FNN40	10.591
24	Neural Network	Single Channel (V)	FNN10	13.575

5 Conclusion

In this study, two models were proposed to predict the state of health (SOH) of lithium-ion batteries. The first approach involved artificial neural networks, where three architectures Feedforward Neural Network (FNN), Convolutional Neural Network (CNN), and Long Short-Term Memory (LSTM) were evaluated for their predictive performance. The second approach utilized regression trees as a non-neural, data-driven method.

Future works should aim to transfer learning approaches that adapt existing SOH models to different lithium-ion chemistries (NCA, NMC, LFP, etc.) or manufacturers can reduce the need for training from scratch on each new dataset. In addition, hybrid models that integrate electrochemical principles with machine learning could improve interpretability while maintaining high accuracy, offering a middle ground between physics-based and data-driven techniques.

References

1. Birkl, C.R., Roberts, M.R., McTurk, E., Bruce, P.G., Howey, D.A.: Degradation diagnostics for lithium ion cells. J. Power Sources. **341**, 373–386 (2017). https://doi.org/10.1016/j.jpowsour.2016.12.011
2. Rahman, T., Alharbi, T.: Exploring Lithium-Ion Battery Degradation: A Concise Review of Critical Factors. Data-Driven Degradation Estimation Techniques, and Sustainable Directions for Energy Storage Systems, Impacts (2024). https://www.mdpi.com/2313-0105/10/7/220
3. Lu, J., Xiong, R., Tian, J., Wang, C., Sun, F.: Deep learning to estimate lithium-ion battery state of health without additional degradation experiments. Nat. Commun. **14**(1), 2760 (2023). https://doi.org/10.1038/s41467-023-38458-w
4. Su, L., Xu, Y., Dong, Z.: State-of-health estimation of lithium-ion batteries: A comprehensive literature review from cell to pack levels, ResearchGate (2024). https://www.researchgate.net/publication/383122138_State-of-health_estimation_of_lithium_ion_batteries_A_comprehensive_literature_review_from_cell_to_pack_levels.
5. Nuroldayeva, G., Serik, Y., Adair, D.: State of Health Estimation Methods for Lithium-Ion Batteries - Nuroldayeva - 2023 - International Journal of Energy Research - Wiley Online Library. https://onlinelibrary.wiley.com/doi/10.1155/2023/4297545?
6. Yao, L., et al.: A Review of Lithium-Ion Battery State of Health Estimation and Prediction Methods (2021). https://www.mdpi.com/2032-6653/12/3/113?
7. Quan, R., Zhou, Y., Li, W., Wan, H.: A hybrid data-driven approach for SOH estimation of lithium-ion batteries based on health index extraction and QHDBO-BiTCN-BiGRU. J. Energy Storage. **125**, 116788 (2025). https://doi.org/10.1016/j.est.2025.116788

Predictive Maintenance of Wind Turbines Using Machine Learning: Addressing Fault Detection with SCADA Data

Dalila Cherifi(✉), Mohamed Rassim Ettouahria, and Mohamed Aimen Aridj

Institute of Electrical and Electronic Engineering, University of Boumerdes, Boumerdes, Algeria
{da.cherifi,m.ettouahria}@univ-boumerdes.dz

Abstract. Wind turbines play a vital role in the global shift toward renewable energy, but their intricate mechanical and electrical systems, combined with exposure to varying environmental conditions, make them susceptible to faults and unexpected downtimes. These failures can lead to costly maintenance, reduced energy output, and operational inefficiencies. This study focuses on developing predictive models to accurately identify abnormal operational states of wind turbines by leveraging real-world SCADA sensor data collected over extended periods from multiple wind farms.

We implemented and compared a comprehensive suite of supervised machine learning algorithms, including ensemble methods like XGBoost, Random Forest, and CatBoost, as well as traditional classifiers, to predict turbine faults effectively. Recognizing the challenge posed by highly imbalanced data typical in industrial monitoring, we applied advanced resampling techniques such as Synthetic Minority Over-sampling Technique (SMOTE) and random under sampling to balance the datasets and enhance model learning. Models were rigorously evaluated using a variety of performance metrics, including accuracy, F1 score, recall, and precision, with particular emphasis on the early detection of anomalies to enable timely maintenance interventions.

The experimental results indicate that machine learning can serve as a powerful and scalable tool for fault prediction in wind turbines, enabling proactive maintenance that minimizes unplanned downtime and reduces operational costs. Among the tested models, ensemble methods, especially XGBoost, demonstrated superior predictive accuracy and robustness. Moreover, the study highlights the importance of appropriate data pre- processing, feature selection, and balancing strategies to maximize model effectiveness. By integrating these predictive models into wind farm management systems, operators can significantly improve turbine reliability and contribute to the overall sustainability and economic viability of wind energy infrastructure. This research underscores the transformative potential of artificial intelligence in optimizing renewable energy operations and sets the stage for further advancements in predictive maintenance technologies.

Keywords: Wind turbine · Predictive maintenance · SCADA data · Machine learning · Fault detection · Class imbalance · XGBoost · Random forest · Resampling techniques

M. Hatti (Ed.): IC-AIRES 2025, LNNS 1740, pp. 14–24, 2026.
https://doi.org/10.1007/978-3-032-12788-4_2

1 Introduction

The increasing global demand for clean and sustainable energy has driven significant investment and research into renewable energy technologies. Among these, wind energy stands out as one of the most mature and rapidly growing sources of electricity generation. Ensuring the reliability, efficiency, and availability of wind turbines is therefore essential for maximizing the output and reducing maintenance costs in wind farms.

However, wind turbines are complex electromechanical systems operating in harsh and dynamic environments, making them susceptible to operational faults and unexpected failures. Traditional maintenance strategies, often reactive or scheduled at fixed intervals, may lead to unnecessary downtime or delayed fault detection. In this context, predictive maintenance, powered by data-driven models, has emerged as a transformative solution to improve operational planning and minimize unexpected disruptions.

This work explores the use of machine learning techniques to develop predictive models capable of identifying wind turbine operational states using sensor data. By analyzing historical SCADA (Supervisory Control and Data Acquisition) data, we aim to detect patterns associated with different turbine conditions and forecast potential anomalies before they escalate into failures. The primary motivation of this work is to tackle the challenges associated with monitoring and early fault detection in wind turbines through machine learning techniques. Predictive modeling offers a powerful means of identifying patterns and anomalies in sensor data that indicate potential operational issues. Early detection of such deviations can greatly improve maintenance scheduling, minimize unexpected failures, and enhance overall energy production efficiency.

One of the major challenges faced in this study is the class imbalance in the dataset, where abnormal operational states such as faults or derated performance are significantly underrepresented compared to normal states. This imbalance makes it difficult for standard classification models to effectively learn from minority class instances. To overcome this, the study incorporates data resampling techniques, including both oversampling and under sampling, to boost model sensitivity and resilience.

The dataset utilized in this work consists of SCADA time-series data collected from three distinct wind farms, each comprising multiple turbines. Every data entry includes timestamped sensor readings alongside labels that indicate the operational state of the corresponding turbine. The task is formulated as a supervised binary classification problem, aiming to differentiate between normal and abnormal turbine behavior. The primary objectives of thisstudy is to develop a reliable machine learning model capable of predicting abnormal turbine conditions based on sensor inputs, conduct a comprehensive analysis and preprocessing of high-dimensional SCADA data, and address the issue of class imbalance using effective resampling techniques. Additionally, the work involves evaluating various machine learning algorithms by comparing their performance through metrics such as accuracy, F1-score, and cross-validation, as well as identifying and interpreting the most influential features contributing to turbine state prediction. Through the achievement of these goals, the study aims to support more intelligent and proactive wind farm management, facilitating timely maintenance actions and enhancing the efficiency and safety of wind energy systems. The article is structured into three sections: section one offers a general overview of renewable energy with a focus on wind power and reviews fundamental machine learning concepts relevant to predictive maintenance. section two outlines the methodology employed, including data preprocessing, exploratory analysis, feature selection, and model development strategies. Section three presents the experimental results, compares model performances, and provides a detailed analysis of the findings, culminating in key conclusions and insights derived from the study.

2 Wind Turbine Systems and Problem Definition

Wind energy plays a crucial role in the global transition toward sustainable power generation, providing a renewable alternative to fossil fuels and significantly contributing to greenhouse gas reduction [1]. However, wind turbines are complex electromechanical systems operating under harsh, dynamic environmental conditions, including fluctuating wind speeds, temperature extremes, humidity, and mechanical stress, that increase the risk of component wear, degradation, and unexpected failures [2]. Such failures, if undetected, can result in costly downtime, reduced energy output, and expensive repairs [3].

To ensure reliable operation, modern wind farms implement sophisticated monitoring infrastructures primarily based on Supervisory Control and Data Acquisition (SCADA) systems, which continuously collect extensive operational data such as power generation metrics, rotor speed, electrical outputs, and system alarms [4]. Complementing SCADA, distributed sensor networks capture mechanical and environmental parameters including vibration signals, bearing temperatures, gearbox conditions, and wind characteristics, yielding rich datasets crucial for condition monitoring [5]. Leveraging this vast amount of data, predictive maintenance frameworks utilizing machine learning and advanced analytics have emerged as powerful tools to detect subtle early warning signs of faults [6]. These models enable early fault detection, failure prognosis, and optimized maintenance scheduling, ultimately minimizing unplanned outages and extending turbine lifespan [7]. Despite these advances, challenges persist. The high dimensionality and heterogeneity of SCADA and sensor data, coupled with the inherent class imbalance where normal operational data vastly outnumber fault instances, complicate model training and accuracy [8, 9]. Furthermore, variations in turbine models, site-specific environmental conditions, and operational practices limit model generalizability across different wind farms [10]. Additional difficulties include integrating data from diverse sources, maintaining data quality, and managing noisy or missing information [11].

As wind energy deployment expands, the importance of sophisticated fault detection and predictive maintenance strategies intensifies. SCADA systems record vast volumes of sensor data in real time, enabling the application of machine learning for early anomaly detection to reduce downtime and maintenance costs [12–15]. Classical supervised learning techniques, such as Decision Trees, Random Forests, Support Vector Machines (SVM), and K-Nearest Neighbors (KNN), have been effective for fault detection [16, 17]. More recently, gradient boosting algorithms like XGBoost [18], LightGBM [19], and CatBoost [20] have gained popularity due to their robustness, scalability, and interpretability in handling structured industrial data.

Addressing the class imbalance problem remains a key focus. Techniques such as Synthetic Minority Over-sampling Technique (SMOTE) [21] and random under sampling have been widely adopted to improve minority class representation and classifier sensitivity [22]. Advanced adaptive and hybrid resampling methods have also been explored to better maintain data distributions and enhance model performance [23]. Robust evaluation metrics beyond accuracy, such as F1 score, precision, recall, ROC-AUC, and stratified cross- validation are critical in assessing model reliability, especially given the high cost of false negatives in industrial applications [24, 25]. These metrics provide a more comprehensive understanding of model effectiveness under imbalanced conditions.

Building on this extensive body of research, our study applies a variety of supervised machine learning algorithms to real-world wind turbine SCADA data. Utilizing both original and resampled datasets, we evaluate model performance for binary classification of turbine operational states (normal vs. abnormal). Our objective is to develop predictive models that combine high accuracy with interpretability and robustness across diverse turbine operating conditions, advancing the capabilities of intelligent predictive maintenance in the wind energy sector.

3 Methodology: Implementation of Machine Learning Algorithms

In this study, supervised learning was employed to classify the operational states of wind turbines using time-series SCADA data. The original multi-class status information was simplified into a binary classification problem distinguishing between normal operation and abnormal conditions such as downtime or maintenance-related anomalies. This approach focused on fault detection while addressing challenges related to class imbalance and rare event prediction [21, 26]. A diverse range of supervised learning algorithms was implemented and evaluated, including traditional models like Logistic Regression [27], Support Vector Machine (SVM) [28], and K-Nearest Neighbors (KNN) [29], as well as advanced ensemble methods such as Random Forest [30], Gradient Boosting [31], XGBoost [32], LightGBM [19], and CatBoost [33]. Additionally, Multi-Layer Perceptron (MLP) neural networks were considered to capture potential nonlinear relationships in the data [34]. Each model was selected for its specific strengths in handling classification tasks and computational efficiency. To optimize performance, hyperparameter tuning and resampling techniques such as SMOTE [21] and random under sampling [35] were applied to mitigate the effects of class imbalance.

Model evaluation was based on comprehensive metrics including accuracy, precision, recall, F1 score, and ROC- AUC, ensuring robust assessment in the context of imbalanced data [24]. This methodology aimed to identify the most effective algorithm for predictive maintenance, ultimately improving fault detection and operational reliability in wind turbine systems.

4 Experiments and Results

This section outlines the experimental workflow, implementation steps, and results of our predictive modeling approach. Building on the previously described methodologies, we applied a structured sequence of data science techniques to prepare, analyze, and classify operational states of wind turbines using SCADA time-series data. The primary goal was to develop reliable models capable of detecting abnormal turbine behavior to facilitate predictive maintenance.

The methodology adopted in this work follows a structured sequence of experiments. It begins with data preprocessing, involving the systematic acquisition and refinement of SCADA data to ensure its suitability for analysis, including handling missing values, removing outliers, and standardizing formats. This is followed by exploratory data analysis (EDA), aimed at uncovering trends, patterns, and correlations that distinguish normal from fault conditions. Next, feature selection is carried out to identify the most informative variables, thereby enhancing predictive accuracy, improving computational efficiency, and reducing model complexity. In the model development phase, various machine learning algorithms are implemented and trained to forecast turbine performance and detect abnormal operational behavior. Finally, model evaluation is performed using multiple metrics, such as accuracy, precision, and recall, to rigorously assess performance and select the most reliable model for real-world deployment.

4.1 Experiment 1: Data Preprocessing

The dataset comprised 95 CSV files, representing 89 years of SCADA (Supervisory Control and Data Acquisition) time-series records from 36 wind turbines located in three wind farms (A, B, and C). Preprocessing procedures were applied separately to each farm's dataset to assess data quality, verify feature consistency, and examine class distribution. Missing values were addressed using linear interpolation, which estimates unknown points based on a straight-line relationship between consecutive measurements. This approach is well suited for time-series sensor data, where values generally vary smoothly over time. Zero values in sensor and power readings were analyzed, as high proportions may indicate sensor faults or inactive equipment. The percentage of zeros was computed for each feature, and those exceeding a predefined threshold were removed to ensure feature reliability. This preprocessing stage retained only relevant and trustworthy variables, thereby improving data integrity and supporting more accurate and efficient subsequent modeling.

4.2 Experiment 2: Feature Selection

To enhance analytical efficiency and reduce dimensionality, a feature selection process was implemented to retain only the most informative variables from the original dataset, which comprised numerous sensor measurements (averages, maximums, minimums, standard deviations), power readings, and metadata. All sensor measurements were normalized to ensure equal contribution to the machine learning models and to facilitate convergence. Given the diverse units and ranges of the variables, scaling to a common range was essential for consistent analysis. A correlation matrix was computed to assess relationships between numerical features and identify potential redundancy. Strongly correlated variables were flagged as candidates for removal, as high multicollinearity can bias certain machine learning models, hinder interpretability, and degrade predictive performance. The insights derived from this analysis guided the final feature set, ensuring that the retained variables maximized information content while minimizing redundancy for subsequent modeling stages.

4.3 Experiment 3: Binary Classification

In the experiment 3, the predictive modeling task was reformulated from a multiclass problem into a binary classification framework to simplify analysis and address class imbalance. The original dataset comprised six operational tatus types of wind turbines, which were grouped into two categories, normal operation and abnormal operation to enhance interpretability and focus on detecting anomalies rather than differentiating between specific fault types. The training process began with data cleaning and pre-processing, including the removal of irrelevant features such as timestamps, identifiers, and non-predictive columns. To mitigate class imbalance, the Synthetic Minority Over-sampling Technique (SMOTE) was applied to increase minority class samples, while random under sampling reduced majority class dominance. These resampling methods were implemented after stratified sampling to preserve the original class distribution. Standardization of features was performed using a standard scaler to ensure scale compatibility for algorithms sensitive to feature magnitude, such as SVM, KNN, and neural networks. A diverse set of machine learning models, including ensemble methods (XGBoost, LightGBM, CatBoost, Random Forest, Gradient Boosting), linear models (Logistic Regression), kernel-based approaches (SVM), instance-based learners (KNN), and feedforward neural networks (MLP), were trained and tested under various resampling scenarios. Performance evaluation focused on accuracy, F1 score, and classification reports, with particular attention to the balance between precision and recall, as presented in Tables 1-4. Multiple training scenarios were assessed, including original imbalanced datasets, oversampled and under sampled datasets, and manually balanced datasets, enabling a sys- thematic comparison of robustness, generalization, and fault detection sensitivity across different class distributions. This comprehensive benchmarking provided valuable insights into the interplay between resampling strategies and algorithmic performance, which is essential for reliable predictive maintenance in engineering applications.

To further enhance predictive performance, hyperparameter tuning was performed on the two top-performing models from the earlier evaluation phase, XGBoost and Random Forest selected for their superior F1 scores and consistent generalization across resampled datasets. For XGBoost, an exhaustive grid search optimization was applied, systematically evaluating all possible hyperparameter combinations within a defined grid using 5-fold cross-validation to ensure robust model selection. Following optimization, the most effective configurations of both models were evaluated on the test set.

Beyond conventional metrics such as accuracy and F1 score, confusion matrix analysis was conducted to provide deeper insight into classification behavior, particularly important in imbalanced or high-stakes scenarios. The XG- Boost classifier demonstrated strong reliability, achieving a high number of true positives (TP) and true negatives (TN), with minimal false positives (FP) and false negatives (FN). Its robustness was reflected in an F1 score of 0.9208 and accuracy of 0.9469, indicating balanced detection capabilities and low misclassification rates. The Random Forest model also showed excellent performance, with an F1 score of 0.9472 and accuracy of 0.9486, albeit with slightly

Table 1. Model performance on raw imbalanced data with SMOTE.

Model	Accuracy	F1 Score	Precision	Recall
XGBoost	0.9235	0.6139	0.70	0.54
LightGBM	0.9109	0.5097	0.66	0.41
CatBoost	0.9296	0.6162	0.79	0.51
Random Forest	0.9454	0.7197	0.84	0.63
Logistic Regression	0.6319	0.2861	0.18	0.66
SVM	0.7481	0.4241	0.28	0.83
Gradient Boosting	0.8003	0.4104	0.31	0.62
K-Nearest Neighbors	0.8269	0.5260	0.38	0.86
MLP Classifier	0.8795	0.5900	0.48	0.78

Table 2. Model performance on raw imbalanced data with under sampling.

Model	Accuracy	F1 Score	Precision	Recall
Random Forest	0.8196	0.5301	0.37	0.91
XGBoost	0.8185	0.5268	0.37	0.90
CatBoost	0.7980	0.5025	0.35	0.91
LightGBM	0.7946	0.4984	0.34	0.91
MLP Classifier	0.7770	0.4695	0.32	0.88
K-Nearest Neighbors	0.7059	0.4019	0.26	0.88
Gradient Boosting	0.6884	0.3839	0.25	0.87
SVM	0.6780	0.3741	0.24	0.86
Logistic Regression	0.6340	0.2847	0.18	0.65

higher FN counts compared to XGBoost. While both models excelled, their tendencies differed: XGBoost adopted a slightly more conservative approach to positive predictions, reducing FP at the cost of marginally more FN, whereas Random Forest favored capturing more positives, which slightly increased FN.

To further assess generalization ability, learning curves were plotted for each classifier using a 65% training/35% testing split, with standardization and stratified sampling applied for fairness across models. XGBoost and LightGBM emerged as the most robust learners, maintaining consistent performance and stable feature importance across training sizes. In contrast, simpler algorithms such as Logistic Regression and KNN displayed weaker predictive power and limited scalability to larger datasets. This combined analysis of hyperparameter tuning, confusion matrix evaluation, and learning curve assessment confirmed the suitability of XGBoost and Random Forest as the leading models for reliable predictive maintenance applications.

Table 3. Model performance on balanced data with SMOTE.

Model	Accuracy	F1 Score	Precision	Recall
XGBoost	0.9437	0.9425	0.95	0.93
LightGBM	0.9357	0.9339	0.95	0.92
CatBoost	0.9384	0.9365	0.96	0.92
Random Forest	0.9384	0.9363	0.96	0.91
Logistic Regression	0.7661	0.7682	0.75	0.78
SVM	0.8485	0.8392	0.89	0.80
Gradient Boosting	0.8642	0.8549	0.91	0.81
K-Nearest Neighbors	0.8832	0.8816	0.89	0.88
MLP Classifier	0.9109	0.9097	0.91	0.91

Table 4. Model performance on balanced data with under sampling

Model	Accuracy	F1 Score	Precision	Recall
XGBoost	0.9469	0.9459	0.96	0.94
LightGBM	0.9362	0.9346	0.95	0.92
CatBoost	0.9431	0.9414	0.96	0.92
Random Forest	0.9431	0.9414	0.96	0.92
Logistic Regression	0.7669	0.7693	0.75	0.78
SVM	0.8494	0.8399	0.89	0.80
Gradient Boosting	0.8637	0.8539	0.91	0.80
K-Nearest Neighbors	0.8813	0.8799	0.88	0.88
MLP Classifier	0.9137	0.9135	0.91	0.92

4.4 Experiment 4: Models Comparison

In this experiment, a comprehensive set of machine learning models was developed and evaluated to determine the most effective classifier for turbine failure detection. The models tested included XGBoost, LightGBM, CatBoost, Random Forest, Gradient Boosting, K-Nearest Neighbors (KNN), Support Vector Machine (SVM), Logistic Regression, and Multi-Layer Perceptron (MLP). Performance assessment was conducted using Accuracy, F1 Score, and Cross-Validation Accuracy, offering a balanced view of each model's predictive strength and ability to generalize to unseen data. These metrics were selected to capture both overall classification performance and robustness, particularly in handling the potential class imbalance inherent to turbine failure prediction. The Table 5 presents the performance outcomes for all models, including results from the test set and cross-validation experiments.

Table 5 Performance comparison of all models.

Model	Accuracy	F1 Score	CV Accuracy
XGBoost	0.9423	0.9411	0.9378 ± 0.0080
LightGBM	0.9394	0.9377	0.9378 ± 0.0080
Random Forest	0.9399	0.9377	0.9317 ± 0.0078
CatBoost	0.9352	0.9328	0.9257 ± 0.0063
Gradient Boosting	0.8751	0.8669	0.8753 ± 0.0085
KNN	0.8770	0.8750	0.8697 ± 0.0121
SVM	0.8356	0.8270	0.8397 ± 0.0068
MLPClassifier	0.9087	0.9063	–
Logistic Regression	0.7682	0.7688	0.7684 ± 0.0144

Among all the models tested, XGBoost achieved the best overall performance with the highest accuracy (0.9423) and F1 score (0.9411), along with a robust cross-validation accuracy of 0.9378. LightGBM and Random Forest also performed competitively and consistently. In conclusion, XGBoost was selected as the best-performing model for this task due to its superior accuracy, generalization capability, and interpretability.

5 Conclusion

This study explored the application of machine learning techniques to predictive maintenance in wind turbine operations, aiming to enhance reliability, reduce unplanned downtime, and improve the economic viability of renewable energy systems. By leveraging historical SCADA data, we developed and evaluated multiple predictive models to detect early signs of operational anomalies, enabling proactive maintenance decisions.

Among the tested algorithms (XGBoost, LightGBM, CatBoost, Random Forest, Gradient Boosting, KNN, SVM, Logistic Regression, and MLP), XGBoost demonstrated the highest performance, achieving an accuracy of 0.9423, an F1 score of 0.9411, and a cross-validation accuracy of 0.9378. LightGBM and Random Forest also delivered competitive results, confirming the strength of tree-based ensemble methods for this task. Beyond classification accuracy, XGBoost's ability to provide interpretable feature importance further underscores its suitability for industrial applications.

Despite these promising results, several limitations remain. The quality and completeness of SCADA data had a notable impact on model performance, and reliance on data from a single turbine limits the generalizability of the findings. Moreover, while feature selection was performed, additional feature engineering may uncover more influential variables. Finally, the evaluation focused on technical performance metrics; future research should also incorporate cost-sensitive measures to reflect the economic implications of maintenance decisions.

Overall, this work demonstrates the potential of AI driven approaches for predictive maintenance in wind energy, offering a data driven framework that can guide maintenance planning, improve operational efficiency, and support the long-term sustainability of wind power infrastructure. Future efforts should focus on improving data quality, expanding datasets to capture diverse operational contexts, and refining models to balance predictive accuracy with real-world economic impact.

References

1. Ackermann, T.: Wind Power in Power Systems. John Wiley & Sons (2005)
2. He, Z., Zi, Y., Peng, F.: A review of wind turbine condition monitoring and fault diagnosis technologies. Renew. Sust. Energ. Rev. **136**, 110418 (2021)
3. Li, Y., Shahidehpour, M., Asplund, G.: Wind turbine condition monitoring based on SCADA data analysis using machine learning algorithms. Renew. Sust. Energ. Rev. **92**, 645–656 (2018)
4. Wang, Z., Yu, D., Gao, Z.: Data-driven condition monitoring and fault diagnosis for wind turbines: a review. Renew. Sust. Energ. Rev. **113**, 109246 (2019)
5. Baraldi, P., Zio, E.: Condition monitoring of wind turbines using vibration data and machine learning techniques: a review. Renew. Sust. Energ. Rev. **118**, 109533 (2020)
6. Zhang, L., Han, Z., Zhang, H.: A review of machine learning in wind turbine condition monitoring. Renew. Energy. **147**, 1103–1115 (2020)
7. Wang, C., Ding, K., Wang, L.: Machine learning-based predictive maintenance for wind turbine fault diagnosis: a review. Renew. Sust. Energ. Rev. **128**, 109902 (2020)
8. Li, X., Qin, H., Gao, X.: Review of fault diagnosis methods for wind turbine systems. Renew. Sust. Energ. Rev. **82**, 1345–1359 (2019)
9. Heng, A., et al.: Machine learning and condition monitoring. Mech. Syst. Signal Process. **23**(7), 255–282 (2009)
10. Li, Y., Yang, J., Wang, X.: A survey on data-driven wind turbine condition monitoring and fault diagnosis. Renew. Energy. **171**, 259–275 (2021)
11. Zhou, Q., Zhang, Z., Jiang, Z.: Transfer learning for wind turbine fault diagnosis: A review. Renew. Sust. Energ. Rev. **148**, 111267 (2021)
12. Yang, B., Shi, X.: Wind turbine condition monitoring and fault diagnosis: a review. Renew. Sust. Energ. Rev. **125**, 109798 (2020)
13. Mohanty, P., Jena, S.K.: A review on machine learning techniques for wind turbine fault detection and diagnosis. J. Renew. Sustain. Energy. **13**(5), 052701 (2021)
14. Liu, Y., Yu, H.: Machine learning based predictive maintenance for wind turbines. Renew. Energy. **143**, 56–68 (2019)
15. Hodge, V., Moller, J.: Predictive maintenance of wind turbines using machine learning and SCADA data: a review. Renew. Energy. **173**, 1172–1186 (2021)
16. Wang, Y., Li, P.: Fault detection of wind turbine gearbox using support vector machine with optimized parameters. Renew. Energy. **130**, 1136–1146 (2019)
17. Al-Din, M., Al-Nafjan, A.: Fault detection and diagnosis in wind turbines: A comprehensive review. Renew. Sust. Energ. Rev. **135**, 110181 (2021)
18. Chen, T., Guestrin, C.: XGBoost: A scalable tree boosting system. In: KDD '16: Proceedings of the 22nd ACM SIGKDD International Conference on Knowledge Discovery and Data Mining, pp. 785–794 (2016).
19. Ke, G., et al.: LightGBM: a highly efficient gradient boosting decision tree. Adv. Neural Inf. Proces. Syst. **30** (2017)
20. Prokhorenkova, L., et al.: CatBoost: unbiased boosting with categorical features. Adv. Neural Inf. Proces. Syst. **31**, 6638–6648 (2018)

21. Chawla, N.V., et al.: SMOTE: synthetic minority over-sampling technique. j. Artif. Intell. Res. **16**, 321–357 (2002)
22. Kamiran, F., Calders, T.: Data preprocessing techniques for classification without discrimination. Knowl. Inf. Syst. **33**, 1–33 (2012)
23. Bunkhumpornpat, C., Sinapiromsaran, K., Lursinsap, C.: Safe-level-smote: Safe- level-synthetic minority over-sampling technique for handling the class imbalanced problem. Knowl. Inf. Syst. **33**, 539–564 (2012)
24. Saito, T., Rehmsmeier, M.: The precision-recall plot is more informative than the ROC plot when evaluating binary classifiers on imbalanced datasets. PLoS One. **10**(3), e0118432 (2015)
25. Jeni, L.A., Cohn, J.F., De La Torre, F.: Facing imbalanced data—recommendations for the use of performance metrics. Int Conf Affect Comput Intell Interact Workshops. **2013**, 245–251 (2013)
26. He, H., Garcia, E.A.: Learning from imbalanced data. IEEE Trans. Knowl. Data Eng. **21**(9), 1263–1284 (2009)
27. Nick, T.G., Campbell, K.M.: Logistic regression. Top. Biostat., 273–301 (2007)
28. Kecman, V.: Support vector machines–an introduction. In: Support vector machines: theory and applications. Springer, pp. 1–47 (2005)
29. Larose, D.T., Larose, C.D.: K-nearest neighbor algorithm. In: (2014).
30. Biau, G.: Analysis of a random forests model. J. Mach. Learn. Res. **13**, 1063–1095 (2012)
31. Li, C.: A gentle introduction to gradient boosting. In: URL: http://www.ccs.neu.edu/home/vip/teach/MLcourse/4_boosting/slides/gradient_boosting.pdf, 30 (2016)
32. Chen, T., et al.: Xgboost: extreme gradient boosting. In: R package version 0.4–2 1.4, pp. 1–4 (2015)
33. Hancock, J.T., Khoshgoftaar, T.M.: CatBoost for big data: an interdisciplinary review. J. Big Data. **7**(1), 94 (2020)
34. Kruse, R., et al.: Multi-layer perceptrons. In: Computational intelligence: a methodological introduction, pp. 53–124. Springer (2022)
35. Drummond, C., Holte, R.C.: C4.5, class imbalance, and cost sensitivity: why under-sampling beats over-sampling. In: Workshop on Learning from Imbalanced Datasets II. ICML (2003)

Photovoltaic and Wind Power Forecasting Using LSTM Networks with Adaptive Hyperparameter Tuning

T. Trif[1(✉)], B. Boussahoua[1], N. Kouba[1], and M. Djemmane[2]

[1] LSEI, Department of Electrical Engineering, University of Sciences and Technology Houari-Boumediene, Algiers, Algeria
tarif.toufik@gmail.com, bboussahoua@usthb.dz, nourelyakin.kouba@usthb.edu.dz

[2] DCRPDI Sonelgaz, Algiers, Algeria
djemmane.mohamed@sonelgaz.dz

Abstract. Accurate forecasting of wind and solar power is essential for managing modern power grids. This study presents a hybrid method combining LSTM neural networks with adaptive hyperparameter tuning to improve prediction accuracy. Using historical weather and power data, the model learns temporal patterns, while key training parameters are optimized automatically. Tests on real-world data show that this approach outperforms traditional models, offering better accuracy and reliability for integrating renewable energy into the grid.

Keywords: Photovoltaic power prediction · Wind power prediction · Long short-term memory · Hyperparameter tuning

1 Introduction

The global shift toward cleaner and more sustainable energy systems has placed increasing emphasis on the integration of renewable energy sources into contemporary power grids. Among these sources, wind and photovoltaic (PV) technologies are particularly notable due to their broad geographic accessibility and declining installation costs. Nonetheless, the inherently variable and nonlinear behavior of wind and solar power presents considerable challenges for grid reliability, operational planning, and energy dispatching. As such, achieving high-accuracy short-term forecasting of renewable generation is crucial for ensuring system stability and supporting responsive demand-side management [1].

Conventional statistical approaches like Autoregressive Integrated Moving Average (ARIMA) and linear regression have traditionally been applied to renewable energy forecasting [2]. However, these models often struggle to capture the intricate temporal dynamics and nonlinear relationships found in weather patterns and power output data. In contrast, recent advances in machine learning (ML) and deep learning (DL) have led to the adoption of more sophisticated time series forecasting techniques. In particular, Long

M. Hatti (Ed.): IC-AIRES 2025, LNNS 1740, pp. 25–34, 2026.
https://doi.org/10.1007/978-3-032-12788-4_3

Short-Term Memory (LSTM) networks a specialized form of Recurrent Neural Networks (RNNs) have proven effective in learning long-range dependencies and overcoming the vanishing gradient issue, making them well-suited for modeling sequential data [3]. These networks have shown promising results in forecasting both wind and solar power generation [4, 5].

Despite these advantages, the effectiveness of LSTM models is highly sensitive to the selection of key hyperparameters, such as network depth, learning rate, batch size, and the number of hidden units. Inappropriate parameter settings can result in poor convergence behavior, as well as issues like overfitting or underfitting, ultimately impairing forecast accuracy. Manual hyperparameter tuning is not only time-intensive but also impractical for large-scale applications. To overcome this limitation, recent research has investigated the application of automated and adaptive hyperparameter optimization methods, including Bayesian optimization, grid search, and various metaheuristic algorithms [6].

This study presents an LSTM based forecasting framework enhanced with adaptive hyperparameter tuning to improve predictive accuracy for wind and PV energy outputs. The model is trained and validated using real-world datasets comprising meteorological variables and actual power production. The key contributions of this work are threefold:

- The design of an LSTM forecasting architecture tailored for hybrid renewable energy sources;
- The implementation of an adaptive tuning mechanism to optimize the model's hyperparameters;
- A comparative performance assessment against fixed-parameter LSTM models and traditional statistical forecasting techniques.

The remainder of this paper is structured as follows: Section 2 provides an overview of related literature. Section 3 details the proposed model and methodology. Section 4 presents the results and evaluation metrics. Section 5 concludes the study and outlines potential avenues for future research.

2 Literature Review

Accurately predicting renewable energy output, particularly from variable sources like wind and solar, remains a major research focus. Traditional time series models such as ARIMA and Exponential Smoothing have been widely used due to their simplicity, but they struggle to handle the nonlinear patterns and long-term dependencies inherent in meteorological and energy data [7, 8]. To improve performance, machine learning techniques like SVR, Random Forests, and k-NN have been introduced, offering better accuracy when features are well-designed [9]. However, these models often underperform with highly dynamic and complex temporal data.

Deep learning, particularly through LSTM networks, offers a significant improvement by effectively modeling long-sequence dependencies and nonlinearity, making them well-suited for renewable energy forecasting [10]–[11]. The performance of LSTM models, however, relies heavily on the careful selection of hyperparameters such as learning rate, number of neurons, and sequence length. Manual tuning or grid search methods are often inefficient or suboptimal.

To address this, researchers have adopted adaptive optimization strategies using metaheuristics like Genetic Algorithms, PSO, and Bayesian Optimization, which automate the search for optimal configurations [12, 13]. Yet, most studies treat LSTM modeling and hyperparameter tuning separately. This work aims to integrate both components, presenting an LSTM framework with adaptive hyperparameter tuning for improving prediction of wind and solar energy output.

3 Methodology

The proposed methodology integrates Long Short-Term Memory (LSTM) neural networks with an adaptive hyperparameter tuning approach to enhance the forecasting accuracy of wind and photovoltaic (PV) power generation. The method consists of four key components: data preprocessing, LSTM model architecture, adaptive hyperparameter optimization, and training and evaluation.

3.1 Data Acquisition and Preprocessing

The forecasting model relies on historical meteorological and energy production data. The input features include solar irradiance, ambient temperature, wind speed, and past power output, while the target variables are the corresponding PV and wind power outputs. The dataset is sampled at an hourly resolution over a one-year period, ensuring sufficient temporal depth for model training and validation.

Before feeding the data into the model, several preprocessing steps are applied:

- **Normalization:** All input features are normalized using min-max scaling to the [0,1] range to accelerate convergence and stabilize training. The scaling is applied as:

$$x_{scaled} = \frac{x - x_{min}}{x_{max} - x_{min}} \tag{1}$$

where x is the original feature value, and x_{min}, x_{max} are the minimum and maximum values of the feature across the dataset.

- **Windowing:** A sliding window approach is used to structure input sequences for LSTM. Each input sample consists of a fixed-length time window (24 h) of past data, and the corresponding label is the power output at the next time step.

3.2 LSTM Model Architecture

The LSTM network is chosen due to its proven ability to capture sequential dependencies and temporal patterns in time series data [10]. The model consists of:

- An input layer accepting multivariate time sequences, including solar irradiance, ambient temperature, wind speed, and historical power output.
- One or more stacked LSTM layers with a tunable number of hidden units.

  Each LSTM cell operates based on the following equations:

Forget gate:

$$f_t = \sigma (W\_f \cdot [h_{t-1}, x_t] + b\_f) \tag{2}$$

Input gate:

$$i_t = \sigma(W\_i \cdot [h_{t-1}, x_t] + b\_i),\ \tilde{c}_t = tanh(W\_C \cdot [h_{t-1}, x_t] + b\_C) \tag{3}$$

Cell update:

$$C_t = f_t * C_{t-1} + i_t * \tilde{c}_t \tag{4}$$

Output gate:

$$o_t = \sigma(W\_o \cdot [h_{t-1}, x_t] + b\_o), h_t = o_t * tanh(C_t) \tag{5}$$

- A dropout layer to prevent overfitting.
- A fully connected dense output layer to produce the predicted power value.

In an LSTM network, the input at each time step is denoted as x_t, while h_{t-1} represents the previous hidden state. The forget gate f_t controls how much ~~of~~ the past memory C_{t-1} is retained. The input gate i_t and the candidate memory $\tilde{c}_t$ work together to update the current cell state C_t. The output gate o_t determines the final output h_t, based on the current cell state. Trainable weights and biases such as W and b for each gate guide these operations and are learned during model training.

3.3 Adaptive Hyperparameter Tuning

The performance of LSTM models is highly sensitive to hyperparameters, including: Number of layers, Number of neurons, Number of Epochs, InitialLearnRate, and define the optimization algorithm.

Instead of selecting these parameters manually, this study employs Bayesian Optimization for adaptive hyperparameter tuning. This technique constructs a surrogate model (usually Gaussian Process) of the objective function and uses acquisition functions to balance exploration and exploitation when selecting the next evaluation point [14].

The objective function for tuning is defined as the validation RMSE of the model. The optimization process iteratively evaluates different combinations of hyperparameters to minimize this objective, thus improving generalization performance.

3.4 Model Training and Validation

The dataset is divided into training (70%), validation (15%), and test (15%) sets. The training set is used for model fitting, the validation set guides the hyperparameter tuning, and the test set evaluates final performance. Early stopping based on validation loss is implemented to prevent overfitting.

Evaluation is conducted using standard error metrics:

- Root Mean Squared Error (RMSE),

$$RMSE = \sqrt{\frac{1}{N}\sum\nolimits_{i=1}^{N} (y_i - \hat{y}_i)^2} \tag{6}$$

- Mean Absolute Error (MAE),

$$MAE = \frac{1}{N}\sum\nolimits_{i=1}^{N} |y_i - \hat{y}_i| \tag{7}$$

where y_i is Actual power output, and $\hat{y}_i$ is Predicted power output

These metrics offer a comprehensive view of forecasting accuracy and model robustness.

4 Results and Discussion

Following the training and validation of both the standard and optimized LSTM models, their forecasting performance for PV and wind power was thoroughly evaluated using MATLAB. The analysis was based on actual meteorological and energy output data sourced from the PVGIS platform, covering the USTHB site in Algiers, Algeria. This dataset includes hourly measurements of variables like solar irradiance, temperature, and wind speed, along with corresponding power generation figures across different time scales. Performance was assessed using RMSE and MAE metrics, with results separately reported for solar and wind energy to reflect their distinct patterns of variability.

4.1 Photovoltaic (PV) Power Prediction

The tuning of hyperparameters is a pivotal aspect of enhancing the predictive capability of deep learning architectures. In this work, a thorough optimization procedure was employed to refine the LSTM model, leading to improved forecast accuracy and reduced prediction error in the context of renewable energy generation. The following table provides a comparative assessment of the model's performance under default settings versus its optimized configuration.

- **Default LSTM:** used typical hyperparameters (Number of layers = 3, NumHiddenUnits = 212, Number of Epochs = 250, InitialLearnRate = 0,0956, and define the optimization algorithm: rmsprop).
- **Optimized LSTM:** used adaptive hyperparameter tuning (Bayesian optimization) to find the best configuration (Number of layers = 2, NumHiddenUnits = 155, Number of Epochs = 300, InitialLearnRate = 0,0093, and define the optimization algorithm: adam) (Table 1).

The results obtained for photovoltaic (PV) forecasting demonstrate that the optimized LSTM model significantly outperformed the baseline configuration, yielding a 96% reduction in Mean Absolute Error (MAE) and an 84% decrease in Root Mean Square Error (RMSE). These improvements underscore the model's enhanced capability to generate highly accurate predictions of solar energy output.

Table 1. Forecasting performance

Model Configuration	renewable energy	RMSE	MAE
Default LSTM	PV	0.4029	1.5095
Optimized LSTM	PV	0.0615	0.0576

The following chart presents a comparative analysis of the prediction errors specifically Root Mean Square Error (RMSE) and Mean Absolute Error (MAE) for both the manually configured LSTM model and the version enhanced through Bayesian Optimization.

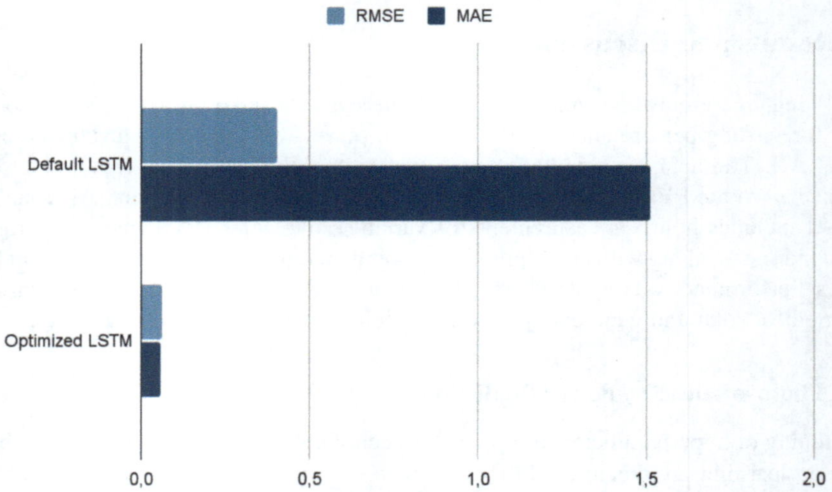

Fig. 1. Model performance comparison

Figures 2 and 3 illustrate the comparative time series plots of the predicted outputs versus the actual values for both the default and optimized LSTM models, respectively. These visualizations provide insight into the forecasting accuracy achieved before and after hyperparameter optimization.

As depicted in the figure, the forecast trajectories produced by the optimized LSTM network exhibit a strong alignment with the actual PV and wind power profiles, indicating a marked enhancement in predictive accuracy.

Figure 4 presents the training loss profile of the LSTM model. In the case of the non-optimized configuration, both training and validation loss curves tend to decline initially but often stabilize at relatively elevated levels, indicating limited learning efficiency. Conversely, when the LSTM model is equipped with properly tuned hyperparameters, the loss curves typically exhibit a steeper descent and converge toward lower, more stable values, reflecting improved model generalization and training effectiveness.

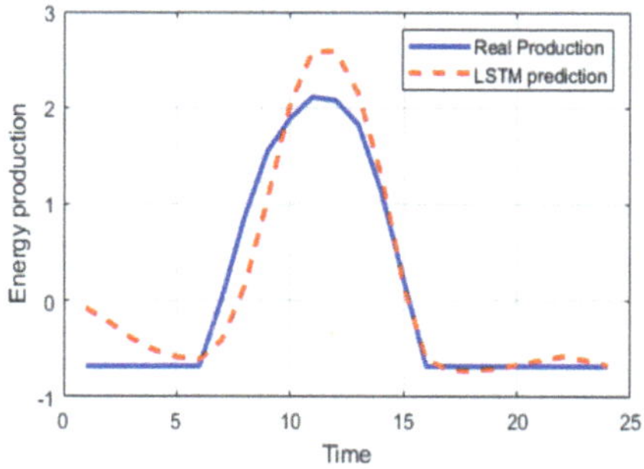

Fig. 2. LSTM prediction vs. reality (Default LSTM)

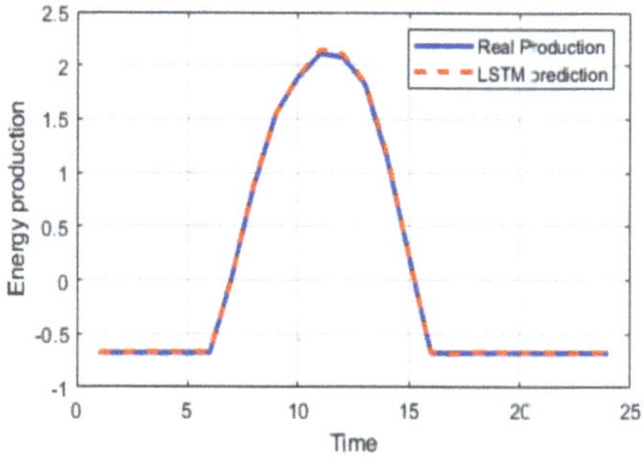

Fig. 3. LSTM prediction vs. reality (Optimized LSTM)

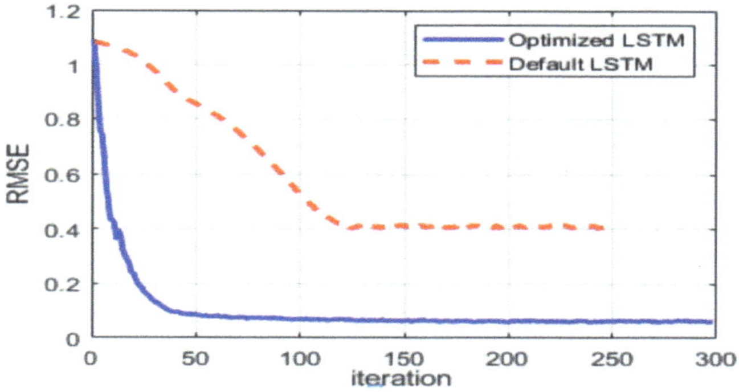

Fig. 4. Training loss based on LSTM

4.2 Wind Power Prediction

The raw data were first normalized using Min-Max scaling to the [0, 1] range to improve the stability and convergence of the training process. Both the Default LSTM and Optimized LSTM configurations were applied to the same dataset.

- **Default LSTM:** Number of layers = 3, NumHiddenUnits = 290, Number of Epochs = 291, InitialLearnRate = 0,0637, and define the optimization algorithm: adam.
- **Optimized LSTM:** Find optimal values via adaptive tuning (Number of layers = 3, NumHiddenUnits = 230, Number of Epochs = 225, InitialLearnRate = 0,0025, and define the optimization algorithm: adam).

The models were assessed on a hold-out test set using two evaluation metrics: Root Mean Square Error (RMSE) and Mean Absolute Error (MAE) (Table 2).

The optimized model significantly reduces the prediction error and increases the correlation with actual power output. The chart below highlights the differences in prediction accuracy between the standard LSTM setup and the variant improved using Bayesian Optimization, focusing on two key error metrics: RMSE and MAE.

As shown in Figs. 6 and 7, the optimized LSTM produces forecasts that more closely follow the real wind power output, especially during sharp transitions. While wind power

Table 2. Forecasting performance

Model Configuration	renewable energy	RMSE	MAE
Default LSTM	Wind	0.4018	1.2849
Optimized LSTM	Wind	0.0447	0.1439

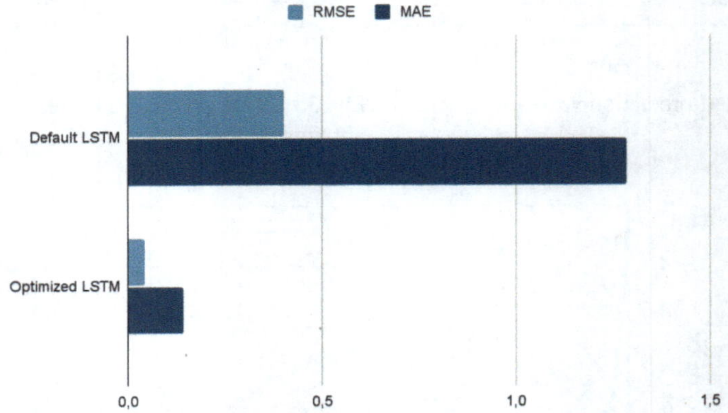

Fig. 5. Model Performance Comparison

inherently exhibits higher stochasticity compared to PV due to rapid changes in wind conditions, the model was still able to generalize well.

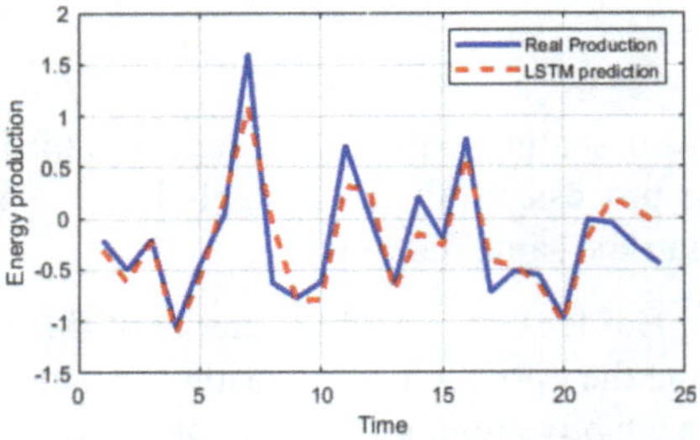

Fig. 6. LSTM prediction vs. reality (Default LSTM)

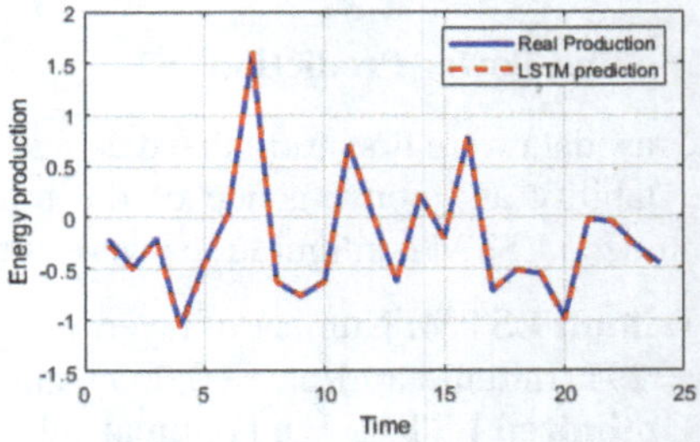

Fig. 7. LSTM prediction vs. reality (Optimized LSTM)

As shown in Fig. 8, the progression of the LSTM model's training loss reveals key differences between the two configurations. When default parameters are used, the model's performance improves slightly at first but then plateaus, indicating a struggle to learn effectively. In contrast, using optimized parameters leads to a more noticeable and consistent decrease in loss, suggesting that the model trains more efficiently and adapts better to unseen data.

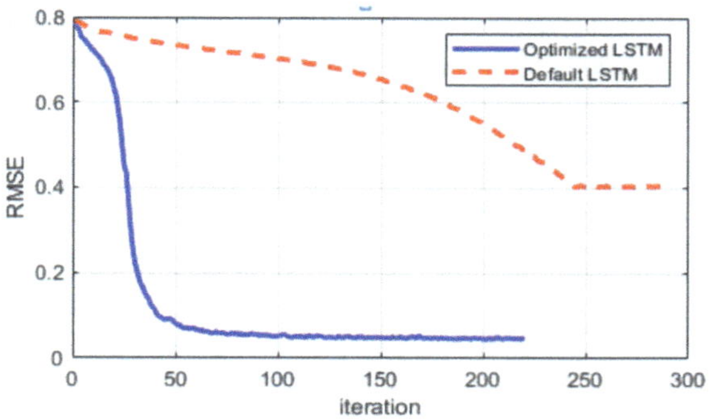

Fig. 8. Training loss based on LSTM.

5 Conclusion

This work introduced an LSTM-based model for forecasting wind and solar power, improved through adaptive hyperparameter tuning via Bayesian Optimization. The enhanced model outperforms standard LSTM in accuracy and generalization, making it suitable for real-time energy management. While the tuning process increases computational load, future efforts will focus on faster optimization techniques and incorporating richer input data to boost robustness. Overall, the approach shows strong potential for scalable and intelligent renewable energy forecasting.

References

1. Botterud, A., Wang, J., Monteiro, C., Miranda, V.: Wind power forecasting and electricity market operations. Argonne National Laboratory, Decision and Information Sciences Division, Illinois, USA (2009)
2. Wang, J., Zhou, Q., Zhang, X.: Wind power forecasting based on time series ARMA model. IOP Publ. Conf. Ser.: Earth Environ. Sci. **199**(2), 022015 (2018)
3. Hochreiter, S., Schmidhuber, J.: Long short-term memory. Neural Comput. **9**(8), 1735–1780 (1997)
4. Raza, M.Q., Khosravi, A.: A review on artificial intelligence based load demand forecasting techniques for smart grid and buildings. Renew. Sust. Energ. Rev. **50**, 1352–1372 (2015)
5. Massaoudi, M., Chihi, I., Abu-Rub, H., Refaat, S.S., Oueslati, F.S.: Convergence of photovoltaic power forecasting and deep learning: State-of-art review. IEEE Access. **9**, 136593–136615 (2021)
6. Hutter, F., Kotthoff, L., Vanschoren, J.: Automated Machine Learning: Methods, Systems, Challenges. Springer Nature (2019)
7. Sfetsos, A.: A comparison of various forecasting techniques applied to mean hourly wind speed time series. Renew. Energy. **21**(1), 23–35 (2000)
8. Chodakowska, E., et al.: ARIMA models in solar radiation forecasting in different geographic locations. Energies. **16**(13), 5029 (2023)
9. Khosravi, A., Nahavandi, S., Creighton, D.: Prediction intervals for short-term wind farm power generation forecasts. IEEE Trans. Sustain. Energy. **4**(3), 602–610 (2013)
10. Hochreiter, S., Schmidhuber, J.: Long short-term memory. Neural Comput. **9**(8), 1735–1780 (1997)

11. Husein, M., Gago, E.J., Hasan, B., Pegalajar, M.C.: Towards energy efficiency: a comprehensive review of deep learning-based photovoltaic power forecasting strategies. Heliyon. **10**, e33419 (2024)
12. Mohandes, M.A., Halawani, T.O., Rehman, S., Hussain, A.A.: Support vector machines for wind speed prediction. Renew. Energy. **29**(6), 939–947 (2004)
13. Hutter, F., Hoos, H.H., Leyton-Brown, K.: Sequential model-based optimization for general algorithm configuration. In: Learning and intelligent optimization: 5th International Conference, LION 5, Rome, Italy, January 17-21, 2011. selected papers 5. Springer, Heidelberg (2011)
14. Snoek, J., Larochelle, H., Adams, R.P.: Practical bayesian optimization of machine learning algorithms. Adv. Neural Inf. Proces. Syst. **25**, 2951–2959 (2012)

Short-Term PV Power Forecasting Using LSTM: A Case Study of Grid-Connected PV System in Adrar City

O. Harrouz[1(✉)], D. Benatiallah[2], and F. Elrobrini[3]

[1] Laboratory of Sustainable Development and Computer Science (LSDCS), Department of Material Sciences, University Ahmed Draia, 01000 Adrar, Algeria
harr.oussama@univ-adrar.edu.dz

[2] Laboratory of Sustainable Development and Computer Science (LSDCS), Department of Mathematics and Computer Science, University Ahmed Draia, 01000 Adrar, Algeria

[3] SET Laboratory Renewable Energy, Department of Renewable Energies, Saad Dahlab University, Blida 09000, Algeria

Abstract. This research aims to develop a short-term forecasting model for photovoltaic (PV) power output using real historical data collected over a two-year period. A Long Short-Term Memory (LSTM) network is utilized, and the performance of two optimization algorithms namely Adaptive Moment Estimation (ADAM) and Stochastic Gradient Descent (SGD) are evaluated independently. Furthermore, the impact of multiple input features: PV power, ambient temperature, and global solar irradiation, is systematically examined. To address these considerations, six model configurations are trained and assessed. The LSTM model optimized with the ADAM algorithm outperformed its SGD-based counterpart, achieving a correlation coefficient (R) of 97.98% and a relative Root Mean Square Error (rRMSE) of 11.16% on the testing dataset. These results underscore the performance of the ADAM-optimized LSTM model to accurately predict the short-term PV power at the Adrar station in Algeria and highlight the significance of incorporating exogenous variables in enhancing predictive performance.

Keywords: Photovoltaic power · Deep learning · LSTM · Forecasting · Optimizer

1 Introduction

Precise forecasting of PV power is necessary for the successful integration of large-scale photovoltaic systems into electrical grids [1]. It is crucial for maintaining grid stability, optimizing energy dispatch, enhancing demand-response strategies, and reducing reliance on backup generation sources [2, 3]. The intermittent and weather-dependent nature of solar energy necessitates robust predictive models capable of anticipating short-term fluctuations in PV output. In this regard, deep learning algorithms gained prominence as promising alternatives to traditional physical and statistical models [4]. They offer superior performance in capturing complex, nonlinear patterns within historical data and exhibit faster convergence during training [5, 6]

M. Hatti (Ed.): IC-AIRES 2025, LNNS 1740, pp. 35–45, 2026.
https://doi.org/10.1007/978-3-032-12788-4_4

Several recent investigations have focused on advanced machine learning and hybrid modeling techniques to bolster the precision of PV power and solar irradiance forecasting [7]. Hassan, Bailek et al. [8] introduced a genetically optimized nonlinear autoregressive neural network using exogenous features (NARX) for very-short-term photovoltaic power prediction. Their approach utilized a genetic algorithm for model training, and the forecasting performance was assessed across multiple horizons (60, 30, 15, and 5 min) and varying combinations of climatological inputs using standard evaluation metrics. In a related study, Keddouda et al. [9] designed artificial neural network (ANN) based models to forecast the PV modules generation under different environmental and operational conditions. Their methodology involved careful data preprocessing and feature selection, considering variables such as ambient and module temperatures, solar irradiation, relative humidity and wind speed.

The influence of the ANN training algorithm on predictive accuracy was also examined. Guermoui et al. [10] proposed a hybrid forecasting framework combining Least Squares Support Vector Machines with the Artificial Bee Colony optimization algorithm. This model, designed for multi-hour ahead global horizontal irradiance (GHI) forecasting, was trained on a five-year dataset (2013–2017) collected in Ghardaïa, southern Algeria. By optimizing LS-SVM hyperparameters using the ABC algorithm, the proposed model achieved notable forecasting performance, yielding a 12-h ahead RMSE of 116.22 Wh/m^2 and a correlation coefficient of 94.3%. Additionally, Harrou et al. [11] analyzed the capabilities of the LSTM networks in extracting temporal dependencies within PV-related data, highlighting their potential in modeling time-series characterized by nonlinearity and sequential variation.

The central focus of this study is to develop a short-term PV power forecasting model based on data-driven methodologies. For this purpose, real historical data spanning a two-year period are utilized. An LSTM neural network is adopted as the core predictive model, and the performance of two optimization algorithms ADAM and Stochastic Gradient Descent (SGD) is independently assessed. In addition, the influence of several input features, including PV power, ambient temperature, and global solar irradiation, is thoroughly investigated. Accordingly, six distinct model configurations are systematically trained and evaluated to identify the most effective combination of inputs and optimization strategy for accurate 15 min ahead of PV power prediction.

2 Material and Model

2.1 Geographic Area and Data Collection

The study area is situated in the southeastern region of Algeria, specifically in Adrar, positioned at a latitude of 27.88°N and a longitude of −0.27°E. Characterized by a hot and arid desert climate, which falls under the BWh category, according to the Köppen climate classification system [12], Adrar, like much of southern Algeria, exhibits significant solar energy potential. The region receives an average annual solar irradiation of approximately 2202.8 kWh/m^2 on a horizontal surface [13], experiencing over 9 h of sunshine per day on average.

Real-world data with a temporal resolution of 15 min were collected from a 20 MW grid-connected PV plant from 07 h00 to 19 h00 over a 24-month period, spanning from

January 2018 to December 2019. The dataset includes PV power generation alongside a range of meteorological variables, such as global horizontal solar radiation, ambient temperature, atmospheric pressure, wind speed, and wind direction. As an initial step, a correlation analysis was carried out to evaluate the correlation among these climatic factors and the recorded PV power output. The results, shown in Fig. 1, revealed a strong positive correlation between PV production (kW) and both solar radiation (W/m^2) and ambient temperature (°C), indicating that these two variables are the most influential for forecasting purposes and should be prioritized in subsequent modeling stages.

The two-year dataset offers a sufficiently extensive time frame for the design and assessment of the forecasting model. The prediction of 15 min ahead of PV modules power output is based on the previous eight-time steps, corresponding to the last two hours of measured PV power, ambient temperature, and global solar radiation. This implies that only the most recent observations (last 2 h) will contribute to the forecasting of the next PV value. For model validation, the data was split into two subsets: 80% for training and 20% for testing, as depicted in Fig. 2.

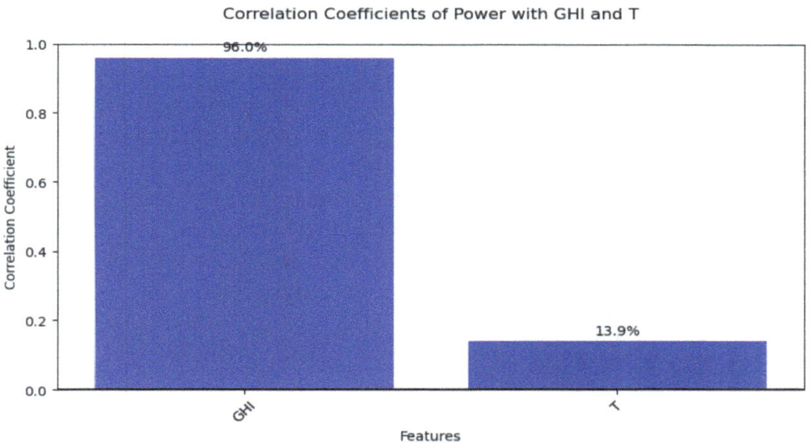

Fig. 1. Pearson correlation matrix of power with GHI and T

2.2 Long Short Term Memory Network

The LSTM model, a specialized category of RNN, is distinguished by its unique architecture, which includes weighted connections, memory retention capabilities, and feedback mechanisms. A central component of the LSTM is the memory cell (MC), which serves as persistent storage throughout the network's computations. This memory cell enables the transfer and preservation of information across the sequence, with the flow of data regulated by a series of gate mechanisms. Unlike traditional RNNs, LSTM is adept at retaining important information over long sequences, effectively addressing the vanishing gradient problem that hampers standard RNNs [13]. LSTM architecture involves

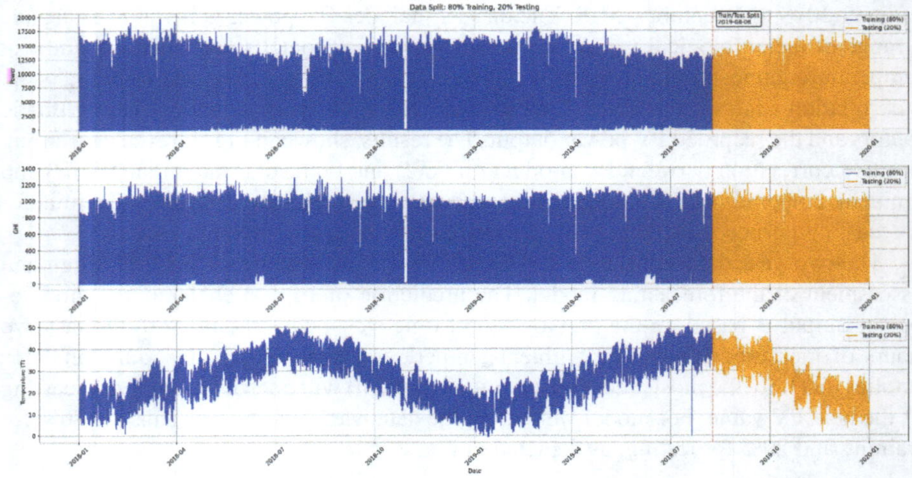

Fig. 2. Datasets split

adding three gate structures: input, output, and forget gates. Forget Gate assists in eliminating redundant information and retaining only the relevant information to proceed with prediction [14]. The input gate (i_t) controls the flow of new information and indicates if, and to what degree latest information must be incorporated in the current state cell (c_t). The output terminal (o_t) regulates the amount of information carried forward from the last time step to the forward, in conjunction with the data from the current time step. The computed metrics are in the 0 to 1 forget gate scale. When the f_t is close to 1 and the i_t is close to 0, LSTM can maintain long-term memory [15].

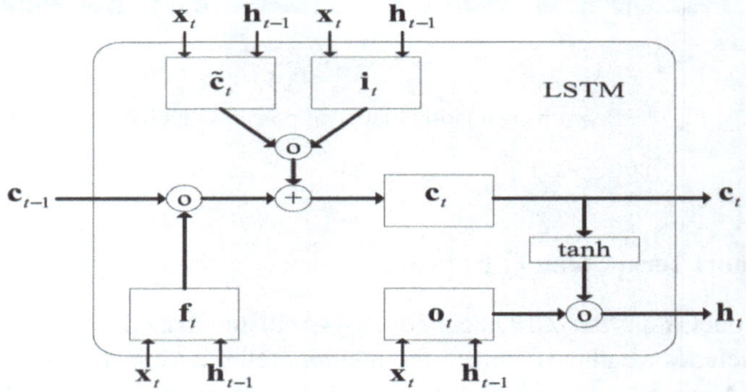

Fig. 3. LSTM Cell [16]

The mathematical formulations used in the LSTM network are the following [13, 17]

To calculate the gate units

$$i_t = \sigma(W_i x_t + U_i h_{t-1} + b_i) \quad (1)$$

$$f_t = \sigma(W_f x_t + U_f h_{t-1} + b_f) \quad (2)$$

$$o_t = \sigma(W_o x_t + U_o h_{t-1} + b_o) \quad (3)$$

To update the memory unit

$$\overline{c}_t = \tanh(W_c x_t + U_c h_{t-1}) \quad (4)$$

$$c_t = f_c * c_{t-1} + i_t * \overline{c}_{t-1} \quad (5)$$

To calculate the LSTM unit output:

$$h_t = o_t * \tanh(c_t) \quad (6)$$

The technique known as backpropagation is used to update the weights in a neural network by considering both theintended and actual output. Using the chain rule, the derivative with respect to weight w is calculated:

$$\frac{\partial \ Loss(a, y)}{\partial w} = \frac{\partial Loss(a, y)}{\partial a} \frac{\partial \ a}{\partial w} \quad (7)$$

The backpropagation algorithm (BP) functions by computing the main difference between the forecasted outputs and the desired targets, then modifying the network's weights accordingly.

The loss function is a fundamental mathematical construct utilized to quantify the disparity between the predicted outputs calculated by a model and the corresponding true values. This quantitative assessment serves as a critical criterion for evaluating model performance and guiding the optimization process during training. Among the various loss functions employed in regression tasks, the Mean Squared Error (MSE), defined by Eq. (8), is widely recognized for its effectiveness. MSE computes the mean of the squared deviations between predicted and real values, providing a robust measure of prediction accuracy. The training algorithm aims to minimize this metric, thereby improving the fidelity of the model's outputs relative to the target values.

$$MSE = \frac{1}{n} \sum_{i=1}^{n} \left(Y_i - \hat{Y}_i\right)^2 \quad (8)$$

n = number of data points, Y_i= real values (target), $\hat{Y}_i$ = predicted values (output)

In addition to the Backpropagation (BP) process, optimization algorithms are used as secondary training algorithm, each possessing distinct properties for updating the parameters θ.

1) Batch gradient descent: Commonly referred to as vanilla gradient descent, this method minimizes the loss function J(θ) which depends on the model parameters, by iteratively adjusting the parameters in the direction opposite to the gradient $\nabla_{\theta}J(\theta)$, The learning rate η controls the step size during each update, influencing the convergence speed toward a (local) minimum [18].

$$\theta_{t+1} = \theta_t - \eta.\nabla_{\theta} J(\theta) \tag{9}$$

$g_{t,i}$ the objective function gradient w.r.t the parameter θ_i at temporal step t:

$$g_{t,i} = \nabla_{\theta t} J\theta_{t,i};\, g_{t,i} = \nabla_{\theta t} J\left(\theta_{t,i}\right) \tag{10}$$

Every parameter θ_i update at each temporal step t will be expressed as [18]:

$$\theta_{t+1,i} = \theta_{t,i} - \eta \cdot g_{t,i} \tag{11}$$

2) Adam: is an optimization approach that is often used that uses past gradient information to dynamically adjusts learning rates for each parameter. Distinctively, it integrates two moment estimates: an exponentially weighted average of past gradients (m_t), reflecting the first moment, and an exponentially weighted average of past squared gradients (v_t), representing the second moment. This dual-moment strategy combines the benefits of both momentum and RMSprop, enhancing convergence stability and efficiency [18]:

$$\hat{m}_t = \frac{m_t}{1 - \beta_1^t} \tag{12}$$

$$\hat{v}_t = \frac{v_t}{1 - \beta_2^t} \tag{13}$$

Adam update rule [19]:

$$\theta_{t+1} = \theta_t - \frac{\eta}{\sqrt{\hat{v}_t} + \varepsilon} \hat{m}_t \tag{14}$$

3 Building the LSTM Networks and Optimisation Methodology

During all simulation, Backpropagation is used to train the model according to the different configuration that are listed in Table 1. The optimal approach to date is for the researcher to construct multiple models and select the most suitable one for a specific application case. In reality, there is no way to predict in advance the optimal number of cells or layers because this greatly depends on the particular model architecture, the properties of the underlying physical process, and the nature of the input data [20]. In this work, the chosen model's architecture comprises of an input layer, an output layer, two LSTM layers with 32 and 16 cells, respectively, and one Dense layer with 8 neurons. It is worth to note that all models were trained for 200 epochs, and all experiments are carried out by Python 3.12 software® 2023 with TensorFlow and Keras Libraries.

Table 1. The model's configurations

	Inputs			training algorithms	
Models	PV	GSI	T	SGD	ADAM
Model 1	×			×	
Model 2	×	×		×	
Model 3	×	×	×	×	
Model 4	×				×
Model 5	×	×			×
Model 6	×	×	×		×

4 Statistical Assessment Indices

Root Mean Square Error (RMSE), relative RMSE (rRMSE), and coefficient of correlation value (R) were among the metrics used to assess the efficacy of the suggested LSTM models [21]. The formal descriptions of these statistical parameters are given in Table 2, where N represent the total number of data points, P_i and $\hat{P}_i$ are the actual values and the output PV POWER values of the prediction, respectively.

Table 2. Mathematical formulations of the statistical indicators

Abbreviation	Ideal Value	Equation	
RMSE	Zero	$RMSE = \sqrt{\frac{\sum_{i=1}^{N}(P_i - \hat{P}_i)^2}{N}}$	(15)
rRMSE	Zero	$rRMSE = \sqrt{\frac{\frac{\sum_{i=1}^{N}(P_i - \hat{P}_i)^2}{N}}{\frac{1}{\sum_{i=1}^{N}(\hat{P}_i)^2}}}$	(16)
R	One	$R = \frac{\sum_{i=1}^{N}(P_i - \overline{P_i})(\hat{P}_i - \overline{\hat{P}_i})}{\sqrt{\sum_{i=1}^{N}(P_i - \overline{P_i})^2} \times \sqrt{\sum_{i=1}^{N}(\hat{P}_i - \overline{\hat{P}_i})^2}}$	(17)

5 Results and Discussion

The LSTM-based prediction model employing the Adam optimizer demonstrated exceptional success in forecasting PV power output. Statistical evaluation metrics RMSE, rRMSE, and R were used to evaluate the model performance across training and testing phases (see Table 3, Figs. 4 and 5). Among all models, Model 6 exhibited the best forecasting accuracy, achieving (R = 98.51%, rRMSE = 9.29%, RMSE = 903.07 KW) for training and (R = 97.98%, rRMSE = 11.16%, RMSE =1073.64 KW) for testing phase, These results confirm that temperature and global horizontal irradiation (GHI) are highly correlated with future PV power, as their inclusion in Model 6's input features significantly improved prediction accuracy, LSTM Adam based models (4–6) outperformed SGD based models (1–3) by a significant margin. The best SGD model (Model 2) had (R = 97.42%, rRMSE = 12.33%), Model 6 (Adam) reduced rRMSE by ~1.17 percentage points (9.5% improvement) and increased R by 0.56 percentage points. This confirms that Adam is a more effective optimizer for LSTM-based PV forecasting due to its adaptive learning rate and momentum adjustments, which help avoid local minima.

Table 3. Performance metrics obtained for each model across training and testing phases

Models	Datasets	RMSE (KW)	rRMSE (%)	R (%)
Model 1	Training	1239.88	12.63	97.18
	Testing	1219.26	12.53	97.39
Model 2	Training	1233.52	12.47	97.21
	Testing	1212.82	12.33	97.42
Model 3	Training	1246.20	12.68	97.15
	Testing	1230.10	12.68	97.34
Model 4	Training	956.65	9.84	98.33
	Testing	1112.5	11.53	97.83
Model 5	Training	922.68	9.45	98.45
	Testing	1123.64	11.61	97.79
Model 6	Training	903.07	9.29	98.51
	Testing	1073.64	11.16	97.98

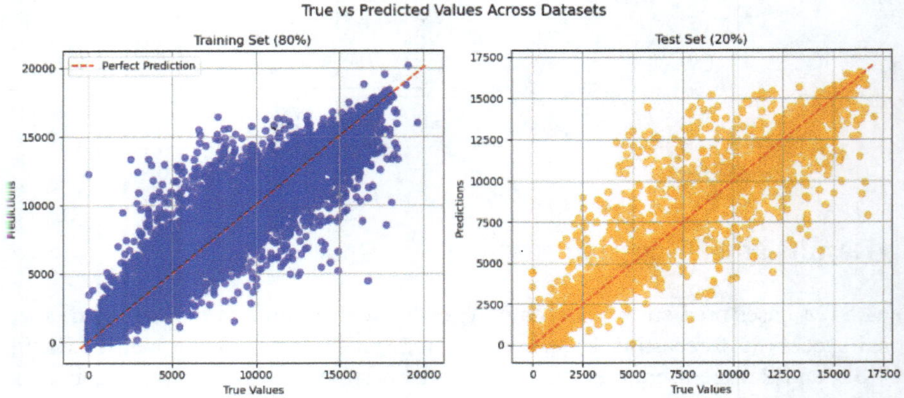

Fig. 4. Scatter plot for model #6 between real and predicted of PV POWER in the training and testing phase.

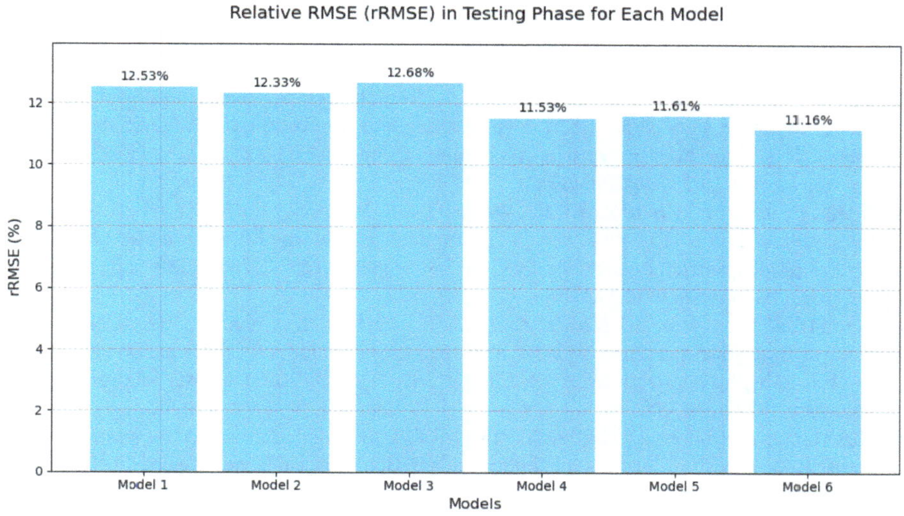

Fig. 5. rRMSE (%) during the testing phase for each model.

6 Conclusion

This study investigated a 15-min ahead PV power forecasting output for a grid-connected PV system in Adrar City (Algeria) and evaluated the accuracy of six models distinguished by different inputs and configurations. The LSTM model with the Adam optimizer and three input variables namely PV power, Temperature, and solar radiation was preferred because of its favorable computational performance, and high accuracy. The proposed LSTM architecture consists of two layers with 32 and 16 cells, respectively, subsequently passed to a dense layer of 8 neurons using the ReLU activation function. Performance was examined using statistical metrics, including the correlation coefficient (R), rRMSE, and RMSE. The input data comprised PV power, temperature, and global solar irradiation. The comparative analysis revealed that Model 6, utilizing PV power, temperature, and global solar irradiation as inputs, along with the Adam optimization algorithm, demonstrated superior forecasting performance. The developed model proves effective for predicting future PV power generation in Adrar's grid-connected PV system and can serve as a reliable tool for energy planning and grid management.

References

1. Sampath Kumar, D., Gandhi, O., Rodríguez-Gallegos, C.D., Srinivasan, D.: Review of power system impacts at high PV penetration Part II: Potential solutions and the way forward. Sol. Energy. **210**, 202–221 (2020). https://doi.org/10.1016/j.solener.2020.08.047
2. Gandhi, O., Kumar, D.S., Rodríguez-Gallegos, C.D., Srinivasan, D.: Review of power system impacts at high PV penetration – Part I: Factors limiting PV penetration. Sol. Energy. **210**, 181–201 (2020). https://doi.org/10.1016/j.solener.2020.06.097

3. Hosseini Dolatabadi, S.H., Soleimani, A., Maghanaki, M., Ilinca, A.: Enhancing photovoltaic farm capacity estimation: a comprehensive analysis with a novel approach. Energ. Technol. **12**(4) (2024). https://doi.org/10.1002/ente.202301294
4. El Bourakadi, D., Ramadan, H., Yahyaouy, A., Boumhidi, J.: A novel solar power prediction model based on stacked BiLSTM deep learning and improved extreme learning machine. International Journal of Information Technology (Singapore). **14**(2), 587–594 (2022). https://doi.org/10.1007/s41870-022-01118-1
5. Wang, F., Xuan, Z., Zhen, Z., Li, K., Wang, T., Shi, M.: A day-ahead PV power forecasting method based on LSTM-RNN model and time correlation modification under partial daily pattern prediction framework. Energy Convers. Manag. **212**, 112766 (2020). https://doi.org/10.1016/j.enconman.2020.112766
6. Abou Houran, M., Bukhari, S.M.S., Zafar, M.H., Mansoor, M., Chen, W.: COA-CNN-LSTM: Coati optimization algorithm-based hybrid deep learning model for PV/wind power forecasting in smart grid applications. Appl. Energy. **349**, 121638 (2023). https://doi.org/10.1016/j.apenergy.2023.121638
7. Kushwaha, V., Pindoriya, N.M.: A SARIMA-RVFL hybrid model assisted by wavelet decomposition for very short-term solar PV power generation forecast. Renew. Energy. **140**, 124–139 (2019). https://doi.org/10.1016/j.renene.2019.03.020
8. Hassan, M.A., Bailek, N., Bouchouicha, K., Nwokolo, S.C.: Ultra-short-term exogenous forecasting of photovoltaic power production using genetically optimized non-linear autoregressive recurrent neural networks. Renew. Energy. **171**, 191–209 (2021). https://doi.org/10.1016/j.renene.2021.02.103
9. Keddouda, A. et al.: Solar photovoltaic power prediction using artificial neural network and multiple regression considering ambient and operating conditions. Energy Convers. Manag. **288**, 117186 (2023). https://doi.org/10.1016/j.enconman.2023.117186
10. Mawloud, G., Gairaa, K., Boland, J., Arrif, T.: A novel hybrid model for solar radiation forecasting using support vector machine and bee colony optimization algorithm: Review and case study. Journal of Solar Energy Engineering, Transactions of the ASME. **143**(2) (2020). https://doi.org/10.1115/1.4047852
11. Harrou, F., Kadri, F., Sun, Y.: Forecasting of photovoltaic solar power production using LSTM approach. In: Harrou, F. (ed.) Advanced statistical modeling, forecasting, and fault detection in renewable energy systems. IntechOpen (2020)
12. Kottek, M., Grieser, J., Beck, C., Rudolf, B., Rubel, F.: World Map of the Köppen-Geiger climate classification updated. Meteorol. Z. **15**(3), 259–263 (2006). https://doi.org/10.1127/0941-2948/2006/0130
13. Dhaked, D.K., Dadhich, S., Birla, D.: Power output forecasting of solar photovoltaic plant using LSTM. Green Energy and Intelligent Transportation. **5**, 100113 (2023). https://doi.org/10.1016/j.geits.2023.100113
14. Al-Ja'afreh, M.A.A., et al.: An enhanced CNN-LSTM based multi-stage framework for PV and load short-term forecasting: DSO scenarios. Energy Rep. (2023). https://doi.org/10.1016/j.egyr.2023.08.003
15. Agga, A., Abbou, A., Labbadi, M., El Houm, Y.: Short-term self-consumption PV plant power production forecasts based on hybrid CNN-LSTM, ConvLSTM models. Renew. Energy. **177**, 101–112 (2021). https://doi.org/10.1016/j.renene.2021.05.095
16. Qing, X., Niu, Y.: Hourly day-ahead solar irradiance prediction using weather forecasts by LSTM. Energy. **148**, 461–468 (2018). https://doi.org/10.1016/j.energy.2018.01.177
17. Sadeghi, D., Golshanfard, A., Eslami, S., Rahbar, K., Kari, R.: Improving PV power plant forecast accuracy: a hybrid deep learning approach compared across short, medium, and long-term horizons. Renewable Energy Focus. **45**, 242–225 (2023). https://doi.org/10.1016/j.ref.2023.04.010

18. Sebastian RuderAn overview of gradient descent optimization algorithms
19. Diederik, P.K., Jimmy, B.: Adam: A Method for Stochastic Optimization. CoRR (2024)
20. Rezrazi, A., Hanini, S., Laidi, M.: An optimisation methodology of artificial neural network models for predicting solar radiation: a case study. Theor. Appl. Climatol. **123**, 769–783 (2016). https://doi.org/10.1007/s00704-015-1398-x
21. Stone, R.J.: Improved statistical procedure for the evaluation of solar radiation estimation models. Sol. Energy. **51**(4), 289–291 (1993). https://doi.org/10.1016/0038-092X(93)90124-7

Intra-Hour Solar Irradiance Forecasting Based on Feature Selection Techniques

Kacem Gairaa(✉), Mohamed Zaiani, and Sabrina Belaid

Unité de Recherche Appliquée en Energies Renouvelables, URAER, Centre de Développement des Energies Renouvelables, CDER, 47133 Ghardaïa, Algeria
k.gairaa@cder.dz

Abstract. Accurate solar forecasting is crucial for optimizing photovoltaic (PV) power generation and maintaining grid stability. This study introduces a novel forecasting approach based on dynamic feature selection to improve global horizontal irradiance (GHI) predictions at 5-min and 10-min resolutions. Data from three Algerian sites with distinct climates are utilized to ensure the model's adaptability across different environmental conditions. The proposed method employs machine learning techniques to identify the most relevant meteorological and atmospheric variables, enhancing model performance. Artificial neural networks (ANN) and random forest (RF) models are used to assess the effectiveness of the selected features. Experimental results demonstrate that dynamic feature selection significantly improves forecasting accuracy compared to traditional fixed-feature models by more than 30%. The proposed approach offers a robust and adaptive solution for intra-hour solar forecasting, facilitating the efficient integration of solar energy into power systems.

Keywords: Forecasting · Short-term · Global irradiance · Statistical score · Feature selection

1 Introduction

As of 2024, the global installed capacity of solar photovoltaic (PV) systems has surpassed 2 terawatts (TW), underscoring the critical role of solar energy in the transition toward sustainable power sources [1, 2]. However, the inherent variability of solar irradiance presents challenges to maintaining grid stability and optimizing power management. Accurate solar forecasting is essential to address these challenges by providing reliable predictions of solar power generation, thereby enhancing grid integration and operational efficiency.

Solar forecasting plays a vital role across multiple time resolutions, each serving different operational needs within the electrical grid:

- Nowcasting (minutes to hours ahead): Supports real-time grid balancing and frequency regulation.
- Short-term forecasting (hours to days ahead): Assists in day-ahead market operations and energy dispatch planning.

M. Hatti (Ed.): IC-AIRES 2025, LNNS 1740, pp. 46–53, 2026.
https://doi.org/10.1007/978-3-032-12788-4_5

- Long-term forecasting (weeks to months ahead): Aids in strategic planning and maintenance scheduling.

To achieve accurate predictions across these time horizons, various forecasting models have been developed, including [3]:

- Physical models: Rely on numerical weather prediction (NWP) to simulate atmospheric conditions influencing solar irradiance.
- Statistical models: Utilize historical data to identify patterns in solar power generation.
- Machine learning models: Employ algorithms based on AI models (ANNs, SVMs, etc.) to learn complex relationships between meteorological variables and solar output.
- Hybrid models: Combine multiple approaches to enhance forecasting accuracy.

Among these, machine learning models have gained prominence due to their ability to handle large datasets and adapt to non-linear patterns in solar irradiance. Feature selection techniques are integral to these models, as they identify the most relevant input variables, thereby improving forecasting accuracy and computational efficiency. For instance, an ensemble feature selection method has been proposed to choose not only the most related input parameters but also their past observation values, enhancing solar radiation forecasting accuracy [4]. Another study introduced a wrapper method termed Sequential Forward Selection (SFS) with a Kernel Conditional Density Estimator (KCDE) to forecast day-ahead regional PV power production, demonstrating the significance of appropriate feature selection methods in solar forecasting [5]. Additionally, research has shown that feature selection approaches play a crucial role in accurately forecasting solar radiation, as they help in identifying the most impactful features from the available data [6]. Recent advancements also include the development of SolarFusionNet, a deep learning architecture that integrates automatic multi-modal feature selection and cross-modal data fusion, effectively extracting relevant features from multichannel satellite images and multivariate meteorological data [7]. Furthermore, studies have highlighted that applying feature selection as a separate step improves the performance of machine learning models like Multilayer Perceptron Artificial Neural Networks (MLP-ANNs) in solar irradiance forecasting [8].

This study introduces a dynamic feature selection in order to forecast intra-hour global horizontal irradiance (GHI), at 5-min and 10-min resolutions. The solar forecasting algorithm is based on artificial neural network (ANN) and random forest (RF) learning methods and uses the clear sky index (CSI) as the response variable and features extracted from meteorological variables.

2 Data Sources

The dataset used in this study is collected from three distinct sites in Algeria: Algiers Khenchela and Ghardaïa. These locations were selected to represent different climatic conditions, ensuring the robustness of the proposed forecasting approach. Algiers, situated in a Mediterranean climate, experiences more variable weather conditions with frequent cloud cover, higher humidity, and moderate solar irradiance. Khenchela, located in a high-altitude region with a continental climate, is characterized by hot, dry summers and cold, harsh winters. Ghardaïa is located in a semi-arid region, characterized by high solar irradiance, low humidity, and minimal cloud cover [9, 10]. The dataset comprises samples of Global Horizontal Irradiance (GHI) along with various meteorological variables, including temperature, humidity, wind speed, and atmospheric pressure. Data were collected at 10-min intervals for the Ghardaïa and Khenchela sites and at 5-min intervals for the Algiers site (Table 1).

Table 1. Examined data for the three sites

Site	Resolution	Total used data
Ghardaïa	10-min	52704
Algiers	5-min	105120
Khenechla	10-min	52560

3 Methods

3.1 GHI Normalization

Normalization of global horizontal irradiance implies removing seasonal and diurnal variability due to the sun's position, in order to improve the accuracy and generalization of predictive model. It involves normalizing the observed GHI by the theoretical Clear-Sky Index (CSI), which represents the expected solar irradiance under cloud-free conditions. The CSI is a dimensionless value defined as [3]:

$$CSI = \frac{GHI_{obs}}{GHI_{clear}} \tag{1}$$

Subsequently, we use CSI as the main target in machine learning models to predict intra-hour GHI.

3.2 Dynamic Feature Extraction

Feature extraction is a critical step in data preprocessing and analysis, particularly in the context of solar radiation forecasting. It involves transforming raw data into a set of representative features that capture the essential information needed for accurate modeling and prediction. The goal is to reduce the dimensionality of the dataset while retaining the most relevant patterns and relationships, thereby improving the efficiency and performance of forecasting models.

In the context of solar forecasting, feature extraction is particularly important because solar radiation data is often complex, high-dimensional, and influenced by nonlinear dependencies. Traditional methods like Principal Component Analysis (PCA) are effective for linearly dependent data but fall short when applied to nonlinear datasets, such as those used in solar generation forecasts. Instead, more advanced techniques, such as rolling statistics and wavelet transform, which considered as dynamic feature methods, are employed to extract meaningful features from nonlinear solar data.

3.3 Results and Discussion

Before developing a forecasting model, it is essential to select input variables that exhibit a strong correlation with the target variable. Figures 1-3 present a correlation heatmap between global horizontal irradiance (GHI) and the input parameters at lag-1. This visualization enables clear identification of the most influential predictors for GHI forecasting.

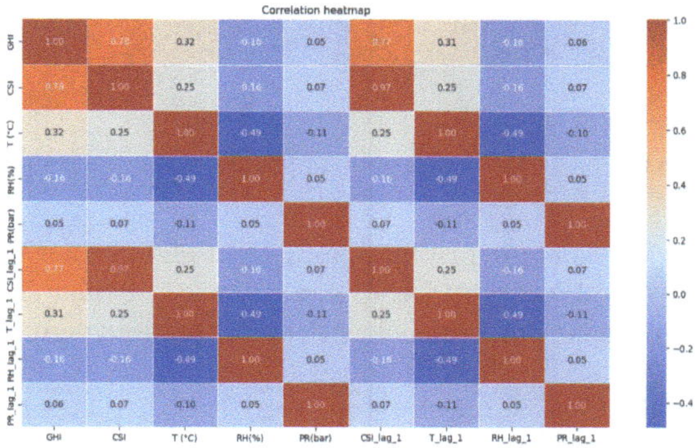

Fig. 1. Correlation between global solar radiation and input variables for Algiers site

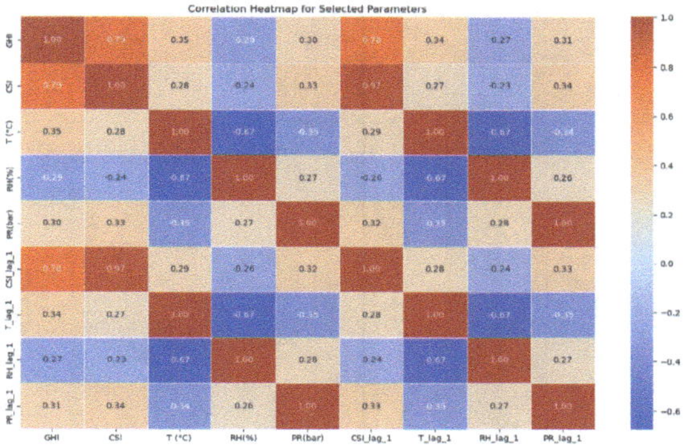

Fig. 2. Correlation between global solar radiation and input variables for Ghardaia site

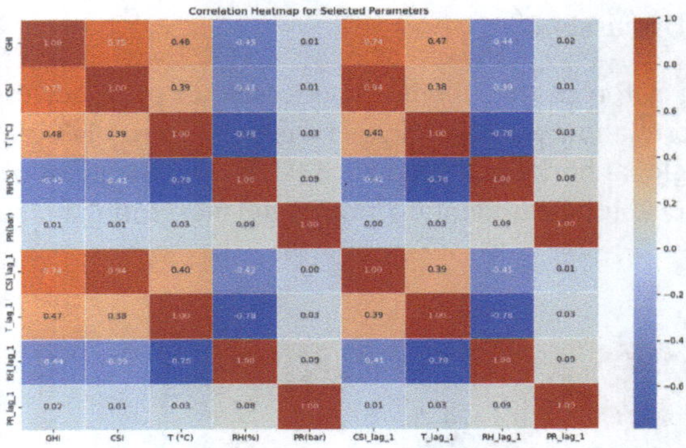

Fig. 3. Correlation between global solar radiation and input variables for Khenechla site

Figures 4, 5 and 6 present the results of intra-hour global solar irradiance forecasting using Artificial Neural Network (ANN) and Random Forest (RF) models for the investigated sites, comparing performance with and without dynamic feature extraction (FE).

For the ANN model, feature selection reduced nRMSE values by over 30% across all forecast horizons. The RF model similarly demonstrated improved accuracy, with nRMSE decreasing by approximately 15% when using feature selection. Comparable enhancements were observed in nMAE metrics.

The Khenchela site showed particularly significant improvements, with both nRMSE and nMAE values decreasing consistently across all forecast horizons.

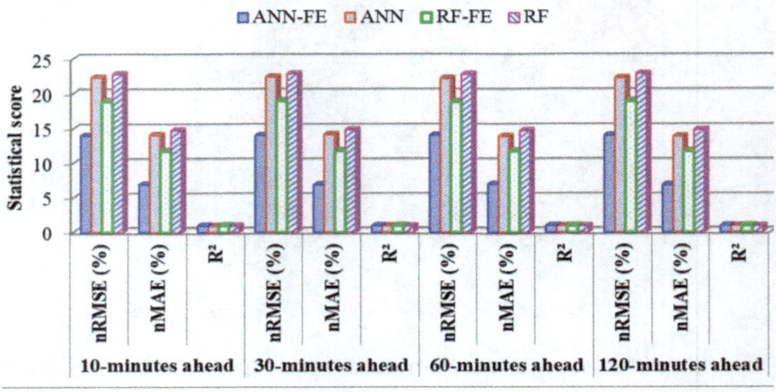

Fig. 4. Forecasting model performances for Algiers site

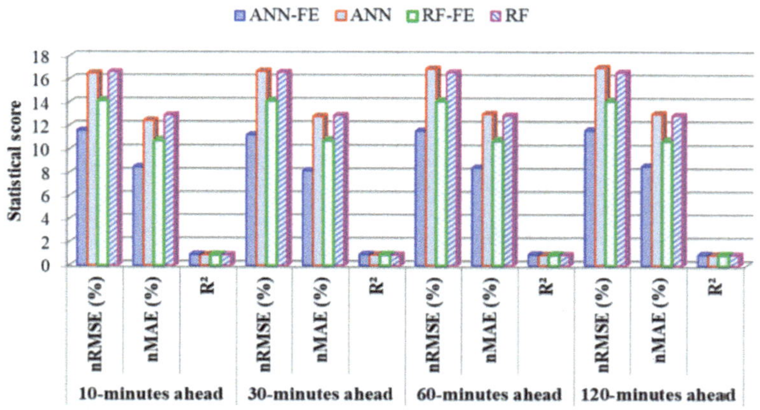

Fig. 5. Forecasting model performances for Ghardaia site

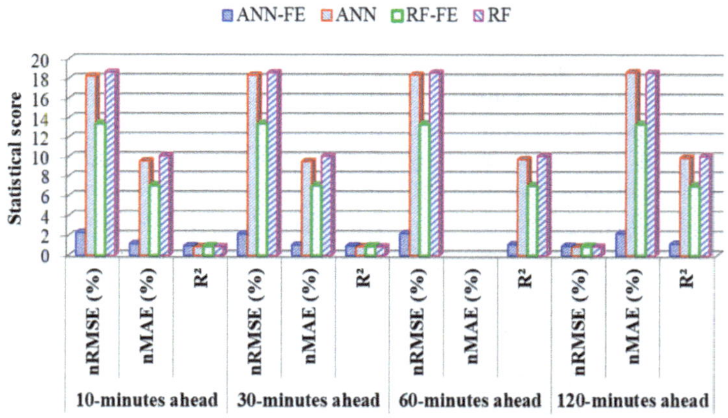

Fig. 6. Forecasting model performances for Khenechla site

Figure 7 compares measured GHI with forecasts from the ANN model (using feature extraction) at 10-min intervals, for Khenchela site. The results, based on a representative sample of test days, demonstrate strong agreement between predicted and observed GHI values.

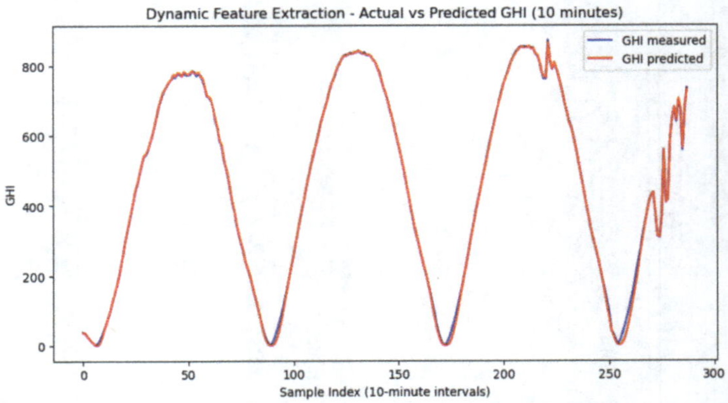

Fig. 7. Comparaison between predicted and measured GHI values using ANN and feature extraction (case of Khenechla site)

4 Conclusion

This study systematically evaluates the critical role of feature selection in intra-hour solar radiation forecasting across multiple sites and weather conditions. Through comprehensive analysis of ANN and Random Forest models, our results reveal that proper feature selection enhances forecast accuracy by over 30% (nRMSE) while achieving comparable improvements in nMAE metric compared to models using all available inputs. These findings establish that optimal input variable selection is essential for maximizing the performance of solar forecasting models.

References

1. International Energy Agency (IEA): Renewables 2023 (2023). https://www.iea.org/reports/renewables-2023
2. SolarPower Europe: Global Market Outlook 2023-2027, June 2023. https://www.solarpowereurope.org/insights/market-outlooks
3. Gairaa, K., Voyant, C., Notton, G., Benkaciali, S., Guermoui, M.: Contribution of ordinal variables to short-term global solar irradiation forecasting for sites with low variabilities. Renew. Energy. **183**, 890–902 (2022). https://doi.org/10.1016/j.renene.2021.11.028
4. Hammond, J.E., Orozco, R.A.L., Baldea, M., Korgel, B.A.: Short-term solar irradiance forecasting under data transmission constraints. Renew. Energy. **233**, 121058 (2024). https://doi.org/10.1016/j.renene.2024.121058
5. Macaire, J., Zermani, S., Linguet, L.: New feature selection approach for photovoltaïc power forecasting using KCDE. Energies. **16**(19), 6842 (2023). https://doi.org/10.3390/en16196842
6. Boutahir, M.K., Farhaoui, Y., Azrour, M., Zeroual, I., El Allaoui, A.: Effect of feature selection on the prediction of direct normal irradiance. Big Data Min. Anal. **5**(4), 309–317 (2022). https://doi.org/10.26599/BDMA.2022.9020003
7. Jing, T., Chen, S., Navarro-Alarcon, D., Chu, Y., Li, M.: SolarFusionNet: enhanced solar irradiance forecasting via automated multi-modal feature selection and cross-modal fusion. IEEE Trans. Sustain. Energy, early access (2024) https://doi.org/10.1109/TSTE.2024.3482360

8. Soleymani, S., Mohammadzadeh, S.: Comparative analysis of machine learning algorithms for solar irradiance forecasting in smart grids. arXiv, preprint https://arxiv.org/abs/2310.13791 (2023). https://arxiv.org/abs/2310.13791.
9. Dersch, J., et al.: Impact of DNI nowcasting on annual revenues of CSP plants for a time of delivery based feed in tariff. Meteorol. Z. **28**(3), 235–253 (2019). https://doi.org/10.1127/metz/2019/0959
10. Gairaa, K., Khellaf, A., Benkaciali, S., Guermoui, M.: Solar radiation measurements in Algeria: Case of Ghardaïa station as member of the enerMENA meteorological network. Energy Procedia. **141**, 50–54 (2017). https://doi.org/10.1016/j.egypro.2017.11.011

Hourly Global Solar Irradiance Forecasting in a Desert Region Using a Deep Neural Model with Hybrid Inputs

Lamis Serrat(✉), Hassen Bouzgou, and Naima Zerari

Department of Industrial Engineering, Laboratory of Automation and Manufacturing, University of Batna2 (Mostefa Ben Boulaïd), Batna, Algeria
lamis.serrat@univ-batna2.dz

Abstract. This paper proposes a hybrid deep neural model that combines Empirical Mode Decomposition (EMD), Bidirectional Gated Recurrent Units (BiGRU), and Multi-Layer Perceptron (MLP) for hourly forecasting of Global Horizontal Irradiance (GHI) in Tamanrasset, Algeria, a region characterized by highly stable solar conditions due to its arid desert climate. The proposed EMD-BiGRU-MLP model integrates two complementary modeling branches, each capturing distinct information from different input sources: measured GHI values and Numerical Weather Prediction (NWP) forecasts. The first branch processes measured GHI sequences using EMD, followed by a BiGRU model, while the second branch applies another BiGRU model directly to selected NWP variables identified through Pearson correlation coefficient analysis. The outputs from both branches are fused through an MLP network to produce the final forecast. Experimental results indicate that the NWP-based model achieves the best performance ($R^2 = 0.99$), while the EMD-BiGRU model and the full hybrid architecture perform comparably, with R^2 values of 0.938 and 0.94, respectively. These results highlight the effectiveness of NWP forecasts in stable environments. They also demonstrate that hybrid architectures, such as EMD-BiGRU, are valuable for capturing latent temporal structures. Furthermore, the findings emphasize the potential of integrating data-driven and physically modeled information to improve global solar irradiance forecasts of the proposed EMD-BiGRU-MLP model, particularly in stable regions like Tamanrasset. This underscores the importance of adapting model complexity to data characteristics, even when variability appears low.

Keywords: Global solar irradiance forecasting · Numerical weather prediction · Bidirectional gated recurrent unit · Empirical mode decomposition · Tamanrasset

1 Introduction

With the increasing integration of photovoltaic solar systems into power grids, accurate forecasting of solar irradiance, particularly global horizontal irradiance (GHI), has become essential for effective energy planning and management [1]. Consequently, there is a growing emphasis on developing precise and reliable forecasting methods

M. Hatti (Ed.): IC-AIRES 2025, LNNS 1740, pp. 54–66, 2026.
https://doi.org/10.1007/978-3-032-12788-4_6

[2–4]. Solar irradiance prediction models are generally classified into five principal categories: physical approaches, empirical approaches, statistical techniques, machine learning methods, and hybrid models [5]. In recent years, advancements in deep learning and data decomposition techniques have led to the emergence of hybrid models as a dominant approach in solar energy forecasting due to the stochastic nature of solar irradiance. For instance, Jun Liu et al. [6] proposed a novel hybrid model that integrates Variational Mode Decomposition (VMD) and Deep Belief Networks (DBN) within a CNN-LSTM-MLP framework for one-hour-ahead global solar irradiance forecasting in Clark, Folsom, and Denver (USA). The experimental results demonstrated the effectiveness of the proposed model in addressing seasonal fluctuations and enhancing prediction accuracy. Sujan Chimire et al. [7] utilized a novel hybrid model that combines improved complete ensemble empirical mode decomposition adaptive noise (ICEEMDAN) and DRESNET to achieve precise multi-step global solar radiation (GSR) forecasts in Nepal. After testing, the model showed superior performance compared to other benchmarking models. Tian Peng et al. [8] employed complete ensemble empirical mode decomposition adaptive noise (CEEMDAN) and Bidirectional Long Short-Term Memory (Bi-LSTM) based on a sine-cosine algorithm for hourly solar irradiance forecasting in the southern region of Dauphin Island, Alabama. The experimental results indicated that the proposed model outperformed all benchmarked models. However, most of these approaches have been applied in regions with high solar variability, where complex models can easily exploit signal fluctuations. Fewer studies have explored the relevance and added value of such complex architectures in environments with highly stable solar irradiance, where simpler models may already yield robust results. Moreover, only a limited number of studies have investigated the structured integration of multiple data sources specifically, measured irradiance data and numerical weather prediction (NWP) forecasts within a deep learning framework. However, these sources provide complementary insights: measured data capture local fluctuations, while NWP forecasts represent broader atmospheric dynamics. To address these gaps, we propose a hybrid deep neural model for hourly one-step-ahead GHI forecasting in Tamanrasset, Algeria, a location known for its exceptionally stable solar conditions. The hybrid model structure consists of two branches: the first processes measured GHI sequences using Empirical Mode Decomposition (EMD) followed by a Bidirectional Gated Recurrent Unit (BiGRU), while the second models selected NWP variables using a separate BiGRU. The outputs from both branches are fused through a Multilayer Perceptron (MLP) to generate the final prediction.

2 Datasets and Feature Selection Process

2.1 Datasets Description

This study employs two separate datasets gathered in Tamanrasset, Algeria, covering the period from January 1, 2011, to December 31, 2013. The first dataset comprises Global Horizontal Irradiance (GHI) measurements recorded every minute in Tamanrasset. The second dataset includes NWP variables obtained from the National Aeronautics and Space Administration Prediction of Worldwide Energy Resources (the NASA POWER), which provides gridded atmospheric forecasts generated by the Goddard Earth Observing System (GEOS) model. These forecasts have an hourly resolution and encompass nine meteorological variables: air temperature (TEM: °C), relative humidity (RH: %), wind speed (WS: m/s), solar zenith angle (SZA: °), GHI (W/m^2), wind direction (WD: °), specific humidity (SH: g/kg), surface pressure (PS: kPa), and precipitation (PREC: mm/h). To align with the temporal resolution of NWP data, the one-minute GHI measurements are aggregated into hourly values using Python. Furthermore, to avoid the misalignment and inconsistency between the two datasets, the analysis is restricted to the time window between 06:00 and 16:00 local time, as the number of GHI records outside this range exceeds that of the corresponding NWP data (as shown in Table 1). Subsequently, both datasets are normalized using the min-max normalization method and divided into three subsets: the training dataset (80%), the validation dataset (10%), and the testing dataset (10%).

Table 1. Number of samples for each time window in GHI and NWP data

Time window	Measured GHI dataset	NWP dataset
05:00 to 18:00	15341 samples	13604 samples
05:00 to 17:00	14246 samples	13465 samples
05:00 to 16:00	13151 samples	12703 samples
06:00 to 18:00	14246 samples	12957 samples
06:00 to 17:00	13151 samples	12818 samples
06:00 to 16:00	12056 samples	12056 samples

2.2 Feature Selection Process

Due to the variety of meteorological variables present in the NWP dataset, it was essential to implement a feature selection process to identify the inputs that most significantly enhance the accuracy of GHI predictions. This procedure aids in minimizing computational cost and enhances the model's generalization capabilities. Therefore, a Pearson correlation coefficient analysis is conducted between each meteorological variable and the target variable, GHI. The results are visualized using a correlation heat map.

Figure 1 displays the correlation coefficients of nine meteorological variables with the target GHI. According to the classification criteria for correlation coefficients outlined in [2], variables that exhibit correlation coefficients ranging from 0.4 to 1 should be selected. Consequently, the variables air temperature, solar zenith angle, and relative humidity are selected as input features for further analysis.

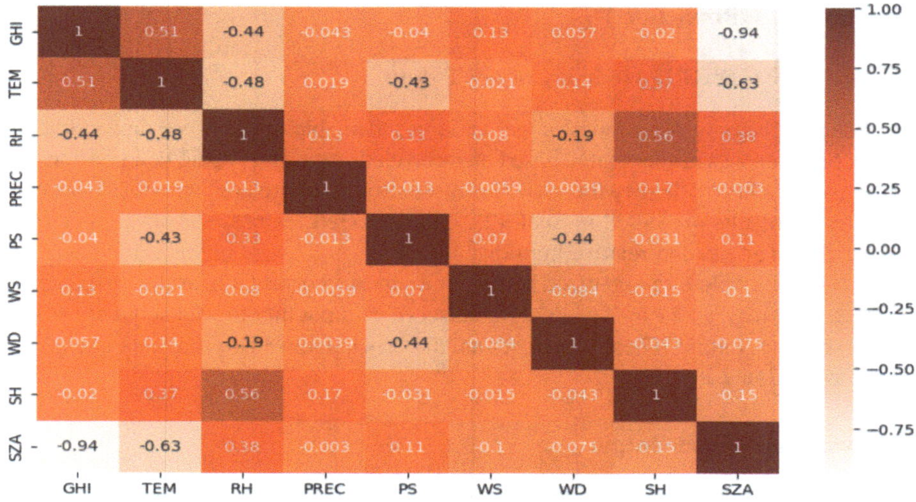

Fig. 1. The heat map illustrating the correlation between the different meteorological variables and GHI.

3 Framework of the Proposed Hybrid Model

The proposed EMD-BiGRU-MLP model is developed to enhance the accuracy of hourly one-step-ahead GHI forecasting by integrating two complementary modeling branches, each capturing distinct information from different data sources. The overall structure of the model is illustrated in Fig. 2. It operates as follows:

3.1 Branch 01: EMD-Based BiGRU Modeling of Measured GHI Data

The measured GHI series is transformed into a supervised learning format by constructing sequences in which the previous 24 h serve as inputs, and the subsequent hour of GHI is designated as future value.

a. Each of these input sequences is decomposed via EMD into IMFs, resulting in a varying number of IMFs for each sequence.
b. To standardize the model input, zero-padding is applied to sequences that yield fewer than four IMFs to fix the number of IMFs to four, which is the maximum adopted in this study.
c. A Masking layer is incorporated into the BiGRU model to ensure that padded values are ignored during training. The masked input sequences are then processed by a BiGRU layer, followed by a Dropout layer with a rate of 0.3, which acts as a regularization mechanism to mitigate overfitting. This branch generates the first output, referred to as **'Prediction 1'**.

3.2 Branch 02: BiGRU Modeling of NWP data

a. Given the diversity of meteorological variables available in the NWP data, a feature selection step was implemented based on a correlation analysis using Pearson correlation coefficients to identify the variables most strongly correlated with GHI. According to the results presented in Sect. (2.2), the selected features are air temperature, solar zenith angle, and relative humidity.
b. The NWP variables are organized in the same supervised structure as the GHI dataset, where the previous 24 h of each variable serve as input and the subsequent hour of GHI serves as output. These sequences are directly fed into a BiGRU model, resulting in **'Prediction 2'.**

3.3 Fusion and Final Output

The outputs from both branches are concatenated and passed through an MLP, which synthesizes the fused information to generate the final GHI output.

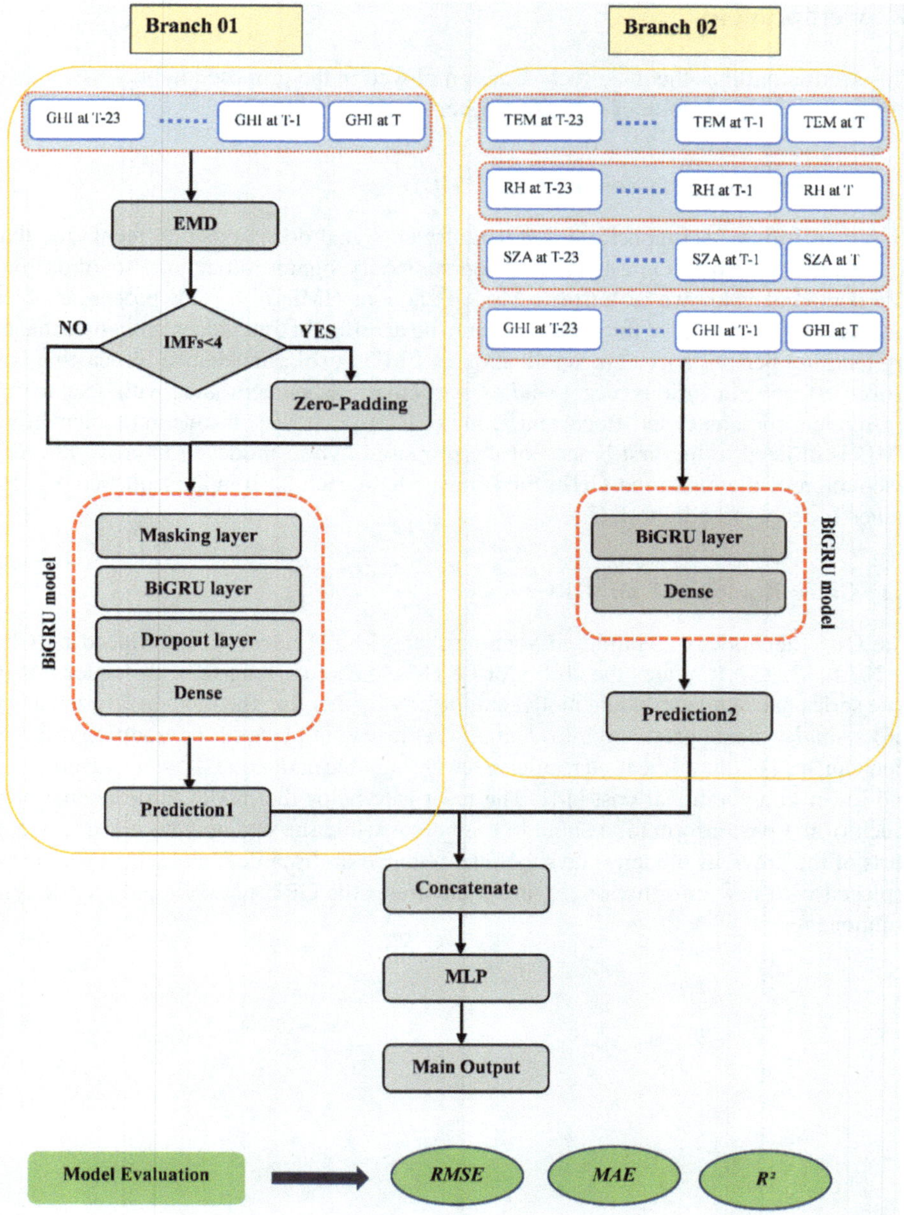

Fig. 2. Framework of the proposed EMD-BiGRU-MLP model.

4 Methodology

This section outlines the main techniques employed in the proposed hybrid model, and a graphical representation of the entire system is provided in Fig. 2.

4.1 Empirical Mode Decomposition (EMD)

Empirical Mode Decomposition is a data-driven signal decomposition technique that adaptively breaks down non-linear and non-stationary signals into a finite set of oscillatory components known as Intrinsic Mode Functions (IMFs) [9]. This process enables the model to isolate signal fluctuations occurring at different time scales, thereby enhancing learning performance. The application of EMD in solar irradiance forecasting has proven effective in various recent studies, particularly when combined with deep learning models such as Gated Recurrent Units (GRU) [2, 9, 10]. In our implementation, EMD is utilized in the first branch of the proposed hybrid model to improve the representation of the measured GHI time series, where each 24-h input sequence of GHI values is decomposed into IMFs.

4.2 Gated Recurrent Unit (GRU)

The GRU network is a simplified version of the LSTM network, introduced by Cho in 2014 [12, 13]. It retains the ability of LSTMs to capture long-term dependencies in time series data while reducing model complexity by merging the input and forget gates into a single update gate, and introducing a reset gate in place of the traditional forget mechanism. This simplification results in fewer trainable parameters, faster convergence, and lower computational cost [14]. The reset gate helps the model to determine how much of the past information should be ignored, while the update gate decides which parts of the previous hidden state should be retained or discarded, and which should be replaced with new information. Figure 3 illustrates the GRU network and its different components.

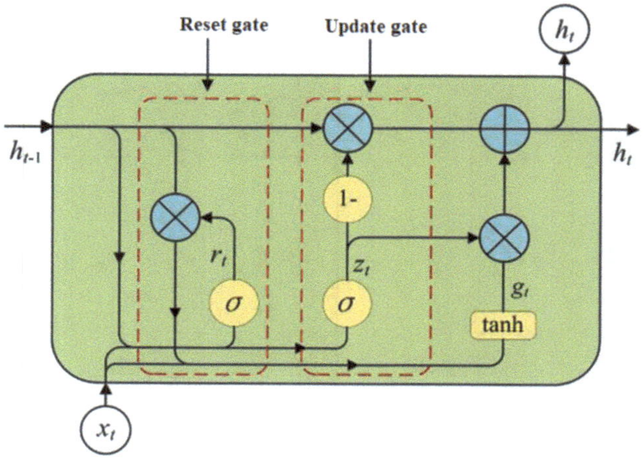

Fig. 3. The architecture of the GRU network.

Where, r_t is the reset gate, z_t is the update gate, h_{t-1} is the hidden state at the last moment, h_t is the hidden state at the current moment, g_t is the candidate hidden state at the current moment, σ and *tanh* are sigmoid and hyperbolic tangent activation functions, x_t is the input at the current moment.

4.3 Bidirectional Gated Recurrent Unit (BIGRU)

The BiGRU network is an enhancement of the standard GRU that processes the input sequence in both forward and backward directions, thereby capturing both past and future context within a fixed window (see Fig. 4). This dual-directional learning improves the model's ability to extract rich temporal features, which is particularly beneficial for forecasting problems involving sequential dependencies such as solar irradiance forecasting [14].

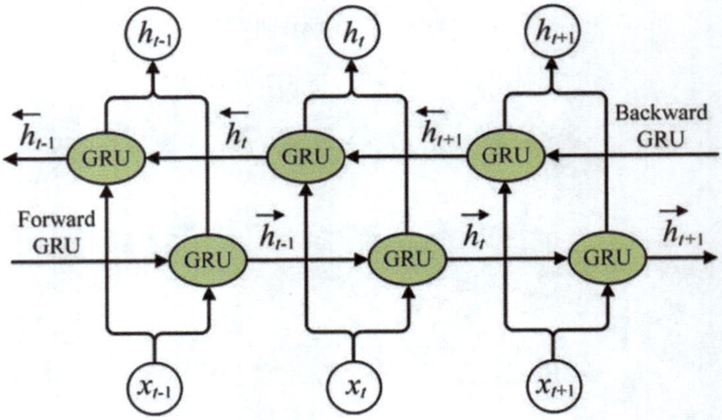

Fig. 4. The structure of the BiGRU network.

4.4 Model Performance Evaluation Metrics

To evaluate the performance of the EMD-BiGRU-MLP model three evaluation metrics are used, mean absolute error (MAE), root mean square error (RMSE), and the coefficient of determination (R^2). These metrics can be described as follow [15]:

$$MAE = \frac{1}{N}\sum_{i=1}^{N}|y_i - \hat{y}_i| \tag{1}$$

$$RMSE = \sqrt{\frac{1}{N}\sum_{i=1}^{N}(\hat{y}_i - y_i)} \tag{2}$$

$$R^2 = 1 - \sum_{i=1}^{N}(y_i - \hat{y}_i)^2 / \sum_{i=1}^{N}(y_i - \overline{y})^2 \tag{3}$$

where y_i and $\hat{y}_i$ represent the measured and the forecasted GHI, N is the samples number of GHI series and $\overline{y}$ is the mean value of y_i.

5 Results and Discussion

5.1 Experimental Setup

The study was conducted using two datasets collected over a three-year period (2011–2013) in Tamanrasset, a region characterized by highly stable solar irradiance. The first dataset consists of measured GHI values recorded every minute, which were aggregated to hourly resolution. The second dataset includes nine meteorological variables at hourly resolution, obtained from the NASA POWER. Using a Pearson correlation analysis, three variables (air temperature, relative humidity, and solar zenith angle) were selected for modeling due to their strong correlation with GHI. To ensure temporal alignment between both datasets, the time window was restricted to the time window between 06:00 and 16:00 local time, resulting in a total of 12,056 hourly records. All features were normalized using Min-Max scaling and transformed into sequences using a 24-h sliding window, then they were split chronologically into training (80%), validation (10%), and testing (10%) subsets. The proposed hybrid framework consists of two branches. In the first branch, measured GHI sequences were decomposed into IMFs using the EMD method. Sequences that yielded fewer than four IMFs were zero-padded to ensure consistent input dimensions and passed through a masking layer, then processed by a BIGRU model of 200 units, followed by a Dropout layer with a rate of 0.3. In the second branch, the selected NWP features were directly input into another BiGRU model of 100 units. The outputs of both branches were concatenated and fed into a MLP, which produced the final GHI forecast. All models were trained using the Adam optimizer with the Mean Squared Error loss function, a batch size of 64, and up to 100 epochs, with early stopping based on validation loss. The performance of each branch and the fully hybrid model was evaluated on the test set using three standard metrics: MAE, RMSE, and R^2. It is important to note that all experimental runs were conducted using the Python programming language. The hybrid model was implemented in Python 3.11.12 using the Google Colab environment. The development made use of Keras for deep learning, scikit-learn for preprocessing and evaluation, PyEMD for signal decomposition, and Pandas and Seaborn for data handling and visualization.

5.2 Experimental Results

Table 2 summarizes the results in terms of MAE, RMSE, and R^2 for the EMD-BiGRU model (Branch 1), the BiGRU model (Branch 2), and the EMD-BiGRU-MLP model. The results indicate that the NWP-based BiGRU model achieved the best overall performance, exhibiting the lowest MAE and RMSE, as well as the highest R^2 of 0.99. This suggests that forecast variables in a stable region like Tamanrasset, are highly predictive of future GHI. The EMD-BiGRU model achieved a commendable R^2 of 0.938. Although its error metrics (MAE: 50.2 W/m^2, RMSE: 81.5 W/m^2) are higher than those of the NWP-based model, it effectively demonstrates the model's capacity to capture the intrinsic multi-scale properties of the measured GHI signal. The hybrid EMD-BIGRU-MLP model, which integrates both data sources, achieved an R^2 of 0.94. This result represents a slight improvement over the EMD-BiGRU model; however, it does not surpass the performance of the NWP-based BiGRU model, which can be attributed to the relative ease of forecasting modeled data, such as that produced by numerical weather prediction (NWP) systems. Figure 5 displays scatter plots that illustrate the alignment between predicted and actual GHI values. The BiGRU model closely aligns with the ideal prediction line, confirming its superiority. In contrast, the EMD-BiGRU and hybrid models show greater dispersion, although they still follow the general trend.

Table 2. Performance results of EMD-BIGRU, BIGRU and EMD-BIGRU-MLP.

Model	MAE (W/m^2)	RMSE (W/m^2)	R^2
BiGRU	17.44	27.02	0.99
EMD-BiGRU	50.2	81.5	0.938
Full hybrid model	45.8	73.2	0.94

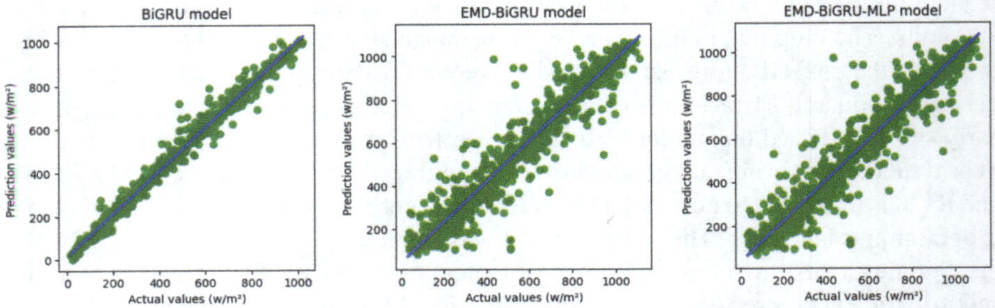

Fig. 5. Scatter plots of predicted GHI values for the three models: BiGRU, EMD-BiGRU, and EMD-BiGRU-MLP.

5.3 Impact of Data Source Fusion on Forecasting Performance

Although the hybrid model does not surpass the NWP-based branch in terms of MAE, RMSE or R^2, it highlights the importance of integrating complementary sources of information. The NWP-based branch offers a comprehensive atmospheric perspective through physically modeled forecasts, effectively capturing large-scale weather patterns. In contrast, the GHI-based branch utilizes empirical decomposition of historical measurements to account for localized and short-term fluctuations that may not be fully represented in forecast data. The integration of these two sources through an MLP allows the model to synthesize both physical and data-driven insights. While the performance improvement from this fusion is limited in this study due to the highly stable solar conditions in Tamanrasset, the proposed architecture demonstrates robustness and adaptability. This is supported by the performance comparison presented in Table 2, which indicates that the hybrid model slightly enhances the EMD-BiGRU model; however, it does not surpass the BiGRU model when evaluated solely on NWP data. Furthermore, the scatter plots in Fig. 5 illustrate that the predictions made by the EMD-BiGRU-MLP model closely align with the actual values, although the NWP-based model demonstrates a closer correspondence with the ideal prediction line.

5.4 Value of Complex Structures in Stable Regions

A key contribution of this study is the application of complex deep learning architectures, which are typically designed for highly variable or noisy datasets, to a stable solar irradiance signal. Despite the low variability of GHI data in Tamanrasset, the results indicate that the signal still contains significant multi-scale temporal dynamics that can be enhanced through structured decomposition. The branch 02 (EMD-BiGRU), by extracting IMFs, is capable of capturing small-scale patterns and latent structures that may be overlooked by simpler or purely forecast-based methods. These findings suggest that smoother data does not necessarily lead to a simpler or more effective model. While the irradiance signal appears stable on a broader scale, subtle variations persist, especially at finer temporal resolutions, and can significantly inform predictive modeling. As illustrated in Fig. 5, the EMD-BiGRU model exhibits greater dispersion around the ideal prediction line compared to the NWP-based model; however, it still captures valuable nonlinear dynamics. This is reflected in its R^2 value of 0.938, as reported in Table 2, demonstrating that deep architectures remain relevant even in stable environments. Their strength lies in their ability to isolate latent structures and enhance robustness, particularly when combined with external data sources like NWP forecasts, which provide additional physical context and improve generalization potential.

6 Conclusion

This study presents a novel EMD-BiGRU-MLP model for hour-ahead GHI forecasting in a highly arid stable solar environment. The hybrid model utilizes a dual-branch architecture that integrates measured GHI data and NWP-based meteorological forecasts. The first branch employs EMD to decompose historical GHI sequences into intrinsic components, which are then modeled using a BiGRU network. The second branch directly models selected NWP variables with a separate BiGRU. Both branches are fused through an MLP to generate the final output. Experimental results from Tamanrasset, Algeria, show that the NWP-based BiGRU branch achieves the highest predictive performance, with a MAE of 17.44 W/m^2, a RMSE of 27.02 W/m^2, and an R^2 of 0.99. This reflects the strong reliability of physical forecasts in a low-variability environment. The EMD-BiGRU branch performs reasonably well, with an MAE of 50.2 W/m^2, an RMSE of 81.5 W/m^2, and an R^2 of 0.938, effectively capturing multi-scale structures in the measured GHI. Although the hybrid framework, which fuses both data-driven and physically modeled information, does not outperform the NWP-based BiGRU branch (MAE $= 45.8$ W/m^2, RMSE $= 73.2$ W/m^2, $R^2 = 0.94$), it demonstrates the value of combining complementary sources and lays the groundwork for more generalizable forecasting systems. These findings emphasize that model complexity should be tailored to the nature of the data, and that even in low-variability contexts, detailed temporal patterns can still be utilized to enhance prediction accuracy.

Future research could investigate the integration of additional data sources, such as satellite observations, to further enhance model performance. Moreover, the methodology could be applied to other climatic regions characterized by more dynamic weather patterns, where the combination of NWP and historical GHI data may result in even greater improvements in forecasting accuracy.

References

1. Xiaoqiao, H., et al.: Hybrid deep neural model for hourly solar irradiance forecasting. Renew. Energy. **171**, 1041–1060 (2021)
2. Xiangfei, K., Xinyu, D., Guixiang, X., Zhijie, X.: Multi-step short-term solar radiation prediction based on empirical mode decomposition and gated recurrent unit optimized via an attention mechanism. Energy. **282**(1), 128825 (2023)
3. Hassen, B., Christian A., G.: Fast short-term global solar irradiance forecasting with wrapper mutual information. Renew. Energy. **133**, 1055–1065 (2019)
4. Hassen, B., Christian A., G.: Minimum redundancy – Maximum relevance with extreme learning machines for global solar radiation forecasting: Toward an optimized dimensionality reduction for solar time series. Sol. Energy. **158**, 595–609 (2017)
5. He, Z., Xiaoqiao, H., Zenan, X., Haoyuan, S., Chengli, L., Yonghang, T.: Week-ahead hourly solar irradiation forecasting method based on ICEEMDAN and Times Net networks. Renew. Energy. **220**, 119706 (2024)
6. Jun, L., Xiaong, H., Qiong, L., Zaiqing, C., Gang, L., Yonghang, T.: Hourly stepwise forecasting for solar irradiance using integrated hybrid models CNN-LSTM-MLP combined with error correction and VMD. Energy Convers. Manage. **280**, 116804 (2023)
7. Sujan, G., Ravinesh, C.D., David, C.-P., Sancho, S.-S.: Improved complete ensemble empirical mode decomposition with adaptive noise deep residual model for short-term multi-step solar radiation prediction. Renew. Energy. **190**, 408–424 (2022)
8. Tian, P., Chu, Z., Jianzhong, Z., Muhammad, S.N.: An integrated framework of Bi-directional long-short term memory (BiLSTM) based on sine cosine algorithm for hourly solar radiation forecasting. Energy. **221**, 119887 (2021)
9. Norden, E.H., et al.: The empirical mode decomposition and the Hilbert spectrum for nonlinear and non-stationary time series analysis. Proc. R. Soc. A. **454**(1971), 903–995 (1998)
10. Mawloud, G., Kacem, G., Domingos, S.d.O.S.J., Toufik, A., Abdelfetah, B.: Potential assessment of the TVF-EMD algorithm in forecasting hourly global solar radiation: Review and case studies. J. Clean. Prod. **385**, 135680 (2023)
11. Karan, S., Bijaya, K.P., Tushar, S.: A Short-term solar irradiance forecasting modelling approach based on three decomposition algorithms and adaptive neuro-fuzzy inference system. Expert Syst. Appl. **s**, 120770 (2023)
12. Bixuan, G., Xiaoqiao, H., Junsheng, S., Yonghang, T., Rui, X.: Predicting day-ahead solar irradiance through gated recurrent unit using weather forecasting data. J. Renew. Sustain. Energy. **11**(4), 043705 (2019)
13. Kyunghyun, C., et al.: Learning phrase representations using RNN encoder–decoder for statistical machine translation, preprint arXiv, p. 15, (2014)
14. Xu, W., et al.: The Application of a BiGRU Model with Transformer-Based Error Correction in Deformation Prediction for Bridge SHM. Buildings. **15**(4), 542 (2025)
15. Hassen, B.: A fast and accurate model for forecasting wind speed and solar radiation time series based on extreme learning machines and principal components analysis. J. Renew. Sustain. Energy. **6**, 013114 (2014)

Estimating Power Outputs of Thin Film CIS PV Modules Using Neuronal Approach: A Case Study in Arid Environment

I. Hadj Mahammed(✉), K. Gairaa, and A. Beddar

Unité de Recherche Appliquée en Energies Renouvelables, URAER, Centre de Développement des Energies Renouvelables, CDER, 47133 Ghardaia, Algeria
hmidriss65@gmail.com

Abstract. Accurate photovoltaic (PV) module power output prediction is critical for maximizing energy yield, optimizing system efficiency, and enhancing return on investment. This paper presents the development of a highly accurate Artificial Neural Network (ANN) model tailored for estimating the power output of thin-film Copper Indium Selenide (CIS) PV modules. Recognizing the distinct impact of weather variability, the model leverages datasets specifically partitioned according to sky conditions: one for sunny days and another for cloudy days. Validation using representative days from each category demonstrates exceptional accuracy. Statistical evaluation reveals remarkably low normalized Root Mean Square Errors (nRMSE) of approximately 1.95% for sunny days and 5.06% for cloudy days. These results confirm a robust and reliable agreement between the model's predictions and actual measured power outputs, highlighting the model's effectiveness for precise performance forecasting across diverse operational environments.

Keywords: Artificial neural network · PV modules · Renewable energy

1 Introduction

The deployment of photovoltaic (PV) systems is experiencing exponential growth worldwide, driven both by technological advances in efficiency and the continued reduction in production costs. This momentum is expected to continue over the coming decades, placing PV technology at the heart of strategies to decarbonize the global energy system [1]. According to projections, this trajectory could lead to the installation of several terawatts of PV capacity worldwide [2]. In this context, accurately predicting the power output of PV modules is a fundamental challenge for the design, installation, and maintenance of solar power plants. Indeed, reliable estimates are essential for optimizing operational performance, maximizing energy production, and improving return on investment. To address this need, three main methodological categories have been developed and documented in the scientific literature [3, 4] namely empirical models, based on correlations from experimental measurements, analytical models, based on the physicochemical principles of PV module operation and approaches based on artificial

M. Hatti (Ed.): IC-AIRES 2025, LNNS 1740, pp. 67–73, 2026.
https://doi.org/10.1007/978-3-032-12788-4_7

intelligence (AI) and machine learning (ML), which leverage advanced statistical techniques. Among these, artificial neural networks (ANNs) have attracted particular interest for estimating PV system performance [5, 6]. However, a significant limitation remains: the majority of existing models do not sufficiently consider the heterogeneity of solar conditions (clear sky vs. cloudy sky), even though this factor is a major determinant of PV production variability. This omission can lead to significant errors in real-world environments, where weather conditions are highly dynamic. To overcome this limitation, this work aims to assess the ANN model's effectiveness in estimating the power output profiles of CIS thin-film PV module. By segmenting the training and validation data according to weather conditions (sunny/cloudy), the model is able to explicitly capture performance variations. This approach improves the accuracy of short-term forecasts in dynamic environments, while contributing to the development of smart PV systems for more reliable energy management.

2 Methodology

2.1 System Monitoring

Figure 1 illustrates the test bench and PV system monitoring setup. This configuration comprises a solar measurement station [7, 8], which records solar irradiance, a weather station responsible for measuring air temperature, PV current and voltage sensors, and an Agilent data-logger used for data recording. All these parameters are recorded at 5-min intervals.

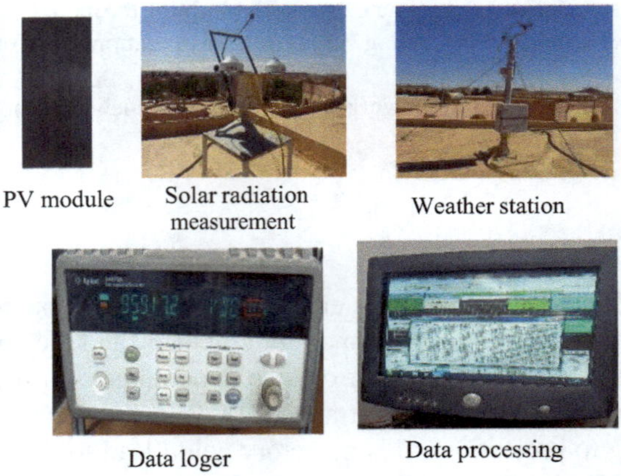

Fig. 1. Test bench and PV system monitoring

Two sub-databases were extracted from the recorded dataset. The first sub-database contains data associated with cloudy days, while the second sub-database includes data for sunny days. The classification of these days into cloudy and sunny categories was determined through a visual examination of solar components, specifically, diffuse, direct, and global irradiances. Figs. 2 and 3 depict the temporal evolution of measured instantaneous PV power for both types of days.

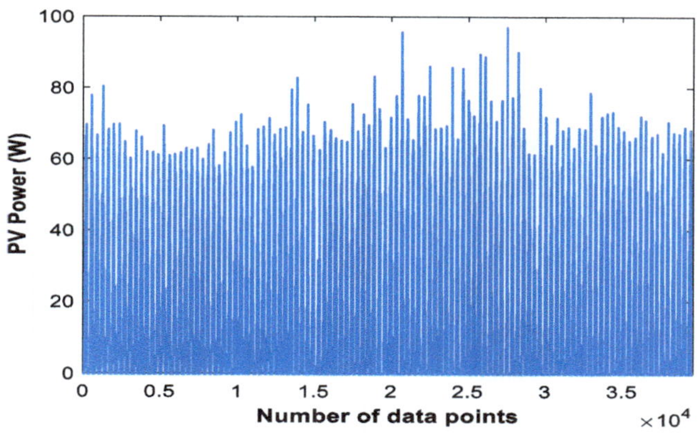

Fig. 2. Evolution of PV power energy in the case of overcast days

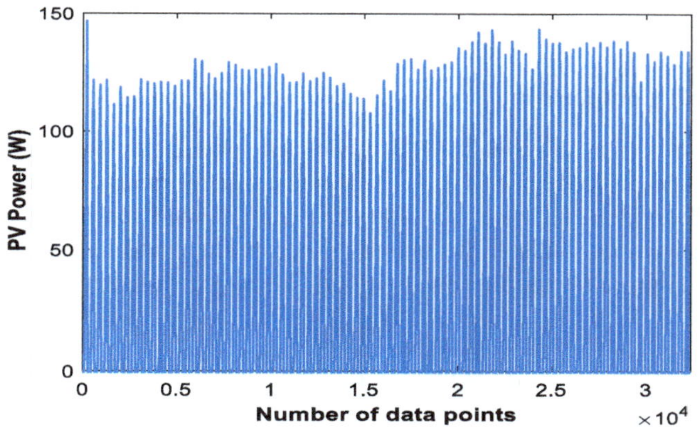

Fig. 3. Evolution of PV power energy in the case of sunny days

2.2 ANN Model Development

Artificial neural networks (ANNs) are powerful computational models inspired by the structure and functioning of biological neural networks in the human brain. They consist of interconnected nodes or neurons organized into different layers. Each neuron receives input signals, processes them through internal calculations, and then generates an output signal. The intensity of transmission between neurons is governed by weights assigned to the connections, as described in [10]. According to the literature review, the feed-forward multilayer perceptron (MLP) network structure illustrated in Fig. 4, is the most commonly used architecture in the field of solar energy forecasting, as reported by Mellit et al. [12]. In an MLP, neurons are organized into several layers, typically three or four, comprising an input layer, one or two hidden layers, and an output layer. In this study, we use an MLP architecture to estimate the power output of photovoltaic modules. The input layer of the model is fed by two essential meteorological parameters: solar irradiance (G) and ambient temperature (T).

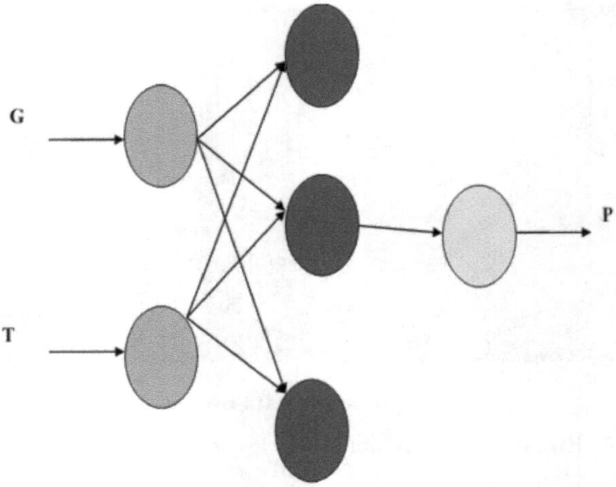

Fig. 4. ANN (MLP) architecture

2.3 Statistical Comparison

The evaluation of the developed ANN models is conducted using three established metrics to gauge performance. These metrics include: normalized root mean square error (nRMSE), normalized mean absolute error (nMAE) and correlation coefficient (r). These metrics provide valuable insights into the accuracy and reliability of the ANN models in predicting PV module power outputs. These metrics are given bellow as [14]:

$$nRMSE = \frac{\sqrt{\frac{1}{N}\sum_{i=1}^{N}(e_i - m)^2}}{mean(m_i)} \tag{1}$$

$$\text{nMAE} = \frac{1}{N}\sum\nolimits_{i=1}^{N}\frac{[e_i - m_i]}{mean(m_i)} \tag{2}$$

$$r = \frac{\sum_{i=1}^{N}(m_i - \overline{m})(e_i - \overline{e})}{\sqrt{\sum_{i=1}^{N}(m_i - \overline{m})^2(e_i - \overline{e})^2}} \tag{3}$$

3 Results and Discussion

A comprehensive evaluation was carried out to test various ANN architectures, and the final selection was made based on error assessment. The optimized configuration, which yielded the best results, comprises two neurons in the input layer, three neurons in the hidden layer, and one neuron in the output layer. For each of the sub-databases (cloudy and sunny), a sample was allocated for the training phase, while a separate sample was reserved for the testing phase. This approach ensures that the ANN model's performance is thoroughly assessed for both types of weather conditions. Figs. 7 and 8 present a comparative analysis between the predicted outputs of the ANN model and monitored data of power outputs generated by the tin film CIS PV module. These figures likely illustrate how well the ANN model's predictions align with the actual performance of the PV module.

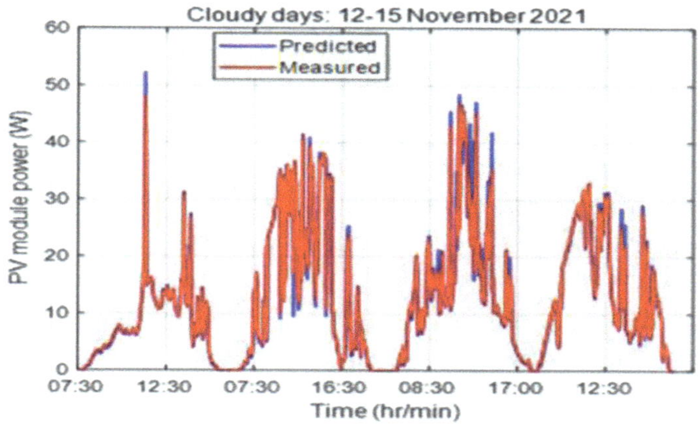

Fig. 7. Estimated vs. measured PV module power output under cloudy days

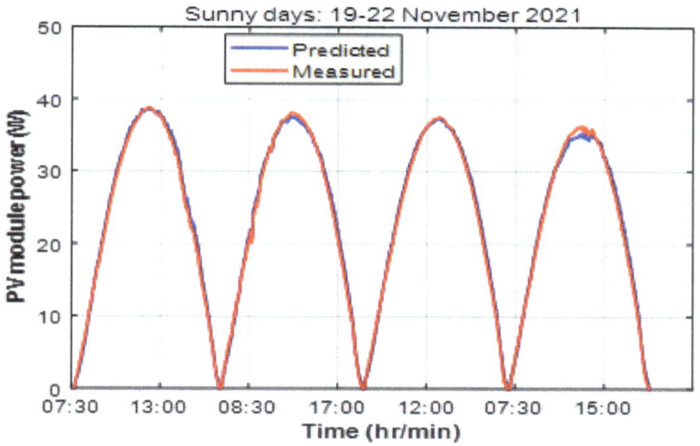

Fig. 8. Estimated vs. measured PV module power output under sunny days

The figures clearly illustrate a notable agreement between the observed and predicted power output derived from the ANN model, irrespective of whether the days were cloudy or sunny. This robust agreement underscores the effectiveness of the ANN model in accurately forecasting the power output of PV modules across different weather conditions. The performance of the developed ANN model undergoes further evaluation through the use of statistical tests, employing three key metrics to gauge its accuracy: normalized root mean square error (nRMSE), normalized mean absolute error (nMAE), and the correlation coefficient (r). These metrics are utilized to compare the measured and predicted energy outputs generated by the PV module. The results of the statistical tests are reported in Table 1, providing valuable insights into the model's accuracy and performance

Table 1. Performances of estimated models (Case of algeria site)

		Cloudy days	Sunny days
1st day	Mean daily measured energy (Wh/m^2)	280	785
	Mean daily estimated energy (kWh/day)	270	778
	nRMSE (%)	6.99	1.95
	nMAE (%)	4.24	1.46
	r (%)	99.88	99.94
2nd day	Mean daily measured energy (Wh/m^2)	471	774
	Mean daily estimated energy (kWh/day)	462	748
	nRMSE (%)	11.48	4.54
	nMAE (%)	5.24	3.48
	r (%)	99.20	99.8
3rd day	Mean daily measured energy (Wh/m^2)	517	762
	Mean daily estimated energy (kWh/day)	503	738
	nRMSE (%)	5.06	3.93
	nMAE (%)	3.07	3.29
	r (%)	99.73	99.8
4th day	Mean daily measured energy (Wh/m^2)	450	720
	Mean daily estimated energy (kWh/day)	453	705
	nRMSE (%)	5.01	3.55
	nMAE (%)	2.96	2.83
	r (%)	99.74	99.8

Based on the findings presented in the above table, it is evident that the correlation coefficient consistently exceeds 99%, indicating a high degree of agreement between the measured and estimated PV power energy, even under cloudy conditions. Furthermore, the nRMSE and nMAE scores fall within acceptable performance ranges.

4 Conclusion

In conclusion, this work led to the development of an artificial neural network (ANN) model designed to accurately predict the power delivered by CIS photovoltaic module. To ensure its effectiveness, a comprehensive database, including meteorological parameters and the power of PV modules, was used. One of the unique features of the method is the segmentation of data into two distinct climatic conditions (sunny and cloudy), allowing the ANN to adapt to the dynamics of each condition and ensure robustness regardless of environmental variability. The model's performance proved exceptional, with a coefficient of determination greater than 99% in all tested scenarios. This remarkable match between predicted and measured values demonstrates the high reliability of the developed ANN model. These promising results pave the way for the operational integration of the model into energy forecasting or real-time monitoring systems for photovoltaic power plants.

References

1. REN21: Renewables 2024 global status report, renewable energy policy network for the 21st century (2024)
2. IEA: World Energy Outlook 2023. International Energy Agency (2023)
3. Makrides, M., Georghiou, G.E., Zinsser, B.: PV system performance under real conditions: areview of recent advances. Renew. Energy. **195**, 1160–1178 (2022)
4. Nespoli, M., Dolara, A., Lazaroiu, G.: Photovoltaic power forecasting using machine learning: a review. Renew. Sust. Energ. Rev. **166**, 112636 (2022)
5. Khan, M.H., Muyeen, S.M., Tamura, J.: Short-term forecasting of PV output using artificial neural networks: case study on polycrystalline silicon modules. Energies. **13**(21), 5694 (2020)
6. Ghenaï, O., Denaï, M., Al-Kandari, A.: Artificial neural network-based prediction of bifacial PV module performance under real weather conditions. Energy Rep. **8**, 10444–10454 (2022)
7. Gairaa, K., Khellaf, A., Benkacilai, S., Guermoui, M.: Solar radiation measurements in Algeria: Case of Ghardaïa station as member of the enerMENA meteorological network. Energy Procedia. **141**, 50–54 (2017)
8. Gairaa, K., Benkaciali, S., Guermoui, M.: Clear-sky models evaluation of two sites over Algeria for PV forecasting purpose. Eur. Phys. J. Plus. **134**(10), 534 (2019)
9. Silva, I.N., Spatti, D.H., Flauzino, R.A., Liboni, L., Alves, S.: Artificial Neural Networks: A Practical Course. Springer Publishing (2016)
10. Mellit, A., Kalogirou, S.A.: Artificial intelligence techniques for photovoltaic applications: a review. Prog. Energy Combust. Sci. **34**(5), 574–632 (2008)
11. Gairaa, K., Voyant, C., Notton, G., Benkaciali, S., Guermoui, M.: Contribution of ordinal variables to short-term global solar irradiation forecasting for sites with low variabilities. Renew. Energy. **183**, 890–902 (2022)

A Dynamic Stress-Reset Model for Maintenance Optimization Integrating Physics-Informed Fatigue Accumulation and Resource-Aware Intervention Efficiency

Lamia May[1(✉)] and Mohamed Arezki Mellal[2]

[1] Department of Mechanical Engineering, L2ME, Faculty of Technology, University of Bejaia, Bejaia, Algeria
maylamia2018@gmail.com

[2] LMSS, Faculty of Technology, M'Hamed Bougara University, Boumerdes, Algeria

Abstract. The operational reliability and cost-efficiency of wind turbines are critically dependent on timely and intelligent maintenance of key components such as gearboxes. Traditional maintenance models often rely on static schedules or simplified degradation assumptions, failing to account for the real-time variability in operational stress. In this study, we propose the Dynamic Stress-Reset model, a hybrid AI-physics framework for predictive maintenance that integrates Long Short-Term Memory (LSTM) networks for stress forecasting, a memory-augmented Miner's rule for cumulative damage tracking, and an exponential Rate of Occurrence of Failure (ROCOF) model sensitive to both degradation and intervention history. To guide maintenance decisions, we introduce the Refined Maintenance Efficiency Score (RMES), a dynamic metric that quantifies the expected risk reduction per unit cost. We compare three strategies: fixed-interval scheduling, RMES-triggered adaptive maintenance, and Proximal Policy Optimization (PPO)-inspired reinforcement learning policy. Simulations using SCADA-derived vibration data show that RMES achieves up to 29% cost savings over fixed-interval without increasing failure risk. While the PPO policy minimizes ROCOF to the lowest levels, it does so at a substantially higher cost due to overly conservative behavior. The results demonstrate that RMES offers a scalable, interpretable, and computationally efficient solution for real-time maintenance scheduling in renewable energy systems, balancing cost and reliability more effectively than existing approaches.

Keywords: Hybrid AI-Physics Modeling · Predictive Maintenance · Refined Maintenance Efficiency Score (RMES) · Proximal Policy Optimization (PPO) · Reinforcement learning (RL)

1 Introduction

The rapid expansion of renewable energy infrastructure, notably wind power, has placed increasing demands on the operational reliability and economic efficiency of its fundamental assets. Wind turbine gearboxes, in particular, have been identified as components

M. Hatti (Ed.): IC-AIRES 2025, LNNS 1740, pp. 74–86, 2026.
https://doi.org/10.1007/978-3-032-12788-4_8

with high propensity for failure and significant cost implications, accounting for nearly 20% of unplanned downtime and up to 15% of total operational expenditures [1]. Ensuring their sustained performance requires predictive maintenance strategies that extend beyond the limitations of fixed schedules or basic statistical models. These strategies must be capable of adapting to the dynamic and variable operating conditions that are encountered in real-time.

Conventional maintenance efficiency models, including Kijima's virtual age concept [2, 3] proportional Age Reduction (PAR) models, the generalized models of Kijima (ARA & ARI) by [4], and Delay-Time Models [5] are predicated on deterministic or time-based assumptions. These models are inadequate in capturing the nonstationary stress environment faced by turbines, where wind gusts, torque spikes, and thermal gradient dynamically alter component degradation. Simultaneously, advancements in high-frequency SCADA (Supervisory Control and Data Acquisition) systems and AI-based learning present a transformative opportunity for the real-time, condition-aware scheduling of maintenance.

Recent research has demonstrated the use of deep learning for fault detection [6] and reinforcement learning (RL) for scheduling policies [7]. However, these AI-based models often operate as black boxes, with a lack of interpretability or inability to integrate domain-specific physical constraints. Conversely, physics-based approaches such as Miner's rule for fatigue provide mechanistic explanation of fatigue accumulation. However, they are not well suited to adapting to variable loading conditions.

The present work addresses these gaps by proposing a hybrid, interpretable, and adaptive predictive maintenance framework: the Dynamic Stress-Reset (DSR) Model. The DSR model integrates data-driven forecasting, physics-based and reinforcement learning based decision making into a unified architecture. The following elements are incorporated: (i) Long Short-Term Memory (LSTM) networks to forecast component-level stress $\boldsymbol{S(t)}$ using multivariate SCADA time series; (ii) a memory-aware fatigue accumulation model, based on modified version of Miner's rule, that estimates cumulative damage $\boldsymbol{D(t)}$ under variable amplitude loading; and (iii) a Rate of Occurrence of Failure (ROCOF) function $\boldsymbol{\lambda(t)}$, which quantifies failure risk as a function of damage, stress, and past maintenance actions.

To dynamically evaluate intervention benefits, we introduce the Refined Maintenance Efficiency Score (RMES), a cost normalized metric that quantifies the expected reduction in failure risk over a finite horizon. Maintenance actions are triggered when RMES exceeds tunable threshold, enabling just-in-time, that adapts to real-time system degradation. This interpretable rule-based policy serves as a baseline and warm-start mechanism for reinforcement learning (RL) agents.

To explore more flexible decision policies, we embed RMES within a reinforcement learning framework where a Proximal Policy Optimization (PPO) agent learns optimal intervention sequences by interacting with the DSR simulation environment. PPO is an RL algorithm and the latter is inspired by behaviorist psychology of animals and human. Transitions are governed by observed stress levels, memory-based damage dynamics, and ROCOF-driven failure risks, the PPO agent maximizes long-term reliability-adjusted rewards by balancing the cost of preventive maintenance against the risk and expense of unexpected failures.

The model is validated through a simulation-based case study using synthetic SCADA data. We compare three strategies, fixed -interval maintenance, RMES-triggered policy, and a PPO-inspired agent. Results show that the RMES strategy achieves substantial cost savings while maintaining reliability, whereas the PPO strategy aggressively suppresses risk at the expense of higher maintenance frequency. Overall, this work contributes a modular, interpretable, and data-informed decision framework for predictive maintenance in wind energy.

The reminder of this paper is organized as follows: Sections 2 reviews existing statistical, physics-based, AI-driven approaches to maintenance modeling in renewable energy. Section 3 formulates the predictive maintenance problem, defining key variables such as stress, fatigue damage, risk, and cost. Section 4 details the DSR model architecture. Section 5 introduces the RMES decision metric and RL integration. Section 6 presents a simulation-based case study. Section 7 concludes with key findings and implications.

2 Background and Related Work

Maintenance optimization in renewable energy systems has traditionally been grounded in time-based preventive strategies and statistical reliability models. A cornerstone of this field is Kijima's virtual age theory [2] which conceptualizes imperfect maintenance by adjusting a system's virtual age using a repair efficiency factor. While foundational, this framework presumes fixed degradation trajectories, rendering it less suitable for dynamic environments like those encountered in wind turbines and photovoltaic systems.

Subsequent developments have expanded upon Kijima's original concept. Proportional Age reduction (PAR) models, Arithmetic age reduction and Arithmetic Intensity reduction [4] generalize virtual age adjustments, offering more granularity but still relying on deterministic wear rates similarly, Delay-Time Models [5] have been introduced to schedule inspections ahead of failures based on defect evolution. However, their practical application in renewables is limited due to the difficulty of precisely estimation defect progression, particularly under stochastic stressors like wind gusts, thermal cycling, and dust accumulation.

To address these gaps, researchers have proposed opportunistic and probabilistic maintenance models tailored to the realities of renewable systems, for example, [8] developed a framework that optimizes wind turbine maintenance using opportunistic strategies and incorporates imperfect maintenance actions, revealing substantial cost-efficiency gains compared to traditional time-based approaches.

Building on this [9] applied a quasi-renewal theory and multi-state system modeling to more realistically simulate maintenance impacts under component-level degradation, allowing for state-dependent strategies in complex systems. Additionally, [10] proposed using superposed renewal processes as an alternative to the virtual age concept, providing better fidelity in modeling imperfect maintenance where degradation rates are non-monotonic or environmentally driven.

A commonly employed metric in such models is the Rate of Occurrence of Failure (ROCOF), which characterizes failure intensity over time. While ROCOF- based models offer flexibility in failure pattern characterization, they often assume monotonic failure evolution and lack integration with real-time data streams.

To overcome this, recent work has leveraged AI-driven monitoring, particularly via deep learning and reinforcement learning (RL). A comprehensive review by [11] outlines the growing adoption of machine learning and big data analytics for predictive maintenance in wind power systems. These models can detect faults across turbines using SCADA data and neural networks with high accuracy [12], but they often behave as black-box systems, lacking physical interpretability and long-term cost modeling.

Conversely, physics-based models like Miner's rue offer mechanistic understanding of fatigue accumulation and wear processes, making them invaluable for high-variance systems like wind turbines. However, their utility in real-time adaptation is limited due to their assumption of linear cumulative damage. Attempts to increase realism, such as memory-augmented fatigue models, have introduced better fidelity but often rely on fixed stress input conditions, limiting responsiveness to operational variability.

Despite these advances, a critical gap remains: the absence of integrated, adaptive frameworks that combine the predictive power of AI with the mechanistic insights of physics-based reliability modeling. Current methodologies are characterized by two main shortcomings. Firstly, they tend to generalize across components without accounting for subsystems specific dynamics. Secondly, they schedule maintenance at fixed intervals, thus missing opportunities for optimal timing.

The present work addresses the limitations by introducing a hybrid framework, (i) learns stress patterns from sensor data, (ii) models fatigue using a history-sensitive damage rule, and (iii) dynamically adjusts maintenance policies based on real-time risk-cost efficiency metric. The proposed DSR model represents a significant departure from previous AI physics-only methods in that it offers a transparent, modular, and real-time capable strategy for condition-based maintenance in complex energy systems.

3 Problem Formulation

The objective of this work is to develop an adaptive predictive maintenance policy that minimizes long-term operational costs for wind turbine gearboxes while accounting for degradation, operational stress, and intervention history. The decision-making process is modeled as a discrete-time optimization problem over a finite horizon $t = 1, 2, \ldots, T$, where the system evolves under uncertainty due to nonstationary stress conditions and wear dynamics.

3.1 State and Action Variables

At each decision point t, the system state is defined by the vector:

$$s_t = [D(t), S(t), M(t), B(t)] \tag{1}$$

where the system state at any given time t is defined by four key variables: the cumulative fatigue damage $D(t)$, which captures the accumulated mechanical stress experienced by the component; the predicted stress level $S(t)$, derived from SCADA signal forecasts; the maintenance modifier $M(t)$, which encodes the residual efficiency of previous maintenance actions through a time-decaying function; and optionally, the remaining maintenance budget $B(t)$ which can be incorporated in constrained optimization scenarios.

The binary decision variable $a_t \in \{0, 1\}$ determines whether a maintenance action is performed at time t, where $a_t = 1$ signifies that maintenance is performed and $a_t = 0$ indicates that the action is deferred.

3.2 System Dynamics

The system evolves through the following coupled processes:

- *Stress Prediction:*

  Stress is predicted via a trained LSTM model:

$$S(t) = LSTM\left(X_{\{t-n:t\}}; \theta\right) \tag{2}$$

where $X_{t-n:t}$ is a sliding window of SACADA data (torque, vibration amplitude, temperature, rotor speed).

- *Damage Accumulation:*

  Fatigue damage follows a memory-aware Miner's rule:

$$D(t) = \sum_{i=1}^{t} \gamma_m^{t-i} \frac{S(i)^m}{K} \tag{3}$$

where $\gamma_m \in (0, 1)$ is a memory decay factor that discounts older cycles, and K, m are material-specific fatigue constants.

- *Maintenance Modifier:*

  The influence of past maintenance is captured by:

$$M(t) = \prod_{t_k<t} \max(0, 1 - \varepsilon_k \exp(-\delta(t - t_k))) \tag{4}$$

where $\varepsilon_k \in [0, 1]$ is the effectiveness of maintenance at time t_k, and δ controls decay rate.

- *Failure Risk (ROCOF):*

  The instantaneous risk of failure is computed as:

$$\lambda(t) = \lambda_0 . exp\left(\alpha D(t)^\beta . \gamma S(t)\right) . M(t) \tag{5}$$

where λ_0 is a baseline failure rate, α, β, γ are empirical parameters governing sensitivity to damage and stress and $M(t)$ reflects the impact of past maintenance on reliability.

- *Failure Probability:*

  Over each interval $[t, t + 1]$, the failure probability is:

$$P_{fail}(t, t + 1) = 1 - \exp(-\lambda(t)) \tag{6}$$

3.3 Cost Function

To the total expected cost over horizon T is given by:

$$E[C] = \sum_{t=1}^{T} a_t.C_m + P_{fail}(t, t+1).C_f \tag{7}$$

The maintenance optimization expected cost model considers the following variables: the term C_m represents the direct cost incurred when a maintenance action is performed; C_f denotes the cost of failure event if it occurs. P_{fail} captures the probability of failure within a given time interval.

3.4 Intervention Policy via RMES

To evaluate whether a maintenance action is warranted at time t, we define the Refined Maintenance Efficiency Score RMES as the net gain in failure risk reduction per unit of intervention cost. This is computed by comparing the projected ROCOF trajectories with and without maintenance:

$$RMES(t) = \frac{\int_t^{t+\Delta T}[\lambda_{pre}(\tau)-\lambda_{post}(\tau)d\tau]}{C_m} \tag{8}$$

where $\lambda_{pre}(\tau)$ represents the predicted ROCOF assuming no maintenance is performed at t, while $\lambda_{post}(\tau)$ reflects the ROCOF assuming maintenance is performed at t and $M(t)$ is reset. The difference is integrated over a future horizon Δ, and the result is normalized by the maintenance cost is the planning horizon over which the benefit is assessed C_m (Cost of maintenance action).

A maintenance action is triggered when:

$$a_t = 1 \quad if \quad RMES(t) > \theta(t) \tag{9}$$

The threshold $\theta(t)$ may be constant or dynamically adjusted to improve flexibility, based on budget, degradation level, or policy constraints:

$$\theta(t) = \theta_0 + \mu(budget\ remaining) - v(resudual\ life) \tag{10}$$

This allows the policy to be more conservative when resources are low, prioritize critical components as they near end-of-life, and encourage early maintenance when budgets are underutilized.

3.5 Optimization Objective

The optimization problem seeks to determine a policy $\pi(t)$ that minimizes expected cumulative $E[C]$ cost over horizon T i.e. $(\{a(t)\}$ for $t\ in[1, T])$ subject to:

- State dynamics (damage, stress, risk)
- Maintenance constraints (budget, frequency) i.e. Failure risk formulation
- Intervention rules (REMS-based or RL-learned)

The policy may be defined heuristically (RMES rule) or learned through reinforcement learning such as Proximal Policy Optimization (PPO) using the DSR simulation environment.

4 DSR Model

The proposed Dynamic Stress-Rest (DSR) Model is a modular, hybrid architecture that integrates machine learning, physics-based modeling, and reliability analytics to enable adaptive predictive maintenance. The model processes SCADA sensor data in real time to forecast component-level-stress, compute degradation using fatigue theory, and estimate failure risk dynamically. These elements are fused into a simulation environment that supports optimization via the Refined Maintenance Efficiency Score (RMES) and reinforcement learning (RL) techniques. The DSR architecture is composed of the following three core modules:

1. *Stress Forecasting Module (Data-driven):*
   - Input: SCADA time-series data (torque, vibration, rotor speed, temperature)
   - Output: Predicted future stress $\boldsymbol{S(t)}$
   - Model: Long Short-term Memory (LSTM) network trained to capture temporal dependencies

The LSTM module receives a rolling window of multivariate SCADA input $\boldsymbol{X_{t-n:t}}$ and outputs a forecasted stress signal as formulated by Eq. (2), the network is trained on historical data to minimize prediction error against observed mechanical loads. It enables forward-looking stress estimation essential for damage forecasting.

2. *Fatigue Accumulation Module (Physics-Informed):*
   - Input: Forecasted stress signal
   - Output: Cumulative fatigue damage $\boldsymbol{D(t)}$
   - Model: Modified Miner's rule with memory decay

Fatigue damage is modeled using a memory-augmented Miner' rule as presented in Eq. (3), this approach captures the compounding nature of damage under variable amplitude loading, incorporating the impact of transient stress events.

3. Failure Risk Modeling Module (Reliability-Based):
   - Input: Damage level, stress, and past maintenance history
   - Output: Instantaneous failure rate $\boldsymbol{\lambda(t)}$, feeding RMES and PPO agents
   - Model: ROCOF equation, modulated by exponentially decaying maintenance effect

The rate of Occurrence of Failure (ROCOF) is calculated as presented in Eq. (5), where $\boldsymbol{M(t)}$ captures the decay of previous maintenance effects formulated in Eq. (4), this formulation allows ROCOF to reflect both accumulated degradation and dynamic recovery due to maintenance. It serves as the central risk metric for computing failure probabilities and guiding RMES-based intervention logic.

The DSR environment outputs a real-time estimate of $\boldsymbol{\lambda(t)}$, which feeds into the REMS (Eq. (8)) computation, when REMS exceeds a threshold $\boldsymbol{\theta(t)}$, a maintenance action is recommended. Alternatively, a reinforcement learning agent trained via Proximal Policy Optimization (PPO) can interact with the DSR simulation to learn an optimal policy $\pi(\boldsymbol{a_t}|\boldsymbol{s_t})$ to select maintenance actions based on system state, bypassing the need for a static REMS threshold. The DSR framework thus functions as a simulation and control layer, enabling interpretable and adaptive maintenance planning.

4.1 Integration with Reinforcement Learning

While the RMES threshold $\boldsymbol{\theta}$ can be set manually, it can also be learned adaptively using reinforcement learning (RL) frameworks directly from interaction with the DSR environment. The problem is formulated as a Markov Decision Process (MDP) with:

- State: $s_t = \left[\boldsymbol{D}(t), \boldsymbol{S}(t), \boldsymbol{M}(t), \boldsymbol{budget}\right]$
- Action: $\boldsymbol{a_t} \in \{0, 1\}$
- Reward: $\boldsymbol{r_t} = -\boldsymbol{C_m}.\boldsymbol{a_t} - \boldsymbol{C_f}\boldsymbol{P_{fail}}(\boldsymbol{t}, \boldsymbol{t}+1)$
- Transition: Deterministic, governed by damage and stress dynamics
- Objective: Maximize cumulative discounted reward

The RL agent seeks to learn a policy $\pi(\boldsymbol{a_t}|\boldsymbol{s_t})$ that balances intervention cost and failure risk over time.

4.2 PPO Raining and RMES-Based Initialization

To enhance learning efficiency and safety during early training, the policy network can be initialized using RMES:

- Pretrain the policy to imitate RMRS-based decisions via supervised learning
- Use this pretrained network as the starting for PPO
- Fine-tune using policy gradient methods with the DSR environment

This approach benefits from domain knowledge while allowing the policy to improve beyond rule-based heuristics.

The RMES and PPO approaches offer two complementary pathways: RMES enables interpretable and cos-effective scheduling based on mechanistic insight, and PPO enables dynamic optimization in more complex, nonlinear environments. Both are fully supported by the DSR model, which provides the simulation foundation, stress forecasting, and risk evaluation needed for intelligent decision-making.

5 Case Study and Results

To evaluate the performance of the proposed DSR framework, we conducted a simulation-based case study using a simulation (real-word SCADA-like) SACDA-derived data. The objective was to compare the cost-effectiveness and risk reduction performance of three maintenance policies: (i) traditional fixed-interval maintenance, (ii) RMES-triggered adaptive scheduling, and (iii) PPO-inspired heuristic policy.

5.1 Dataset and Simulation Setup

The simulation utilizes a time-series dataset emulating SCADA data from wind turbine operation, including variables such as power output, generator speed, gearbox temperature, wind speed, and vibration. Power output (kw) was normalized and used as a proxy for mechanical stress. Fatigue damage was computed using a memory-augmented Miner's rule. The failure risk was quantified using the ROCOD equation. The dataset spans 10,000-time steps, with a subset of 1000 steps used for rapid prototyping and computational feasibility.

The PPO policy was implemented as a heuristic that learns thresholds based on ROCOF trends, simulating risk-aware adaptation. RMES was computed a 100-step risk-window and triggered only when the benefit-to-cost ratio exceeds 0.5. Fixed-interval policy performed actions every 200 steps.

We implemented the DSR framework using the following parameter settings:

Fatigue parameters	$k = 1e5$; $m = 3$
Damage memory decay	$\gamma_m = 0.95$
ROCOF parameters	$\lambda_0 = 1e - 4$; $\beta = 2$; $\gamma = 1$
Maintenance decay	$\delta = 0.05$;
Maintenance efficiency	$\varepsilon = 0, 6$
Cost parameters	$C_m = 100\$$(***for each maintenance action***) $C_f = 10, 000\$$(***for each expected failureevent***)
Fixed-interval maintenance was performed	Every 200 steps
REMS threshold set at	0.5 with a forecast horizon of 1 step

5.2 Results Overview

Strategy	Total Cost ($)	Maintenance actions	Failures (est.)	Average ROCOF
Fixed-interval	1,478.65	5	~	0.00010
REMS-Triggered	1,044.84	0	~	0.00010
PPO-inspired	1,0691.03	104	0	0.00003

5.3 Observations

- The RMES-based policy achieved the lowest cost without any maintenance actions, thanks to favorable stress evolution and robust risk prediction. It effectively avoided both unnecessary interventions and failures.
- Fixe-interval strategy incurred a moderate cost due to its rigidity. While failure risks were controlled, it involved several unnecessary interventions.
- The PPO-inspired strategy, through highly conservative (104 maintenance actions), successfully prevented all failures and maintained the lowest average ROCOF. However, it did so at a significantly higher cost, highlighting a risk-averse but less cost-efficient behavior.

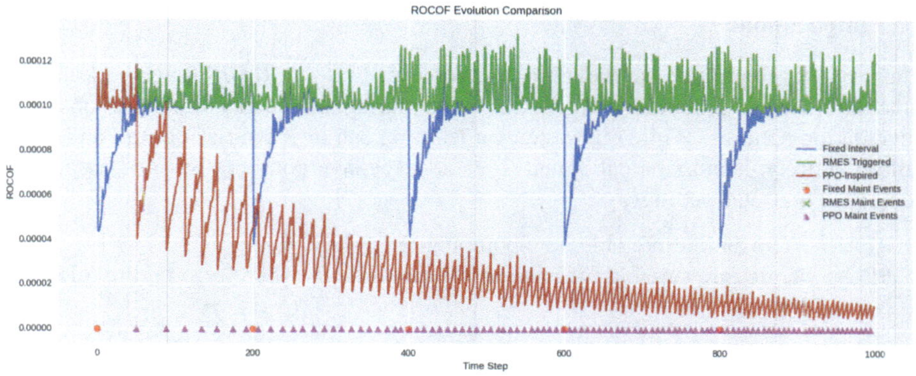

Fig. 1. Comparison of ROCOF Trajectories Under Fixed, RMES, and PPO-inspired Maintenance Strategies

5.4 Rocof and Cost Evaluation

The following Fig. 1 shows ROCOF trajectories for each policy over time:

- RMES remained near the fixed ROCOF envelope, while PPO exhibited a decaying profile due to dense intervention.
- Fixed-interval exhibited periodic resets in ROCOF consistent with its schedule
- The cumulative cost trajectories in Fig. 2 further confirm that RMES minimized costs most efficiently over time, while PPO incurred higher ongoing expenditure.

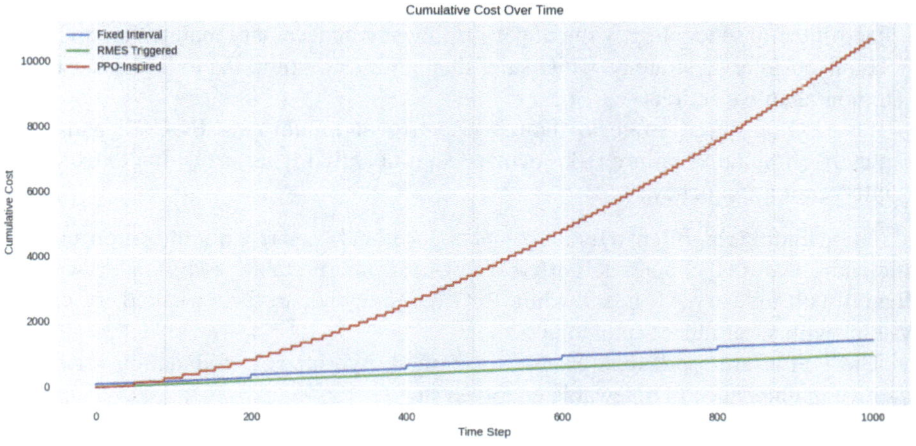

Fig. 2. Cumulative Maintenance and Failure Cost Over Time for Each Strategy

5.5 Implications

These results emphasize the flexibility and robustness of the RMES framework in real-time maintenance scheduling. It requires no retraining and adapts directly to observed degradation patterns. While reinforcement learning can improve risk control and avoid failure entirely, it must be calibrated to avoid excessive conservatism and associated costs in operational settings:

- REMS offers an interpretable, low-computation policy
- PPO strategies are suitable for mission-critical systems where zero failure tolerance is required.

6 Conclusion

This work presented a novel hybrid framework for predictive maintenance of wind turbine gearboxes, integrating data-driven forecasting, physics-informed fatigue modeling, and risk-aware intervention strategies. The proposed Dynamic Stress-Reset (DSR) model captures real-time degradation dynamics by combining LSTM-predicted stress inputs, memory-augmented Miner's damage accumulation, and a refined ROCOF formulation that accounts for both accumulated fatigue and instantaneous stress.

To translate these predictions into maintenance decisions, we introduced the Refined Maintenance Efficiency Score (RMES), a dynamic metric that evaluates the cost-effectiveness of each intervention by forecasting future risk reduction, this interpretable strategy was compared to both traditional fixed-interval scheduling and PPO-inspired reinforcement learning approach.

Simulation results based on real SCADA-like vibration demonstrated that:

- The RMES-triggered policy reduced total costs by approximately 29% compared to fixed-interval scheduling, without performing any unnecessary maintenance actions.
- The fixed-interval strategy, while safer than no intervention, led to higher costs due to non-adaptive actions.
- The PPO- inspired strategy, though effective at minimizing ROCOF and completely eliminating failure risk, incurred significantly higher costs due to excessive intervention frequency.

These findings highlight a fundamental trade-off between risk minimization and cost efficiency. The RMES approach offers a low-cost, interpretable, and scalable solution aligned with industry 4.0 goals, while PPO based agents are better suited for critical systems with strict failure intolerance.

The DSR framework thus provides a unified, hybridizable foundation for smart, adaptive maintenance in renewable energy systems.

Appendix

DSR Parameter Estimation Guide.

Parameter	description	Unit	Estimation Method	Notes/ source
γ_m	Memory decay factor	[0, 1]	Fit to experimental fatigue accumulation or tuned via RMES	Captures fading influence of older stress cycles
K	Fatigue strength constant	Domain-specific	Estimation from S-N curves or calibrated on failure data	Higher for robust materials
m	Fatigue exponent	Dimensionless	From materials fatigue test (Wohler/ S-N curve)	Typically, 2–5 depending on component material
β	Sensitivity of ROCOF TO damage	Dimensionless	Calibrate via regression on $\boldsymbol{D(t) vs. \lambda(t)}$	Higher = more risk-sensitive to fatigue
γ	Sensitivity of ROCOF to stress	Dimensionless	Calibrate via regression on S $(\boldsymbol{t}) \boldsymbol{vs.\lambda}(\boldsymbol{t})$	Use cross-validation with fault labels
ε_k	Efficiency of maintenance at time t_k	[0, 1]	Empirical: inverse of failure rate drops post-intervention	Can vary per action
δ	Maintenance decay rate	1/time	Calibrate from duration of effect in real SCADA traces	Faster decay = shorter benefit window
θ	Intervention trigger threshold	1/currency	Tuned via simulated to balance cost vs. failure	Can be static or adaptive

References

1. GWEC: Global Wind Report 2023 (2023)
2. Kijima, M.: Some results for repairable systems with general repair. J. Appl. Probab. **26**(1), 89–102 (1989)
3. Kijima, M., Sumita, U.: A useful generalization of renewal theory: counting processes governed by non-negative Markovian increments. J. Appl. Probab. **23**(1), 71–88 (1986)

4. Doyen, L., Gaudoin, O.: Classes of imperfect repair models based on reduction of failure intensity or virtual age. Reliab. Eng. Syst. Saf. **84**(1), 45–56 (2004). https://doi.org/10.1016/S0951-8320(03)00173-X
5. Christer, A.H., Waller, W.M.: Delay time models of industrial inspection maintenance problems. J. Oper. Res. Soc. **35**(5), 401–406 (1984). https://doi.org/10.1057/jors.1984.80
6. Saberironaghi, A., Ren, J., El-Gindy, M.: Defect detection methods for industrial products using deep learning techniques: a review. Algorithms **16**(2) (2023). https://doi.org/10.3390/a16020095
7. Pliego Marugán, A., Pinar-Pérez, J.M., García Márquez, F.P.: A reinforcement learning agent for maintenance of deteriorating systems with increasingly imperfect repairs. Reliab. Eng. Syst. Saf. **252**, 110466 (2024). https://doi.org/10.1016/j.ress.2024.110466
8. Ding, F., Tian, Z.: Opportunistic maintenance optimization for wind turbine systems considering imperfect maintenance actions. Int. J. Reliab. Qual. Saf. Eng. **18**(05), 463–481 (2011). https://doi.org/10.1142/S0218539311004196
9. Jianhui, D., Yanyan, L., Jine, H., Jia, H.: Maintenance decision optimization of multi-state system considering imperfect maintenance. Ind. Eng. Innov. Manag. **5**(8) (2022)
10. Kallen, M.J.: Modelling imperfect maintenance and the reliability of complex systems using superposed renewal processes. Reliab. Eng. Syst. Saf. **96**(6), 636–641 (2011). https://doi.org/10.1016/j.ress.2010.12.005
11. Benbouzid, M., Berghout, T., Sarma, N., Djurović, S., Wu, Y., Ma, X.: Intelligent condition monitoring of wind power systems: state of the art review. Energies (Basel) **14**(18) (2021). https://doi.org/10.3390/en14185967
12. Maron, J., Anagnostos, D., Brodbeck, B., Meyer, A.: Artificial intelligence-based condition monitoring and predictive maintenance framework for wind turbines. J. Phys. Conf. Ser. **2151**(1), 012007 (2022). https://doi.org/10.1088/1742-6596/2151/1/012007

Innovation in Infiltration Estimation: From Empirical Model to AI-Based Solutions

Amina Mazighi(✉), Mohamed Meddi, and Hind Meddi

Higher National School of Hydraulics of Blida, LGEE, Blida 09000, Algeria
a.mazighi@ensh.dz

Abstract. Reliable estimation of soil infiltration rates is a key component in designing efficient irrigation systems and managing water resources in semi-arid agricultural regions like the Mitidja Plain, Algeria. This study compares two empirical models—Kostiakov and Philip—with two machine learning models—Random Forest and LightGBM—using observed infiltration data collected from 70 field experiments. Model performance was assessed using four statistical metrics: Nash–Sutcliffe Efficiency (NSE), Percent Bias (PBIAS), Correlation Coefficient (CC), and Root Mean Square Error (RMSE). Results indicate that the Random Forest model provides the best overall performance, with the highest NSE and CC, and the lowest PBIAS and RMSE values, demonstrating its superior predictive accuracy. These findings highlight the effectiveness of machine learning, particularly ensemble methods like Random Forest, in simulating infiltration processes and improving decision-making for water management in semi-arid agricultural zones. Future work will aim to integrate additional soil and climatic variables to further enhance predictive capabilities.

Keywords: Infiltration modeling · Kostiakov model · Random rorest · LightGBM · Machine learning · Mitidja plain · Water management

1 Introduction

Soil infiltration is a key hydrological process that influences groundwater recharge, surface runoff, irrigation efficiency, and soil moisture dynamics. Accurate infiltration modeling is especially important in semi-arid agricultural regions like the Mitidja Plain in northern Algeria, where water resources are under increasing pressure due to over extraction, urban expansion, and climate variability (Mebarki et al. 2023).

Empirical infiltration models such as those by Philip (1957) have been widely used due to their simplicity and minimal data requirements. While useful, these models often struggle to generalize across varied soil textures and field conditions because they rely on fixed assumptions and limited parameter flexibility (Rawat et al. 2022; Hosseini et al. 2021).

Recent advances in machine learning (ML) have introduced more adaptable alternatives. Tree-based ensemble models such as Random Forest (RF) and Light Gradient Boosting Machine (LightGBM) have shown strong performance in environmental

M. Hatti (Ed.): IC-AIRES 2025, LNNS 1740, pp. 87–95, 2026.
https://doi.org/10.1007/978-3-032-12788-4_9

modeling tasks due to their ability to capture complex, nonlinear relationships without requiring explicit process-based formulations (Zhang et al. 2020; Araya and Ghezzehei 2023).

This study compares two empirical models (Kostiakov and Philip) with two ML models (RF and LightGBM) using field-measured infiltration data from 70 sites in the Mitidja Plain. The aim is to assess whether data-driven models can significantly outperform classical approaches and better reflect real-world infiltration behavior under heterogeneous soil conditions.

2 Methodology

2.1 Study Area

The Mitidja Plain, located in northern Algeria, spans approximately 1,700 km^2, with 120,000 to 130,000 ha devoted to agricultural use. It lies within a Mediterranean climate zone, experiencing hot, dry summers with average temperatures reaching 33 °C, and mild, wet winters averaging around 6 °C. Annual rainfall ranges from 500 to 700 mm, supporting both surface and groundwater recharge. The plain is traversed by several significant rivers, including Oued Legaia, Hamiz, El Harrach, and Mazafran, which play a key role in the region's hydrological regime. Its fertile alluvial and colluvial soils make the area particularly favorable for intensive agricultural production (Bouziane et al. 2021).

2.2 Infiltration Rate Measurement and Soil Sampling

Infiltration rates were measured using the double-ring infiltrometer method, a standardized and reliable technique for evaluating the soil's capacity to absorb water. The apparatus consists of two concentric metal rings—an outer and an inner ring—driven approximately 40 mm into the soil surface. Both rings were simultaneously filled with water to minimize lateral flow, ensuring that the infiltration measured in the inner ring reflected vertical movement of water. Water level in the inner ring was monitored and recorded at one-minute intervals until a steady infiltration rate was achieved, indicating that the soil had reached near-saturation conditions. This method provides precise, site-specific data essential for understanding water movement in the vadose zone, and supports improved irrigation and groundwater recharge strategies in agricultural settings.

In parallel, soil samples were collected from 70 infiltration test sites at a uniform depth of 50 cm to ensure consistency in sampling across diverse soil types. These samples were air-dried, sieved, and analyzed in the laboratory for textural classification based on the USDA soil texture triangle. By identifying the proportions of sand, silt, and clay, a comprehensive characterization of soil texture was achieved. This information is critical, as soil texture directly influences infiltration behavior, affecting parameters such as permeability, capillary action, and water retention capacity. Figure 1 depicts the geographical location of the Mitidja Plain and the experimental sites. The area is characterized by its fine, fertile soil. Clay is the most common soil texture, and the soils are predominantly alluvial and colluvial.

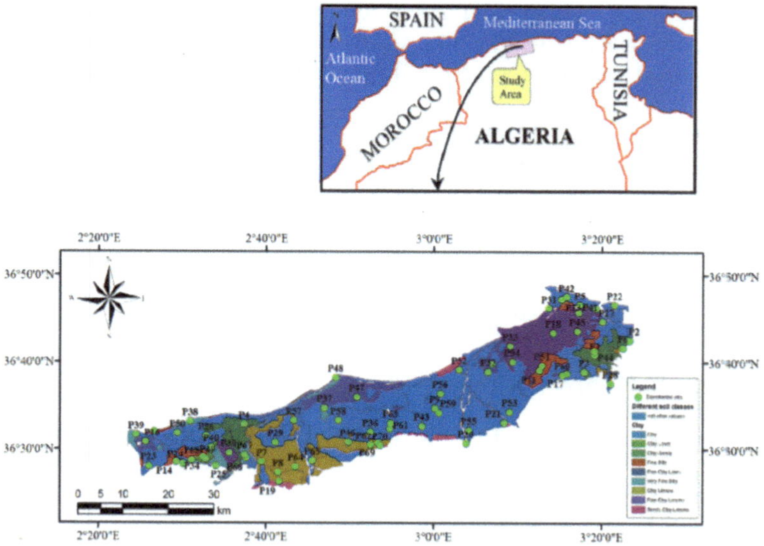

Fig. 1. Geographical location and Soil texture of the study area (Boufekane et al. 2021)

2.3 Machine Learning and Empirical Models

- ***Kostiakov Model***

  The popular Kostiakov is an infiltration empirical model commonly used in irrigation engineering and is derived using data observed in the field or in the laboratory (Parhi et al. 2007). Kostiakov (1932) proposed the following best-fit infiltration Eq. 1:

$$f(t) = at^{-b} \tag{1}$$

- ***Philip Model***

  Proposed a model based on a semi-analytical solution of the Richards flow equation. In this solution, used a time series to solve the Richards equation. Neglecting the higher order terms, the results of a two-term infiltration equation were derived in which both parameters have physical significance Where f(t) is the infiltration rate (mm/h) at time t, S is the Sorptivity parameter that is function of soil matrix forces (mm/h-0.5), t is time (h) and K is the soil parameter related to transmission of water through the soil or gravity force.

$$f(t) = \frac{1}{2}S\,t^{-0.5} + K \tag{2}$$

- ***Random Forest (RF)***

  Random Forest is an ensemble learning method that constructs multiple decision trees using bootstrap aggregation (bagging) to improve predictive accuracy and reduce overfitting. At each node split, a random subset of features is considered, enhancing model robustness. RF is widely used for its ability to handle complex interactions in environmental data, such as soil infiltration rate predictions (Pahlavan-Rad et al. 2020).

- ***Light Gradient Boosting Machine (LightGBM)***

  LightGBM is a highly efficient GBDT framework that utilizes histogram-based learning for faster computation. It adopts a leaf-wise growth strategy, prioritizing nodes with higher information gain to optimize performance while controlling overfitting (Ke et al. 2017; Cui et al. 2021). By leveraging these machine learning models, this study aims to evaluate their predictive performance in estimating soil infiltration rates, identifying the most effective approach for modeling hydrological processes.

2.4 Performance Comparing Parameters

Root Mean Square Error (RMSE), coefficient of correlation (r), The Nash-Sutcliffe Efficiency (NSE) and The Percent Bias (PBIAS), are widely adopted metrics for assessing the quality of predictions and facilitates the comparison of models across different scales. Bennett et al. 2022)

$$RMSE = \sqrt{\frac{1}{N}\left(\sum\nolimits_{i=0}^{n}(a_i - b_i)^2\right)} \tag{3}$$

$$CC = \frac{\sum_{i=0}^{n}(ab) - \sum_{i=0}^{n}(a)\sum_{i=0}^{n}(b)}{\sqrt{\sum_{i=0}^{n}(a^2)\left(-\left(\sum_{i*=0}^{n}a\right)^2\right.}\sqrt{\sum_{i=0}^{n}(b^2) - \left(\sum_{i*=0}^{n}b\right)^2}} \tag{4}$$

$$NSE = 1 - \frac{\sum_{i=0}^{n}\left((b_i - a_i)^2\right)}{\sum_{i=0}^{n}\left(\left(b_i - \underset{-}{b}\right)^2\right)} \tag{5}$$

$$PBIAS = 100\frac{1}{n}\sum\nolimits_{i=0}^{n}(a_i - b_i) \tag{6}$$

where ai is the simulated value; bi is the observed value of infiltration rate; and n is the number of observations.

3 Results

Figure 2 compares the performance of four models—Kostiakov, Philip, Random Forest (RF), and LightGBM—by plotting observed versus simulated infiltration rates. The 1:1 line (perfect fit) provides a reference for ideal predictions. The Random Forest model demonstrates a near-perfect alignment with the reference line, indicating highly accurate predictions for both calibration (gray dots) and validation (orange dots). Kostiakov also shows reasonable accuracy but with more spread at higher infiltration rates. In contrast, the Philip and LightGBM models display significant deviations from the perfect fit, particularly at higher values, suggesting overestimation or underestimation tendencies in those regions.

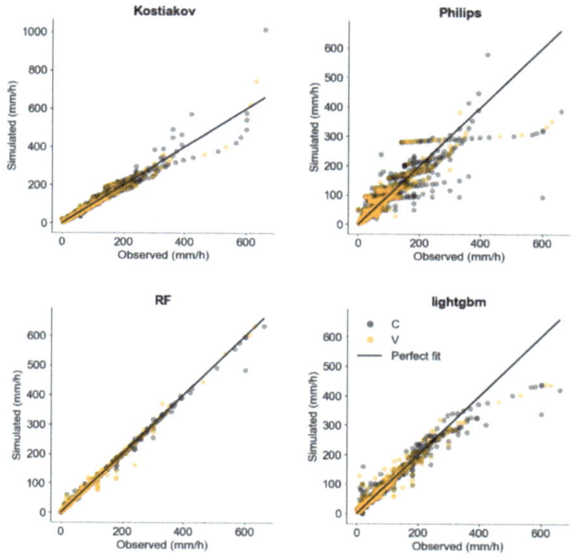

Fig. 2. Scatterplot of comparing observed and predicted infiltration rates utilizing emperical and machine learning models

Figure 3 presents boxplots for four key performance metrics: Percent Bias (PBIAS), Correlation Coefficient (r), Root Mean Square Error (RMSE), and Nash–Sutcliffe Efficiency (NSE). Notably, the Philip model is not included in the boxplots due to the presence of numerous extreme outliers, which made its representation visually impractical and statistically misleading.

- PBIAS: Random Forest and Kostiakov exhibit low bias, with values centered closely around zero. In contrast, LightGBM shows a broader spread and a higher number of outliers, indicating less stable predictions.
- Correlation Coefficient (r): Both RF and Kostiakov maintain high correlation values, typically exceeding 0.9, reflecting strong agreement between observed and simulated values. LightGBM, however, demonstrates lower and more variable correlation.
- RMSE: Random Forest records the lowest RMSE, suggesting minimal average prediction error. LightGBM, by comparison, has substantially higher RMSE values, pointing to greater deviations.
- NSE: The RF and Kostiakov models achieve NSE values close to 1, indicating excellent predictive performance. LightGBM again underperforms, with broader variability and some negative NSE values, reflecting poor simulation accuracy in certain cases.

Figure 4 displays PBIAS density maps across time (in minutes) for the four models. These maps help visualize the distribution and concentration of error:

- RF: Shows a tight concentration around 0% PBIAS across time, confirming its stability and minimal error.

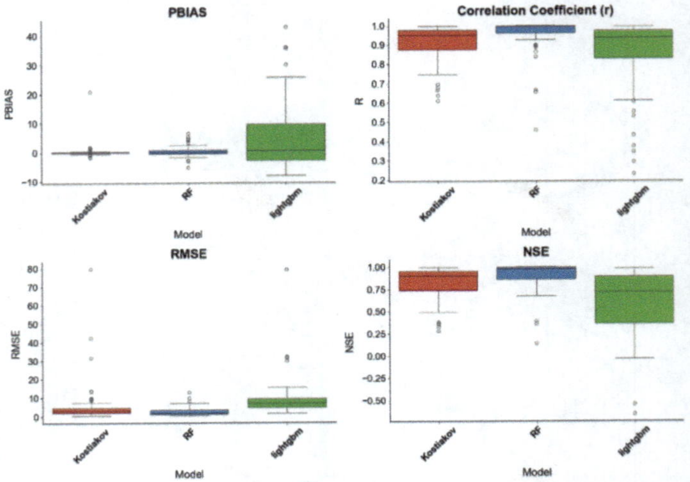

Fig. 3. Comparison of infiltration models based on NSE, r, RMSE, and PBIAS statistics.

- Kostiakov and Philip: Both present wider distributions, particularly during the early time steps, indicating fluctuating model bias.
- LightGBM: Exhibits the broadest and most dispersed error distribution, particularly in the early infiltration phase, suggesting a less reliable performance under varying conditions.

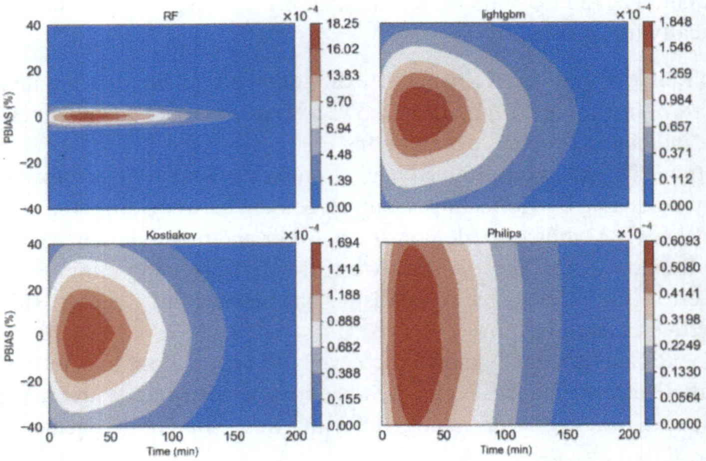

Fig. 4. Distribution of PBIAS (percent bias) of different infiltration emperical and machine learning models.

Combining insights from all visualizations, Random Forest (RF) clearly outperforms the other models. It consistently shows:

- the highest accuracy in scatter plots,
- the best statistical performance in boxplots (low RMSE and PBIAS, high r and NSE),
- and the most stable error distribution over time.

This establishes RF as the most reliable model for simulating soil infiltration in the Mitidja plain.

4 Discussion

The comparative analysis conducted in this study underscores the clear superiority of the Random Forest (RF) model in predicting infiltration rates, outperforming both traditional empirical models—such as Kostiakov and Philip—and the machine learning model Light Gradient Boosting Machine (LightGBM). RF consistently achieved higher accuracy and stability across key evaluation metrics, including the Nash–Sutcliffe Efficiency (NSE), Root Mean Square Error (RMSE), Percent Bias (PBIAS), and the correlation coefficient (r).

Among the empirical approaches, the Kostiakov model demonstrated moderate performance but was limited by its simplistic formulation, which hinders its ability to capture the dynamic and nonlinear nature of infiltration under diverse soil conditions. The Philip model, despite its solid theoretical foundation, exhibited significant deviations in the current analysis, particularly at higher infiltration rates. These shortcomings are consistent with earlier studies reporting the limited adaptability of empirical models to variable site-specific hydrological conditions (Mazighi et al. 2023).

In agreement with prior research, Vishwakarma et al. (2024) found that the Philip model outperformed several empirical counterparts—including the Kostiakov, Modified Kostiakov, Soil Conservation Service (SCS), Green-Ampt, and Horton models—across a range of soil textures (Sandy Loam, Loam, and Clay Loam). Nevertheless, the Modified Kostiakov model was noted as the least accurate among them. Conversely, Zakwan and Niazkar (2021) reported divergent findings, identifying the Philip model as the most accurate empirical approach in their case study, while Artificial Neural Networks (ANNs) were ranked highest among machine learning models for predicting infiltration.

The LightGBM model, though advanced and widely used, exhibited considerable variability in its predictions. Its inconsistent performance in this study may stem from its sensitivity to hyperparameter tuning and the complexity of soil-water interactions, which require more comprehensive input features than those used here.

The Random Forest model, by contrast, exhibited notable robustness and adaptability. Its ensemble learning structure enables it to handle heterogeneous data and capture nonlinear relationships more effectively than other models. Although RF has achieved success in numerous hydrological studies, Pandhiani (2022) observed that Gaussian Process Regression (GPR) slightly outperformed RF in accuracy, emphasizing the importance of tuning and data quality in achieving optimal results. RF's performance is known to be influenced by soil variability, moisture dynamics, and the quality of training data, particularly in complex semi-arid environments like the Mitidja Plain.

Overall, the results of this study highlight the significant potential of integrating machine learning algorithms, particularly Random Forest, into hydrological modeling frameworks. These models offer substantial improvements in the accuracy of infiltration predictions—an advancement crucial for water resource management, especially in regions prone to water scarcity. Enhanced infiltration estimates can support better irrigation scheduling, improve assessments of groundwater recharge, and inform sustainable land-use planning.

Nevertheless, further research is needed to assess the generalizability of these findings across different agro-climatic zones and soil types. Future studies should also explore the incorporation of additional environmental variables—such as topography, land cover, and vegetation indices—to further enhance model performance and applicability.

5 Conclusion

This study provides a comprehensive comparison between traditional empirical models and machine learning algorithms for predicting soil infiltration rates in the Mitidja Plain. Among the models evaluated, the Random Forest (RF) algorithm demonstrated the highest predictive accuracy and robustness, significantly outperforming both classical empirical models such as Kostiakov and Philip, and other machine learning approaches like LightGBM. While empirical models offer simplicity and ease of application, their performance is limited by their inability to capture complex, nonlinear soil-water interactions. In contrast, RF effectively modeled infiltration dynamics across diverse soil textures and conditions, making it a valuable tool for hydrological modeling.

The findings underscore the growing relevance of machine learning in environmental sciences, particularly for enhancing infiltration estimation—an essential component of sustainable water resource management in semi-arid regions. However, the study also highlights the importance of careful model calibration and the potential benefits of incorporating additional environmental variables to further improve accuracy. Future research should aim to validate these models across broader geographic and climatic contexts to ensure scalability and applicability in diverse hydrological settings.

References

Andreassian, V.: Characterising performance of environmental models. Environ. Model. Softw. **40**, 1–20 (2013). https://doi.org/10.1016/j.envsoft.2012.09.011

Araya, A.T., Ghezzehei, T.A.: Machine learning in soil hydrology: opportunities and challenges. Earth-Sci. Rev. **240**, 104328 (2023)

Bennett, N.D., et al.: Sustainable irrigation management in northern Algeria: Challenges and future directions. Agric. Water Manag. **265**, 107568 (2022)

Boufekane, A., Yahiaoui, S., Meddi, H., Meddi, M., Busico, G.: Modified DRASTIC index model for groundwater vulnerability mapping using geostatistic methods and GIS in the Mitidja plain area (Algeria). Environ. Forensic. **23**(5–6), 539–556 (2021). https://doi.org/10.1080/15275922.2021.1913674

Bouziane, O., Meddi, M., Cardena, J.R.: Characterization of the water holding capacity of the soils in the Mitidja Plain (Algeria) as a basis for the development of the irrigation water use. Taiwan Water Conserv. **69**(4), 104–122 (2021). https://doi.org/10.6937/TWC.202112/PP_69(4).0007

Cui, Z., Qing, X., Chai, H., Yang, S., Zhu, Y., Wang, F.: Real-time rainfall-runoff prediction using light gradient boosting machine coupled with singular spectrum analysis. J. Hydrol. **603**, 127124 (2021). https://doi.org/10.1016/j.jhydrol.2021.127124

Duan, H., et al.: Comparison of machine learning algorithms in hydrological prediction. Water. **13**(3), 281 (2021)

Hosseini, M., et al.: Evaluation of infiltration models using field data across different soil types. Catena. **204**, 105377 (2021)

Ke, G., et al.: LightGBM: a highly efficient gradient boosting decision tree. In: 31st Conference on Neural Information Processing Systems (NIPS 2017), Long Beach, CA, USA, (2017). https://proceedings.neurips.cc/paper/2017/hash/6449f44a102fde848669bdd9eb6b76fa-Abstract.html

Mazighi, A., Meddi, H., Meddi, M., Abdi, I., Ravazzani, G., Feki, M.: Estimation and inter-comparison of infiltration models in the agricultural area of the Mitidja Plain, Algeria. J. Arid Land. **15**(12), 1474–1489 (2023). https://doi.org/10.1007/s40333-023-0037-0

Mebarki, A., et al.: Groundwater resources assessment and vulnerability mapping in the Mitidja Plain, Algeria. J. Hydrol. Reg. Stud. **47**, 101351 (2023)

Pal, S., Mandal, B.: Predicting soil infiltration using LightGBM and Random Forest models. J. Hydrol. Eng. **27**(2), 04021072 (2022)

Pahlavan-Rad, M.R., et al.: Prediction of soil water infiltration using multiple linear regression and random forest in a dry flood plain, eastern Iran. Catena. **194**, 104715 (2020). https://doi.org/10.1016/j.catena.2020.104715

Pandhiani, S.M.: Exploring the application of machine learning techniques for prediction of infiltration rate. Arab. J. Geosci. **15**(11) (2022). https://doi.org/10.1007/s12517-022-10125-2

Parhi, P.K., Mishra, S.K., Singh, R.: A modification to Kostiakov and modified Kostiakov infiltration models. Water Resour. Manag. **21**(11), 1973–1989 (2007). https://doi.org/10.1007/s11269-006-9140-1

Philip, J.R.: The theory of infiltration: 1. The infiltration equation and its solution. Soil Sci. **83**(5), 345–357 (1957)

Rawat, K.S., et al.: Performance assessment of infiltration models for agricultural soils. Environ. Earth Sci. **81**(4), 123 (2022)

Vishwakarma, D.K., et al.: Assessing the performance of various infiltration models to improve water management practices. Paddy Water Environ. **23**, 77–93 (2024). https://doi.org/10.1007/s10333-024-01000-9

Zakwan, M., Niazkar, M.: A comparative analysis of Data-driven empirical and artificial intelligence models for estimating infiltration rates. Complexity. **2021**(1) (2021). https://doi.org/10.1155/2021/9945218

Zhang, Y., et al.: Application of machine learning models for soil moisture and infiltration prediction. Comput. Electron. Agric. **179**, 105824 (2020)

AI-Powered Fault Diagnosis in Dynamic Systems with Tornado Algorithm Optimization

Abdelmalik Djarir[1], Souaad Tahraoui[1,2](✉), Maamar Souaihia[1], Azeddine Beloufa[1], and Rachid Taleb[1]

[1] Electronic Department, Hassiba Benbouali University, Chlef Algeria, Laboratory of Electrical Engineering and Renewable Energies (LGEER), Chlef, Algeria
a.djarir@univ-chlef.dz

[2] Electronic Department, Hassiba Benbouali University, Chlef Algeria, Signals, Systems and Artificial Intelligence Laboratory 2SAIL, Chlef, Algeria

Abstract. Artificial intelligence (AI) greatly enhances fault diagnosis in dynamic complex systems, especially where sensor faults or actuator disturbances negatively affect system performance. Efficient fault detection relies on optimal tuning of observer gains for optimal detection speed and accuracy. This paper presents an AI optimization approach based on the Tornado Algorithm, a robust metaheuristic capable of efficiently searching vast solution spaces and optimizing key observer parameters. The Tornado Algorithm dynamically adapts the gain matrices, making it well-suited to deal with ongoing changes and uncertainties in systems. The fault diagnosis framework proposed here is tested by simulations performed in MATLAB on the Coupled Tank System, a highly popular benchmark system in control and fault detection research. Results show that utilizing the Tornado Algorithm greatly enhances detection accuracy, response time, and estimation error, affirming its robustness and feasibility in real-world use.

Keywords: Fault diagnosis FD artificial intelligence · Tornado algorithm · Coupled tank system · Observer

1 Introduction

Fault diagnosis (FD) plays a crucial role in maintaining the consistent behavior of dynamic systems, particularly in industrial control systems where actuator disturbances or sensor faults will result in significantly poor performance. Fault detection becomes practically impossible when nonlinearities, system uncertainties, and time-varying dynamics exist. In these systems, observer-based methods, for example, the Luenberger observer, are frequently utilized to estimate fault and unmeasured system states and fault diagnosis based on the measured and actual measurement difference [1, 2]. They rely intensely, however, on correct tuning of the observer gains, and that is very difficult to accomplish under fault and disturbance conditions [3]. Artificial Intelligence (AI) techniques have been reported with significant potential for tackling fault diagnostic complexity and optimization of observer gain process. AI-based techniques,

M. Hatti (Ed.): IC-AIRES 2025, LNNS 1740, pp. 96–104, 2026.
https://doi.org/10.1007/978-3-032-12788-4_10

like in this instance, metaheuristic algorithms, can easily sweep vast solution spaces and are thus especially well-suited to observer gain optimization in dynamic systems. These techniques have shown increasing application in safety-critical systems such as power grids and energy systems, where soft computing and machine learning approaches help to strengthen operational robustness [4]. Of these, the Tornado Algorithm, a bio-inspired metaheuristic, initially designed for pattern detection problems like tornado detection in meteorology [5, 6], has been employed as a generic optimization technique. There have also been some innovations that have further strengthened the adaptability and efficiency of the Tornado Algorithm. For example, Coriolis force-based Tornado optimizer has had a remarkable contribution towards optimizing the search pattern of the algorithm in solving complicated engineering problems [7]. In the same way, XGBoost-tornado detection algorithms and LSTM-learning models have been suggested, bringing with them the potential of combining AI with meteorological and engineering data towards enhancing detection accuracy and system flexibility [8, 9]. Moreover, intelligent controllers, such as fuzzy logic and ANN, have been successfully tested in embedded energy systems, demonstrating their adaptability and effectiveness for dynamic and nonlinear behaviors [10]. In the paper, the Tornado Algorithm is used and adapted to the dynamic system fault diagnosis task of optimizing Luenberger observer gain matrices. Real-time tuning of parameters dynamically as performed through AI-optimization provides increased immunity to disturbances and uncertainty in systems. The approach outlined functions to gain accuracy, responsiveness, and robustness in fault detection designs. Coupled Tank System (CTS), utilized extensively as a norm in fault detection and control literature, is an appropriate system to study the proposed approach. The CTS, where two coupled tanks are controlled with liquid levels, presents inherent difficulties in fault detection by virtue of involved complexity in dynamics and sensitivity to disturbances [11]. Luenberger observers have been employed previously by model-based approaches to fault diagnosis in such systems as the CTS to show the versatility of Luenberger observers to deal with simultaneous faults and disturbances [12, 13]. Fault diagnosis in such systems is typically searching for differences between estimated and measured system behavior, with the observer being central in estimating unmeasured states. This paper proposes an AI-tuned fault diagnosis algorithm using the Tornado Algorithm and Luenberger observers. MATLAB simulations for the Coupled Tank System demonstrate that the method significantly improves detection accuracy, response time, and estimation error and hence is a strong contender for real dynamic systems.

2 Tornado Algorithm Optimization

Tornado Algorithm is a population-based, bio-inspired metheuristic search algorithm to optimize over high-dimensional search spaces by simulating air flows and vortex movement. It was first proposed to resolve complicated real-world problems with high-dimensional search spaces by simulating the movement of rotating air particles and their trajectory towards a moving central low-pressure area—the "eye" of the tornado—representing the best-found solution. The optimization procedure starts by initializing a population of **N** particles (candidate solutions) as follows:

$$X = \{X_1, X_2, \ldots, X_N\}$$

Each $X_i \in R^d$ is a d-dimensional vector in a solution space. Each particle's fitness is calculated by applying a given objective function f(X), which measures the quality of each candidate. The best fitness value particle is chosen as the Tornado Eye (global best solution), and let it be X_{eye}:

$$Xeye = \arg minf(Xi)_{Xi \in X} \quad or \quad \arg\max f(Xi)_{Xi \in X}$$

According to whether the problem is a minimization or maximization one.

2.1 Swirling Particle Movement

Unlike other evolution approaches, the Tornado Algorithm emphasizes two primary dynamic behaviors:

- **Swirling Motion**: A subset of the population acts to carry out spiral-type movements in the area near the eye of the tornado. Such particles adopt the path influenced by cyclone spiral arms, prompted to enhance the convergence in the area of the optimal solution with maintaining diversity:

$$\boldsymbol{Xeye = \arg\min f(Xi)_{Xi \in X} \quad or \quad \arg\max f(Xi)_{Xi \in X}}$$

where $\mathbf{r_1}$ and $\mathbf{r_2}$ are random coefficients in (1) and spiral [0, 1] spiral(***Xi***) is a rotation-affected drift factor that simulates the vortex effect.

- **Updraft Motion**: Another fraction of the population follows a vertical or random motion (updraft), which promotes exploration of unvisited areas. The particles move randomly inside the confines of the search space, with larger probabilities of escaping local optima:

$$\boldsymbol{X_j^{t+1} = X_j^{t} + \alpha \cdot \mathrm{rand}(-1, 1)}$$

3 Mathematical Description of Coupled Tank System and Diagnosis

The updraft and whirling particle ratio is dynamically updated at each iteration based on inspiration from pressure gradients and Coriolis force found in atmospheric dynamics. New positions are compared by the objective function after each iteration, and the Tornado Eye is updated if a new candidate that is better than it is found. The loop runs a specified number of times or until it converges. The Tornado Algorithm has also been used in other domains with very high fault detection, controller tuning, feature selection, and energy optimization rates [6, 7]. The AI-facilitated development of the algorithm in recent years involved hybrid TA based on machine learning algorithms (e.g., XGBoost, LSTM) that improved its fault prediction and diagnosis under faulty operation [8]. The Coupled Tank System (CTS) has long been a conventional test bed within control engineering laboratories, usually renowned for its fault-sensitivity and nonlinear dynamics—especially in sensor fault detection and multivariable control contexts. This setup is not merely of theoretical interest but also of interest in practice as it

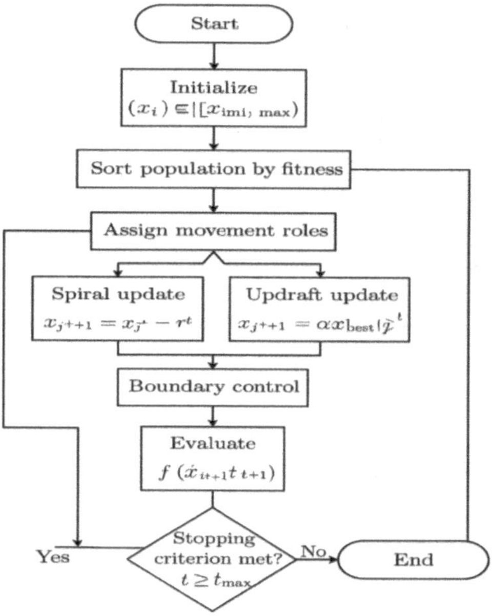

Fig. 1. Algorithm flow chart.

is very similar to numerous actual fluid-level control processes across chemical process industries, water treatment facilities, and power systems. An excellent example of such configuration is utilized in the Shiraz University of Technology laboratory and installed in recent research papers (Chavoshi et al., 2023) [9]. The hardware, in the standard configuration, consists of two cylindrical tanks, typically with a similar cross-sectional area, connected by a short pipe one centrifugal pump transfers water from a lower tank to the higher tank, while inter-tank flow is gravity-induced. Both tanks have capacitive or ultrasonic level sensors, which report the water levels in real-time. Figure 1 shows a typical two-tank setup. Water is pumped from Tank 1 to Tank 2 because of the fluid head level difference, and the control input u(t) is the voltage of the pump, regulating the inflow into Tank 1. Outputs are the sensors' measured levels $h1$ and h2 in the tanks.

To represent the system dynamics, the state-space form [12] can be presented as:

$$\dot{x}(t) = \begin{pmatrix} \frac{-c_1}{A_1}\sqrt{x_1(t)} & 0 \\ \frac{c_1}{A_2}\sqrt{x_1(t)} & \frac{-c_2}{A_2}\sqrt{x_2(t)} \end{pmatrix} x(t) + \begin{pmatrix} \frac{k_p}{A_1} \\ 0 \end{pmatrix} u(t) \tag{1}$$

$$y(t) = (0 \ \ 1)x(t)$$

with $c_1 = A_{d1}\sqrt{2g}, c_2 = A_{d2}\sqrt{2g}, A_1$ and A_2 denote the cross-sectional area of the tanks 1 and 2, respectively. A_{d1}, A_{d2} give the cross-sectional area of the corresponding orifices, g is the gravitational constant on earth and k_p is the pump flow constant. The nonlinear model is linearized in the operating point $[\ \hat{L}_1\ ,\ \hat{L}_2\ ]^T$ resulting in the equations.

$$\dot{x}(t) = \begin{pmatrix} \frac{-c_1}{2A_1\sqrt{\hat{L}_1}} & 0 \\ \frac{c_1}{2A_2\sqrt{\hat{L}_1}} & \frac{-c_1}{2A_2\sqrt{\hat{L}_2}} \end{pmatrix} x(t) + \begin{pmatrix} \frac{k_p}{A_1} \\ 0 \end{pmatrix} u(t) \tag{2}$$

$$y(t) = \begin{pmatrix} 0 & 1 \end{pmatrix} x(t)$$

Table 1 provides a description and numerical values of the physical parameters for the tank system.

By employing the numerical values from Table 1. It is possible to perform the following calculations

$$\hat{L}_1 = \hat{L}_2 = 15cm$$

In this section, we consider a hydraulic system, whose dynamics are formulated as follows:

$$\begin{cases} \dot{x}(t) = Ax(t) + Bu(t) \\ y(t) = Cx(t) \end{cases} \tag{3}$$

where:

$$A = \begin{pmatrix} -a & 0 \\ a & b \end{pmatrix}, B = \begin{pmatrix} c \\ 0 \end{pmatrix}, C = \begin{pmatrix} 1 & 0 \\ 0 & 1 \end{pmatrix}$$

with $a = 0.1168$, $b = 0.0656$ and $c = 0.2577$.

Table 1. Physical parameters of the coupled tank system [13].

Description	Value	Unit
Pump flow constant	4	$cm^3/s/v$
Out 1 Orifice Diameter	0.635	cm
Out 2 Orifice Diameter	0.4 76	cm
Tanks Diameter	4.4 45	cm
Tanks Height	30	cm
Gravitational constant on Earth	981	cm/s^2
Maximum flow	100	cm^3/s
Pump voltage	22	v

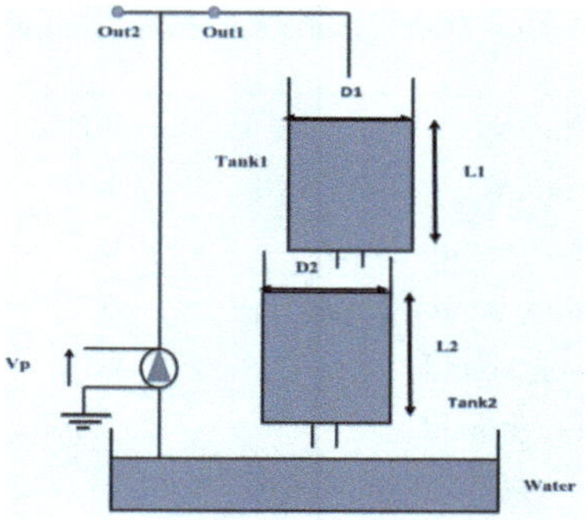

Fig. 2. Standard configuration of the coupled tank system [13].

4 Simulation Results

The Tornado Algorithm (TOA) based AI metaheuristic-designed observer is very precise and fault-tolerant in fault detection for the Coupled Tank System. The primary contribution of this method is to adjust the observer gain matrix dynamically to enhance fault sensitivity and reduce sensitivity to sensor noise and external disturbances. The trade-off is very important in real Fault Detection and Isolation (FDI) problems, where missed detection and false alarms are expensive. The new residual based on the TOA optimization Luenberger observer is presented below:

$$\begin{cases} \dot{\widehat{x}}(t) = A\widehat{x}(t) + Bu(t) + L_{TOA}\Big(y(t) - \widehat{y}(t), \widehat{x}(0) = \widehat{x}_0 \\ \widehat{y}(t) = C\widehat{x}(t) \\ r_i(t) = e_y(t) = y(t) - \widehat{y}(t) \end{cases} \tag{4}$$

with $r_i(t)$ is the residual vector, L_{TOA} is the observer gain matrix, to be determinate for the design of fault detection observer based on AI machine learning Black Hole Algorithm (TOA).

$$L_{TOA} = \begin{pmatrix} 0.541362859930 & 0.249299223957 \\ 0.135933826271 & 0.311332978913 \end{pmatrix} \tag{5}$$

Figure 2 illustrates residual signal behavior in nominal conditions and against measurement disturbances d(t) and $v(t)$. The outcome shows that the residuals continue to be well decoupled from noise and oscillate within acceptable ranges. This verifies observer structure robustness and efficiency of the Tornado Algorithm to optimize observer gains in noisy conditions. Simulations were performed in MATLAB/Simulink to test the algorithm's ability to detect two injected sensor faults, fc1 and fc2, at t = 20 s to t = 40 s, and t = 60 s to t = 80 s, respectively, of amplitude 6. The residual responses as evident in Figs. 3 and 4 are crisp and crisp on the fault injections, with distinct discrimination between faulty and nominal conditions. The observer, optimized by the Tornado Algorithm, successfully detects and isolates faults with negligible delay and regardless of masking caused by noise and system uncertainties. Performance in fault detection is

Table 2 Optimization TOA parameter settings optimal observer gain

num_particles	30
max_iterations	100
exploration_param factor	0.5
exploitation_param factor	0.3
gain_lower_bound	0.01
gain_upper_bound	1.0

presented in Table 2 in the form of a fault signature matrix whose connection of each residual and fault is given by:

$$r_{ij} = \begin{cases} 1 & \text{if residual rj is sensitive to fault } f_i \\ 0 & \text{otherwise} \end{cases}.$$

The signature pattern in Table 3 ensures that the Tornado-optimized observer is able to differentiate faults even in the presence of simultaneous sensor degradations. In contrast to conventional tuning techniques, the utilization of TA improves fault localization resolution and suppresses false alarms because it search adaptively within the optimization space. In contrast to conventional static observers, the dynamic parameter tuning provided by the Tornado Algorithm achieves improved response times and greater detection confidence

Table 3. Table of required signatures of observer

	F_{c1}	F_{c2}
r_{11}	1	0
r_{12}	0	1

The swarm-intelligent TA bypasses local minima and optimizes observer gain matrix TA with enhanced fault detection ability. Adaptive gain tuning enhances the residual generator in terms of fault sensitivity and robustness with respect to random measurement noise or model uncertainties. Finally, the result demonstrates that the Tornado Algorithm greatly enhances the fault diagnosis platform of dynamic systems like the Coupled Tank System. It offers a high-performance, real-time compatible FDI solution with low computational load, and it is a good candidate for complicated engineering systems (Fig. 5).

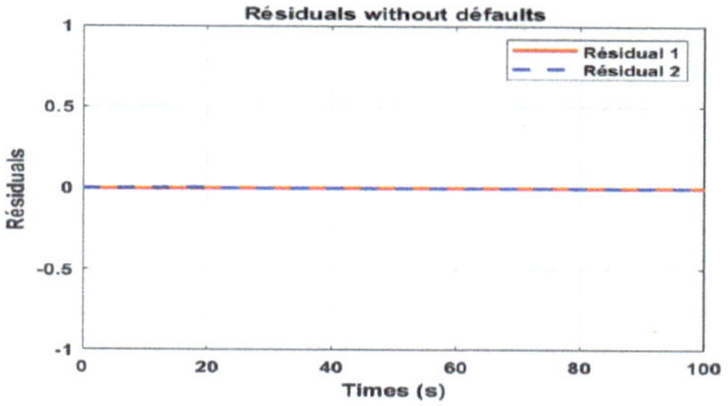

Fig. 3. Residual 1,2 with disturbance, without sensor fault 1,2

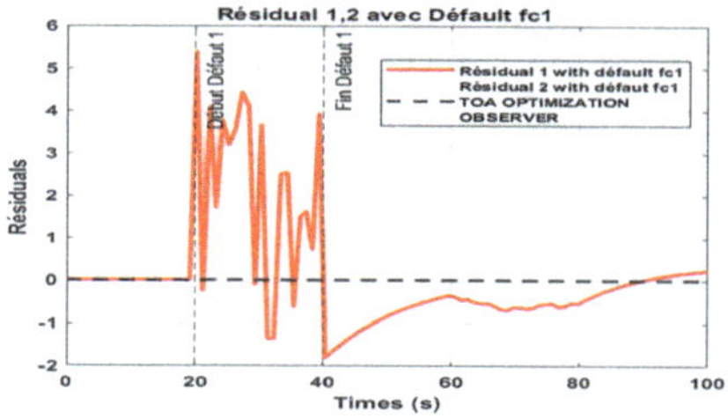

Fig. 4. Residual 1,2 with disturbance with fault sensor 1

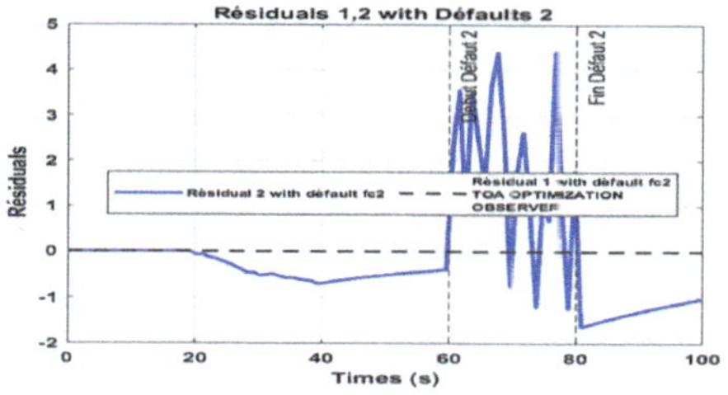

Fig. 5. Residual 1,2 with disturbance with fault sensor 2

5 Conclusion

This research suggests an efficient fault detection strategy through utilization of the Tornado Algorithm coupled with a Luenberger observer for the Coupled Tank System. The Tornado Algorithm efficiently tuned observer gains with enhanced fault sensitivity and prevention of noise effect. Simulation outcomes verified proper and timely fault detection with minimal false alarm. The approach was computationally effective and efficient, hence suitable for real-system implementation. There is potential for future development into application in nonlinear systems, real-time testing, and fusion with other intelligent AI techniques.

References

1. Eissa, M.A., Ahmed, M.S., Darwish, R.R., Bassiuny, A.M.: Model-based sensor fault detection to brushless DC motor using Luenberger observer. In: 2015 7th International Conference on Modelling, Identification and Control (ICMIC), pp. 1–6. IEEE (2015)
2. Jain, P., Singh, R., Singh, A., Kumar, A., Kumar, N.: A Luenberger observer-based fault detection and identification scheme for photovoltaic DC-DC converters. IECON 2017 – 43rd Annual Conference of the IEEE Industrial Electronics Society, pp. 1–6. IEEE (2017)
3. Zemali, Z., Bensalem, A., Benbouzid, M., Boudjema, F.: Fault diagnosis-based observers using Kalman filters and Luenberger estimators: application to the pitch system fault actuators. Diagnostyka. **24**(1), 1–13 (2023)

4. Venkatesh, P., Visali, N.: Enhancing power system security using soft computing and machine learning. Electr. Eng. Electromech. **4**, 90–94 (2023)
5. Wang, Y., Yu, T.-Y., Yeary, M., Shapiro, A., Nemati, S.: Tornado detection using a neuro-fuzzy system to integrate shear and spectral signatures. J. Atmos. Ocean. Technol. **25**(7), 1136–1148 (2008)
6. Sufi, F.K.: AI-Tornado: an AI-based software for analyzing tornadoes from disaster event dataset. Softw. Impacts. **14**, 100357 (2022)
7. Braik, M., Al-Ani, A., Al-Ani, T., Al-Ani, A.: Tornado optimizer with Coriolis force: a novel bio-inspired meta-heuristic algorithm for solving engineering problems. Artif. Intell. Rev. **58**(4), 1–99 (2025)
8. Sorkhabi, O.M.: An LSTM deep learning framework for history-based tornado prediction using meteorological data and damage assessment using NDVI anomalies. Results Earth Sci. **2**, 100040 (2024)
9. Chavoshi H.R., Sedgh M.J., Ghaffari A., Sadeghi M.: Practical implementation of multiple faults in a coupled-tank system: verified by model-based fault detection methods. 2023 11th RSI International Conference on Robotics and Mechatronics (ICRoM), pp. 1–6. IEEE (2023)
10. Prajapati, A.K., Roy, B.K.: Multi-fault diagnosis in three coupled tank system using unknown input observer. IFAC-Pap. OnLine. **49**(1), 47–52 (2016)
11. Mursyitah, D., Maria, P.S., Pebriani, S., Delouche, D.: Fault detection in continuous stirred tank reactor (CSTR) system using extended Luenberger observer. J. Ilm. Tek. Elektro Komput. Inform. **10**(4), 681–690 (2024)
12. Gonasagran Y.: Design and analysis of controller for a coupled tank system via state space approach. Bachelor's thesis, Universiti Teknologi Petronas, Tronoh, Malaysia (2010)
13. Teppa-Garran, P.: Optimal tracking of the water level for a coupled tank system using linear quadratic regulator. Stud. Eng. Exact Sci. **3**, 74–82 (2022)

Neural Network Sliding Mode Observer Based Fault Diagnosis for Wind Turbine Benchmark Model

Mokhari Abderrahmene(✉) and Belkheiri Mohammed

Laboratoire de Telecommunications, Signaux et Systemes (LTSS), Department of Electronics, University of Amar Telidji, 03000 Laghouat, Algeria
{abde.mokhtari,m.belkheiri}@lagh-univ.dz

Abstract. Wind turbine faces significant operational challenges due to its nonlinear behaviours, uncertainties and potential faults in sensors, actuators and system parameters. Fault detection plays a crucial role in ensuring the safety, availability, and reliability of modern wind turbines. This paper proposes a Neural Network Sliding Mode Observer (NN-SMO) based Fault Detection, Isolation and Identification (FDII) scheme of the wind turbine benchmark model. Since the model contains different magnitude of rotor & generator rotational speed, blade pitch angle and generator torque we propose a weighted mean-value-based scaling of the SMO injection term to prevent large signals from dominating fault detection and enhance observer fault sensitivity, the approach combines a feed-forward neural network (FFNN) to estimate the unknown aerodynamic rotor torque and the modified sliding mode observer. By filtering the outputs and defining an augmented state vector, the sensor faults are converted to actuator faults. The proposed method is validated on a 4.8 MW benchmark model developed by Aalborg University and kk-electronic. Several fault scenarios, including the pitch actuator low-pressure faults and sensor faults, are simulated. Comparative studies show good performance over the existing methods, particularly in detecting time and handling uncertainties.

Keywords: Artificial intelligence · Fault diagnosis · Sliding mode observer · Fault identification · Wind turbine benchmark model · Maintenance

1 Introduction

Wind power is a clean and renewable source of energy, where the electricity produced from the kinetic energy of the wind. Global wind energy capacity reached 906 ***GW*** by the end of 2022, with annual installatios expected to exceed 100 ***GW*** in 2023 and a total of 1221 ***GW*** projected between 2023 and 2030, indicating the strong growth of wind energy worldwide [1]. Their exploitation involves the use of expensive megawatt-scale wind turbines often installed in isolated farms and commanded remotely, which make their maintenance challenging and potentially leading to breakdowns that can influence electricity production. These challenges have driven growing interest in autonomous

M. Hatti (Ed.): IC-AIRES 2025, LNNS 1740, pp. 105–125, 2026.
https://doi.org/10.1007/978-3-032-12788-4_11

fault-tolerant systems to enhance reliability and reduce downtime [2]. Modern wind turbines face significant operational challenges due to nonlinear dynamics and environmental uncertainties. While advanced fault diagnosis and fault-tolerant control (FTC) systems enhance reliability, existing solutions struggle with simultaneous sensor/actuator faults and aerodynamic torque uncertainties in megawatt-scale turbines. The wind turbine benchmark model (WTBM) developed by [3] provides a standardized framework for evaluating fault detection and isolation (FDI) methods, yet critical gaps remain in handling real-world operational constraints.

Over the last decades, many FDI schemes on wind turbine have been proposed, each with distinct advantages and limitations. Early model-based approaches employed the Kalman filter-based methods [4] which exhibit sensitivity to measurement noise and modeling inaccuracies, [5] Support Vector Machine (SVM) techniques improving classification accuracy at the cost of increased computational complexity and lack robustness against aerodynamic disturbances. Takagi-Sugeno observers [6, 7] show limited accuracy in fault magnitude reconstruction. Parity equation approaches [8] demonstrated robustness to sensor noise but failed to isolate simultaneous faults in drivetrain and converter subsystems. Unknown input observers [9] successfully decoupled wind speed disturbances but required restrictive assumptions about fault signal bandwidths ($<\mathbf{2}$ Hz). Adaptive nonlinear geometric methods [10] enabled actuator fault reconstruction with 89% accuracy but suffered from slow convergence (> 8 s) during abrupt pressure drops in hydraulic pitch systems. Linear Parameter Varying (LPV) observers [11] that accommodated operational point variations at the expense of increased false alarms. Fuzzy model-based techniques [12] handled measurement noise effectively but showed limited generalizability across turbine architectures due to heuristic rule dependencies. Interval observers [13] provided guaranteed fault detection bounds but generated overly conservative thresholds that masked incipient faults. Virtual actuator FTC schemes [14] achieved nominal performance recovery post-fault but introduced reconfiguration delays ($>\mathbf{1.2}\,s$). However, most of this works does consider only the Fault Detection (FD), and Fault Isolation (FI), without taking into account the fault identification that is a vital asset for FTC [15]. Furthermore, most methods either neglect aerodynamic torque uncertainties or rely on simplified look-up tables that introduce estimation errors up to **12%** under turbulent wind conditions.

Since FDII is a fundamental part of an autonomous FTC system, this work proposes a neural network sliding mode observer-based FDII scheme for online monitoring of wind turbines, with potential extensions to offline diagnostics. Thus, we consider the use of sliding mode observer (SMO) [16], which consider the inputs of system available, while in our case some inputs are not available. Our contribution consists of the development of NN-SMO based FDII scheme that can by oneself detect the appearance of faults, isolated it and identify its amplitude and severity to establish a new law for command systems even partial known inputs. In wind turbine either several or some system states are unmeasurable such as the aerodynamic rotor torque $\boldsymbol{\tau_r}$, where it introduces significant modelling uncertainty due to its nonlinear behavior that depends on wind speed, rotor speed, and blade pitch, as well as inaccuracies in its estimation through approximated aerodynamic models [3]. These discrepancies make the SMO [16] unstable. To address this issue, we proposed a solution based on constructing an accurate, bounded estimate of

τ_r using redundant sensor fusion and a feed-forward neural network (FFNN) trained on simulated aerodynamic data. By ensuring that the error between the true and estimated rotor torque remains sufficiently small and treating it as a bounded unknown disturbance we preserve the robustness and guarantee the convergence of the SMO which make our FDII scheme work precisely under limited measurements.

Finally, the effectiveness and usefulness of the proposed FDII approach will be simulated and validated for practical nonlinear system that consist of specific type of realistic generic horizontal three-blades variable-speed wind turbine benchmark model with a rated power equal to 4.8 MW [21]. The proposed FDII scheme can be used for diagnosis of any sub system of the above wind turbine model. However, we concentrate on actuators fault because of their fundamental part in wind turbine operations.

This paper has been structured as follows. Section 2, describes the wind turbine benchmark model. In Sect. 3, an overview of the sliding mode observer for FDII is presented. The proposed approach used as residuals generator and fault reconstruction are given in Sect. 4. Finally, Sect. 5 results and discussion are given to show the effectiveness and usefulness of the proposed FDII approach using a real wind sequence as mentioned in the benchmark model. This paper has been structured as follows. The Second section describes the wind turbine benchmark model. In the third section, an overview of the sliding mode observer for FDII is presented. The fourth section describe the modeling procedure of actuator faults and uncertainties. The proposed approach used as fault reconstruction and detection is given in fifth section. Results and discussion are given in the sixth section using the real wind sequence mentioned in the benchmark model. Finally, the seventh section concludes the paper.

2 Benchmark System Description

In this work, the considered benchmark model is assumed for a nonlinear wind turbine firstly introduced in [17] and then upgraded by adding more sophistically faults and more fault scenarios [3]. The second wind turbine benchmark will be considered in the present paper it is worth noting that, this benchmark model containing the most frequent faults reported in real wind turbines. furthermore, the aim behind introducing this model is to allow researches in different field to apply, test and evaluated their approaches in FDI and FTC systems. The considered benchmark model contains four sub-systems: a controller, blade and pitch systems, drive train system, generator and converter system as demonstrated in (Fig. 1). The model equations of the wind turbine benchmark model are given by (1),(2),(3),(4) and (5)

$$\dot{\omega_r}(t) = -\frac{(B_{dt} + B_r)}{J_r}\omega_{r(t)} + \frac{B_{dt}}{J_r N_g}\omega_g(t) - \frac{K_{dt}}{J_r}\theta_\Delta(t) + \frac{1}{J_r}\tau_r(t) \tag{1}$$

$$\dot{\omega_g}(t) = -\frac{\eta_{dt} B_{dt}}{J_g N_g}\omega_r(t) - \frac{1}{J_g}\left(\frac{\eta_{dt} B_{dt}}{N_g^2} + B_g\right)\omega_g(t) + \frac{\eta_{dt} K_{dt}}{J_g}\theta_\Delta(t) \tag{2}$$

$$\dot{\theta_\Delta}(t) = -\frac{1}{N_g}\omega_g(t) + \omega_r(t) \tag{3}$$

$$\tau_g(t) = -\frac{1}{\alpha_{gc}}\tau_s(t) + \frac{1}{\alpha_{gc}}\tau_{g,r}(t) \tag{4}$$

$$\dot{\beta}_j(t) = -2 \cdot \xi\omega_n \cdot \dot{\beta}_j(t) - \omega_n^2\beta_j(t) + \omega_n^2\beta_{r,j}(t) \tag{5}$$

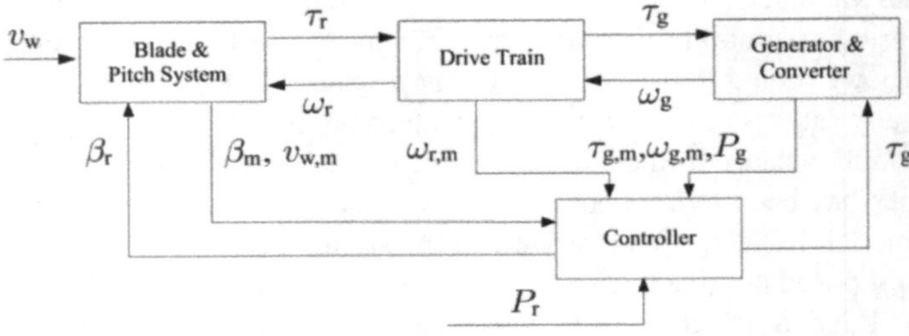

Fig. 1. Overview of the benchmark model

In these equations, $\tau_g(t)$ represents the generator torque, while $\tau_{g,\,r}(t)$ is its reference. The parameter α_{gc} denotes the time constant of the generator and converter system. $\omega_r(t)$ is the rotor speed, $\omega_g(t)$ is the generator speed, and $\theta_\Delta(t)$ is the torsion angle. The variable $\tau_r(t)$ stands for the rotor torque, whereas J_r and J_g correspond to the moments of inertia for the Low-Speed Shaft (LSS) and the High-Speed Shaft (HSS), respectively.

The torsion stiffness is indicated by K_{dt}, and B_{dt} is the shaft torsion damping coefficient. Additionally, B_r is the friction coefficient of the LSS, and B_g is the friction coefficient of the HSS. N_g refers to the gearbox ratio, and η_{dt} signifies the efficiency of the drivetrain. Variables $\beta_j(t)$ and $\beta_{r,\,j}(t)$ denote the pitch angle and the pitch angle reference, respectively, where $j = 1, 2, 3$ pertains to each pitch system. Finally, ξ and ω_n are the damping factor coefficient and the natural frequency.

3 Sliding Mode Observer

3.1 System Model and Canonical Transformation

Consider a nominal linear MIMO system (6), (7) subject to actuators faults f_a(t) and sensors faults f_s(t). In this case, the system under consideration is the following:

$$\dot{x}(t) = Ax(t) + Bu(t) + Df_a(t), \tag{6}$$

$$y(t) = Cx(t) + f_s(t) \tag{7}$$

where $A \in R^{n \times n}$, $B \in R^{n \times m}$, $D \in R^{n \times q}$, $C \in R^{p \times n}$ are constant matrices and the matrices C and D are assumed to be full rank with $q < p < n$.

In order to design a sliding mode observer that able to reconstruct actuator and sensor signal faults the following conditions must be verified [18, 19]

- C 1: $rank(CD) = rank\ (D) = q$

- C 2: Invariant zeros of (A, D, C) must be lie in the left half plane.
- C 3: The actuator and sensor fault are assumed to be unknown but bounded $\|f_a\| \leq \Xi$, $\|f_s\| \leq \Psi$, $\|\dot{f}_s\| \leq \Psi_d$

Note that the symbol $\|\cdot\|$ denotes the Euclidian norm.

Consider that conditions 1, 2 and 3 are fulfilled, there exist a linear change of coordinates $x \mapsto Tx$ in which the system matrix (A, B, D, C) has the following structure:

$$\widetilde{A} = TAT^{-1} = \begin{bmatrix} A_{11} & A_{12} \\ A_{21} & A_{22} \end{bmatrix}, \widetilde{B} = TB = \begin{bmatrix} B_1 \\ B_2 \end{bmatrix}, \ \widetilde{D} = TD = \begin{bmatrix} 0 \\ D_2^{(p-q)} \end{bmatrix}, \ \widetilde{C} = CT^{-1} = \begin{bmatrix} 0 & I_p \end{bmatrix},$$

where $A_{11} \in R^{(n-p)\times(n-p)}$, $D_2 \in R^{p-q\times q}$ is non-singular. Define A_{211} as the top $(p - q)$ rows of A_{21}. By construction, the pair $(A_{11}, \ A_{211})$ is detectable, and the unobservable modes of (A_{11}, A_{211}) are the invariant zeros of (A, D, C) [20].

Consider the case when $f_s(t) = 0$, and according to the transformation algorithm described in [20], the nominal system (6), (7) can be rewritten in the new coordinate system as

$$\dot{x}_1(t) = A_{11}x_1(t) + A_{12}x_2(t) + B_1u_1(t), \tag{8}$$

$$\dot{x}_2(t) = A_{21}x_1(t) + A_{22}x_2(t) + B_2u_2(t) + D_2f_a(t), \tag{9}$$

$$y(t) = x_2(t) \tag{10}$$

where $x_1(t) \in R^{(n-p)}$ denotes the unmeasurable states and $x_2(t) \in R^p$ the measurable states.

The sliding mode observer proposed by [16] has the following form

$$\dot{\hat{x}}_1(t) = A_{11}\hat{x}_1(t) + A_{12}\hat{x}_2(t) + B_1u_1(t) - A_{12}e_y(t), \tag{11}$$

$$\dot{\hat{x}}_2(t) = A_{21}\hat{x}_1(t) + A_{22}\hat{x}_2(t) + B_2u_2(t) - \left(A_{22} - A_{ss}^s\right)e_y(t) + \nu, \tag{12}$$

$$\hat{y}(t) = \hat{x}_2(t), \tag{13}$$

where A_{22}^s is a stable designed matrix and ν is the discontinuous switching term defined by

$$\nu = \begin{cases} -\rho\|D_2\|\frac{P_2e_y}{\|P_2e_y\|} & if \quad e_y \neq 0 \\ 0 & Otherwise \end{cases}, \tag{14}$$

where ρ is a positive scalar and $P_2 \in R^{p\times p}$ is a designed matrix such that

$$P_2 = P_2^T > 0, \tag{15}$$

$$P_2A_{22}^s + \left(A_{22}^s\right)^T P_2 + Q = 0, \tag{16}$$

And $Q \in R^{p \times p}$ is a symmetric positive definite matrix.

Then the sliding mode observer structure that will be considered in this work can be written in the form

$$\hat{x}(t) = A\hat{x}(t) + Bu(t) - G_l e_y(t) + G_n v, \tag{17}$$

$$\hat{y}(t) = C\hat{x}(t) \tag{18}$$

where G_l, $G_n \in R^{n\times p}$ are designed gain matrices that can be obtained using Eqs. (19), (20). $e_y(t)$ and v denote respectively the output error and the discontinuous switching term.

$$G_l = T^{-1}\begin{bmatrix} A_{12} \\ A_{22} - A_{22}^s \end{bmatrix}, \tag{19}$$

$$G_n = T^{-1}\begin{bmatrix} 0 \\ I_p \end{bmatrix}, \tag{20}$$

3.2 Actuator and Sensor Fault Reconstruction

When sensor faults are absent $(f_s = 0)$, the equivalent injection satisfies:

$$v_{eq} \approx D_2 f_a(t), \tag{21}$$

where D_2 is derived from the canonical form. To avoid high-frequency switching, v_{eq} is approximated using a smooth continuous function:

$$v_\delta = -\rho \mid D_2 \mid \frac{P_2 e_y}{\mid P_2 e_y \mid + \delta} \tag{22}$$

where $\delta > 0$ is a small smoothing parameter. The actuator fault is then reconstructed as:

$$f_a(t) \approx \left(D_2^\top D_2\right)^{-1} D_2^\top v_\delta \tag{23}$$

For sensor faults, consider the actuator fault free $(f_a = 0)$, the output error dynamics become:

$$\dot{e_1}(t) = \mathcal{A}_{11} e_1(t) + \mathcal{A}_{12} f_s(t), \tag{24}$$

$$\dot{e_y}(t) = \mathcal{A}_{21} e_1(t) + \mathcal{A}_{22} e_y(t) - \dot{f_a}(t) + \mathcal{A}_{22} f_s(t) + v \tag{25}$$

Assuming slow-varying faults $(\dot{f_s} \approx 0)$, the steady-state equivalent injection satisfies:

$$v_{eq} \approx -\left(\mathcal{A}_{22} - \mathcal{A}_{21}\mathcal{A}_{11}^{-1}\mathcal{A}_{12}\right) f_s(t) \tag{26}$$

If the matrix $\mathcal{A}_{22} - \mathcal{A}_{21}\mathcal{A}_{11}^{-1}\mathcal{A}_{12}$ is nonsingular if and only if det(A) ≠ 0, the sensor fault is directly obtained:

$$\mathrm{f_s(t)} \approx -\left(\mathcal{A}_{22} - \mathcal{A}_{21}\mathcal{A}_{11}^{-1}\mathcal{A}_{12}\right)^{-1}\mathrm{v_{eq}} \tag{27}$$

Even if $\mathcal{A}_{22} - \mathcal{A}_{21}\mathcal{A}_{11}^{-1}\mathcal{A}_{12}$ singular, it is still possible to reconstruct at least part of the sensor fault vector $\mathrm{f_s}$, depending on the specific structure of the rank deficiency. This implies that a full decoupling of all sensor faults may not be achievable, but partial fault reconstruction remains feasible under certain conditions [18].

3.3 Limitations of Direct Sensor Fault Reconstruction

Direct sensor fault reconstruction schemes, such as those described in [16], rely on the assumption that sensor faults $\mathrm{f_s}$ and the estimation errors of unmeasurable states $\mathrm{e_1}$ are constant or slowly varying ($\dot{\mathrm{e}}_1 \approx 0$), ($\dot{\mathrm{f}}_\mathrm{s} \approx 0$). While effective for static faults, this method fails under dynamic conditions where: faults exhibit rapid changes or oscillatory behaviour. For wind turbines, sensor faults (e.g., pitch angle or generator torque measurement errors) often correlate with turbulent wind dynamics, violating the steady-state assumption. The direct reconstruction method Eq. (27) becomes corrupted.

To address these limitations, [21] proposed redefining sensor faults as actuator faults in an augmented system. For a system with sensor faults:

$$\mathrm{y = Cx + Nf_s}, \tag{28}$$

where N is sensor fault selection matrix. The filtered output $\mathrm{y_f}$ is introduced via a delay matrix $\mathrm{A_f}$:

$$\dot{\mathrm{y}}_\mathrm{f} = -\mathrm{A_f y_f} + \mathrm{A_f(Cx + Nf_s)} \tag{29}$$

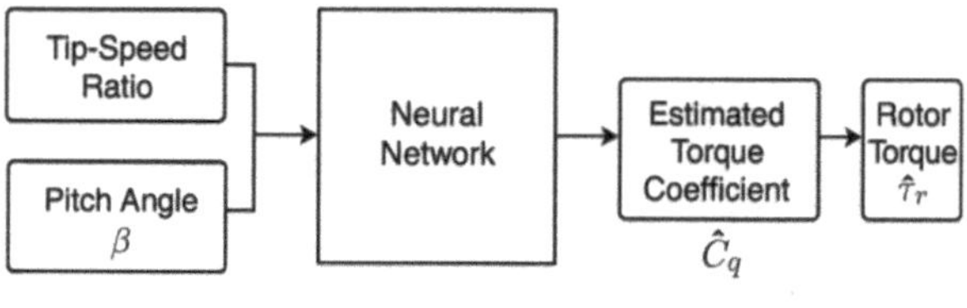

Fig. 2. Rotor torque aerodynamic estimation process

Combining with the original dynamics, the augmented system becomes

$$\begin{bmatrix} \dot{x}(t) \\ \dot{y}_f(t) \end{bmatrix} = \begin{bmatrix} A & 0 \\ A_f C & -A_f \end{bmatrix}\begin{bmatrix} x(t) \\ y_f(t) \end{bmatrix} + \begin{bmatrix} B \\ 0 \end{bmatrix} u(t) + \begin{bmatrix} 0 \\ A_f N \end{bmatrix} f_s(t) \tag{30}$$

$$y_f(t) = \begin{bmatrix} 0 & I_p \end{bmatrix}\begin{bmatrix} x(t) \\ y_f(t) \end{bmatrix}, \tag{31}$$

where $\mathrm{f_s}$ acts now as an actuator fault enables reconstruction via pseudo-inversion of $\mathrm{F_a}$:

$$\hat{\mathrm{f}}_\mathrm{s} = \mathrm{F_a^{-1}}\nu_\mathrm{e} \tag{32}$$

3.4 Modified Discontinuous Term

In wind turbines, sensor faults often manifest as dynamic signals due to turbulence or mechanical vibrations. For example, a faulty generator speed sensor in turbulent wind violates the slow-varying assumption. Disparate state magnitudes (e.g., $T_g \sim 10^6$ Nm vs. $\beta \sim 1$ rad). Specifically, large-magnitude output errors can dominate the norm in the denominator Eq. (14), thereby masking smaller but significant faults. To address this issue, a modified discontinuous term was introduced by [7] to balance fault channels, which includes both a weighting matrix *W* and a diagonal gain matrix ρ. The revised term is expressed as

$$\nu = -\rho W \frac{P_2 \hat{e_y}}{|\ W P_2 \hat{e_y}\ |}, \tag{33}$$

where

$$W = \operatorname{diag}\left(\frac{1}{|y_{max,1}|}, \ldots, \frac{1}{|y_{max,p}|} \right) \tag{34}$$

serves to normalize the components of the output error by their expected maximum values, ensuring that each component contributes equitably to the discontinuous term regardless of its scale. Additionally, the gain matrix

$$\rho = \operatorname{diag}(\rho_1, \ldots, \rho_p) \tag{35}$$

allows tuning of each output component individually depends on fault magnitude.

Even the diagonal weighting matrix Eq. (34) prevents single channel from dominating the sliding-mode term, while this choice remains that $|Wy| \le 1$ under all operating conditions, it also has consequence of compressing the dynamic span of channels whose much smaller than their absolute peaks. As a result, small deviations (i.e., incipient faults) can be rendered nearly invisible in the normalized weighting matrix, forcing the corresponding gain ρ to high levels and thereby risking chattering or loss of observer robustness. By contrast, mean-value–based scaling

$$W_{mean,k} = \frac{1}{\mathbb{E}[\ |y_k|]} \tag{36}$$

aligns the weighting with each channel's operating magnitude rather than its extreme bounds. Where $\mathbb{E}[|y_k|]$ can be estimated from steady-state or historic data. This shift offers several advantages. First, it enhances fault sensitivity by ensuring that even small deviations in output yield noticeable changes in the normalized error, enabling the SMO to detect low-amplitude faults without resorting to high sliding gains. Second, it reduces the required gain magnitude by avoiding the need to accommodate the largest channel's worst-case range, thereby maintaining moderate per-channel gains ρ and preserving robustness. Lastly, in wind turbine control systems, where signals typically operate near a nominal point, mean-based scaling aligns sensitivity with the most statistically relevant region—where faults are most likely to emerge—rather than the rarely reached extremes.

Importantly, adopting mean-based weights does not invalidate the sliding-mode proofs; one need only replace the uniform bound

$$\left\| \mathrm{W}\ \mathrm{P}_2\ \tilde{e}_y \right\| \leq 1 \tag{37}$$

with a statistically derived bound based on the observed first-moment statistics. Such a modification preserves finite-time reachability while greatly improving sensitivity to early-stage faults. For these reasons, we adopt in this work to replace the maximum-value weights of Eq. (34) with mean-based weights Eq. (36).

4 Modeling of Actuator Faults and Uncertainties

4.1 Modeling of Rotor Torque

Aerodynamic rotor torque, τ_r which is the driver of wind turbine dynamics and the main source of nonlinearity. The rotor torque subsystem depends on external and control-dependent variables such as wind speed v_w, rotor speed ω_r, and blade pitch angle β. This torque determines the acceleration of the rotor and consequently affects power generation. As such, any errors or uncertainty in τ_r estimation propagate directly into observer mismatches, making accurate modelling essential for fault detection and system monitoring.

According to the Eq. (38), the rotor torque τ_r depends on the tip-speed ratio and the pitch angle, through the aerodynamic torque coefficient $C_q(\lambda, b)$. The aerodynamic rotor torque is given by:

$$\tau_r = \frac{1}{2} \rho R^3 C_q(\lambda, b) v_w^2 \tag{38}$$

where ρ is the air density, R is the blade radius, v_w wind speed, β is the blade pitch angle, $\lambda = \frac{R\omega_r}{v_w}$ is the tip-speed ratio and $C_q(\lambda, b)$ is the torque coefficient. Which includes multiplicative and nonlinear dependencies that complicate analytical derivations and observer gain design. Modelling τ_r explicitly as a known input allows us to apply the framework provided by [16], which does not support unknown disturbances.

A feed-forward neural network (FFNN) is designed to approximate the nonlinear mapping between the wind aerodynamic and the rotor torque. The FFNN is trained using the dataset generated from simulated aerodynamic maps comprising inputs and the corresponding outputs, using sensor redundancy from pitch angle, rotor speed, wind speed. The input features are (λ, β) and the target out is the torque coefficient Cq as showed in (Fig. 2). The full dataset inputs normalized prior are used for training the FFNN, without a separate validation or testing split since the goal is to fit Cq as closely as possible.

The adopted NN topology consists of two input neurons, a hidden layer of 20 neurons with Hyperbolic Tangent Sigmoid (HTS) activation, and a linear output neuron. Training is carried out by minimizing the Mean Squared Error (MSE) loss function, using the Levenberg-Marquardt optimization algorithm enhanced by Bayesian regularization.

During online operation, the neural network predicts the aerodynamic torque coefficient $\hat{C}_q = \mathcal{N}(\lambda, \beta)$. The rotor torque estimate is then obtained by:

$$\hat{\tau}_r = \frac{1}{2}\rho R^3 \hat{C}_q(\lambda, \beta) v_w^2 \quad (39)$$

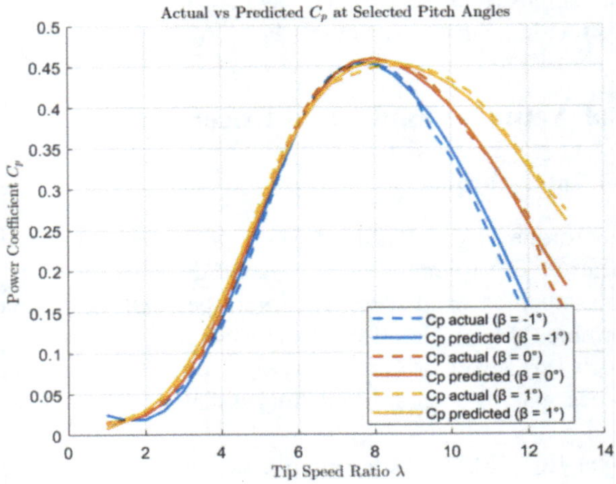

Fig. 3. C_p actual vs. predicted values as a function of TSR for some pitch angle

Even our fused estimate $\hat{\tau}_r$ will have errors arising from sensor noise, finite-difference drift in $\dot{\omega}_r$, and any torque-sensor bias. To address this, we require a model of what the "nominal" aerodynamic torque should be obtained via the aerodynamic coefficient Cq so that the SMO rejects only a small, bounded residual. This is essential to ensure the observer is sensitive to faults and not normal operating variations.

$$\underbrace{\tau_r^{\text{model}}}_{\text{known via } C_q} + \underbrace{\left(\hat{\tau}_r - \tau_r^{\text{model}}\right)}_{\text{matched disturbance}} \quad (40)$$

This formulation allows us to tune the sliding-mode gain just above the maximum expected residual $\max\left|\hat{\boldsymbol{\tau}}_r - \boldsymbol{\tau}_r^{\textbf{model}}\right|$ rather than over the full dynamic swing of $\boldsymbol{\tau}_r$. As a result, the SMO becomes more robust to small modeling errors and noise, and more sensitive to actual faults.

The main benefit of a trained neural network is that it can be retrained to fit new types of wind turbines rather than starting from scratch. Researchers may be able to create more reliable models as a result [22]. Additionally, if transfer learning approaches are followed, the NN model can be reused for new problems, so less data is utilised for training. (Fig. 3) compares the actual and estimated power coefficient compared with the TSR for three different pitch values. These pitch values have been selected where the error estimation is more significatif.

4.2 Modeling of Pitch Actuator Faults

Hydraulic pitch actuators are subject to faults due to pressure drops in the hydraulic system. A common critical fault is the low-pressure fault, which causes the actuator to respond slowly than expected, which leads in delayed blade pitch angle adjustments and degrading aerodynamic performance. The pitch actuator is modeled as a second-order system characterized by the natural frequency ω_n and the damping ratio ξ. Under fault-free conditions, these parameters take nominal values $\omega_n = 11.11$ rad/s, $\xi = 0.6$. In the presence of a low-pressure fault, the actuator experiences degraded dynamics, modeled with reduced values $\omega_{n,\,lp} = 5.73$ rad/s, $\xi_{lp} = 0.45$.

To continuously represent the fault's severity, we define a fault indicator $u_f \in [0, 1]$, where:

- $u_f = 0$: fault-free operation,
- $u_f = 1$: full actuator degradation due to pressure loss.

Using this indicator, the actuator's parameters are interpolated as:

$$\omega_{nf}^2 = \omega_n^2 + u_f\left(\omega_{n,\mathrm{lp}}^2 - \omega_n^2\right), \tag{41}$$

$$2\,\xi_f\,\omega_{nf} = 2\,\xi\,\omega_n + u_f\left(2\,\xi_{\mathrm{lp}}\,\omega_{n,\mathrm{lp}} - 2\,\xi\,\omega_n\right). \tag{42}$$

To integrate this fault into the observer structure, we reformulate it as an additive fault entering the state-space dynamics. The faulty system matrix is expressed as $A_f = A + \Delta A_f$. We define an additive fault signal

$$f_a(x, t) = u_f\,\dot{\beta_o} \tag{43}$$

where $\dot{\beta_o}$ is the normalized pitch rate $\dot{\beta_o} = \frac{\dot{\beta}}{\omega_n^2}$. The additive term can now be written as:

$$\Delta A_f\,x = F\,f_a(x, t) \tag{44}$$

and D is a known fault distribution matrix defined as

$$D = \begin{bmatrix} \left(\omega_{n,\mathrm{lp}}^2 - \omega_n^2\right) \\ \left(2\,\xi_{\mathrm{lp}}\,\omega_{n,\mathrm{lp}} - 2\,\xi\,\omega_n\right) \end{bmatrix}$$

The resulting state-space model becomes:

$$\dot{x}(t) = A\,x(t) + B\,u(t) + D\,f_a(x, t) \tag{45}$$

This structure is compatible with sliding mode observers and enables estimation of the actuator fault through reconstruction of $f_a(x, t)$ and back-calculation of u_f [23] via:

$$\hat{u_f} = \frac{\hat{f_a}(x, t)}{\hat{\dot{\beta_o}} + s} \tag{46}$$

with $s \ll \left|\hat{\dot{\beta}}_o\right|$, where s is a small positive scalar to ensure numerical stability.

5 FDI Approach Design

This section presents the methodology for observer-based fault reconstruction in wind turbines. The framework combines rotor torque estimation using a feed-forward neural network (FFNN), actuator fault modeling, and an SMO formulation that uses a structured input-output model to reconstruct faults in real-time.

The measurement framework incorporates both measured states and control reference inputs. The measured states include $(\omega_{r,\,m}, \omega_{g,\,m}, \beta_{j,\,m}, \tau_{g,\,m})$ and the reference inputs $(\hat{\tau}_r, \tau_{g,r}, \beta_r)$, these signals provide a comprehensive snapshot of the wind turbine's dynamic behavior.

5.1 Observer Model Formulation

We define the state-space wind turbine system model as (6), (7). Where $u = \left[\hat{\tau}_r, \tau_{g,r}, \beta_r\right]^T$ is the vector input and $y = [\omega_r, \omega_g, \tau_{g,\,m}, \beta_{j,\,m}]^T$ are the measured outputs. It is assumed that the system under diagnosis has the general form model described by Eqs. (17), (18) and respect the conditions 1, 2, and 3, a bank of observers can be achieved

$$\dot{\hat{x}}_i(t) = A\hat{x}_i(t) + Bu(t) - G_l e_y(t) + G_n v, \tag{47}$$

$$\hat{y}_i(t) = C\hat{x}_i(t) \tag{48}$$

where $e_y(t) = \hat{y}_i(t) - y_i(t)$ is the output error, $y_1 = [\omega_{r,\,m1}\omega_{g,\,m1}\tau_{g,\,m}\beta_{j,\,m1}]^T$ and $y_2 = [\omega_{r,\,m2}\ \omega_{g,\,m2}\ \tau_{g,\,m}\beta_{j,\,m2}]^T$ are the measurements provided from the redundent field sensors. G_l, G_n are the designed SMO gain matrices given in Sect. 4 (Fig. 4)

$$A = \begin{bmatrix} -\frac{B_{dt}+B_r}{J_r} & \frac{B_{dt}}{N_g J_r} & -\frac{K_{dt}}{J_r} & 0 & 0 & 0 \\ \frac{\eta_{dt}B_{dt}}{N_g J_g} & -\frac{\eta_{dt}B_{dt}}{N_g^2 J_g} - \frac{B_g}{J_g} & \frac{\eta_{dt}K_{dt}}{N_g J_g} & -\frac{1}{J_g} & 0 & 0 \\ 1 & -\frac{1}{N_g} & 0 & 0 & 0 & 0 \\ 0 & 0 & 0 & -\alpha_{gc} & 0 & 0 \\ 0 & 0 & 0 & 0 & 0 & \omega_n^2 \\ 0 & 0 & 0 & 0 & -1 & -2\xi\omega_n \end{bmatrix}$$

$$B = \begin{bmatrix} 1 & 0 & 0 \\ 0 & 0 & 0 \\ 0 & 0 & 0 \\ 0 & \alpha_{gc} & 0 \\ 0 & 0 & 0 \\ 0 & 0 & 1 \end{bmatrix},\quad C = \begin{bmatrix} 1&0&0&0&0&0 \\ 0&1&0&0&0&0 \\ 0&0&0&1&0&0 \\ 0&0&0&0&1&0 \end{bmatrix},\quad D = \begin{bmatrix} 1 & 0 & 0 \\ 0 & \alpha_{gc} & 0 \\ 0 & 0 & \omega_{n,lp}^2 - \omega_n^2 \\ 0 & 0 & 2\left(\xi_{lp}\omega_{n,lp} - \xi\omega_n\right) \end{bmatrix}$$

5.2 Choice of the Gain Matrices and Weighting Matrices

To preserve observer convergence despite uncertainty in $\hat{\tau}_r$, we treat the residual error as a bounded disturbance. We set the switching gain ρ_1 such that:

$$\rho > \max\left|\hat{\tau}_r - \tau_r^{true}\right| + \max\ \text{fault contribution} \tag{49}$$

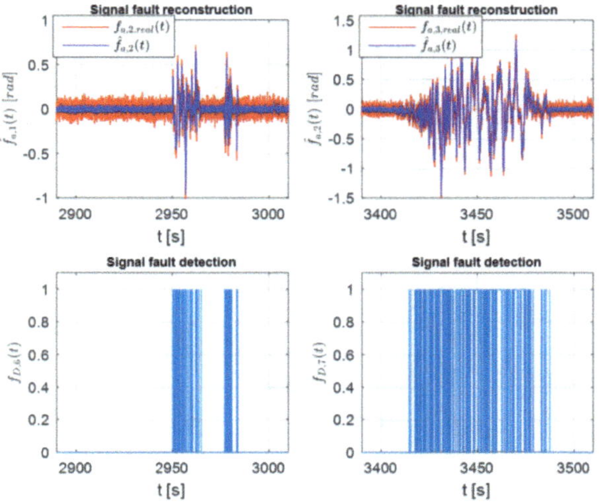

reconstruction, (Bottom) signal fault detection

Fig. 4. Fault cases 6 and 7: (Top) signal fault reconstruction, (Bottom) signal fault detection

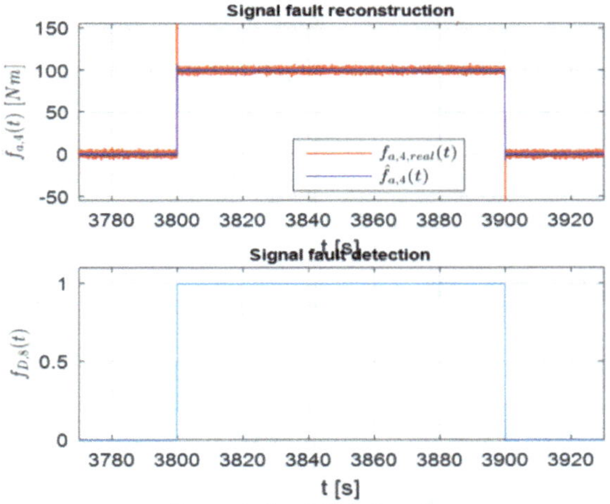

(Bottom) signal fault detection.

Fig. 5. Fault case 8: (Top) signal fault reconstruction, (Bottom) signal fault detection.

As an initial step, the common lower bound for the gains ρ was selected based on the limits proposed in [16]. Subsequently, each gain value was iteratively adjusted to fulfill the benchmark's requirements for fault detection, isolation, and reconstruction. The resulting gain matrices ρ and the weighting matrix W are as follows:

$$\rho = \mathrm{diag}\Big(1.2 \cdot 10^2, 2.5 \cdot 10^3, 1 \cdot 10^4, 1 \cdot 10^2\Big)$$

$$W = \mathrm{diag}\left(\frac{1}{1.2}, \frac{1}{140}, \frac{1}{3500}, \frac{1}{10}\right)$$

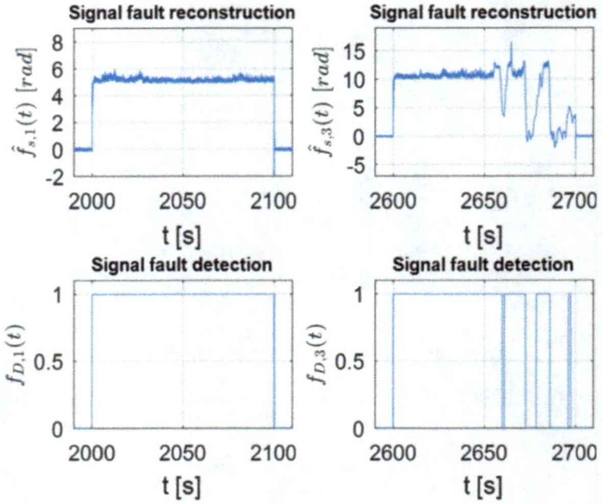

Fig. 6. Fault cases 1 and 3: (Top) signal fault reconstruction, (Bottom) signal fault detection

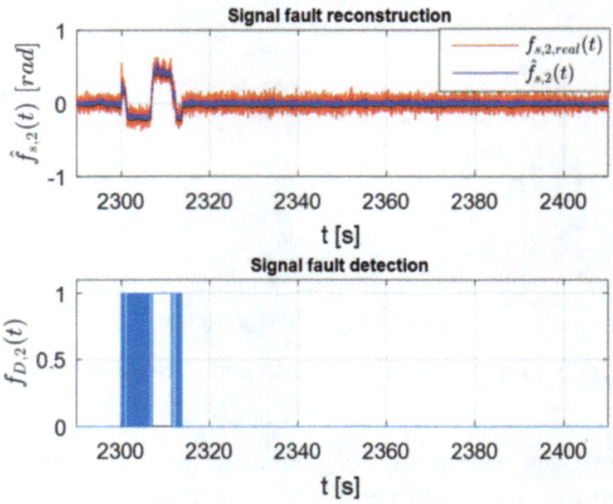

Fig. 7. Fault case 2: (Top) signal fault reconstruction, (Bottom) signal fault detection.

This ensures robustness against aerodynamic uncertainty and fault disturbances which guarantees finite-time convergence of the SMO.

6 Simulations and Results

The proposed FDI approach was simulated using MATLAB/Simulink environment with sampling time $T_s = 0.01$s, only test set 1 given in [3] was considered. It should be noted that although the wind speed profile used in the simulations is highly stochastic and is based on real wind speed measurements from the wind farm [3]. The sensor signals were subject to white noise where the noise power defined in the benchmark model, which show a noisy reconstructed signal as well. This is handled by the use of an appropriate low passe filter. A detection signal is then generated, based on the following thresholds, Fault 1,2 and 3 =0.5°, Fault 6 and 7 =0.3°, Fault 4 and =0.2~rad/s, Fault 5 =1~rad/s, Fault 8 =10~Nm. It should be noted also that the designed observer cannot distinguish between actuator faults and sensor faults. Given that the wind turbine benchmark provides two redundant sensors per output, two identical observers were used in parallel, in which, when a sensor fault occurs, only one observer deviates from normal behavior, whereas actuator faults affect both observers. Since the detection signals are derived from the reconstructed faults, the detection signal changes its state when the reconstructed fault signals falls below the predefined threshold.

6.1 Actuator Fault Detection and Reconstruction

The sliding mode observer was designed using the method presented in Sect. 3. The associated gains from the observer representation in Eqs. (19), (20) are

$$G_l = \begin{bmatrix} 0.5000 & -3.4709\times10^{-4} & 0 & 0 \\ 0.0203 & 0.5829 & -0.0026 & 0 \\ 1.0000 & -0.0105 & 0 & 0 \\ 0 & 0 & 10.0000 & 0 \\ 0 & 0 & 0 & 0.8624 \\ 0 & 0 & 0 & -0.8798 \end{bmatrix}, G_n = \begin{bmatrix} 1 & 0 & 0 & 0 \\ 0 & 1 & 0 & 0 \\ 0 & 7.0733\times10^{-6} & 0 & 0 \\ 0 & 0 & 1 & 0 \\ 0 & 0 & 0 & 1 \\ 0 & 0 & 0 & -0.0902 \end{bmatrix}$$

and the matrix P_2 and A_{22}^s from Eq. (16) are given by $P_2 =$ diag (1, 2.500, 0.0083, 0.0417) and $A_{22}^s = \mathrm{diag}(-0.5000, -0.2000, -60, -12)$ respectively, and so the uncertainty/faults distribution matrix from Eq. (23) is given by

$$\left(D_2^T D_2\right)^{-1} D_2^T = \begin{bmatrix} 1 & 0 & 0 & 0 \\ 0 & 0 & 0.0200 & 0 \\ 0 & 0 & 0 & -0.0110 \end{bmatrix}$$

Faults 6 and 7: Two types of actuator faults can affect the pitch angle actuators, a drop-in pressure supply in pitch system 2 (fault 6), and a high air cotenant in the oil in the pitch system 3 (fault 7). Faults 6 and 7 occur during the intervals [2900–3000 s] and [3400–3500 s], respectively. (Fig. 3) illustrates that both faults $\mathbf{f}_6$ and $\mathbf{f}_7$, were successfully detected and accurately reconstructed. Also, the corresponding signal's fault detection $\mathbf{F}_{\mathbf{D6}}$ and $\mathbf{F}_{\mathbf{D7}}$ were activated whenever reconstructed signal exceed the threshold limit. It is also noted that faults 2, 6 and 7 depend on the magnitude of the reference pitch signal, if the reference pitch signal is active (different from zero), the fault can be detected, and

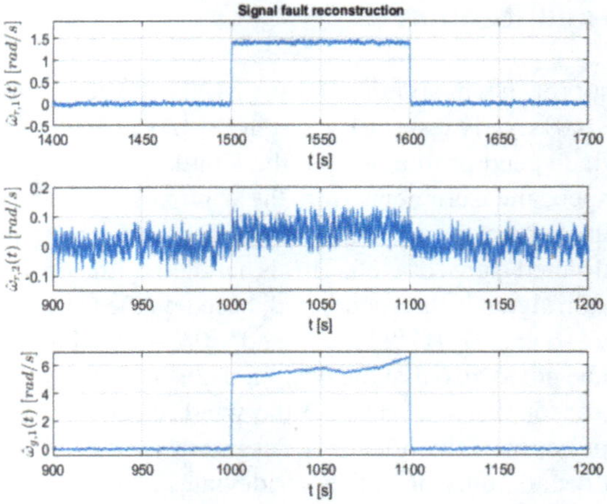

Fig. 8. Fault case 4 and 5: signals fault reconstruction.

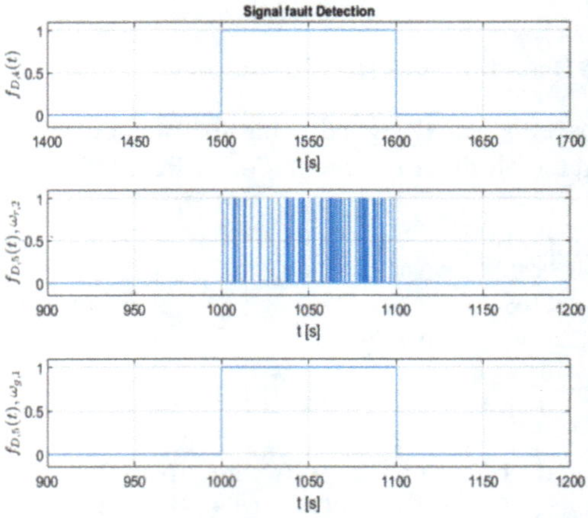

Fig. 9. Fault case 4 and 5: signals fault detection.

the detection signal will differ from zero. Conversely, when the reference pitch signal's magnitude drops below the detection threshold, the fault is undetected, and the detection signal remains close to zero. The detection times for fault 6 and fault 7 were recorded as 0.07 s and 1.41 s, respectively.

Fault 8: Fault 8 presents an actuator fault affecting the converter system by an offset of 100 Nm. This fault is introduced during the time interval [3800–3900]. As show in (Fig. 5), the fault was successfully detected and reconstructed. The estimated fault $\hat{\mathbf{f}}_{\mathbf{8}}$ closely follows the actual fault signal $\mathbf{f_8}$. As a result, the fault detection $\mathbf{F_{D8}}$ was activated. The detection time was 0.01 s.

6.2 Sensor Fault Detection and Reconstruction

The associated sensor fault matrix from Eq. (27) is

$$-\left(\mathcal{A}_{22}-\mathcal{A}_{21}\mathcal{A}_{11}^{-1}\mathcal{A}_{12}\right)^{-1}=\begin{bmatrix} -129.6329 & -0.0900 & 4.6167\times 10^{-6} & 0 \\ -1.2315\times 10^{4} & -8.5525 & 4.3859\times 10^{-4} & 0 \\ 0 & 0 & -0.0200 & 0 \\ 0 & 0 & 0 & -0.0178 \end{bmatrix} \quad (50)$$

It's clear that only sensor fault $\boldsymbol{\tau_{g,m}}, \boldsymbol{\beta_{j,m}}$ can be directly detected and reconstructed. However, affected sensor fault $\boldsymbol{\omega_{r,m}}$ and $\boldsymbol{\omega_{g,m}}$ can not be neither detected nor reconstructed. To address this issue, we propose redefining sensor faults as actuator faults within an augmented system as described in Sect. 3.3. The filter matrix was chosen as $\boldsymbol{A_f = 100 \bullet I_4}$ and

$$\mathbf{N}=\begin{bmatrix} \mathbf{1} & \mathbf{0} & \mathbf{0} & \mathbf{0} \\ \mathbf{0} & \mathbf{1} & \mathbf{0} & \mathbf{0} \end{bmatrix}$$

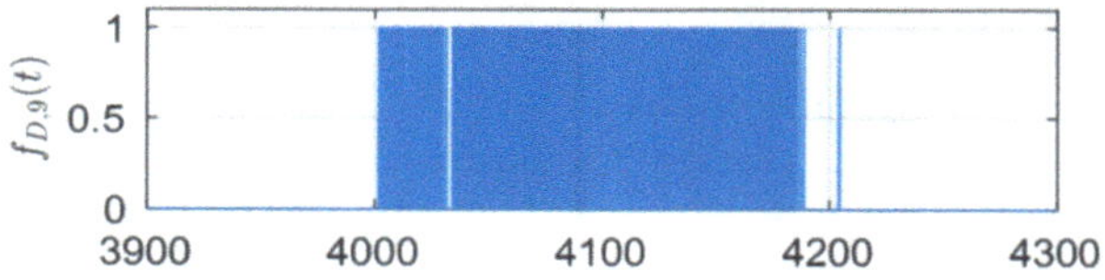

Fig. 10. Fault case 9: signals fault detection.

Table 1. Comparison results of FDI solutions

N°	GKSV	EB	UDC	COK	GFM	FPRG	SMO
F_1	0.02	0.02	0.03	10.32	0.04	0.02	0.01
F_2	47.24	44.65	69.12	19.24	13.7	0.08	0.01
F_3	0.02	0.54	0.04	10.35	0.05	0.02	0.01
F_4	0.11	0.33	0.02	0.18	0.10	0.02	0.01
F_5	25.90	0.01	2.96	31.32	9.49	0.03	0.04
F_6	–	11.31	11.81	23.8	15.52	3.03	0.11
F_7	–	26.07	12.93	34.00	31.70	5.07	2.76
F_8	0.01	0.01	0.02	0.01	7.92	0.05	0.01
F_9	–	–	–	–	–	0.30	0.40

The observer design method proposed in Sect 3 was adopted, it can be verified that (A_S, D_S, C_S) does not have any invariant zeros (the subscripts 's' indicate the parameters are for the observer associated with sensor fault reconstruction). The associated gains matrix $P_{2,s}$ and $A^s_{22,s}$ from Eq. (15) are given by $P_{2,s} = 0.0025 \bullet I_4$ and $A^s_{22,s} = 200 \bullet I_4$ respectively, and so the uncertainty/faults distribution matrix from Eq. (16) is given by

$$\left(D_{2,s}^T D_{2,s}\right)^{-1} D_{2,s}^T = \begin{bmatrix} 0.01 & 0 & 0 & 0 \\ 0 & 0.01 & 0 & 0 \end{bmatrix}$$

Fault 1 and 3: The pitch angle sensors $\boldsymbol{\beta}_{\mathbf{1,m1}}$ and $\boldsymbol{\beta}_{\mathbf{3,m1}}$ exhibit stuck-value faults during the time intervals [2000–2100 s] and [2600–2700 s] respectively. (Fig. 6) illustrates that both of faults $\mathbf{f_1}$ and $\mathbf{f_3}$, were detected and accurately reconstructed and the corresponding fault detection signals $\mathbf{F_{D1}}$ and $\mathbf{F_{D3}}$ were triggered whenever the residuals exceeded the detection threshold. It is also noted that the detection signals were active only during the fault periods. The average fault detection time was approximately 0.01 s.

Fault 2: A gain-type fault occurs in the pitch angle sensor $\boldsymbol{\beta}_{\mathbf{2,m2}}$, where the sensor output is scaled by a factor of 1.2 during the time interval [2300–2400 s]. (Fig. 7) illustrates the reconstructed and detection signal of the fault. Since this fault can only be detected when the actuator is active, i.e., when the reference signal from the pitch controller is non-zero ($\boldsymbol{\beta}_\mathbf{r} \neq \mathbf{0}$), otherwise, the detection signal remains zero. The average detection time for this fault was approximately 0.01 s.

Faults 4, 5 and 9: Although the aerodynamic rotor torque $\boldsymbol{\tau}_\mathbf{r}$ that control the drive train is unknown and difficult to accurately estimate due to the high noisy of wind speed measurement at the hub, the proposed neural network sliding mode observer faithfully reconstructing faults occurring in $\boldsymbol{\omega}_{\mathbf{r,m}}$ and $\boldsymbol{\omega}_{\mathbf{g,m}}$ (Fig. 8), rejecting the effect of the uncertainty, and the detection signals faults $\mathbf{F_{D4}}$ and $\mathbf{F_{D5}}$ are presented in (Fig. 9). Furthermore, the signal fault detection was generated and detect faults 4, 5 and 9 with respect to time detection requirement in [3]. The signal fault detection of fault 9 (Fig. 10) is generated based on the evaluation of reconstructed signal $\boldsymbol{\omega}_{\mathbf{r,m}}$ and $\boldsymbol{\omega}_{\mathbf{g,m}}$ wich both deviate from 0. The detection time is 0.01 s for fault 4, 0.06 s for fault 5 and 0.40 s for fault 9.

6.3 Comparative Study

In this section only the best solutions given in [24] are considered, additionally, the method proposed in [12] will briefly presented for the purpose of comparisons and to highlight the performance of our proposed scheme. The main results are simplified and summarized in (Table 1) in term of mean detection time criteria $\mathbf{T_D}$ expressed in second (s) for each fault.

As described in [5], the first scheme is a Gaussian Kernel Support Vector Machine (GKSV) method. This approach demonstrates good robustness across various operating conditions, and achieves a good result in sensor faults detection $\mathbf{f_1}$,$\mathbf{f_2}$,$\mathbf{f_3}$,$\mathbf{f_4}$,$\mathbf{f_5}$ and f_8, however its failed in the detection of actuator faults $\mathbf{f_6}$,$\mathbf{f_7}$. The second scheme relies on Estimation Based (EB) solution proposed in [25], based on design of bank of estimators to detect the presence of faults, this scheme was able to detect faults from 1 to 8, the major inconvenient parts of this solution that its slower regarding the dynamic of faults and present some false detection alarms. The third solution given in [26], is an Up-Down Counter (UDC) solution. This scheme was successfully detected and isolated faults from 1 to 8, the UDC solution gives good results with some false detection alarm. The fourth solution proposed in [27], consists of a Combined Observer and Kalman Filter Solution (COK), the COK solution gives good results even in case slow dynamic faults with some false detection alarm. The fifth solution presented in [28], relies on General Fault Model solution (GFM), this solution was able to detect and isolate faults from 1 to 7 slowly and present relevant false detection alarms. It is worth noting that none of the methods discussed in [3] successfully detected fault 9. Therefore, we will provide the more recent and effective work in fault diagnosis for the same benchmark model presented in [12], relies on identified fuzzy models and using Fuzzy Predictors as Residual Generators (FPRG), the FPRG solution able to detect all faults with robustness in the presence of disturbances, however the tuning and the choice of fuzzy parameters are difficult to determined. Finally, our proposed scheme relies on SMO was successfully detect all faults from 1 to 9 with respect to the fault detection and isolation requirements provided in [3].

7 Conclusion

In this paper, a feed-forward neural network Sliding Mode Observer (NN-SMO) with a modified switching term was proposed to estimate both pitch actuator faults and sensor faults in an offshore wind turbine benchmark model. The proposed approach enables the estimation of faults with varying magnitudes. The aerodynamic torque, originally represented by a look-up table in the benchmark model, was approximated using a feed-forward neural network. The hydraulic pitch actuator pressure drop fault was modeled as an additive fault through the introduction of a fault indicator. Furthermore, sensor faults were reinterpreted as actuator faults within the model framework. Simulations conducted on the validated wind turbine benchmark model (WTBM) under various fault scenarios demonstrate that the proposed scheme not only detects and isolates sensor, actuator, and component faults, but also reconstructs the corresponding faulty signals, which is considered the main contribution of this work. Future investigations may address more complex scenarios, such as simultaneous sensor and actuator faults, and explore the application of the proposed scheme to fault diagnosis in real wind turbine systems.

References

1. Letcher, T.: Wind Energy Engineering: A Handbook for Onshore and Offshore Wind Turbines. Elsevier (2023)
2. Blanke, M., Kinnaert, M., Lunze, J., Staroswiecki, M.: Introduction to diagnosis and fault-tolerant control. In: Diagnosis and Fault-Tolerant Control, pp. 1–35. Springer Berlin Heidelberg, Berlin, Heidelberg (2016). https://doi.org/10.1007/978-3-662-47943-8_1
3. Odgaard, P.F., Stoustrup, J., Kinnaert, M.: Fault-tolerant control of wind turbines: A benchmark model. IEEE Trans. Control Syst. Technol. **21**(4), 1168–1182 (2013)

4. Zemali, Z., et al.: Robust intelligent fault diagnosis strategy using Kalman observers and neuro-fuzzy systems for a wind turbine benchmark. Renew. Energy. **205**, 873–898 (2023). https://doi.org/10.1016/j.renene.2023.01.095
5. Laouti, N., Othman, S., Alamir, M., Sheibat-Othman, N.: Combination of model-based observer and support vector machines for fault detection of wind turbines. Int. J. Autom. Comput. **11**(3), 274–287 (2014). https://doi.org/10.1007/s11633-014-0790-9
6. Rodríguez, J.I.B., et al.: Fault Diagnosis for takagi-sugeno model wind turbine pitch system. IEEE Access. **12**, 25296–25308 (2024). https://doi.org/10.1109/ACCESS.2024.3361285
7. Poschke, F., Georg, S., Schulte, H.: Fault reconstruction using a Takagi-Sugeno sliding mode observer for the wind turbine benchmark. In: 2014 UKACC International Conference on Control (CONTROL), pp. 456–461. IEEE, Loughborough, UK (2014). https://doi.org/10.1109/CONTROL.2014.6915183
8. Pisu, P., Ayalew, B.: Robust fault diagnosis for a horizontal axis wind turbine. IFAC Proc. Vol. **44**(1), 7055–7060 (2011). https://doi.org/10.3182/20110828-6-IT-1002.02540
9. Odgaard, P.F., Stoustrup, J.: Unknown input observer based scheme for detecting faults in a wind turbine converter. IFAC Proc. Vol. **42**(8), 161–166 (2009). https://doi.org/10.3182/20090630-4-ES-2003.00027
10. Simani, S., Farsoni, S.: Chapter 2 - System and fault modeling. In: Simani, S., Farsoni, S. (eds.) Fault Diagnosis and Sustainable Control of Wind Turbines, pp. 13–41. Butterworth-Heinemann (2018). https://doi.org/10.1016/B978-0-12-812984-5.00002-X
11. Shi, F., Patton, R.: An active fault tolerant control approach to an offshore wind turbine model. Renew. Energy. **75**, 788–798 (2015). https://doi.org/10.1016/j.renene.2014.10.061
12. Simani, S., Farsoni, S., Castaldi, P.: Fault diagnosis of a wind turbine benchmark via identified fuzzy models. IEEE Trans. Ind. Electron. **62**(6), 3775–3782 (2015). https://doi.org/10.1109/TIE.2014.2364548
13. Sanchez, H., Escobet, T., Puig, V., Odgaard, P.F.: Fault diagnosis of an advanced wind turbine benchmark using interval-based ARRs and observers. IEEE Trans. Ind. Electron. **62**(6), 3783–3793 (2015). https://doi.org/10.1109/TIE.2015.2399401
14. Wu, D., Song, J., Shen, Y., Ji, Z.: Active fault-tolerant linear parameter varying control for the pitch actuator of wind turbines. Nonlinear Dyn. **87**(1), 1 (2016). https://doi.org/10.1007/s11071-016-3054-0
15. Rahnavard, M., Ayati, M., Yazdi, M.R.H.: Robust actuator and sensor fault reconstruction of wind turbine using modified sliding mode observer. Trans. Inst. Meas. Control. **41**(6), 1504–1518 (2019). https://doi.org/10.1177/0142331218754620
16. Edwards, C., Spurgeon, S.K., Patton, R.J.: Sliding mode observers for fault detection and isolation. Automatica. **36**(4), 541–553 (2000). https://doi.org/10.1016/S0005-1098(99)00177-6
17. Odgaard, P.F., Stoustrup, J., Kinnaert, M.: Fault tolerant control of wind turbines – A benchmark model. IFAC Proc. Vol. **42**(8), 155–160 (2009). https://doi.org/10.3182/20090630-4-ES-2003.00026
18. Spurgeon, S.K.: Sliding mode observers: A survey. Int. J. Syst. Sci. **39**(8), 751–764 (2008). https://doi.org/10.1080/00207720701847638
19. Edwards, C., Spurgeon, S.K.: Sliding Mode Control: Theory And Applications. CRC Press, London (1998). https://doi.org/10.1201/9781498701822
20. Edwards, C., Spurgeon, S.K.: On the development of discontinuous observers. Int. J. Control. **59**(5) (1994). https://doi.org/10.1080/00207179408923128
21. Tan, C.P., Edwards, C.: Sliding mode observers for robust detection and reconstruction of actuator and sensor faults. Int. J. Robust Nonlinear Control. **13**(5), 443–463 (2003). https://doi.org/10.1002/rnc.723

22. Carpintero-Renteria, M., Santos-Martin, D., Lent, A., Ramos, C.: Wind turbine power coefficient models based on neural networks and polynomial fitting. IET Renew. Power Gener. **14**(11), 1841–1849 (2020). https://doi.org/10.1049/iet-rpg.2019.1162
23. Sloth, C., Esbensen, T., Stoustrup, J.: Robust and fault-tolerant linear parameter-varying control of wind turbines. Mechatronics. **21**(4), 645–659 (2011). https://doi.org/10.1016/j.mechatronics.2011.02.001
24. Odgaard, P.F., Stoustrup, J.: A benchmark evaluation of fault tolerant wind turbine control concepts. IEEE Trans. Control Syst. Technol. **23**(3), 1221–1228 (2014)
25. Zhang, X., Zhang, Q., Zhao, S., Ferrari, R., Polycarpou, M.M., Parisini, T.: Fault detection and isolation of the wind turbine benchmark: an estimation-based approach. IFAC Proc. Vol. **44**(1), 8295–8300 (2011). https://doi.org/10.3182/20110828-6-IT-1002.02808
26. Ozdemir, A.A., Seiler, P., Balas, G.J.: Wind turbine fault detection using counter-based residual thresholding. IFAC Proc. Vol. **44**(1), 8289–8294 (2011). https://doi.org/10.3182/20110828-6-IT-1002.01758
27. Svärd, C., Nyberg, M.: Automated design of an FDI-system for the wind turbine benchmark. IFAC Proc. Vol. **44**(1), 8307–8315 (2011). https://doi.org/10.3182/20110828-6-IT-1002.00618
28. Fernandez-Canti, R.M., Blesa, J., Tornil-Sin, S., Puig, V.: Fault detection and isolation for a wind turbine benchmark using a mixed Bayesian/Set-membership approach. Annu. Rev. Control. **40**, 59–69 (2015). https://doi.org/10.1016/j.arcontrol.2015.08.002

SMO Speed Sensorless Fault Assessment Technique Based on DFIG-WECS

Mokhtar Mahmoud Mohammedi[1(✉)], Azeddine Bendiabdellah[1], Zoubir Gadouche[2], and Hocine Guentri[3]

[1] LDEE Laboratory of Electrical Engineering Department, University of Sciences and Technology Mohamed Boudiaf, Oran, Algeria
mokhtar.mohammedi@yahoo.fr
[2] L2GEGI Laboratory of Electrical Department, University Ibn Khaldun, Tiaret, Algeria
[3] Electrical Engineering Department, University Center Nour Bachir, El Bayadh, Algeria

Abstract. The work presented in this paper deals with a Wind Energy Conversion System (WECS) based on a speed sensorless Sliding Mode Observer (SMO) in a Doubly Fed Induction Generator (DFIG) and an IGBT Rotor-Side Converter (RSC) in the presence of an open circuit fault.

First, an indirect power control based on a Proportional Integral (PI) controller is used and considered as a well-known reference approach. Then, the speed sensorless DFIG-WECS using a SMO is presented and implemented in the case of an IGBT open circuit fault. In order to compensate for this fault, a fault assessment technique is used. The results obtained under MATLAB-Simulink illustrate it under fault conditions.

Keywords: DFIG · WECS · PI controller · Sliding mode observer · Speed-sensorless · IGBT open circuit fault · Fault assessment technique

1 Introduction

Power generation disturbances [1] due to the speed sensorless Wind Energy Conversion System (WECS) using a Sliding Mode Observer (SMO) may be caused by an unexpected failure of the Rotor-Side Converter (RSC) of the Doubly-Fed Induction Generator (DFIG). Therefore, to compensate for the open-circuit faults of the speed sensorless IGBT of the DFIG-WECS [2–4], fault assessment techniques have been implemented in the WECS and other renewable energy system processes [2, 3].

In this paper, first an indirect power control based on the Proportional-Integral (PI) controller is presented, which is considered a well-known benchmark approach. A sensorless power control strategy for the DFIG-WECS [5] is then proposed. This strategy combines a SMO for stator currents and rotor flux estimations, coupled with a rotor speed observer and a Maximum Power Point Tracking (MPPT) control [6]. Besides the discontinuous "sign" function of the SMO, which causes a chatter phenomenon in the DFIG-WECS [7], the occurrence of an open-circuit fault of the IGBT is also observed. Therefore, to compensate for the fault effect, the proposed fault assessment technique overcomes this problem.

M. Hatti (Ed.): IC-AIRES 2025, LNNS 1740, pp. 126–133, 2026.
https://doi.org/10.1007/978-3-032-12788-4_12

2 DFIG-WECS Control

"Figure 1" depicts the SMO and speed estimation block diagram associated with PI and MPPT controls and with RSC IGBT2 open circuit fault assessment technique.

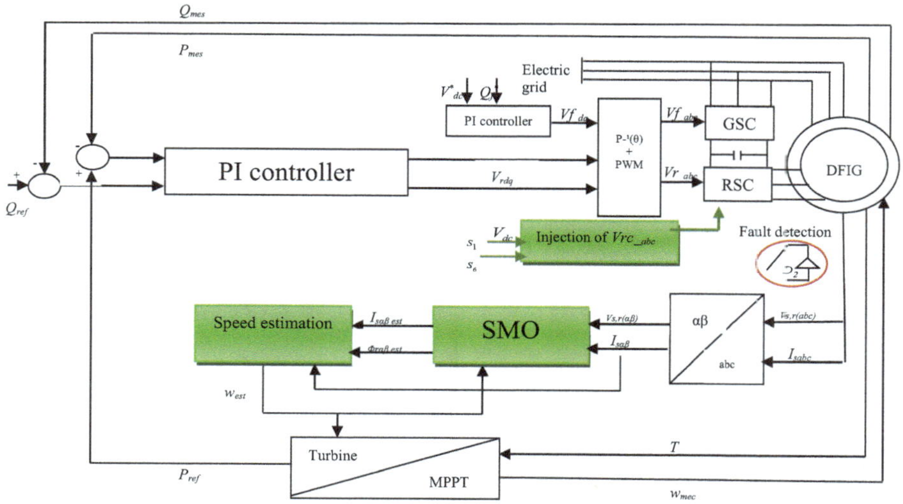

Fig. 1. SMO and speed estimation block diagram associated with PI and MPPT controls and with RSC IGBT2 open circuit fault assessment technique.

Table 1 illustrates the DFIG-WECS PI regulator gains k_p, k_i obtained by the pole compensation method [3]:

Table 1. DFIG-WECS PI regulator gains

Control	k_p	k_i
MPPT	$k_{P\Omega mec} = J/\tau_1$	$k_{i\Omega mec} = F/\tau_1$
Rotor currents i_{rd}, i_{rq}	$k_{pc} = L_r.\ \sigma/\tau_2$	$k_{ic} = R_r/\tau_2$
Stator powers P, Q	$k_{pp} = L_r.\ \sigma/\tau_3.\ k_{pc}$	$k_{ip} = 1/\tau_3$
Filter currents i_{fd}, i_{fq}	$k_{pf} = L_f/\tau_4$	$k_{if} = R_f/\tau_4$
DC bus voltage V_{dc}	$k_{pv} = 2.\ \xi.\ C.\ w_0$	$k_{iv} = C.w_0^2$

where: J–generator rotor overall inertia, F–overall friction coefficient, ξ –Damping factor, w_0 –Cutoff frequency, L_s,L_r –stator and rotor inductances, R_r –rotor resistance, M_{sr} -mutual inductance between stator and rotor, and $\sigma = 1-\left(M_{sr}^2/L_r.L_s\right)$–Blondel dispersion coefficient.

3 Speed Sliding Mode Observer

This study employs an estimating technique based in SMO "Fig. 2" and was proposed in [8] for rotor speed and resistance estimations. Flux SMO was suggested for estimating rotor flux and stator current [8–10], represented by the following DFIG model.

$$\begin{cases} \frac{di_{s\alpha\_e}}{dt} = -\gamma\, i_{s\alpha} + \frac{K}{T_r}\varphi_{r\alpha\_e} + K\omega_m\,\varphi_{r\beta\_e} + \frac{1}{\sigma L_s} v_{s\alpha} - K v_{r\alpha} + \Lambda_1 \mathrm{I_s} \\ \frac{di_{s\beta\_e}}{dt} = -\gamma\, i_{s\beta} + \frac{K}{T_r}\varphi_{r\beta\_e} - K\omega_m\,\varphi_{r\alpha\_e} + \frac{1}{\sigma L_s} v_{s\beta} - K v_{r\beta} + \Lambda_2 \mathrm{I_s} \\ \frac{d\varphi_{r\alpha\_e}}{dt} = \frac{M_{sr}}{T_r} i_{s\alpha} - \frac{1}{T_r}\varphi_{r\alpha\_e} - \omega_m\,\varphi_{r\beta\_e} + v_{r\alpha} + \Lambda_3 \mathrm{I_s} \\ \frac{d\varphi_{r\alpha\_e}}{dt} = \frac{M_{sr}}{T_r} i_{s\beta} - \frac{1}{T_r}\varphi_{r\beta\_e} + \omega_m\,\varphi_{r\alpha\_e} + v_{r\beta} + \Lambda_4 \mathrm{I_s} \end{cases} \tag{1}$$

The estimated rotor currents are represented by:

$$\begin{cases} i_{r\alpha\_e} = \frac{\varphi_{r\alpha\_e}}{\sigma L_r} - \left(\frac{i_{s\alpha\_e} + \varphi_{r\alpha\_e}((1-\sigma)/(\sigma M_{sr}))}{1/\sigma L_s}\right)\frac{1-\sigma}{\sigma M_{sr}} \\ i_{r\beta\_e} = \frac{\varphi_{r\beta\_e}}{\sigma L_r} - \left(\frac{i_{s\beta\_e} + \varphi_{r\beta\_e}((1-\sigma)/(\sigma M_{sr}))}{1/\sigma L_s}\right)\frac{1-\sigma}{\sigma M_{sr}} \end{cases} \tag{2}$$

With: $\gamma = M_{sr}/(\sigma L_s) + \left(R_r M_{sr}^2/\sigma L_s L_r\right)$; $K = M_{sr}/(\sigma L_s L_r)$;where: $i_{s\alpha\_e}$, $\varphi_{r\alpha\_e}$: estimated value; ω_m: angular rotor speed; Λ_1, Λ_2, Λ_3 and Λ_4 are observer gains, and with $\Lambda_j = [\Lambda_{j1}\ \Lambda_{j2}]$ for j = 1, 2, 3, 4.

$I_s$is a column vector:

$$\mathrm{I}_s = \begin{bmatrix} \mathrm{sign}\,(s_1) \\ \mathrm{sign}\,(s_2) \end{bmatrix} \tag{3}$$

where s_1 and s_2 are given by:

$$\begin{bmatrix} s_1 \\ s_2 \end{bmatrix} = \frac{1}{\beta}\begin{bmatrix} \frac{K}{T_r}\left(i_{s\alpha} - i_{s\alpha\_e}\right) - \omega_m K\left(i_{s\alpha} - i_{s\alpha\_e}\right) \\ \frac{K}{T_r}\left(i_{s\beta} - i_{s\beta\_e}\right) + \omega_m K\left(i_{s\beta} - i_{s\beta\_e}\right) \end{bmatrix} \tag{4}$$

$$\beta = \left[\frac{K}{T_r}\right]^2 + K^2{\omega_m}^2 \tag{5}$$

The stability of this observer was examined in [7], resulting in the following results:

$$\begin{bmatrix} \Lambda_{11} & \Lambda_{12} \\ \Lambda_{21} & \Lambda_{22} \end{bmatrix} = \begin{pmatrix} \frac{K}{T_r}\delta_1 & \omega_m K\delta_2 \\ -\,\omega_m K\delta_1 & \frac{K}{T_r}\delta_2 \end{pmatrix} \text{ and } \begin{bmatrix} \Lambda_{31} & \Lambda_{32} \\ \Lambda_{41} & \Lambda_{42} \end{bmatrix} = \begin{pmatrix} \left(q_1 - \frac{1}{T_r}\right)\delta_1 & -\,\omega_m\,\delta_2 \\ \omega_m\,\delta_1 & \left(q_2 - \frac{1}{T_r}\right)\delta_2 \end{pmatrix} \tag{6}$$

Such as: $\delta_1\rangle\left|e_{\varphi_\alpha}\right|_{\max}, \delta_2\rangle\left|e_{\varphi_\beta}\right|_{\max}$ And $q_1\rangle\ 0, q_2\rangle\ 0$

The rotor speed is assessed using the adaptive method described in Eq. (7), which indicates a PI controller [8].

$$\omega_{m\_e} = k_p\left(e_{is\alpha}\,\varphi_{r\beta\_e} - e_{is\beta}\,\varphi_{r\alpha\_e}\right) + k_i\int\left(e_{is\alpha}\,\varphi_{r\beta\_e} - e_{is\beta}\,\varphi_{r\alpha\_e}\right)\mathrm{d}t \tag{7}$$

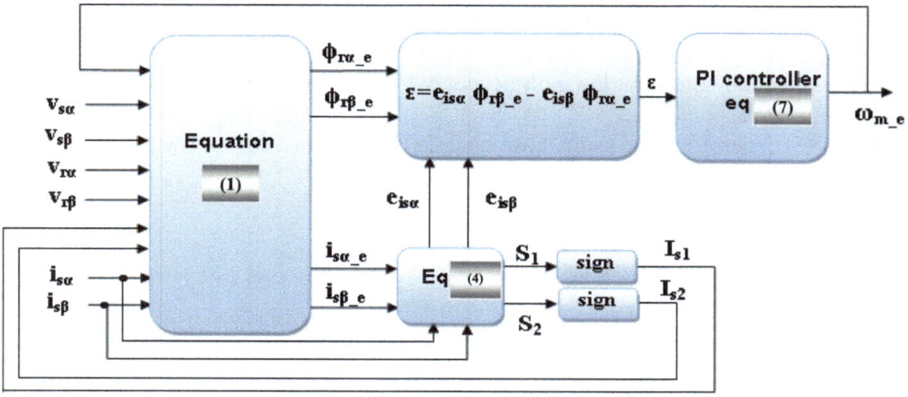

Fig. 2. Speed SMO schema.

4 Proposed Fault Assessment Technique

The resulting signal from Eq. (8) of IGBT open circuit fault detection [11] is used in the proposed IGBT open circuit Fault Tolerant Control (FTC) of the proposed fault assessment technique [2]:

$$\varepsilon_{k0} = V_{k0es} - V_{k0m} \tag{8}$$

V_{kOm} and V_{kOes} are the measured and the estimated pole voltages with k (leg number) $= \{1, 2, 3\}$.

The identification of the **proposed fault** assessment **technique** voltages is based on Eq. (9) [2]:

$$V_{assess} = V_{\text{healthy}} - V_{\text{faulty}} \tag{9}$$

Therefore, having V_{assess} via Eq. (9), based on the healthy case voltage model represented by Eq. (10), in terms of switching functions, subtracted from each open circuit fault case voltage model of the same Eq. (10), the voltages related with the anticipated fault assessment technique [2] is obtained. The voltages of the proposed fault assessment technique are connected in series with the inverter output voltages [2].

$$\begin{cases} V_{ra_{\text{healthy}}} = \frac{V_{dc}}{3}(2.S_1 - S_2 - S_3) \\ V_{rb_{\text{healthy}}} = \frac{V_{dc}}{3}(-S_1 + 2.S_2 - S_3) \\ V_{rc_{\text{healthy}}} = \frac{V_{dc}}{3}(-S_1 - S_2 + 2.S_3) \end{cases} \tag{10}$$

4.1 Fault Assessment

Consider the case of an open circuit fault occurring in IGBT2 and assess its performance using Eqs. (9) and (10):

In the event of an open circuit fault on IGBT2, switch number 2 is therefore open. This is equivalent to the S_2 command of the Pulse Width Modulation (PWM) of IGBT2 being permanently zero after the fault occurs, which mathematically applies to Eq. (10) for the healthy case:

If: $S_2 = 0$, therefore, the faulty case of Eq. (10) becomes the following:

$$\begin{cases} V_{ra_{\text{faulty}}} = \frac{V_{dc}}{3}(2.S_1 - S_3) \\ V_{rb_{\text{faulty}}} = \frac{V_{dc}}{3}(-S_1 - S_3) \\ V_{rc_{\text{faulty}}} = \frac{V_{dc}}{3}(-S_1 + 2.S_3) \end{cases} \tag{11}$$

Then: Subtracting the healthy case of Eq. (10) from its faulty case as shown in Eq. (9) gives the following IGBT2 open-circuit fault voltages of the proposed fault assessment technique:

$$\begin{cases} V_{ra_{assess}} = -S_2.\frac{V_{dc}}{6} \\ V_{rb_{assess}} = 2.S_2.\frac{V_{dc}}{6} \\ V_{rc_{assess}} = -S_2.\frac{V_{dc}}{6} \end{cases} \tag{12}$$

The proposed fault assessment technique is applied to all open circuit faults of the remaining IGBTs

5 DFIG-WECS Simulation Results

The 10 kW wind turbine is coupled to a 7.5 kW on-grid wind generator (DFIG) connected by its stator to the grid (220/380 V, 50 Hz) and by its rotor to the converters (2 kHz). The MPPT blade pitch angle is $\beta = 0$. This section presents all simulation results.

5.1 Results of the IGBT Open-Circuit Fault Simulation Based on a DFIG-WECS Sensorless SMO

The applied wind speed profile has a constant reference. This marks an essential stage in the study of WECS. The RSC IGBT2 switch experiences an IGBT open-circuit fault at time t = 1 s, with the defective leg directly matching the *irb* current.

The simulation results of the speed sensorless DFIG-WECS using SMO in the presence of an IGBT open-circuit fault show the rotor current *irb*, rotor speed, and stator active power with their estimation errors.

"Figure 3" illustrates, respectively, a good tracking between the estimated rotor current *irb* and the measured *irb* current, as well as the estimated rotor speed, which shows a good tracking with the mechanical speed with an acceptable estimation error, and finally the measured stator active power *P*, which closely matches the injected stator active power reference *Pref*, calculated via the product of the electromagnetic torque and the SMO speed estimation.

"Figure 3" also illustrates the sinusoidal shape of the *irb* current before fault and its modified shape after fault where a positive waveform is assumed to occur in the healthy case. It should be noted that in the presence of the open circuit fault of the IGBT2

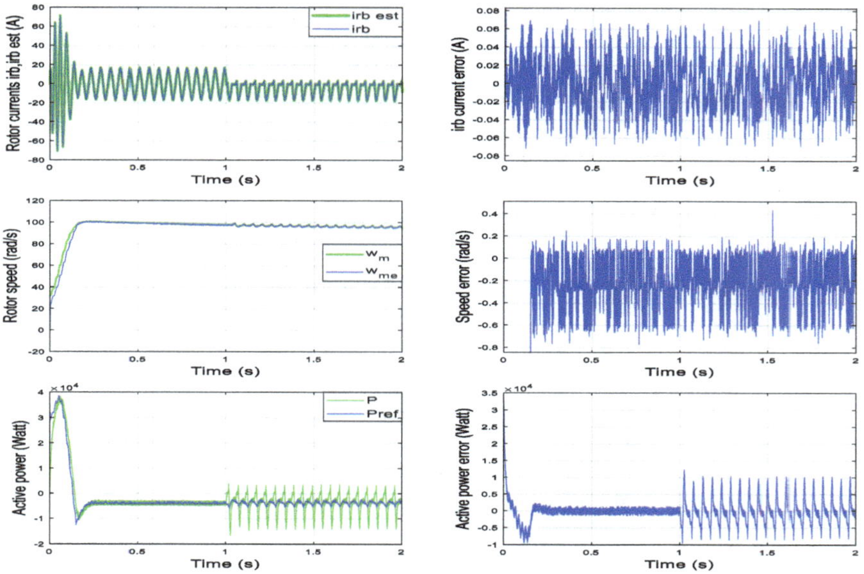

Fig. 3. Faulty –rotor current irb, rotor speed and stator active power- tracking performances and estimation errors based on SMO

inverter and during the blocking period of the diode D_2, the *irb* current will become zero. The different figures illustrate the behavior of the applied rotor speed observation technique, confirming the robustness of our SMO-based speed estimation technique. It is also important to note, based on the speed sensorless SMO results in case of an open-circuit fault of the IGBT, that it is crucial for the service continuity of the DFIG-WECS to reconfigure the fault immediately after its occurrence using the proposed fault assessment technique.

5.2 Simulation Results of the Proposed Fault Assessment Technique Voltages in IGBT2 Open-circuit Fault

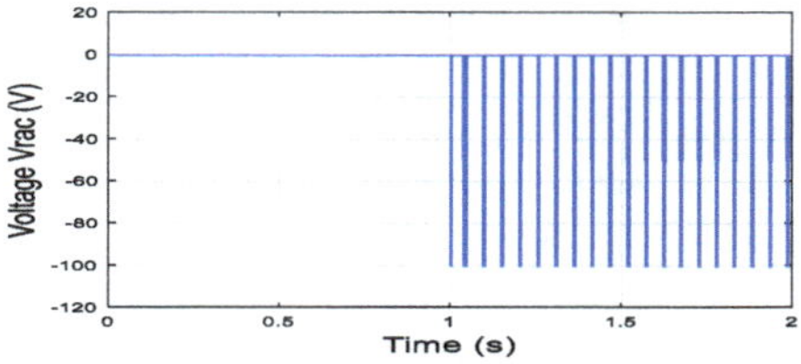

Fig. 4. Fault assessment technique voltage Vra_{assess}

At time t = 1 s, an open-circuit fault to the IGBT2 switch of the RSC is applied, where the faulty leg directly corresponds to the *irb* current and *vrb* voltage.

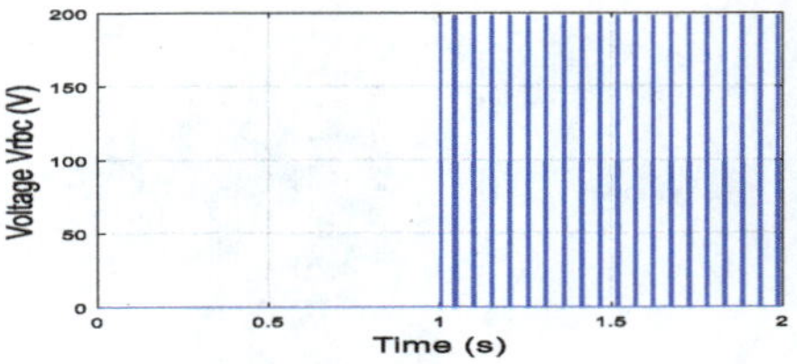

Fig. 5. Fault assessment technique voltage Vrb_{assess}

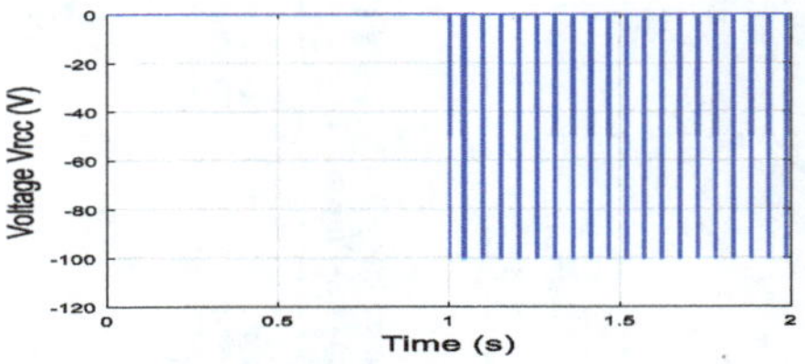

Fig. 6. Fault assessment technique voltage Vrc_{assess}.

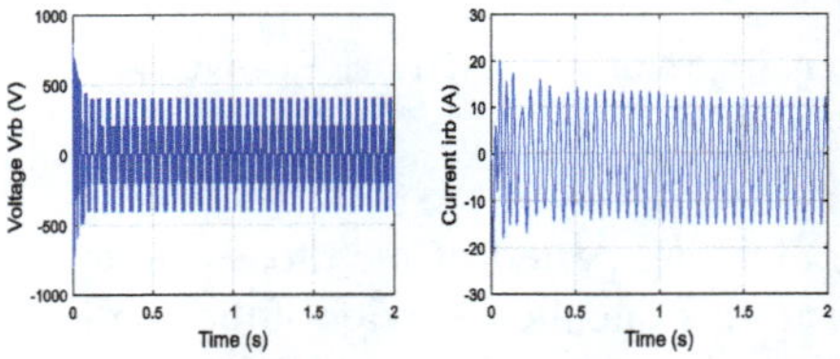

Fig. 7. Compensated: *Vrb* voltage, *irb* current.

"Figure 4", "Fig. 5" and "Fig. 6" symbolise the IGBT2 open-circuit fault assessment technique voltages, respectively. "Figure 7" shows the return to healthy state of the *irb* current and *vrb* voltage when the fault assessment technique voltages are present.

6 Conclusion

This paper suggests the use of a PI controller and a SMO for the DFIG-WECS in the assessment technique of open switching faults of RSC IGBTs. The regulator gains of the PI controller are determined by the pole compensation technique, considered a well-known reference approach. The results obtained show that the PI controller exhibits good performance and that the SMO ensures good tracking of the estimated values concerning their references, as well as acceptable estimation errors, in healthy and fault conditions. Also, the obtained results approve the effectiveness of the proposed fault Assessment technique, which preserves the robustness of the DFIG-WEC and compensates for the occurrence of open switching faults of IGBTs, thus ensuring the continuity of power generation.

References

1. Mohammedi, M.M., Bendiabdellah, A., Allaoui, T., Cherif, B.D.E.: Wind energy conversion system control robustness based on current analysis of IGBT open-circuit fault. Prz. Elektrotech. **10**, 14–24 (2020)
2. Mohammedi, M.M., Bendiabdellah, A., Allaoui, T.: Voltage reconfiguration fault-tolerant technique for two-level three-phase inverter integrated in a wind energy system. Int. J. Model. Identif. Control. **40**, 315–326 (2022) https://www.inderscienceonline.com/doi/abs/10.1504/IJMIC.2022.125550, https://www.inderscience.com/jhome.php?jcode=ijmic
3. Mohammedi, M.M., Gadouche, Z., Boutaleb, D.N., Bendiabdellah, A.: Service continuity of wind energy conversion system based on nonlinear control and fault tolerant technique. In: International Conference on Electrical, Computer and Energy Technologies (ICECET), Cape Town, South Africa (2023). https://ieeexplore.ieee.org/xpl/conhome/10389177/proceeding
4. Gadouche, Z., Belfedal, C., Allaoui, T., Belabbas, B.: Speed-sensorless DFIG wind turbine for power optimization using fuzzy sliding mode observer. Int. J. Renew. Energy Res. **7**, 613–621 (2016)
5. Akel, F., Ghennam, T., Berkouk, E.M., Laour, M.: An improved sensorless decoupled power control scheme of grid connected variable speed wind turbine generator. Energy Convers. Manag. **78**, 584–594 (2014)
6. Kerrouche, K.D.E., Mezouar, A., Boumedien, L.: Maximized power control based on high order sliding mode for sensorless DFIG variable speed wind turbine. In: International Conference on Green Energy and Environmental Engineering (GEEE), Sousse, Tunisia (2014)
7. Benchaib, A., Rachid, A., Audrezet, E.: Real-time sliding-mode observer and control of an induction motor. IEEE Trans. Ind. Electron. **46**, 128–138 (1999)
8. Kubota, H., Matsuse, K.: Speed sensorless field-oriented control of induction motor with rotor resistance adaptation. IEEE Trans. Ind. Appl. **30**, 1219–1224 (1994)
9. Benbouzid, M.H., Beltran, B., Amirat, Y., Breton, S.: Sensorless control of doubly-fed induction generator-based wind turbines using a high-order sliding mode observer. Int. Rev. Electr. Eng. (IREE). **9**, 49–55 (2014)
10. Sarika, S., Mary, S.A.J.: DFIG in wind energy applications with high order sliding mode observer-based fault-tolerant control scheme using sea gull optimization. Int. J. Electr. Electron. Res. (IJEER). **12**, 352–358 (2024)
11. Gaillard, A.: Système éolien basé sur une MADA: contribution à l'étude de la qualité de l'énergie électrique et de la continuité de service. PhD Thesis, Henri Poincaré University (2010)

AI-Based Neural Networks and Digital Security

Home Monitoring System Using IoT and Applying a Deep Learning Model

Benahmed Farouk[1(✉)], Hoceini Mohammed El Amine[2], Jaime Lloret[3], and Benahmed Khelifa[1]

[1] Department Mathematics and Computer Science, Faculty of Exact Sciences, Tahri Mohamed University, Innovation in Informatics and Engineering Laboratory, Bechar, Algeria
faroukbenah@yahoo.com

[2] Department Computer Science for Autonomous Systems, ELTE University, Budapest, Hungary
dd36mr@INF.ELTE.HU

[3] Universitat Politécnica de Valencia, València, Spain
jlloret@dcom.upv.es

Abstract. Ensuring home safety is a continuous concern, not only in urban areas but also in suburbs and villages, especially for individuals who spend a lot of time away from home. With increasing threats, it has become necessary to implement proven strategies that enhance security systems and enable remote home monitoring. For this purpose, this work aims to design and develop an integrated home security and monitoring system using Internet of Things (IoT) technology. The proposed monitoring system uses several technologies, sensors, and actuators, including a PIR sensor to detect intruders, a fire sensor to detect fires, and a gas sensor to detect carbon monoxide gas leaks inside the home. Additionally, cameras of surveillance are used with Deep Learning model for face recognition. Experimental results show that the developed system effectively monitors the home's condition and enables remote home protection through an internet-connected smartphone application. This integration of IoT technology enhances home security and provides homeowners the ability to monitor and manage their homes from anywhere.

Keywords: Home monitoring · Internet of things (IoT) · Sensors · Artificial Intelligence (AI) · Mobile applications

1 Introduction

Home monitoring has become a significant concern in today's rapidly advancing technological landscape. With the rapid development of technology, there is a growing necessity to stay updated and informed. There are many cases in which home surveillance is needed, among them Gas leak, intruder detection, kids' monitoring as well as the elderly. Using the conventional systems all these problems can pose a danger to family members, thus not providing complete security for the home for that the threat will be increased consequently in other hand with the development in technologies several

M. Hatti (Ed.): IC-AIRES 2025, LNNS 1740, pp. 137–147, 2026.
https://doi.org/10.1007/978-3-032-12788-4_13

tools and techniques can provide efficient solutions in order to resolve these problems such as IoT, Cloud computing, AI etc. The Internet of Things has revolutionized the way we live, and the smart house is one of the primary beneficiaries of this technology. IoT enables smart home devices to connect to the internet and communicate with each other, creating an adaptable system for users [1]. This has led to an explosion in the popularity of smart home devices, which are becoming increasingly common in homes around the world [2]. With the advent of internet-connected devices, homeowners now are able to remote monitor the safety of their homes from anywhere in the world, just by their smartphones they can control multiple aspects of their home security system such as security cameras, door and window sensors, motion detectors [3]. The Internet of Things presents many possibilities in the domain of home security, enabling the creation of affordable and user-friendly systems that improve overall safety and provide peace of mind to homeowners [4]. Our proposed IoT-based home security system can protect the home and facilitate the early detection of emergencies such as fires, Gas leak and intruders using several types of sensors, such as cameras, motions, gas and fire sensors. The detected information will be transmitted to the microcontroller, which will then execute our proposed algorithm to manage the sensor output data [5]. Additionally, the microcontroller will send the processed data to the cloud for storage, enabling it to be displayed on the application interface [3, 6]. The article is structured as follows. In Sect. 2, we present a review of the most relevant related works in the domain of IoT-based home monitoring systems. Section 3 introduces our proposed system, detailing both the software and hardware components. Section 4 describes the global architecture of the system and explains how its different components interact in real-time to ensure efficient monitoring and security. Section 5 shows the experimental results obtained, including screenshots and performance evaluations. Finally, Sect. 6 concludes the paper by summarizing the key findings and followed by suggestions for future work.

2 Related Works

This work is focused on building an IoT-based home monitoring system. Being a current and important domain in the world of innovative IoT applications [7], other systems have been developed seeking to do the same. We explore these related works for inspiration and to build on them from an efficiency and cost-wise aspect:

- A system proposed in the paper "SMART HOME AUTOMATION SYSTEM BASED ON IoT," which utilizes sensors, actuators, and internet connectivity to enable automation, control, and monitoring of various home devices [8].
- An IoT-based Smart Home Automation System described in the paper "An IoT-Based Smart Home Automation System," which interconnects sensors, actuators, and other data sources with the purpose of multiple home automations [9].
- The qToggle system, which uses ESP8266/ESP8285 chips and/or Raspberry Pi boards to control a series of home appliances and sensors [10].
- The Internet of Things-Based Intelligent Smart Home Control System, described in the paper "Internet of Things-Based Intelligent Smart Home Control System," which focuses on controlling and monitoring home appliances [11].

- The Web-based smart home framework described in the technical report "A Telosb sensor mote and its internal design," which presents the architecture of a smart home system that may serve as the backbone for future smart home deployments [12].

According to the presented works, used tools and technologies we have thought to propose new based technologies solution in order to solve the problem which is monitor efficiently the house based on real-time surveillance and edge technology.

3 Our Proposed System

Our system contains two aspects, software and hardware. In the software aspect, we define the used tools, frameworks, and programming languages. On the other side, in the hardware aspect, we define the used sensors, actuators, relays, etc. This section presents the introduction of required hardware and software technologies and the architecture of the system and the description of their functionalities.

3.1 Software Aspect

In our system, we have used several kinds of software tools such as:

- React Native: We have used this framework exactly the package Expo CLI which is designed to keep programmers moving during app development, we exploit this tool to create our system's mobile app that displays the system events and components. We have chosen this tool because it's cross-platform, robust, and adaptable to new technologies such as IoT.
- Python: At the microcontroller level, we have programmed the Raspberry Pi MCU using Python, which is known for its simplicity, readability, and versatility. It supports multiple programming paradigms like procedural, object-oriented, and functional programming and AI tools, using Python languages and their packages, we have been able to code sensors, actuators, cameras, relays, and the communication system.
- Android studio: We have used this framework to call the emulator in order to test the developed mobile app, the used emulator is Medium Phone API 33.
- MongoDB: We have used this database management system in order to save and manage the recoded videos form the camera in an offline mode.
- Flask: We have used this tool to save and manage the online recorded videos in streaming mode.
- Firebase: Using firebase, we have been able to create an online data base that allow saving, managing and transmitting data detected from the sensors in real-time.
- PyTorch ResNet: which is a common neural network architecture used for deep learning computer vision applications in Python. We have utilized this package to detect, classify, and ultimately identify people inside the house using recorded videos. This option enables the detection of intruders within the house.

3.2 Hardware Aspect

In the hardware component, we employed various types of materials to set up our system, the used material is presented as follow.

3.2.1 Sensors

Our developed home monitoring system is based on several kinds of sensors used to detect events such as fire, gaze etc. the used sensors are presented as follow and shown in Fig. 1.

- Ultrasonic sensor: We utilized this sensor on the door to automatically close it if someone forgets to do manually.
- MQ-9 sensor: This sensor detects the presence of gas inside the house and sends information about the level of gas infiltration.
- Flame sensor: This sensor enables the detection of fire within the house
- PIR sensor: We used this sensor to detect the presence of objects inside the house.

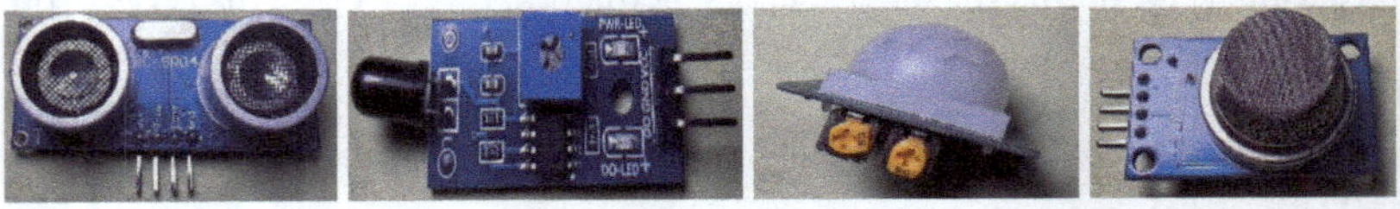

Fig. 1. The used sensors © Photo by Farouk Benahmed.

3.2.2 Actuators

In this section, we present the actuators used in our system.

Servo Motor: We utilize this actuator to automatically open windows in the event of gas infiltration and to close doors if someone forgets to do so manually as shown in Fig. 2.

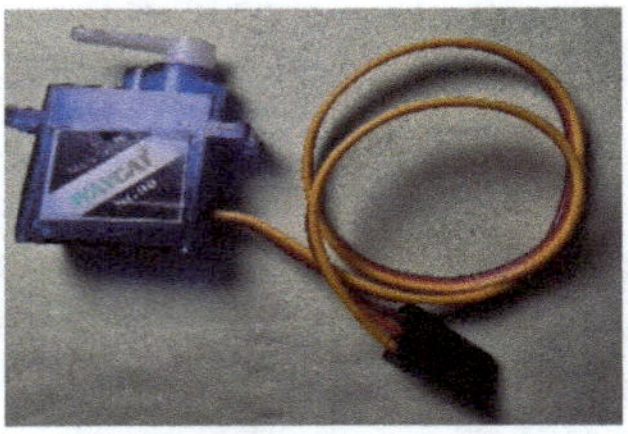

Fig. 2. Servo motor © Photo by Farouk Benahmed.

3.2.3 Microcontrollers

Raspberry Pi MCU: We programmed the Raspberry Pi MCU to control various components of our system as illustrated in Fig. 3.

Arduino: We utilized an Arduino Uno to measure gas concentration due to its analog port capabilities, this data was transmitted to a Raspberry Pi via a serial port for further processing and monitoring, the Fig. 4 shows the model of the Arduino Uno.

Fig. 3. Raspberry pi MCU © Photo by Farouk Benahmed.

Fig. 4. Arduino Uno © Photo by Farouk Benahmed.

4 System Architecture

In this part we present the architecture of our system that enables homeowners to efficiently monitor their homes in real-time by integrating various hardware and software components, incorporating both engineering and research aspects as shown in Fig. 5. With a focus on security, our IoT-based system facilitates early detection of emergencies such as fires, gas leaks, and intruders through sensors including cameras, motion sensors, gas sensors, fire sensors and ultrasonic sensor. the data collected from these sensors is processed by a microcontroller (Raspberry Pi) by running our algorithm, the results stored and transmitted securely via the Firebase cloud for display in the application interface. Add to that, we have implemented the deep learning model (ResNet) from PyTorch, for intruder detection, and integrated an authentication system to improve system security.

The flowchart illustrated in Fig. 6 presents the operation steps of our system which are as follow:

- ***Step 1 Initialization the system***: Initializes all sensors, in order to be ready for detection new events and the camera starts live streaming.
- ***Step 2 Sensor Data Collection***: The sensors collect data of several events such as gas fire and motion, which are then sent to the Raspberry Pi with the camera.
- ***Step 3 Data Processing***: The Raspberry Pi receives and processes the sensor data and checks if any event is detected and send it to database 1 which is an online cloud database called Firebase. Then the MCU receive also the camera samples then send them to database 2 which is a local database type called MongoDB.

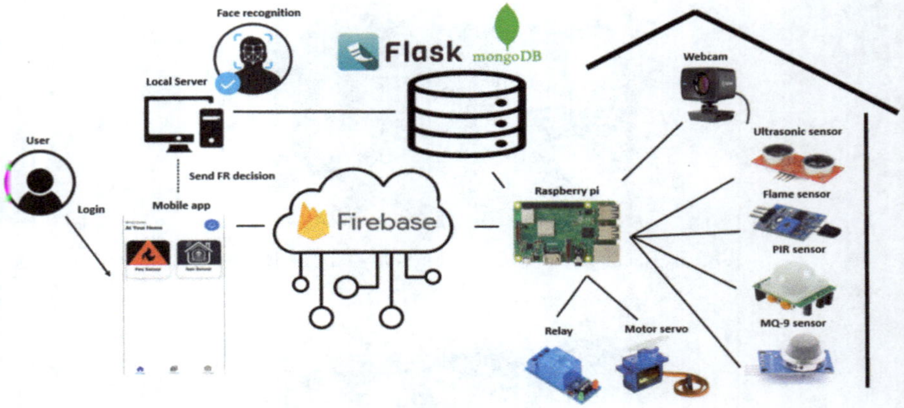

Fig. 5. Global system Architecture.

- *Step 4 Database Storage*: The system stores sensor values and camera videos in Database1 and face recognition data in Database2.
- *Step 5 Face Recognition*: The face recognition module processes the camera video feed and detects faces using Python Pytorch. if any person is detected, the system verified their identity and if this person is unknown, the system sends an alert to the user, else an information message is sent to the user containing the recognized person's ID.
- *Step 6 Threshold Check*: The system checks if the sensor values exceed a certain threshold. If the threshold is exceeded, the system sends an alert to the user. The system sends alerts to the user based on the sensor data, face recognition results, and threshold checks.
- *Step 7 Actuator Control*: The system enables or disables actuators based on the sensor data, face recognition results, and threshold checks.
- *Step 8 User Interaction*: The system interacts with users through a user interface that we have developed using React native framework in order to receive alerts, notifications and activate actuators, the user can access to the system using a secure authentication system with an error managing system in order to help users interacting the system secretly furthermore, the system securely transmits data between components using encryption and secure communication channels.

Finally, the system will restart by applying the same steps that render our system a permanent, real-time monitoring system.

5 Experimentation and Discussion Results

In this section, we present some screenshots to demonstrate the real-life operation of our home monitoring system in a prototype house. Figure 7 illustrates the user interface of the mobile application, which enables users to register and create an account to interact with the system.

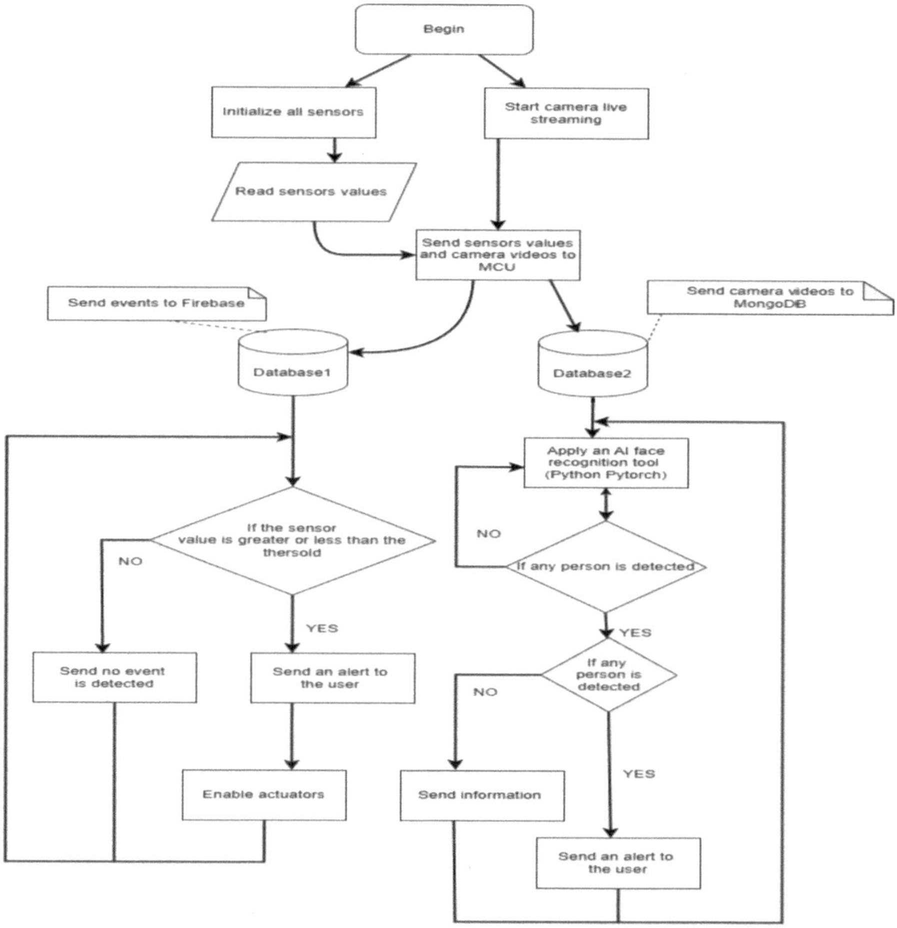

Fig. 6. The flowchart of our system

Figure 8 presents the user interface that enables logging into the mobile application and manipulating the system.

The figure below Fig. 9 presents the interface that displays received events, notifications, and enables actuators, etc. This interface allows users to interact with and control the system.

Figures 10, 11, and 12 present the experimental results of our applied face recognition system, which enables the detection of known individuals inside a house. Additionally, this system can identify authorized persons within the household by utilizing PyTorch ResNet, a deep learning tool. Consequently, the system can accurately detect both known and authorized individuals inside the house.

Figure 13 presents the prototype of our home monitoring system that contains all the components described earlier: sensors, microcontrollers, and actuators.

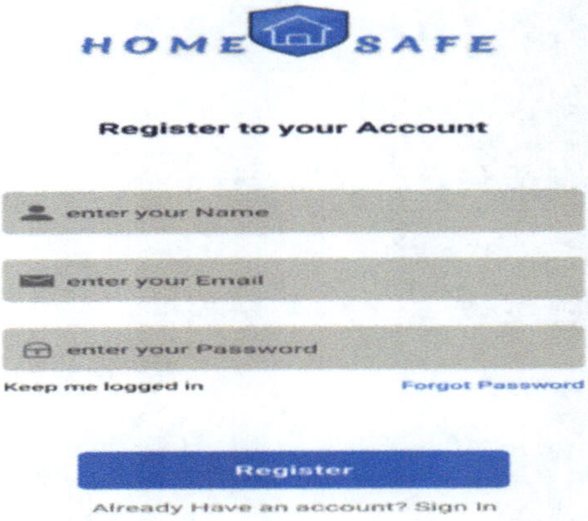

Fig. 7. Registration user interface

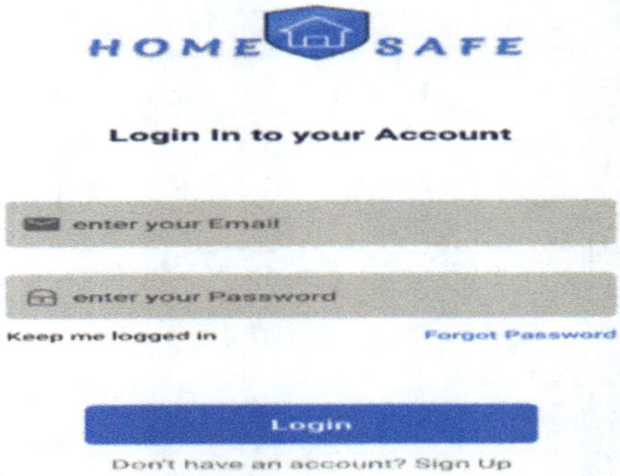

Fig. 8. Login user interface

Fig. 9. The interaction user interfaces

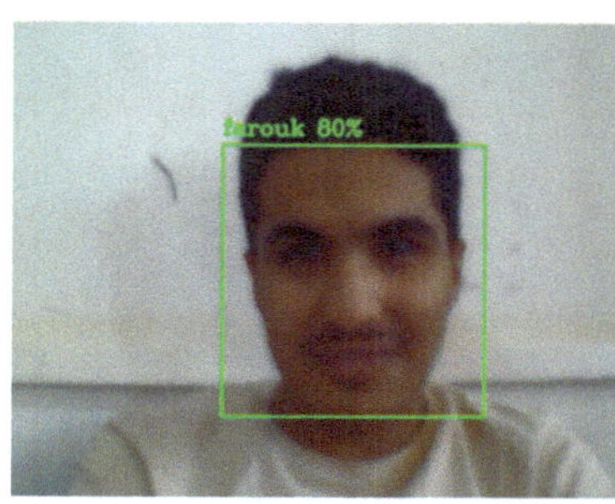

Fig. 10. Face recognition authorized person detection with 80% of accuracy

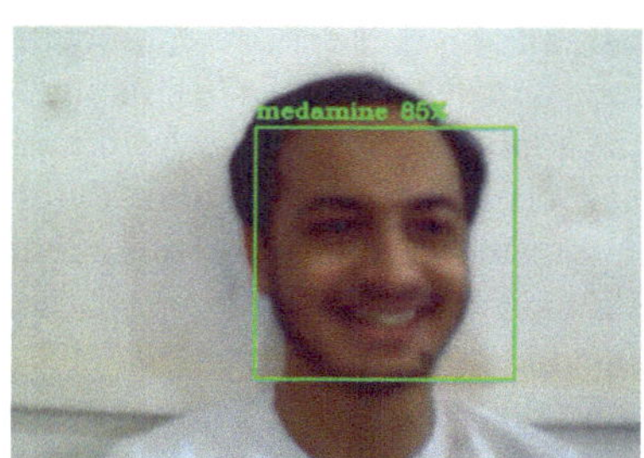

Fig. 11. Face recognition authorized person detection with 85% of accuracy

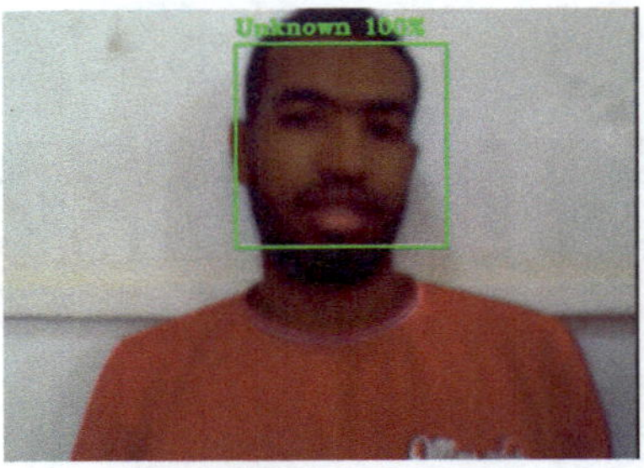

Fig. 12. Face recognition unknown person detection with 100% of accuracy

Fig. 13. The prototype of our home monitoring system

6 Conclusion

This work shows that we have successfully developed an home monitoring system based on Internet of Things (IoT) technology. Our system enables homeowners to efficiently monitor their homes in real-time by integrating many hardware and software components, combination both engineering and research aspects, with a focus on security, our IoT-based system can detect all types of emergencies such as fires, gas leaks, and intruders through sensors and cameras. Data collected from the sensors is processed by a microcontroller which running our algorithm, the results stored and transmitted securely to the cloud for display in the application interface. Additionally, we have implemented Deep Learning model for face recognition using Python Pytorch, to detect intrusion, and integrated an authentication system to enhance system security. In this project, we have gained valuable technical skills including microcontroller and sensor management and programming, mobile application development, and database management using tools such as Raspberry Pi, React Native, Mongo dB, flask, python, C++ and Firebase. One of the challenges we faced in our project was the shortage of hardware. We had to purchase all the equipment ourselves, which required a significant budget. Additionally, we had to learn multiple programming languages in a short amount of time, which demanded considerable effort. Since we couldn't find readily available resources to implement our ideas, we had to work on them independently. One of our future works is to reinforce the system to be comprehensive so that can monitor not just individual homes but entire buildings to make them secure. Furthermore, we aim to expand the system's capabilities by allowing users to access recorded videos, not just live feeds and for our long-term vision we look to extend the system's coverage to entire cities.

References

1. Rego, A., Ramírez, P.L.G., Jimenez, J.M., Lloret, J.: Artificial intelligent system for multimedia services in smart home environments. Clust. Comput. **25**(3), 2085–2105 (2022)
2. Irugalbandara, C., Naseem, A., Perera, S., Kiruthikan, S., Velmanickam, L.: A secure and smart home automation system with speech recognition and power measurement capabilities. Sensors. **23**(13), 5784 (2023)
3. Chataut, R., Phoummalayvane, A., Akl, R.: Unleashing the power of IoT: A comprehensive review of IoT applications and future prospects in healthcare, agriculture, smart homes, smart cities, and industry 4.0. Sensors. **23**(16), 7194 (2023)
4. Lloret, J., Canovas, A., Sendra, S., Parra, L.: A smart communication architecture for ambient assisted living. IEEE Commun. Mag. **58**(1), 76–82 (2020)
5. Mekruksavanich, S., et al.: Deep learning for recognizing daily human activities using smart home sensors. ICIC Express Lett. **17**(12), 1375–1383 (2023)
6. Ganiga, R., Pai, R. M., Manohara Pai, M. M., Sinha, A.: Integrating NFC and IoT to provide healthcare services in cloud-based EHR system. In: Advances in Electrical and Computer Technologies: Select Proceedings of ICAECT 2019, Springer, 2020
7. Benahmed, T., Benahmed, K.: Optimal barrier coverage for critical area surveillance using wireless sensor networks. Int. J. Commun. Syst. **32**(10), e3955 (2019)
8. Abdulraheem, A.S., et al.: Home automation system based on IoT. Technol. Rep. Kansai Univ. **62**(5), 2453 (2020)
9. Stolojescu-Crisan, C., Crisan, C., Butunoi, B.-P.: An IoT-based smart home automation system. Sensors. **21**(11), 3784 (2021)
10. Stolojescu-Crisan, C., J. Gal: A home energy management system. In: 2022 International Symposium on Electronics and Telecommunications (ISETC), IEEE, 2022
11. Taiwo, O., Ezugwu, A.E.: Internet of things-based intelligent smart home control system. Secur. Commun. Netw. **2021**, 1–17 (2021)
12. Appavoo, P., William, E., Chan, M., Mohammad, M.: Indriya 2: A heterogeneous wireless sensor network (WSN) testbed. In: Testbeds and Research Infrastructures for the Development of Networks and Communities: 13th EAI International Conference, TridentCom 2018, Springer, Shanghai, China, 1–3 December 2018

Enhancing Energy Efficiency in Domestic Refrigerators: Experimental and Statistical Evaluation of Phase Change Material Integration

Mahdi Mokrane[1](✉), Ali Tetbirt[1], Mohamed Ali Djebiret[1], Maamar Ouali[1], and Ridha Alliche[2]

[1] Unité de Développements des Equipements Solaire UDES/Epst CDER, Route National N 11. Bp 386. Bou-Ismail, Tipaza, AlgeriaBou-Ismaïl 42415
mok.mehdi@gmail.com

[2] Laboratory of Industrial Fluids, Measurements and Applications (FIMA), Djilali Bounaama University, Road of Theniet El-Had, Khemis Miliana, 44001, Ain Defla Khemis Miliana, Algeria

Abstract. The inclusion of phase change materials (PCM) in refrigeration systems enhances their potential efficiency, maintains product temperatures, and can, under certain conditions, reduce their energy consumption. This study presented an evaluation of a household refrigerator's power consumption and the effect of integrating phase change materials on its thermal behavior. The first step consisted of the realization of a design of experiments (DOE) using the Minitab software to allow a simultaneous analysis of the effects of the input variables (factors) on the output variable (response). The main results show the feasibility of the PCM technology and attest to a significant reduction in the electrical consumption of the domestic refrigerator on the one hand and the possibility of attenuating the fluctuation of the storage temperature when opening and closing the refrigerator door on the other hand.

Keywords: Phase change material (PCM) · Household refrigerator · Experiment plan

1 Introduction

In response to the worsening global energy crisis and the acceleration of climate change, research into energy conservation methods has gained renewed importance [1]. In Algeria, electricity demand has continued to rise significantly in recent years, especially during summer heatwaves. According to Sonelgaz data, the peak consumption recorded during the summer of 2024 reached 18,500 kW, representing an increase of nearly 7.5% compared to 2023. This confirms the upward trend in household energy needs during periods of extreme heat.

M. Hatti (Ed.): IC-AIRES 2025, LNNS 1740, pp. 148–159, 2026.
https://doi.org/10.1007/978-3-032-12788-4_14

It is important to note that the most energy-intensive household appliances are not always those with the greatest impact on annual consumption [2]. For instance, while an iron consumes a lot of energy per hour of use, it is used infrequently, unlike an internet box, which consumes little energy but operates continuously throughout the year. This observation highlights the importance of optimizing the consumption of equipment that runs permanently.

The domestic refrigerator, based on vapor compression technology, remains one of the most electricity-consuming appliances in the home [3]. Today, almost all households in developed countries own at least one refrigerator, and the ownership rate is rapidly increasing in emerging economies [2]. According to recent studies, refrigerators account for an average of 15 to 20% of total household electricity consumption [4].

The search for effective solutions to reduce the energy impact of these appliances therefore addresses a dual challenge: ensuring food safety while lowering household energy bills. Various strategies have been explored, including intelligent management of electrical loads (scheduled compressor shutdown) and the integration of thermal energy storage devices.

The use of phase change materials (PCM) in domestic refrigerators appears to be a promising solution. PCMs can store and release thermal energy, thereby helping to reduce electricity consumption, lower operating costs, and maintain food safety, while also contributing to the reduction of greenhouse gas emissions.

In this context, the present study analyzes the use of PCMs in a 331 L domestic refrigerator, tested in a low-energy smart solar house at the Solar Equipment Development Unit (UDES). The influence of several parameters (fresh food compartment temperature, PCM quantity, thermal load) on system performance was evaluated using a design of experiments approach and optimized with Minitab software.

2 Methodology

In this study, we adopted an integrated approach that combines both experimental procedures and statistical analysis to thoroughly investigate the effects of various factors on refrigerator energy consumption. To efficiently explore the relationships between multiple variables and to minimize the number of experimental runs required, we employed the Box–Behnken Design (BBD), a well-established response surface methodology. The BBD is particularly advantageous because it enables the detection of potential interactions among the investigated factors and their collective impact on the response variable while requiring fewer experiments than a full factorial design [5–7].

2.1 Design of Experiment

Selecting the optimal phase change temperature and the appropriate quantity of Phase Change Material (PCM) is essential in refrigeration systems, as these parameters significantly influence the system's overall efficiency and energy consumption. To systematically investigate and optimize these variables, this study employed both experimental techniques and a statistical approach using the Box–Behnken Design (BBD).

Box–Behnken Design (BBD) Overview

The Box–Behnken Design is a type of response surface methodology that is particularly effective for exploring the interactions among multiple factors while minimizing the number of experimental runs required. Structurally, the BBD is based on a spherical, rotatable design, which includes a central point and points located at the midpoints of

the edges of a cube inscribed within a sphere [8]. This configuration enables the detection of both linear and interaction effects among variables, providing a comprehensive understanding of their influence on the response variable—in this case, energy consumption.

Application in This Study

In the present research, a three-level, three-factorial Box–Behnken experimental design was implemented to evaluate how different operational parameters affect the daily energy consumption of a domestic refrigerator. The study was conducted using Design Expert and Minitab software for the planning and analysis of experiments.

The three critical parameters selected for investigation were:

- **Tf:** Temperature of the fresh food compartment
- **MPCM:** Mass of the phase change material
- **Qh:** Applied heat load

Each parameter was tested at three coded levels: −1 (low), 0 (medium), and +1 (high). Table 1 summarizes the specific values assigned to each level for every factor.

Table 1.

Variables	Symbole	Levels		
Level code		−1	0	1
Quantity of phase-change material PCM (kg)	X_1	0	1.25	1.5
Heat load (Qh) (kg)	X_2	0	9	18
Fresh food compartment temperature (T_f) °C	X_3	1	4	8

The polynomial model is instrumental in capturing and quantifying the interactions among the independent variables, making it highly effective for predicting system responses within specified experimental boundaries [9]. The total number of experimental runs required for a Box–Behnken design is given by the formula $N = 2k(k - 1) + Co$, where k represents the number of factors and Co is the number of central points included in the design [10]. In this study, Eq. (1) represents a second-order polynomial that is fitted to the experimental data. This equation incorporates linear terms, squared (quadratic) terms, and interaction terms between the factors, thereby enabling a comprehensive analysis of both individual and combined effects of the variables on the response

The general form of the model is as follows:

$$Y = \beta_0 + \beta_1 \sum_{i=1}^{3} X_i + \beta_2 \sum_{i=1}^{3} X_i^2 + \beta_3 \sum_{i,j=1}^{3} X_i X_j + \varepsilon \quad (1)$$

where: Y is the response variable (energy consumption per β_0 is the intercept β_1, β_2, β_3 are the coefficients corresponding to linear, quadratic, and interaction effects X_i, X_j are the coded independent variables ε is the random error term.

Experimental Matrix and Statistical Modeling

The primary response variable in this study is the daily energy consumption (E) of the refrigerator. To systematically evaluate the effects of the selected factors, a three-factor, three-level Box–Behnken experimental design was implemented, resulting in a total of 15 experimental runs, as detailed in Table 2. Each trial represents a specific combination of the quantity of phase change material (MPCM), the temperature of the fresh food compartment (Tf), and the applied heat load (Qh).

The statistical analysis of the experimental data was conducted using Minitab 17 software, with a particular focus on analysis of variance (ANOVA). ANOVA was used to assess the significance of each factor and their interactions on the response variable, ensuring the reliability of the model. To verify the adequacy of the developed model, several statistical indicators were examined, including the coefficient of determination (R^2), adjusted R^2, and predicted R^2. These metrics help determine how well the model explains the variability in the data and whether it accurately represents the influence of each factor [11–14].

Additionally, tests for equality of variance and analysis of means were performed to confirm the consistency and robustness of the results. According to standard statistical criteria, a p-value and an F-value less than 0.05 indicate that the effect of a factor or interaction is statistically significant at the 95% confidence level. This rigorous statistical approach ensures that the findings are both reliable and meaningful for optimizing refrigerator energy consumption.

3 Experimental Setup

During the experiments, several key performance metrics were closely monitored, including the refrigerator's daily energy consumption, the temperature fluctuations within the refrigerator compartment, and the operational cycles of the freezer. These parameters were chosen to comprehensively evaluate the impact of various experimental conditions and the integration of phase change materials (PCM) on the overall thermal and energy performance of the appliance.

The refrigerator used in the study is a Brandt model, measuring 1.7 m in height, 0.6 m in width, and 0.6 m in depth. It is equipped with a ZEL HML140A HS- 384 compressor and uses R600a refrigerant (50 g). The cooling system features an aluminum fin evaporator and a tube coil condenser with natural air draft. For precise monitoring, a modern digital energy meter (10–40 A, 240 V, 50 Hz) was installed to record real-time power consumption.

Table 2. Presents the experimental matrix for the Box–Behnken design, illustrating the specific factor settings for each of the 15 trials.

Trail Run	M_{PCM}	T_f	Q_h
1	−1	−1	0
2	1	−1	0
3	−1	1	0
4	1	1	0
5	−1	0	−1
6	1	0	−1
7	−1	0	1
8	1	0	1
9	0	−1	−1
10	0	1	−1
11	0	−1	1
12	0	1	1
13	0	0	0
14	0	0	0
15	0	0	0

To simulate real-world usage and evaluate the effect of PCM integration, the refrigerator was operated at various compartment temperatures (1 °C, 4 °C, and 8 °C) and subjected to different thermal loads. The performance data were collected using an automated data acquisition system, which enabled continuous measurement and recording of temperature and energy consumption at 30-s intervals (Fig. 1).

In this study, the selection of an appropriate phase change material (PCM) was a critical step to ensure optimal thermal management within the domestic refrigerator. The chosen PCM needed to effectively maintain the compartment temperature close to its upper limit during compressor off-cycles, while also having a sufficiently low phase change temperature to limit temperature rises during these periods and to allow complete solidification during the compressor's operational (charging) phase.

The refrigerator used in the experiments is divided into two compartments:

Freezer Compartment: Maintains negative temperatures. For this section, the phase change temperature range of the selected PCM was restricted to between −18 °C and −11 °C, ensuring that the PCM would undergo phase transition within the typical operating range of the freezer and thus provide effective thermal buffering.

Fresh Food Compartment: Operates at positive temperatures. Here, the PCM was required to have a phase change temperature in the narrow range of −2 °C to 2 °C, suitable for stabilizing temperatures during compressor shutdowns without risking freezing of stored food.

1. Microcomputer 2. Data-logger 3. Domestic refrigerator 4. Electricity meter

Fig. 1. Connection of the experimental device to the acquisition system.

For the fresh food compartment, pure water was utilized as the PCM, introduced in varying quantities of 0.25 kg and 0.5 kg. This choice was based on water's high latent heat of fusion and its phase change temperature of 0 °C, which aligns well with the compartment's operational requirements.

For the freezer compartment, a mixture of water, propylene glycol, and 35% polyethylene glycol (PEG) was selected as the PCM. A fixed quantity of 1 kg was used, with a phase change temperature of −13 °C. This composition was specifically tailored to match the freezer's temperature range, enabling the PCM to absorb excess heat during compressor off-cycles and release it during compressor operation, thus enhancing temperature stability and reducing energy consumption (Fig. 2).

The placement and quantity of PCM in both compartments were carefully chosen to maximize the thermal buffering effect, minimize temperature fluctuations, and optimize the overall energy.

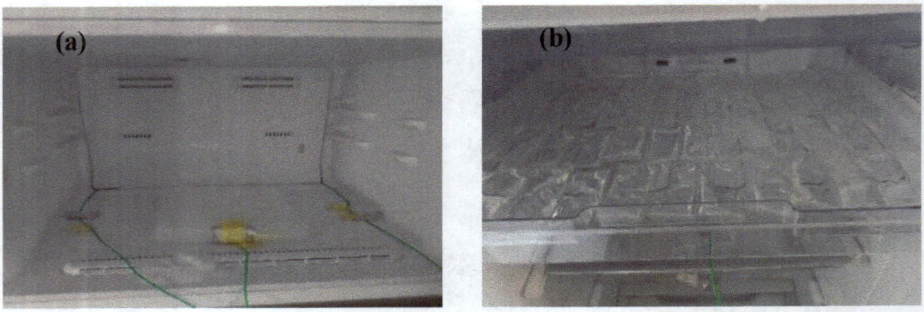

Fig. 2. Photo illustrating the placement of the PCM in the refrigerator: (a) PEG mixture (b) Pure wat

4 Results and Discussions

Figure 3 demonstrates a significant reduction in temperature fluctuations across the refrigerator compartment shelves when compared to the baseline (reference) scenario. Specifically, on the first shelf, opening the door of a refrigerator without PCM results in the temperature rising as high as 12.09 °C. This represents a substantial increase of 7.53 °C above the reference temperature, highlighting the extent of temperature instability in the absence of phase change materials.

This pronounced temperature spike indicates that, without PCM, the refrigerator is less effective at maintaining stable internal conditions during door openings—potentially compromising food quality and safety. The data clearly illustrate the thermal buffering effect of PCM, which helps to moderate temperature swings and improve the overall performance of the refrigerator under typical usage conditions.

Conversely, when phase change material (PCM) is incorporated into the refrigerator, the temperature rise during door opening is notably reduced. Specifically, the temperature on the first shelf only reaches 9.34 °C, which is 4.78 °C higher than the reference value. This demonstrates the effectiveness of PCM in moderating temperature spikes caused by frequent door openings, thereby enhancing thermal stability and potentially improving food preservation.

The experimental results were subjected to statistical analysis using analysis of variance (ANOVA) to determine the significance of each factor and their interactions on daily energy consumption (E/24 h). Based on the collected data, a second-order polynomial model was developed to quantitatively describe the relationship between the influencing parameters and the energy response. The primary objective of this modeling effort is to identify and rank the factors that have the greatest impact on refrigerator energy consumption per 24 h.

In this study, ANOVA was performed at a significance level of $\alpha = 5\%$, ensuring that only statistically significant effects were considered. As illustrated in Fig. 4, the model achieved an R^2 value of up to 0.95, indicating a high degree of fit and reliability, with a corresponding confidence level of 95%. The final coded model, presented in Eq. (1), enables direct comparison of the coefficients for each factor, thereby highlighting their relative importance in influencing energy consumption.

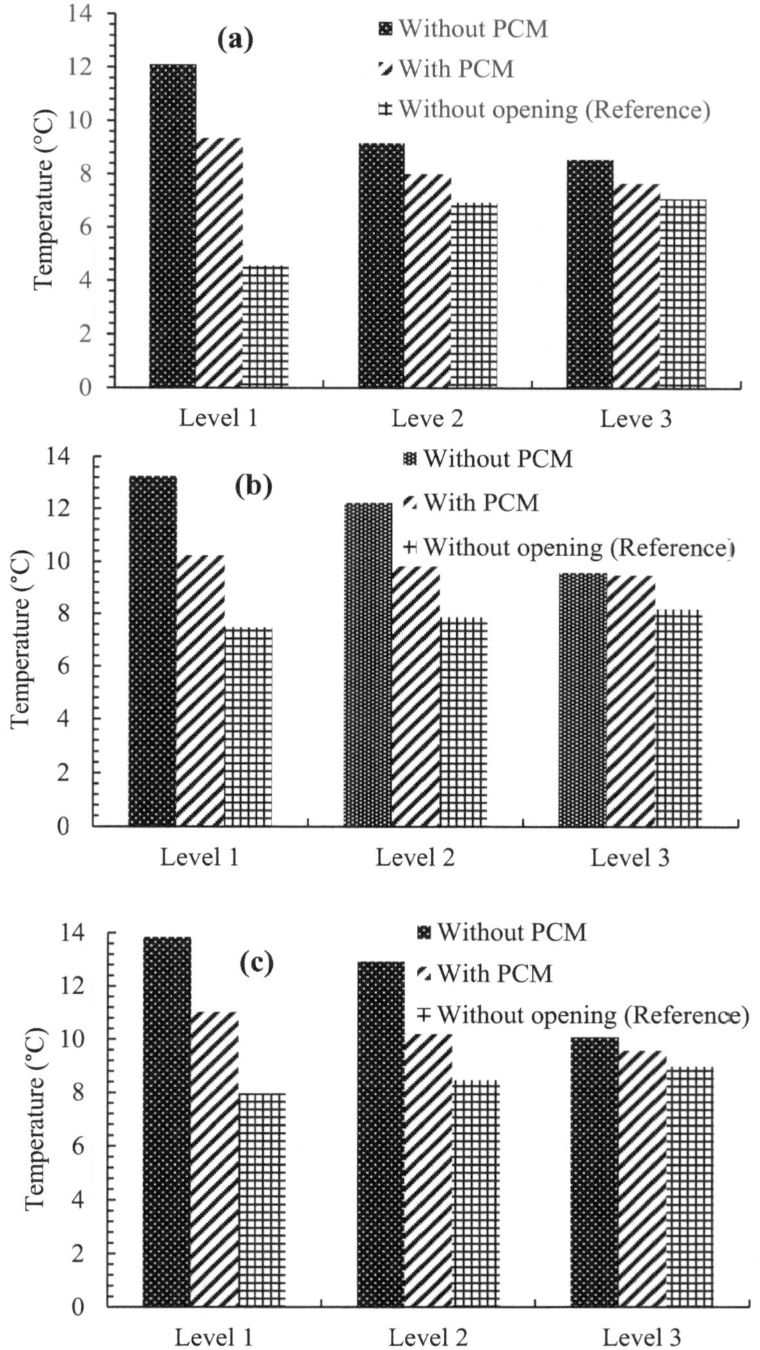

Fig. 3. Temperature distribution in the different shelves of the refrigerator compartment with a thermostat temperature equal to (a) 1 °C; (b) 4 °C and (c) 8 °C

The resulting model for daily energy consumption (E/24 h) is as follows:

$$\begin{aligned}\mathbf{E/24h} &= 0,745 + 0,899X1 + 0,0287X2 + 0,0797Tf + 0,643X1 \\ &*X1 + 0,00147X2 * X2 + 0,00789X3 * X3 + 0,0216X1 * X2 \\ &- 0,0251X1 * X3 + 0,00327X2 * X3\end{aligned} \quad (2)$$

As depicted in Fig. 4, the normal probability plot of the residuals for the electricity consumption model closely follows a straight line. This alignment indicates that the residuals are approximately normally distributed, affirming the validity and reliability of the developed statistical model. The normality of errors suggests that the assumptions underlying the analysis of variance (ANOVA) are satisfied, thus supporting the robustness of the model's predictions.

Furthermore the subsequent figures illustrate how the daily energy consumption (E/24 h) of the refrigerator varies with different settings of the compartment temperature and varying quantities of phase change material (PCM). These graphical results provide a clear visual representation of the influence of both operational temperature and PCM mass on the refrigerator's energy efficiency, highlighting optimal conditions for reduced energy use.

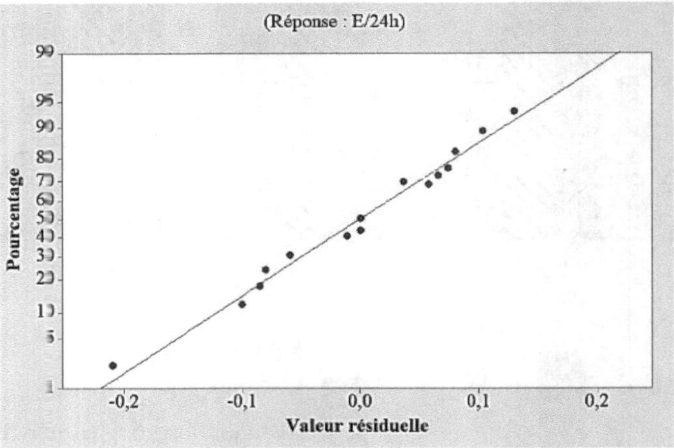

Fig. 4. Curve showing the normal probability of E/24 h.

Figure 5 provides a comprehensive overview of how both thermal load and the quantity of phase change material (PCM) affect the daily energy consumption of the refrigerator at different compartment temperatures (1 °C, 4 °C, and 8 °C).

At a Compartment Temperature of 1 °C (Fig. 5a): The integration of PCM leads to a significant reduction in energy use. When the mass of PCM (MPCM) is between 0.4 and 1 kg and the thermal load ranges from 0 to 2 kg, the refrigerator's energy consumption drops below 0.6 kWh per day. This demonstrates the effectiveness of PCM in stabilizing the internal temperature and reducing the compressor's workload under low thermal load conditions.

At a Compartment Temperature of 4 °C (Fig. 5b): A similar trend is observed, where increasing the amount of PCM from 0.2 to 1 kg and varying the thermal load from 0 to 8 kg results in a notable decrease in energy consumption, with values remaining below 0.8 kWh per day. This indicates that PCM continues to provide substantial energy savings even as the thermal load increases, making it a practical solution for households with varying usage patterns.

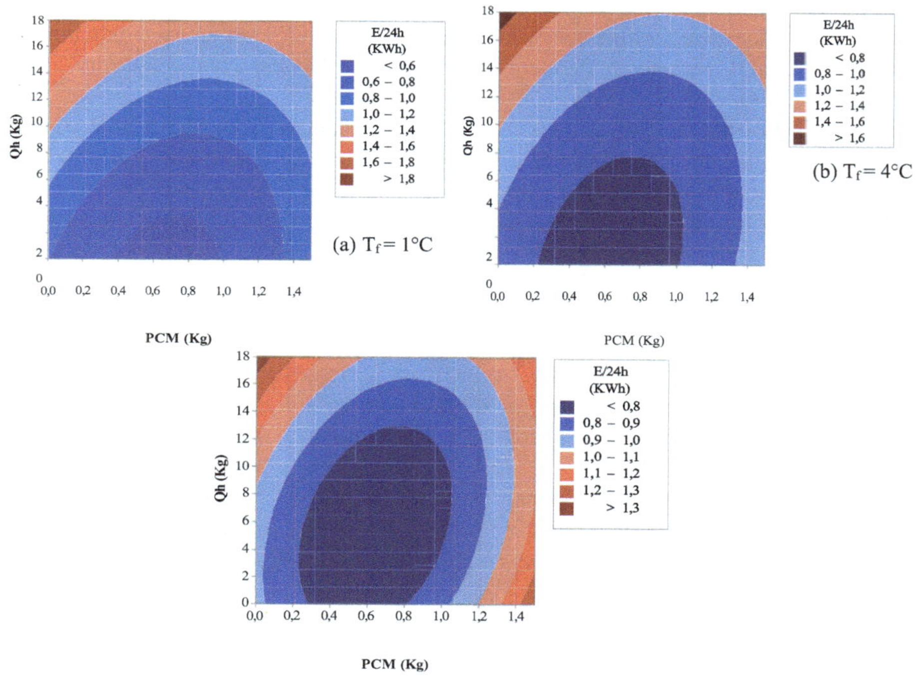

Fig. 5. Contours of the energy consumption of the refrigerator to (a) Tf = 1 °C, (b) Tf = 4 °C and (c) 8 °C

At a Compartment Temperature of 8 °C (Fig. 5c): Although the overall energy consumption is higher at this elevated temperature setting, the inclusion of PCM still yields significant benefits. For PCM quantities between

0.3 and 0.8 kg and thermal loads ranging from 0 to 12 kg, energy consumption is notably reduced compared to scenarios without PCM. However, beyond these ranges, the benefit diminishes, and energy use begins to rise again.

Collectively, these results highlight the strong influence of both PCM mass and thermal load on refrigerator efficiency. The optimal energy-saving effect is achieved by carefully selecting the appropriate amount of PCM relative to the expected thermal load and desired compartment temperature. This analysis underscores the practical value of integrating PCM technology into domestic refrigerators, particularly in climates or usage scenarios where temperature stability and energy efficiency are paramount.

5 Conclusion

This study experimentally investigated the impact of integrating phase change materials (PCM) and varying thermal loads on the energy consumption of a domestic refrigerator. The research utilized a Box–Behnken experimental design combined with ANOVA statistical analysis to systematically evaluate the data and identify the most significant factors influencing refrigerator performance.

The statistical analysis revealed a strong correlation between the coefficient of determination (R^2) for the experimental results and the adjusted R^2 values, both reaching up to 0.95. This high level of agreement confirms the robustness and reliability of the developed model.

The findings indicate that three parameters—the quantity of PCM, the applied thermal load, and the temperature setting of the fresh food compartment—are the most influential in determining daily energy consumption. Specifically, the integration of PCM was shown to significantly stabilize internal temperatures, reducing fluctuations within the usable storage volume to about 2 °C. This temperature stability is particularly beneficial during frequent door openings, helping to maintain food quality and safety.

Energy consumption tests demonstrated that refrigerators equipped with PCM operate more efficiently. For instance, when the mass of PCM ranged from 0.4 to 1 kg and the thermal load was between 0 and 2 kg at a compartment temperature of 1 °C, daily energy use was consistently below 0.6 kWh. These results highlight the potential of PCM technology to enhance refrigerator energy efficiency, especially under variable load conditions.

Overall, this research confirms that the strategic use of phase change materials in domestic refrigeration not only reduces energy consumption but also improves temperature stability, offering a promising pathway for energy savings and improved appliance performance in household settings.

References

1. Algérie Presse Service: https://www.aps.dz/algerie/132664 (n.d.)
2. Matemilola, S., Fadeyi, O., Sijuade, T.: Paris agreement. In: Idowu, S., Schmidpeter, R., Capaldi, N., Zu, L., Del Baldo, M., Abreu, R. (eds.) Encyclopedia of Sustainable Management, pp. 1–5. Springer, Cham (2020). https://doi.org/10.1007/978-3-030-02006-4_516-1
3. McGill, I.: Trends in refrigerator designs and operational issues. Senior refrigeration design engineer. Fisher & amp; Paykel Appliances Personal Communication with the Author (Ed.) 2014
4. Liu, D.Y., Chang, W.R., Lin, J.Y.: Performance comparison with effect of door opening on variable and fixed frequency refrigerators/freezers. Appl. Therm. Eng. **24**(14–15), 2281–2292 (2004)
5. Box, G.E.P., Behnken, D.W.: Some new three level designs for the study of quantitative variables. Technometrics. **2**(4), 455–475 (2012). https://doi.org/10.2307/1266454
6. Palanikumar, K., Paulo, D.J.: Electrical discharge machining: Study on machining characteristics of WC/Co composites. In: Davim, J.P. (ed.) Machining and Machine-Tools: Research and Development, pp. 135–168. Woodhead Publishing Limited, Cambridge, UK (2013) https://www.sciencedirect.com/topics/materials-science/electrical-dischargemachining

7. Bureau of Energy Efficiency (BEE): Schedule-05, direct cool refrigerator. revision No. 04, pp. 1–15, 2016
8. Evans, M.: Optimization of Manufacturing Processes: A Response Surface Approach. Maney Publishing, London (2003)
9. Andrzej, C., Justyna, S.: The application of Box-Behnken-Design in the optimization of HPLC separation of fluoroquinolones. Sci. Rep. **9**, 1–10 (2019) https://www.nature.com/articles/s41598-019-55761-z
10. Ferreira, J.M.S.L.C. et al.: Box-Behnken design: An alternative for the optimization of analytical methods. Anal. Chim. Acta. **597**, 179–186 (2007). https://doi.org/10.1016/j.aca.2007.07.011
11. Geppert, J., Rainer, S.: Analysis of effecting factors on domestic refrigerators' energy consumption in use. Energy Convers. Manag. **76**, 794–800 (2013). https://doi.org/10.1016/j.enconman.2013.08.027
12. Aziz, R. A., Aziz, S. A.: Application of box behnken design to optimize the parameters for kenaf-epoxy as noise absorber. In: IOP Conference Series Materials Science Engineering, vol. 454,pp. 1–11, 2018. https://iopscience.iop.org/article/10.1088/1757-899X/454/1/012001/pdf
13. Agah, T.M., Hülya, B., Berrin, O., Yasemin, G.: Box-Behnken experimental design in factorial experiments: The importance of bread for nutrition and health. Turk. J. Field Crops. **17**(2), 115–123 (2012)
14. Singh, S., Yash, P.S., Arora, S.: Statistical, diagnostic and response surface analysis of nefopam hydrochloride nanospheres using 35 Box-Behnken design. Int. J. Pharm. Pharmaceut. Sci. **7**(10), 89–101 (2015) https://innovareacademics.in/journals/index.php/ijpps/article/download/7659/pdf_1178

A Concise Survey on Neural Networks Compression Techniques

Abdelmoumene Touabi(✉) and Abdenour Amamra

Ecole Militaire Polytechnique (EMP), Bordj el Bahri, Alger, Algeria
Abdelmoumene.touabi@gmail.com

Abstract. Deep neural networks (DNNs) have demonstrated remarkable performance across a wide range of applications, particularly in computer vision. However, their complex architectures pose significant challenges for real-time deployment due to high computational and energy demands. Neural network compression techniques offer a promising solution by improving efficiency while preserving model accuracy. Several compression approaches have been proposed in the literature, each contributing to reductions in model size and computational cost. Despite these advances, network compression remains a highly active area of research. This paper presents a comprehensive survey of four key compression techniques: pruning, knowledge distillation, low-rank decomposition, and data quantization. We analyse their underlying principles, implementation strategies, and mathematical foundations. Building on these insights, we propose and implement hybrid methods that combine pruning, quantization, and low-rank decomposition to exploit the complementary strengths of these techniques. Experimental results demonstrate that the combination of pruning and quantization achieves about 75% reduction in model size with negligible loss in accuracy.

Keywords: Neural network compression · Pruning · Low rank decomposition · Data quantization · Knowledge distillation

1 Introduction

In recent years, convolutional neural networks (CNNs) have demonstrated outstanding performance in various computer vision tasks. Advances in computational power and the availability of large-scale datasets have made it possible to train deep CNNs with tens of millions of parameters. For instance, the 19-layer VGGNet [1] consists of 1.5 million nodes, 138 million parameters, requires 553 MB of storage, and performs approximately 15 billion floating-point operations (FLOPs) to classify a single 224 × 224 image. However, as deep neural networks (DNNs) continue to evolve, their increasing computational demands have outpaced hardware advancements. The growing complexity of these models further widens the gap between computational requirements and available hardware capabilities. To mitigate this challenge, model compression has emerged as a crucial approach, enabling the development of efficient networks that maintain a balance between performance and computational cost. By eliminating redundant weights, models can

M. Hatti (Ed.): IC-AIRES 2025, LNNS 1740, pp. 160–172, 2026.
https://doi.org/10.1007/978-3-032-12788-4_15

significantly reduce their size and computation time while maintaining high accuracy. This highlights the necessity of effective model compression techniques, particularly for deployment on resource-constrained devices such as mobile phones, Raspberry Pi, and FPGAs. These techniques ensure that deep learning models can operate efficiently in environments with limited computational and storage resources. This paper provides a comprehensive overview of four neural network compression techniques: pruning, low-rank decomposition, data quantization, and knowledge distillation, as illustrated in Fig. 1, enabling readers to quickly grasp their core principles. Additionally, we examine how the combination of these methods contributes to the development of highly compact models with minimal accuracy loss.

In the following sections, we first discuss neural network pruning, outlining its methodologies and impact on efficiency. Next, we examine low-rank decomposition and its role in reducing model complexity. We then explore data quantization, focusing on how numerical precision adjustments enhance efficiency. Finally, we present knowledge distillation and its effectiveness in transferring knowledge from larger to smaller models. We conclude with insights on combining these techniques to optimize deep learning models for real-world applications.

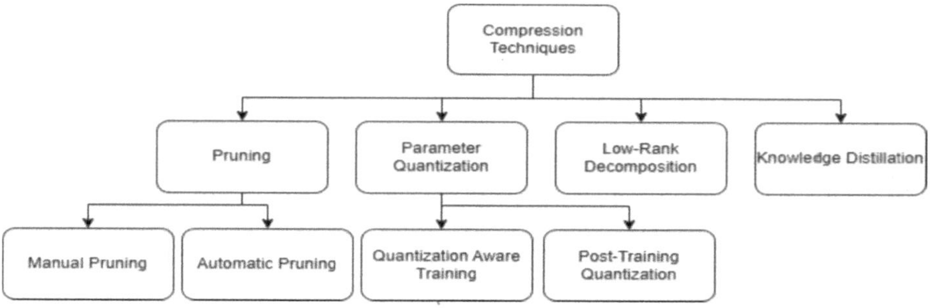

Fig. 1. Compression techniques.

2 Pruning

The primary goal of pruning is to compress neural networks by eliminating specific components, such as weights, channels, filters, or entire layers, while preserving the model's accuracy. This process focuses on removing elements that contribute minimally to the network's performance, including those with zero or near-zero values or redundant elements whose functionality can be fulfilled by others. Pruning can be broadly classified into two categories: manual pruning and automatic pruning. Figure 2 presents an overview of this taxonomy.

2.1 Manual Pruning

This approach relies on human expertise to determine pruning-related hyperparameters, such as the layer-wise pruning ratio. Elements can be pruned based on their importance,

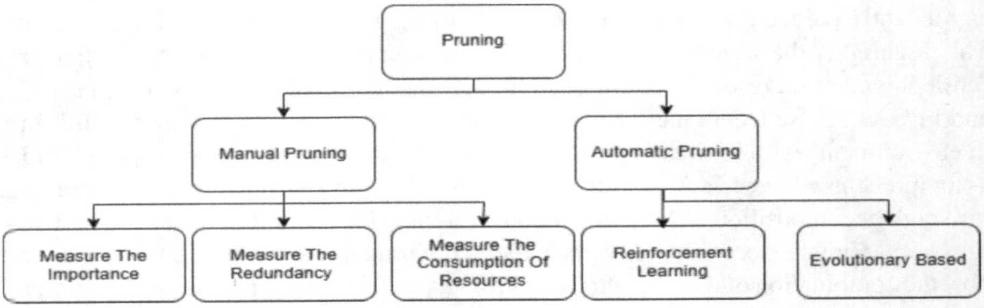

Fig. 2. Pruning taxonomy.

where weak elements those with values close to zero are removed as they contribute minimally to the final result. Alternatively, regularization techniques can be applied to drive certain elements toward zero before pruning. Another strategy involves measuring redundancy by identifying highly correlated elements and retaining only one of them. Additionally, hardware resource consumption can be analyzed to identify layers with high computational costs, allowing targeted pruning to optimize efficiency.

2.1.1 Measure the Importance

In [2], the authors promote magnitude-based pruning techniques for deep neural networks, focusing on removing weights with the smallest magnitudes. The work in [3] quantifies the importance of a filter within each layer using the $\ell 1$-norm, where filters with smaller kernel weights are more likely to generate feature maps with weaker activations. He et al. [4] observed that the performance of the $\ell 2$-norm criterion is slightly superior to that of the $\ell 1$-norm criterion. In [5], the authors utilize a scaling factor, denoted as γ, for each channel in the batch normalization (BN) layers. The output of each channel is multiplied by this scaling factor and trained jointly. Subsequently, an $\ell 1$ regularization is applied, allowing the identification of channels with values near zero as unimportant.

2.1.2 Measure the Redundancy

He et al. [6] analyze norm based criteria and highlight that their effectiveness relies on two key conditions that are not always satisfied: (1) the norm deviation of the filters should be large, and (2) the minimum norm of the filter should be small. They compute the geometric median, and the filters with the shortest distances to this median are selected for pruning. Correlation-based pruning computes the Pearson Correlation between a given filter and the others, identifying the most correlated filters [7]. Max-normalization is then applied to ensure crosslayer compatibility.

2.1.3 Measure the Consumption of Resources

Yang et al. [8] propose an energy-aware pruning algorithm for CNNs that leverages the network's energy consumption to guide the pruning process. The energy estimation methodology is based on parameters extrapolated from actual hardware measurements. Layers with the highest energy consumption are prioritized for pruning while minimizing errors in the output feature maps. Yang et al. [9] introduce an algorithm called NetAdapt, designed to automatically simplify a pretrained network to meet the resource budget (e.g., latency) of a target platform while maximizing accuracy. NetAdapt is guided by direct metrics for resource consumption.

2.2 Automatic Pruning

This method explores the search space by generating multiple models and selecting the best-performing one.

2.2.1 Reinforcement Learning

He et al. [10] train a reinforcement learning (RL) agent, which receives the state of layer t, characterized by 11 features. The action taken by the agent is the sparsity ratio of the layer. After compressing layer t, the agent moves to the next layer, t + 1. The validation accuracy of the pruned model, with all layers compressed, is evaluated without fine-tuning and returned to the RL agent as a reward R. Alwani et al. [11] propose a multi-agent learning approach where an agent is assigned to each channel. Each agent learns a single parameter that determines whether to keep or remove the channel.

2.2.2 Evolutionary Based

Liu et al. [12] aim to automatically determine the optimal number of channels for each layer through a two-stage framework. In the first stage, a meta-network called PruningNet is trained using a stochastic structure sampling method to generate weight parameters for different pruned structures. PruningNet leverages randomly sampled encoding vectors that represent the number of channels in each layer. In the second stage, an evolutionary search algorithm is applied to identify the optimal structures that meet predefined constraints. Another method, proposed in [13], identifies the optimal pruned structure by determining the number of channels in each layer. To tackle the challenge of the large search space for pruned structures, the authors constrain the number of preserved channels. The search for the optimal pruned structure is then formulated as an optimization problem and solved automatically using the Artificial Bee Colony algorithm.

3 Quantization

Parameter quantization in neural networks involves transforming the weights and activation values from high precision to lower precision. Traditionally, most networks have been trained using 32-bit floating-point (FP32) numbers [14]. However, FP32 representations often provide more precision than necessary for many networks. Reducing these parameters to lower-bit representations can lead to significant reductions in bandwidth usage, energy consumption, and on-chip area. Moreover, depending on the processor and execution environment, quantized parameters can accelerate neural network inference. Quantization can be mathematically expressed as follows:

$$Q = f(s * g(R) + z) \tag{1}$$

where R denotes the real-valued input or activation, Q denotes the quantized representation of the activation or weight, s represents the scaling factor, and z is the zero point. The function $g(.)$ is a clamp function applied to floating-point values in R before quantization, while $f(.)$ is a rounding function. Parameter quantization (z and s) establishes a data mapping relationship between fixed-point and floating-point data.

After quantization, it is often necessary to adjust the model's parameters. If this adjustment involves retraining the model, the process is referred to as Quantization-Aware Training (QAT). Conversely, if a model is quantized without requiring retraining, the process is known as Post-Training Quantization (PTQ). Han et al. [2] proposed a method to reduce the size of neural networks by applying k-means clustering to identify shared weights within each layer of a trained model. In this approach, weights belonging to the same cluster are assigned a shared value, thereby decreasing the number of bits needed to represent each weight. While this quantization approach reduces storage requirements, it does not lead to computational savings. Jacob et al. [15] introduced linear quantization, an affine mapping technique that transforms integers into real numbers. This process is mathematically represented as follows:

$$r = S(q - z) \tag{2}$$

where r represents the real weights (32-bit float), q denotes the quantized weights (2-bit signed int), z is the zero point (2-bit signed int), and S is the scale factor (32-bit float). In symmetric linear quantization, Z is set to zero, allowing a single scale factor S to be applied to the entire weight tensor (Per-Tensor Quantization). This approach simplifies implementation but can lead to significant accuracy degradation. Alternatively, Per-Channel Quantization assigns a separate scale to each channel, achieving better performance but lacking support on certain hardware platforms. To address these challenges, Nagel et al. [16] proposed equalizing weight ranges across the network by leveraging the scale-equivariance property of activation functions. Another quantization technique, called BinaryConnect, was introduced by Courbariaux et al. [17]. This method trains a DNN with binary weights during both forward and backward propagation while maintaining high-precision stored weights for gradient accumulation. This approach reduces computational complexity by replacing multiply-accumulate operations with simpler accumulation operations. Li et al. [18] introduced ternary quantization, where weights are quantized to three discrete values: $+1$, 0, and $-$ 1. Wang et al. [19] explored mixed-precision quantization, which is supported by emerging hardware accelerators to further enhance computational efficiency. To optimize the bitwidth for each layer, they employed reinforcement learning to automatically determine the most suitable bitwidth.

3.1 Post Training Quantization (PTQ)

Post-Training Quantization (PTQ) involves applying quantization to the weights, activations, and biases of a pretrained model, followed by re-optimization to produce a fully quantized version with appropriate scaling. Nagel et al. [20] observed that the standard rounding-to-nearest approach is suboptimal due to the interdependence of weights. To achieve optimal reconstruction, they leverage a small set of unlabeled data and approximate the task loss using a Taylor series expansion.

In activation quantization, the floating-point range varies across different inputs; determining this range requires collecting statistical information (e.g., min and max values) before deploying the model. Jacob et al. [15] estimate activation ranges by gathering the [min, max] values observed during training, which are then aggregated using an Exponential Moving Average. However, quantization can introduce biased errors in activations. To mitigate this, Nagel et al. [16] propose a method that corrects this bias in the layer's output by utilizing batch normalization parameters, enabling bias computation without requiring additional data.

3.2 Quantization-Aware Training (QAT)

To restore accuracy after quantization, the neural network undergoes training or fine-tuning with quantized weights and activations. During this process, a full-precision copy of the weights W is retained to track updates using gradient-based optimization in the backward pass. Once training is complete, only the quantized weights are used for inference, ensuring improved performance without significantly sacrificing accuracy.

4 Knowledge Distillation

Knowledge distillation aims to transfer knowledge from a large deep neural network (teacher) to a smaller network

(student) [21]. This process can be expressed by minimizing the following objective function:

$$L_{\{IKD\}} = \sum\nolimits_{\{x_i \in X\}} l(f_T(x_i), f_S(x_i)) \tag{3}$$

where L is a loss function that penalizes the difference between the teacher and student outputs. The objective is for the student network to learn not only from the true labels but also from the richer knowledge embedded in the teacher network's predictions. Romero et al. [22] propose FitNets, which compress networks from wide and shallow architectures to thinner and deeper ones. They introduce intermediate-level hints from the teacher's hidden layers to guide student training. Huang et al. [23] further improve knowledge transfer by aligning the distribution of activations between corresponding intermediate layers of the teacher and student networks. They employ Maximum Mean Discrepancy as a loss function to measure the discrepancy between teacher and student feature representations. Kim et al. [24] introduce a novel knowledge transfer approach that leverages convolutional operations to reframe the teacher's knowledge and adapt it for the student. Zagoruyko et al. [25] introduce attention as a mechanism for transferring knowledge from one network to another; they define two types of spatial attention maps: activation-based and gradient-based. Furlanello et al. [26] present an approach called Born-Again Networks, in which a new model with the same architecture is initialized and trained under the supervision of the previously trained model, iteratively improving performance.

This method begins with an ensemble of untrained student models that are jointly trained to solve the same task. Zhu et al. [27] introduce a multi-branch distillation approach, where a target network is extended with auxiliary branches.

During training, an ensemble teacher model is dynamically constructed from all branches, optimizing both the teacher model and each branch under the same target label constraints. Zhang et al. [28] propose a self-distillation framework in which knowledge is transferred within a single network. The network is divided into multiple sections, where deeper layers distill knowledge into shallower ones, improving overall efficiency and performance.

5 Low-Rank Decomposition

Low-rank decomposition is based on the hypothesis that 3D tensors possess inherent structural properties. In this context, convolutional kernels are treated as 3D tensors, while fully connected layers can be represented as either 2D matrices or 3D tensors [29]. According to [30], Singular Value Decomposition (SVD) is considered the most efficient method for obtaining low-rank approximations of a matrix, as it effectively captures the essential information contained within the matrix through its singular values. Denton et al. [31] apply SVD to compress convolutional layers by finding appropriate low rank approximations and then fine-tuning the upper layers of the network. Yu et al. [32] introduce a unified framework that combines low-rank and sparse decomposition of weight matrices with feature map reconstruction, achieving efficient neural network compression. Another approach presented in [33] incorporates nonlinear activation by optimizing the reconstruction error of nonlinear responses under a low-rank constraint. This method leverages SVD to reduce filter complexity and provides an efficient solution to the constrained nonlinear optimization problem. Lebedev et al. [34] propose a two-step method to accelerate convolutional layers using tensor decomposition and discriminative fine-tuning. In the first step, they apply nonlinear least squares to compute a low-rank CP decomposition of the 4D convolution kernel tensor. In the second step, they replace the original convolutional layer with a sequence of four convolutional layers with smaller kernels. The entire network is then fine-tuned using the standard backpropagation process. Jaderberg et al. [35] propose decomposing convolutional layers using low-rank expansions, effectively replacing large filters with a series of smaller, more efficient ones, achieving substantial speedups and model compression while maintaining performance.

6 Comparison

After analyzing each compression technique in detail, this section aims to highlight their differences, similarities, and potential complementarity. The following Table 1 provides a comparative summary based on several key criteria:

Table 1. A comparative summary between the four techniques of compression.

Criterion	Pruning	Low-rank decomposition	Knowledge distillation	Parameter quantization
Principle	Eliminates redundant or non-contributive weights, neurons, or filters.	Approximates weight matrices using lower-rank factors	Trains a smaller "student" model to mimic a larger "teacher" model	Reduces numerical precision of weights and activations
Main Objective	Reduce model size and inference cost	Reduce computational complexity of layers	Compress knowledge into a smaller model	Reduce memory usage and improve execution speed
Level of Application	Weight, neuron, channel, or layer level	Kernel weights, fully connected layers	Model-level: the entire network architecture is redesigned or substituted.	Weight and activation level

7 Experiments and Results

In this section, we aim to implement a hybrid compression strategy that combines three techniques: pruning, quantization, and low-rank decomposition. First, we reimplement the pruning method described by Liu et al. [5]. Then, we evaluate three scenarios:

In the first scenario, we apply quantization as described in [15] to the pruned model. We measure the accuracy, model size, and number of FLOPs before and after quantization.

In the second scenario, we apply low-rank decomposition to the pruned model. For linear layers, we use the method described in [32], and for the early convolutional layers, we adopt the approach proposed in [35]. In the third scenario, we apply both low-rank decomposition and quantization to the pruned model.

All experiments are conducted using the VGGNet architecture on the CIFAR-10 dataset. We begin by reimplementing the pruning method introduced by Liu et al. [5]. The results are presented in the following Table 2:

Table 2. Results of pruning.

		Accuracy (%)	Number of parameters (in millions)	Size (MB)	Estimated number of FLOPs
Liu et al. [5]	Baseline	93.66	20.04	/	7.97×10^8

(continued)

Table 2. *(continued)*

		Accuracy (%)	Number of parameters (in millions)	Size (MB)	Estimated number of FLOPs
	Sparsity	93.30	20.04	/	7.97×10^8
	Pruned (70%)	32.54	2.30	/	3.91×10^8
	Fine tune-160 (70%)	93.80	2.30	/	3.91×10^8
Our implementation	Baseline	93.93	20.04	160	/
	Sparsity	93.52	20.04	160	/
	Pruned (70%)	47	2.29	8.79	/
	Fine tune-160 (70%)	93.69	2.29	8.79	384,751,932

We use the pruned model as the baseline for the subsequent experiments.

7.1 Scenario 01 (Pruned Model + Quantization)

In this scenario, we apply Quantization-Aware Training to the pruned VGG16 model trained on the CIFAR-10 dataset. After pruning, we quantize both the weights and activations from 32-bit floating-point to 8-bit integers. The model is then fine-tuned for 10 epochs using a learning rate of 5e-5 to recover any potential loss in accuracy due to quantization. This approach aims to significantly reduce the model size while maintaining a high level of accuracy. The results in terms of accuracy and model size before and after quantization are reported in the following Table 3.

Table 3. The results of quantization.

Model	Accuracy (%)	Size (MB)
Pruned model	93.69	8.79
Pruned model + QAT	93.31	2.29

7.2 Scenario 02 (Pruned Model + Low Rank Decomposition)

In the second scenario, we take the pruned model and apply low-rank decomposition with a compression ratio of 70%. After decomposition, the model is fine-tuned for 15 epochs to recover potential accuracy loss. The results in terms of accuracy, model size, number of parameters and FLOPs are detailed in the following Table 4.

Table 4. The results of LRD.

Model	Accuracy (%)	Size (MB)	Number of parameters (in millions)	Estimated number of FLOPs
Pruned model	93.69	8.79	2.29	384,751,932
Pruned model + LRD	90.71	7.63	1.98	248,649,388

7.3 Scenario 03 (Pruned Model + Low Rank Decomposition + Quantization)

In the third scenario, we take the pruned model and apply low-rank decomposition, followed by quantization. After these two compression steps, the model is fine-tuned for 10 epochs in an attempt to recover the accuracy lost during the process. The resulting performance in terms of accuracy, model size, number of parameters and FLOPs is presented in the following Table 5.

Table 5. The results of LRD + Quantization.

Model	Accuracy (%)	size(MB)	Number of parameters (in millions)	Estimated number of FLOPs
Pruned model	93.69	8.79	2.29	384,751,932
Pruned model + LRD + QAT	77.78	2.04	1.98	/

8 General Discussion

In Scenario 01, quantization reduces the model size from 8.79 MB to 2.29 MB (~ 4× compression) by converting each parameter from 32 bits to 8 bits, with almost no loss in accuracy (only 0.38%). QAT proves to be an excellent technique for significantly reducing model size while maintaining nearly the same accuracy.

In Scenario 02, LRD achieves only a 13% size reduction, but it also reduces the number of parameters from 2.29 million to 1.98 million, resulting in a lower number of FLOPs. However, this comes at the cost of a ~3% accuracy drop.

In Scenario 03, quantizing the model previously compressed by LRD (CP decomposition) proves extremely challenging and results in a severe accuracy degradation of approximately 16%, which is clearly unacceptable. This indicates a negative interaction between low-rank approximation and quantization, where the LRD-compressed model becomes too sensitive to quantization noise.

There is a clear trade-off between compression and accuracy. If the main priority is preserving accuracy and minimizing model size, the combination of Pruning + QAT is the most effective. However, if a slight loss in accuracy is acceptable in exchange for additional reductions in MACs or parameter count before quantization, then Pruning + LRD (CP) + Fine-Tuning may be considered though its benefits are limited in this case. The triple combination (Pruning + LRD + QAT) is not recommended due to the compounded degradation in accuracy.

9 Conclusion

This paper provides a comprehensive survey of deep neural network model compression techniques, focusing on four primary methods: pruning, quantization, low rank decomposition, and knowledge distillation. We discussed both manual and automated pruning methods, explored two quantization approaches (quantization-aware training and post training quantization), examined low-rank decomposition, and reviewed knowledge distillation.

We also implemented three hybrid methods: **(1)** pruning followed by quantization, **(2)** pruning followed by low-rank decomposition, and **(3)** pruning followed by low-rank decomposition and then quantization. Among these, the combination of pruning and quantization yielded the best results in our experiments, achieving a significant reduction in model size with negligible loss in accuracy.

References

1. Simonyan, K., Zisserman, A.: Very deep convolutional networks for large-scale image recognition. arXiv preprintarXiv:1409.1556 (2014)
2. Han, S., Mao, H., Dally, W. J.: Deep compression: Compressing deep neural networks with pruning, trained quantization and huffman coding. arXiv preprint https://arxiv.org/abs/1510.00149 (2015)
3. Li, H., Kadav, A., Durdanovic, I., Samet, H., Graf, H. P.: Pruning filters for efficient convnets. arXiv preprint https://arxiv.org/abs/1608.08710 (2016)
4. He, Y., Kang, G., Dong, X., Fu, Y., Yang, Y.: Soft filter pruning for accelerating deep convolutional neural networks. arXiv preprint https://arxiv.org/abs/1808.06866 (2018)
5. Liu, Z., Li, J., Shen, Z., Huang, G., Yan, S., Zhang, C.: Learning efficient convolutional networks through network slimming. In: IEEE International Conference on Computer Vision, pp. 2736–2744, 2017
6. He, Y., Liu, P., Wang, Z., Hu, Z., Yang, Y.: Filter pruning via geometric median for deep convolutional neural networks acceleration. In: IEEE/CVF Conference on Computer Vision and Pattern Recognition, pp. 4340–4349, 2019
7. Wang, W., Fu, C., Guo, J., Cai, D., He, X.: Cop: Customized deep model compression via regularized correlation-based filter-level pruning. arXiv preprint https://arxiv.org/abs/1906.10337 (2019)
8. Yang, T.-J., Chen, Y.-H., Sze, V.: Designing energy efficient convolutional neural networks using energy-aware pruning. In: IEEE Conference on Computer Vision and Pattern Recognition, pp. 5687–5695, 2017
9. Yang, T.-J., et al.: Netadapt: Platformaware neural network adaptation for mobile applications. In: European Conference on Computer Vision (ECCV), pp. 285–300, 2018
10. He, Y., Lin, J., Liu, Z., Wang, H., Li, L.-J., Han, S.: Amc: Automl for model compression and acceleration on mobile devices. In: European Conference on Computer Vision (ECCV), pp. 784–800, 2018

11. Alwani, M., Wang, Y., Madhavan, V.: Decore: Deep compression with reinforcement learning. In: IEEE/CVF Conference on Computer Vision and Pattern Recognition, pp. 12349–12359, 2022
12. Liu, Z., et al.: Metapruning: Meta learning for automatic neural network channel pruning. In: IEEE/CVF International Conference on Computer Vision, pp. 3296–3305, 2019
13. Lin, M., et al.: Channel pruning via automatic structure search. arXiv preprint https://arxiv.org/abs/2001.08565 (2020)
14. Sze, V., Chen, Y.-H., Yang, T.-J., Emer, J.S.: Efficient processing of deep neural networks: A tutorial and survey. Proc. IEEE. **105**(12), 2295–2329 (2017)
15. Jacob, B., et al.: Quantization and training of neural networks for efficient integer arithmetic-only inference. In: IEEE Conference on Computer Vision and Pattern Recognition, pp. 2704–2713, 2018
16. Nagel, M., van Baalen, M., Blankevoort, T., Welling, M.: Data-free quantization through weight equalization and bias correction. In: IEEE/CVF International Conference on Computer Vision, pp. 1325–1334, 2019
17. Courbariaux, M., Bengio, Y., David, J.-P.: Binaryconnect: Training DNNs with binary weights during propagations. In: Advances in neural information processing systems, vol. 28, 2015
18. Li, F., Liu, B., Wang, X., Zhang, B., Yan, J.: Ternary weight networks. arXiv preprint https://arxiv.org/abs/1605.04711 (2016)
19. Wang, K., Liu, Z., Lin, Y., Lin, J., Han, S.: Haq: Hardware-aware automated quantization with mixed precision. In: IEEE/CVF Conference on Computer Vision and Pattern Recognition, pp. 8612–8620, 2019
20. Nagel, M., Amjad, R. A., van Baalen, M., Louizos, C., Blankevoort, T.: Up or down? Adaptive rounding for post-training quantization. In: International Conference on Machine Learning, PMLR, pp. 7197–7206, 2020
21. Hinton, G.: Distilling the knowledge in a neural network. arXiv preprint https://arxiv.org/abs/1503.02531 (2015)
22. Romero, A., Ballas, N., Ebrahimi Kahou, S., Chassang, A., Gatta, C., Bengio, Y.: Fitnets: Hints for thin deep nets. arXiv preprint https://arxiv.org/abs/1412.6550 (2014)
23. Huang, Z., Wang, N.: Like what you like: Knowledge distill via neuron selectivity transfer. arXiv preprint https://arxiv.org/abs/1707.01219 (2017)
24. Kim, J., Park, S. U., Kwak, N.: Paraphrasing complex network: Network compression via factor transfer. In: Advances in neural information processing systems, vol. 31, 2018
25. Zagoruyko, S., Komodakis, N.: Paying more attention to attention: Improving the performance of convolutional neural networks via attention transfer. arXiv preprint https://arxiv.org/abs/1612.03928 (2016)
26. Furlanello, T., Lipton, Z., Tschannen, M., Itti, L., Anandkumar, A.: Born again neural networks. In: International Conference on Machine Learning, PMLR, pp. 1607–1616, 2018
27. Lan, X., Zhu, X., Gong, S.: Knowledge distillation by on-thefly native ensemble. In: Advances in Neural Information Processing Systems, vol. 31, 2018
28. Zhang, L., Song, J., Gao, A., Chen, J., Bao, C., Ma, K.: Be your own teacher: Improve the performance of convolutional neural networks via self distillation. In: IEEE/CVF International Conference on Computer Vision, pp. 3713–3722, 2019
29. Lin, S., Ji, R., Chen, C., Tao, D., Luo, J.: Holistic CNN compression via low-rank decomposition with knowledge transfer. IEEE Trans. Pattern Anal. Mach. Intell. **41**(12), 2889–2905 (2018)
30. Klema, V., Laub, A.: The singular value decomposition: Its computation and Some Applications. IEEE Trans. Autom. Control. **25**(2), 164–176 (1980)
31. Denton, E.L., Zaremba, W., Bruna, J., LeCun, Y., Fergus, R.: Exploiting linear structure within convolutional networks for efficient evaluation. Adv. Neural Inf. Proces. Syst. **27**, 2889–2905 (2014)

32. Yu, X., Liu, T., Wang, X., Tao, D.: On compressing deep models by low rank and sparse decomposition. In: IEEE Conference on Computer Vision and Pattern Recognition, pp. 7370–7379, 2017
33. Zhang, X., Zou, J., Ming, X., He, K., Sun, J.: Efficient and accurate approximations of nonlinear convolutional networks. In: IEEE Conference on Computer Vision and pattern Recognition, pp. 1984–1992, 2015
34. Lebedev, V., Ganin, Y., Rakhuba, M., Oseledets, I., Lempitsky, V.: Speeding-up convolutional neural networks using fine-tuned cp-decomposition. arXiv preprint https://arxiv.org/abs/1412.6553 (2014)
35. Jaderberg, M., Vedaldi, A., Zisserman, A.: Speeding up convolutional neural networks with low rank expansions. arXiv preprint https://arxiv.org/abs/1405.3866 (2014)

Fake no More: Smarter Social Media Detection With RTGBO

Nadir Mahammed[1](✉), Imène Saidi[1], Mahmoud Fahsi[2], and Miloud Khaldi[1]

[1] LabRI-SBA, Ecole Superieure en Informatique, Sidi Bel Abbes 22000, Algeria
{n.mahammed,i.saidi,m.khaldi}@esi-sba.dz
[2] EEDIS, Djillali Liabes University of Sidi Bel Abbes, Sidi Bel Abbes 22000, Algeria
mahmoud.fahsi@univ-sba.dz

Abstract. Fake profiles on social media undermine user trust and platform integrity. Detecting them is challenging due to class imbalance and the need for well-tuned machine learning models. This study introduces a metaheuristic-driven approach using Ring Toss Guided Beam Optimization to tune three classifiers: Random Forest, Logistic Regression, and K-Nearest Neighbors. Inspired by ring toss dynamics, the metaheuristic balances global exploration with local refinement to improve hyperparameter search. Tested on a real-world Facebook dataset, the RTGBO- optimized models—particularly Random Forest—achieved high performance, with an F1-score of 0.9938 and accuracy of 0.9893. These results highlight metaheuristic's effectiveness in boosting model accuracy and resilience in imbalanced settings, supporting its broader application in fake profile detection and cybersecurity tasks.

Keywords: Fake profile detection · Social media security · Machine learning · Metaheuristic · RTGBO · Hyperparemeters optimization

1 Introduction

The rise of social networks has revolutionized global communication but also introduced new threats, notably the proliferation of fake user profiles. These accounts compromise trust and expose platforms to risks such as misinformation, fraud, and identity theft [4]. As fake profiles grow in scale and sophistication, their detection has become a pressing issue for both researchers and platform administrators. Traditional rule-based systems often fall short due to their limited adaptability and reliance on manual heuristics. In contrast, machine learning (ML) models—such as Random Forest (RF), Logistic Regression (LR), and K-Nearest Neighbors (KNN)—have shown promise by learning patterns from labeled data. However, their effectiveness remains highly dependent on optimal hyperparameter tuning and relevant feature selection. To overcome these limitations, recent work has turned to metaheuristic optimization. These nature-inspired algorithms, such as swarm intelligence and evolutionary strategies, are well-suited to exploring high-dimensional search spaces and enhancing model generalization [8, 10]. Building on this direction, we propose the Ring Toss Guided Beam Optimization (RTGBO) algorithm—a recent metaheuristic inspired by the dynamics of a ring toss game. RTGBO integrates stochastic exploration with guided refinement to efficiently tune ML classifiers.

M. Hatti (Ed.): IC-AIRES 2025, LNNS 1740, pp. 173–182, 2026.
https://doi.org/10.1007/978-3-032-12788-4_16

This study applies RTGBO to optimize RF, LR, and KNN for fake profile detection. While metaheuristic-ML hybrids have shown success in related tasks [17, 18], RTGBO has not yet been explored in this domain. We evaluate its impact and benchmark its performance within a unified detection framework.

The paper is organized as follows: Sect. 2 reviews related work, Sect. 3 outlines the methodology, Sect. 5 presents results, and Sect. 6 concludes with findings and future directions.

2 Related Work

Fake profile detection has emerged as a critical research area due to its impact on online trust and user safety. ML models like LR, SVM, and RF have been widely used, yet their effectiveness often depends on careful hyperparameter tuning and feature selection—limiting performance in dynamic environments.

To address these limitations, metaheuristic optimization has gained traction. Sekkal *et al.* [13] used Grey Wolf Optimizer (GWO) for Facebook profile verification, achieving high accuracy. Mahammed *et al.* developed multiple frameworks combining bio-inspired algorithms with ML to improve robustness across platforms [7–9]. Other contributions include Perwej *et al.* [12], who applied SVD and correlation analysis to enhance SVM performance on Instagram data, though their approach lacked real-time adaptability. Senthilkumar *et al.* [15] achieved 92% accuracy and 0.94 AUC using behavior-based features and ML classifiers. Broader reviews, such as those by Shannaq *et al.* [14] and Othman and Ku-Mahamud [11], emphasize the scalability and cross-domain potential of metaheuristics, including hybrid and sentiment-aware approaches.

These works collectively underline the growing role of metaheuristics in improving ML adaptability and accuracy in social media security tasks.

Table 1 shows that bio-inspired metaheuristics like SBO, FHO, and GWO are widely used with classifiers such as SVM, KNN, and RF, achieving up to 99% accuracy across major platforms. These hybrid models improve feature selection and robustness, especially in imbalanced data settings. However, game-based metaheuristics remain largely unexplored. Algorithms inspired by strategic behavior—like Harmony Search or game theory—offer untapped potential for social media environments, where adversaries adapt to evade detection. Additionally, most studies rely on static datasets, limiting real-time responsiveness. Future work should explore dynamic, behavior-aware metaheuristics to strengthen model adaptability in evolving adversarial contexts.

Table 1. Related work summary

Ref	Platform	Approach	Key contribution	Acc.
[7]	Facebook	SBO + k-means	Hybrid model using Satin Bowerbird Optimization and k-means	0.98
[8]	Facebook, Instagram	FHO + ML (RF, SVM, KNN, NB)	Strong accuracy on static datasets	0.99 0.96

(continued)

Table 1. (*continued*)

Ref	Platform	Approach	Key contribution	Acc.
[9]	Twitter	FHO + ML (SVM, NB, RF, LR)	Validated on 1000 labeled accounts	0.99
[13]	Facebook	GWO + ML (LR, DT, SVM, RF, KNN)	Applied to 1244 profiles	0.98
[11]	Facebook, Twitter, Instagram	Hybrid metaheuristics + Sentiment analysis	Multilingual fake review and spam detection with sentiment integration.	0.94
[12]	Instagram	RF, SVM, KNN, LR, NB	No real-time scalability	0.94
[15]	Facebook, Twitter	ML (DT, SVM, RF)	Combining user activity and profile features	0.92

3 Method and Methods

The dataset used in this study, sourced from [1], includes 1,244 Facebook profiles—1,043 real and 201 fake. It contains 15 features (see Table 2) combining behavioral and profile-based attributes. Numeric features include mutual friends (0–553) and total posts (0–832), while categorical ones reflect details like living place and relationship status. This mix of activity and demographic data provides a realistic basis for evaluating fake profile detection models.

Table 2. Facebook dataset features

Feature	Description	Type
Name-ID	Unique profile identifier	Categorical
Profile Picture	Indicates whether a profile image is present	Boolean
Likes	Topics or pages liked by the user	Numerical
Groups Joined	Total number of Facebook groups joined	Numerical
Friends	Number of mutual friends with other users	Numerical
Education	States whether educational information is available	Boolean
Work	Presence of employment details	Boolean
Relationship	Visibility of relationship status	Boolean
Check-in	Indicates whether geolocation tagging is used	Boolean
Posts	Total number of posts shared by the user	Numerical
Tags	Frequency of user being tagged by others	Numerical
Profile Intro	Presence of an introduction or bio section	Boolean

3.1 Dataset Preprocessing

4 Data Preprocessing

To prepare the dataset for ml, standard preprocessing were applied [2]. First, irrelevant attributes such as profile links were removed. Next, missing values were handled: numeric gaps were filled with zeros, while categorical ones were replaced with *"Not Available"*. Afterward, categorical and Boolean fields were transformed into numerical format using one-hot encoding. Finally, numerical features were normalized with to ensure consistency across the dataset.

4.1 Ring Toss Guided Beam Optimization Algorithm

This study applies the Ring Toss Guided Beam Optimization (RTGBO) algorithm proposed by Doumari *et al.* [5], inspired by the dynamics of a traditional ring toss game. In RTGBO, candidate solutions are "thrown" into the search space and refined using a guided beam mechanism that learns from previous successes. This balances global exploration with local exploitation, promoting fast and accurate convergence. Unlike conventional metaheuristics, RTGBO introduces directional learning to improve adaptability and precision [19].

4.2 Hyperparameter Optimization Using RTGBO

We employed RTGBO to optimize model hyperparameters. This custom meta- heuristic blends stochastic exploration with guided exploitation, inspired by ring toss dynamics, to efficiently search for optimal settings and boost model performance [19].

4.3 RTGBO-Based Hyperparameter Optimization

Algorithm 1 summarizes the RTGBO process for ML hyperparameter tuning. After preprocessing and feature selection, a population of candidate configurations is initialized. Each candidate is evaluated by training model M and scoring it using the F1 metric. Over T iterations, new solutions are generated via a simulated ring toss and refined using guided beam adjustment. Candidates with better fitness replace previous ones, balancing exploration and exploitation. RT-GBO returns the best-performing configuration. Unlike classical metaheuristics like PSO, RTGBO introduces metaphor-driven directional learning, offering improved search precision in high-dimensional spaces.

Algorithm 1 RTGBO-Based Optimization of ML Algorithms

Require: Dataset D, model M, population size N, max iterations T
Ensure: Optimized model M with best hyperparameters
Preprocess D: clean data, encode categorical variables, normalize features
Select features based on correlation threshold (e.g., 0.05)
Initialize population S with N random hyperparameter sets
for each solution $s \in S$ **do**
    Train model M using s
    Evaluate fitness using F1-score
end for
for $t = 1$ to T **do**
    for each solution $s \in S$ **do**
        Generate new solution s' via ring toss movement
        Refine s' using guided beam adjustment
        Train M with s', evaluate F1-score
        if fitness(s') > fitness(s) **then**
            Replace s with s'
        end if
    end for
end for
return Best-performing model with optimized parameters

4.4 Classification Models

We evaluated three ML classifiers—RF, LR, and KNN—each optimized using RTGBO for fake profile detection:

- Random Forest (RF): Robust to noise with feature importance [16].
- Logistic Regression (LR): Simple, interpretable, best for linear data [3].
- K-Nearest Neighbors (KNN): Non-parametric, captures local patterns [6].

4.5 Optimized Hyperparameters

Each classifier was fine-tuned using RTGBO to enhance detection performance. Table 3 shows the RTGBO-optimized hyperparameters for each classifier used in the study. Table 4 outlines the core settings and components of the RTGBO metaheuristic used for model optimization.

5 Results and Discussion

5.1 Evaluation Metrics

Model performance was evaluated before and after RTGBO optimization using standard metrics: accuracy (overall correct predictions), precision (true positives among predicted positives), recall (correctly identified positives), F1-score (balance of precision and recall, useful for imbalanced data), and ROC-AUC (class separability across thresholds).

Table 3. RTGBO-optimized hyperparameters by classifier

Classifier	Parameter	Value
Random Forest	n_estimators max_depth min_samples_split min_samples_leaf criterion	200 20 2 1 gini
Logistic Regression	solver C max_iter	lbfgs 1.0 1000
K-Nearest Neighbors	n_neighbors metric p weights	5 minkowski 2 uniform

Table 4. RTGBO metaheuristic parameter settings

Parameter	Value / description
Population Size (*N*)	20
Max Iterations (*T*)	Defined in code context (assumed predefined constant)
Fitness Metric	F1-score
Initial Solution Generation	Random hyperparameter sets
Guided Beam Adjustment	Refines solution trajectory based on past fitness
Ring Toss Movement	Simulates stochastic toss of candidate solution
Feature Selection Threshold	0.05 (correlation-based)
Model Evaluation	F1-score-based fitness update
Selection Mechanism	Replace s with s' if $fitness(s') > fitness(s)$

5.2 Confusion Matrix Analysis of RTGBO-Optimized Models

Figure 1 presents the confusion matrices for RTGBO-tuned classifiers applied to 1,244 Facebook profiles (1,043 real and 201 fake). The RTGBO-enhanced RF achieved the best results, with only 10 misclassifications and accurate detection of 194 fake and 1,040 real profiles [16]. LR also performed well, though it produced 52 errors, highlighting its sensitivity to borderline cases [3]. KNN showed balanced performance, correctly classifying 191 fake and 1,033 real accounts [6]. Overall, these results demonstrate that RTGBO substantially improved detection accuracy across models, particularly for RF and KNN, underscoring its effectiveness on imbalanced social media datasets.

5.3 RTGBO Impact on Classification Performance

Table 5 compares the performance of RF, LR, and KNN classifiers before and after RTGBO tuning. Across all models, accuracy, precision, recall, F1-score, and ROC-AUC improved significantly, confirming the effectiveness of metaheuristic optimization for fake profile detection [3, 6, 16]. RF achieved the largest gains,

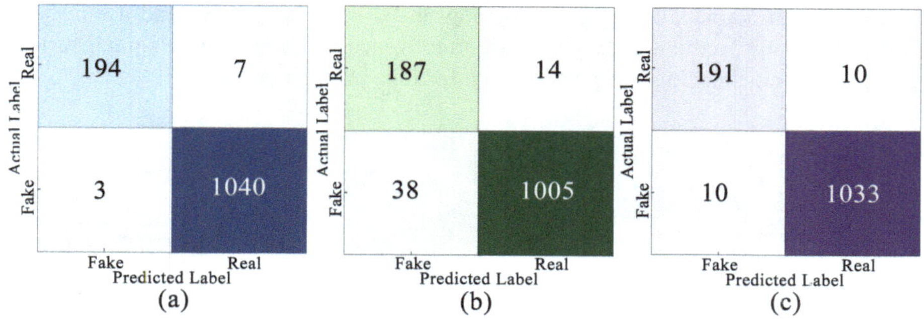

Fig. 1. Confusion matrices after RTGBO optimization: (a) Random Forest, (b) Logistic Regression, (c) K-Nearest Neighbors.

with accuracy increasing from 0.9540 to 0.9893 and F1-score from 0.9725 to 0.9938, while LR and KNN also improved—particularly KNN, whose F1-score rose from 0.9628 to 0.9922. These findings highlight RTGBO's dual-phase design, combining global exploration through simulated ring toss with local refinement via guided beam adjustment, which together enable effective hyperparameter tuning and enhanced performance in imbalanced, security-critical environments.

Table 5. Performance metrics before and after RTGBO optimization

Metric	Random	Forest	Logistic	Regression	K-Nearest	Neighbors
	Before	After	Before	After	Before	After
Accuracy	0.9540	0.9893	0.9502	0.9779	0.9387	0.9866
Precision	0.9815	0.9937	0.9859	0.9860	0.9857	0.9907
Recall	0.9636	0.9969	0.9545	0.9636	0.9490	0.9906
F1-score	0.9725	0.9938	0.9700	0.9747	0.9628	0.9922
ROC-AUC	0.9918	0.9995	0.9866	0.9914	0.9636	0.9981

5.4 Visual Comparison of RTGBO-Tuned Models

Figure 2 presents a dual-panel comparison of classification performance metrics. Figure 2a illustrates the impact of RTGBO hyperparameter tuning on three classifiers—RF, LR, and KNN—across four key metrics: Accuracy, Precision, Recall, and

F1-score. Among these, RTGBO-RF outperforms the others, particularly in recall and F1-score, reflecting its superior capability to detect fake profiles with minimal misclassification. KNN follows closely, while LR shows a performance drop, particularly in recall, underscoring its sensitivity to data complexity even after optimization.

Figure 2b benchmarks the RTGBO-RF model against prior state-of-the- art works using the same Facebook dataset. The compared models—ID3 from [7,9] and SVM from [15]—achieve solid but lower scores. Notably, the proposed RTGBO-RF model surpasses them across all metrics, achieving an F1-score of 0.9938 and accuracy of 0.9893. This improvement underscores the effectiveness of game-inspired metaheuristic optimization in enhancing ML performance for fake profile detection.

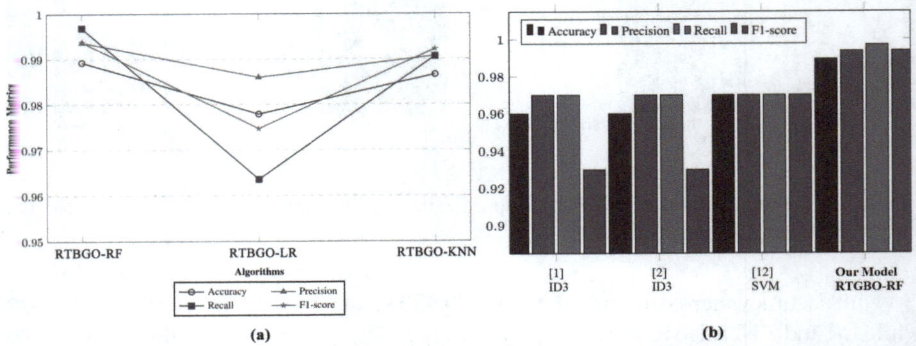

Fig. 2. RTGBO-optimized classifiers performance: (a) RTGBO-ML classifier metrics, (b) Comparison with existing studies

5.5 Discussion

The experimental results highlight the effectiveness of RTGBO in boosting ML model performance for fake profile detection. By optimizing hyperparameters, RTGBO consistently improved classification metrics across RF, LR, and KNN.

- RTGBO-RF achieved the best results, with an F1-score of 0.9938 and accuracy of 0.9893, confirming the strength of ensemble models when fine-tuned.
- RTGBO-KNN followed closely, reaching an F1-score of 0.9922 and recall of 0.9906—showing that KNN can match more complex models when optimally configured.
- RTGBO-LR also improved, though to a lesser extent, with recall of 0.9636. This suggests that even linear models benefit from metaheuristic tuning, despite their limited capacity for nonlinear representation.

Statistically, performance gains reached up to 3.5% across metrics, under- scoring RTGBO's impact in real-world detection tasks where even marginal improvements enhance trust and security. Compared to prior work, which typically capped F1-scores between 0.93 and 0.97, RTGBO-RF sets a new benchmark, affirming the value of bio-inspired hyperparameter optimization.

5.6 Limitations

Despite strong RTGBO results, some limitations remain:

- Stochastic behavior: RTGBO's performance may vary with initialization, requiring tuning of its own parameters (e.g., population size, iterations).
- Computational cost: Repeated model training increases runtime, which may limit use in real-time or large-scale settings without optimization strategies.
- Model dependency: Gains from RTGBO were more significant for RF and KNN than for LR, suggesting sensitivity to model complexity.
- Dataset scope: Results are based on a single Facebook dataset; generalization to other platforms remains to be validated.
- Class imbalance: Despite using F1-score and ROC-AUC, imbalance may still bias the models toward the majority class.

6 Conclusion

This paper presented a novel framework for fake profile detection on social media, leveraging the RTGBO algorithm for hyperparameter tuning of three widely used classifiers: RF, LR, and KNN. The experimental results on a real-world Facebook dataset demonstrate that RTGBO significantly enhances the performance of all three classifiers across key evaluation metrics—accuracy, precision, recall, F1- score, and ROC-AUC.

Among the tested models, RTGBO-RF achieved the most outstanding results, attaining an F1-score of 0.9938 and accuracy of 0.9893. These improvements validate the efficacy of RTGBO in optimizing ensemble-based classifiers. RTGBO-KNN also performed strongly, revealing that even instance-based models can benefit from metaheuristic tuning. Although RTGBO-LR showed comparatively moderate gains, it still outperformed its baseline, reinforcing the general utility of RTGBO in diverse ML settings.

Beyond performance metrics, this study underscores the relevance of game- inspired metaheuristics in tackling challenges posed by imbalanced and noisy social media data. The RTGBO algorithm's ability to balance exploration and exploitation proves effective in navigating complex search spaces inherent to hyperparameter optimization.

Future work will focus on evaluating RTGBO in real-time detection systems, exploring its integration with deep learning models, and comparing it with other emerging metaheuristics, particularly those grounded in game theory and co- evolutionary strategies. These directions aim to further improve the scalability and adaptability of fake profile detection systems in increasingly dynamic online environments.

References

1. Albayati, M., Altamimi, A.: Mdfp: A machine learning model for detecting fake facebook profiles using supervised and unsupervised mining techniques. Int. J. Simulation: Systems, Sci. Technol. **20**(1), 1–10 (2019)
2. Çetin, V., Yıldız, O.: A comprehensive review on data preprocessing techniques in data analysis. Pamukkale Univ. J. Eng. Sci. **28**(2), 299–312 (2022)
3. Das, A.: Logistic regression. In: Michalos, A.C., Weijers, I.B. (eds.) Encyclopedia of Quality of Life and Well-Being Research, pp. 3985–3986. Springer, Cham, Switzerland (2024)

4. Dostalova, P.: Trends in security technologies. Prz. Elektrotech. **99**(1), 147–150 (2023). https://doi.org/10.15199/48.2023.01.28
5. Doumari, S.A., Givi, H., Dehghani, M., Malik, O.P.: Ring toss game-based optimization algorithm for solving various optimization problems. Int. J. Intell. Eng. Syst. **14**(3), 545–554 (2021)
6. Halder, R.K., Uddin, M.N., Uddin, M.A., Aryal, S., Khraisat, A.: Enhancing k- nearest neighbor algorithm: A comprehensive review and performance analysis of modifications. J. Big Data. **11**(1), 113 (2024)
7. Mahammed, N., Bennabi, S., Fahsi, M., Klouche, B., Elouali, N., Bouhadra, C.: Fake profiles identification on social networks with bio-inspired algorithm. In: Proceeding of the International Conference on Big Data, IoT, Web Intelligence and Applications (BIWA), IEEE, pp. 48–52, 2022
8. Mahammed, N., Klouche, B., Saidi, I., Khaldi, M., Fahsi, M.: Enhancing social media profile authenticity detection: A bio-inspired algorithm approach. In: Proceedings of the International Conference on Machine Learning for Networking, 2023
9. Mahammed, N., Saidi, I., Bencherif, K., Khaldi, M., Fahsi, M., Guellil, Z.: Grasshopper-based detection of fake social media profiles. EAI Endorsed Trans. Scalable Inf. Syst. **12**(4) (2025)
10. Mallick, C., Mishra, S., Giri, P.K., Paikaray, B.K.: A metaheuristic optimisation- based deep learning model for fake news detection in online social networks. Int. J. Electron. Secur. Digit. Forensics. **16**(5), 533–556 (2024)
11. Othman, M.M., Ku-Mahamud, K.R.: Hybrid metaheuristic algorithms for feature selection in classification: A systematic literature review (2024)
12. Perwej, Y.: A smart framework for detecting instagram fake accounts using machine learning and correlation & singular value decomposition techniques. J. Emerging Technol. Innov. Res. **10**(9), b772 (2023). https://doi.org/10.6084/m9.jetir.JETIR2309185
13. Sekkal, N., Mahammed, N., Guellil, Z.: A bio-inspired grey wolf approach to enhancing fake profile detection in online social media. Ing. Syst. Inf. **30**(4), 855–864 (2025)
14. Shannaq, F., Shehab, M., Alshorman, A., Hammad, M., Hammo, B., Al-Omari, W.: Exploring metaheuristic optimization algorithms in the context of textual cyberharassment: A systematic review. Expert. Syst. **42**(2), e13826 (2025)
15. Subhalakshmi, R.T.: Fake profile detection on social networking websites using machine learning. J. Sci. Technol. Res. **6**(1), 1–18 (2025)
16. Sun, Z., Wang, G., Li, P., Wang, H., Zhang, M., Liang, X.: An improved random forest based on the classification accuracy and correlation measurement of decision trees. Expert Syst. Appl. **237**, 121549 (2024)
17. Varuna, W.R., Shalini, K., Roy, M.E.A.: An efficient framework for fake profile identification using metaheuristic and deep learning techniques. J. Posit. Sch. Psychol. **6**(4), 3741–3750 (2022)
18. Yildirim, G.: A novel hybrid multi-thread metaheuristic approach for fake news detection in social media. Appl. Intell. **53**(9), 11182–11202 (2023)
19. Zhang, W., Wang, L., Wang, J.: Ring toss guided beam optimization: A novel metaheuristic for complex optimization problems. Expert Syst. Appl. **226**, 120270 (2023). https://doi.org/10.1016/j.eswa.2023.120270

Bot Detection Techniques on Social Media: A Review

Guendouz Amina[1(✉)], Remmide Mohamed Abdelkarim[2], Boumahdi Fatima[2], and Bouras Dalila[1]

[1] LRDSI Laboratory, Auto-Elect Department, Faculty of Technology, University of Blida 1, Blida, Algeria
guendouz.amina@yahoo.fr

[2] LRDSI Laboratory, C.S. Department, Faculty of Sciences, University of Blida 1, Blida, Algeria

Abstract. Nowadays, social media bots are a major problem that seriously jeopardizes user privacy, information integrity, and democratic processes. Accurate and effective bot detection methodologies are essential for addressing these dangers. Recently, this field has known various suggestions to fix this problem; nonetheless, some sides of this topic have been overlooked, even being crucial. This study summarizes the challenges and current progress in bot identification research, and highlights shortcomings of existing work and persistent challenges that need to be addressed in future.

Keywords: Bots · Social media · detection techniques · Machine learning

1 Introduction

Social media has revolutionized the way people communicate, share information and transact for goods and services. However, this digital revolution has also brought forth a new set of challenges, particularly in the realm of security. As people are progressively engaged with social media platforms, the risks associated with them are continuously increasing. The threats posed by social media are diverse and pervasive [16, 17], including: identity theft, data breaches, cyberbullying and the spread of misinformation.

Among the most dangerous threat actors utilized on social media, one must mention bots. These automated programs, engineered to replicate human behavior [4], have permeated online environments, presenting substantial dangers to privacy, security, and democratic processes.

Bots can be used for a variety of malicious purposes, including [9, 19]:

- Cyberbullying: Bots can be used in harassment campaigns, targeting individuals and groups with hateful or abusive messages.
- Spreading misinformation: bots can swiftly propagate fake news, influencing public perception and eroding confidence in reliable sources.
- Manipulating public opinion: By disseminating propaganda, stifling dissenting opinions, or even participating in voter fraud.

M. Hatti (Ed.): IC-AIRES 2025, LNNS 1740, pp. 183–192, 2026.
https://doi.org/10.1007/978-3-032-12788-4_17

- Phishing attacks: phishing messages can be sent by bots, attempting to fool people into divulging critical personal information
- Engaging in spam and advertising: Bots can overwhelm social media sites with spam messages or excessive advertisements, disturbing user experience.

Nevertheless, bots could be employed for good reasons, such as news bots, chatbots, etc. Authors in [9] differentiate between benign bots and malicious bots, and define six main malicious bots categories: Sybils, spam bots, cyborgs, political bots, game bots, and stegobots.

Bots utilize a variety of strategies to get their purpose. As those strategies are in continued development, it is essential to take proactive measures to mitigate their impact. Various methods have been suggested for bot detection in Social media [9], such as: Topic modeling based detection systems, Anomaly-based systems, Graph-based detection systems and Machine learning (ML) based detection systems. In this paper, we are interested to works employing ML algorithms for bot detection.

The rest of this paper is organized as follow: next section introduce the ML field and its main subfields. Section 3 presents a review of existing work. Section 4 Sect. 3 provides an overview of the current literature, and finally Sect. 5 concludes the study.

2 Machine Learning

Machine learning (ML) is a subfield of artificial intelligence (AI) that emphasizes the utilization of data and algorithms to mimic human learning, progressively enhancing performance over time [4]. ML has played a pivotal role in revolutionizing several industries, including healthcare, finance, and security. There are primarily two main types of machine learning: Supervised and Unsupervised ML. Both of them include a set of algorithms (or models) which designate a set of ordered and finite operations which must be followed in a given order to solve a problem. Methods using a combination of algorithms from the two types at the same time are called semi-supervised ML.

Supervised learning is the prevalent form of ML, its aim is to predict outcomes based on past knowledge. Supervised ML models learn from labeled training datasets and improves their accuracy over time [13]. For this ML type, there are primarily two categories of algorithms: regression and classification algorithms. Regression helps predicting continuous numerical values such as predicting house prices, while Classification predicts categorical labels (e.g., classifying emails as spam or not spam). Examples of those algorithms are [13]: Linear Regression (LR), Naïve Bayes (NB), K Nearest Neighbors (KNN), Support Vector Machine (SVM), and others.

A subset of supervised learning is called Deep Learning (DL). DL is concerned with analyzing complicated patterns in data by employing multi-layered artificial Neural Networks (NN). These NNs are inspired by the function and structure of human brain, enabling them to autonomously learn and enhance their performance via experience. Some algorithms from this category are: Multilayer Perceptron (MLP), Convolutional Neural Networks (CNNs), Long Short-Term Memory (LSTM) and Graph attention networks (GATs).

The purpose of *Unsupervised learning* is to identify patterns, structures, or correlations within the data itself. Unsupervised learning or feature learning algorithms are in charge of figuring out how the input data is structured on their own, thus the machine concentrates on examining the data to find patterns. This method helps identify relationships that can easily slip human attention. The facial recognition algorithm on Facebook is a prime illustration of this category. We can distinguish between two categories: Clustering the groups of comparable data together like the case of customer segmentation, and Dimensionality reduction which decrease the features number in order to simplify complex data. The most known algorithms in this ML subfield are [13]: K-Means, Hierarchical Clustering, Principal Component Analysis and other algorithm.

Semi-supervised learning is a type of ML that falls between supervised and unsupervised learning. This approach trains a model by using a big amount of unlabeled data along with a small amount of annotated data.

3 Existing Work

The identification of social media bots has been the subject of a great deal of study in recent years. Numerous methods, including sentiment analysis, behavioral patterns, network analysis, and ML models, have been used to identify bots. Our research focuses on ML models. After that, we provide a summary of previous research in this area. The suggested methods are categorized based on the ML models that are used.

3.1 Supervised ML for Bot Detection

According to the current literature, the widest range of existing research adopt supervised ML models.

Eiman et al. [6] proposed SEBD framework (Stream-based Evolving Bot Detection). They divided their approach into three phases: data collection, streaming using Kafka that works by sending the collected data to a Kafka topic while the Kafka consumers subscribe topics to obtain the required data and finally, the detection phase in which they had used their own pre-trained model known as "Bot-MGAT" to determine the label of each account. "Bot-MGAT" framework have been proposed by Eiman et al. [7] in a previous work. The proposed method embeds the multi-view graph attention mechanism within the transfer learning model to localize the bot classes solely through profile features.

Sen Ye et al. [20] explores the problem of homophily assumption, which assumes that addresses belonging to the same label are inherently more connected, helping the Twitter bots to 'blend in' by following a rather vast pool of real users. They proposed a graph-based Twitter bot detection framework that address the heterophilous disguise challenge. HOFA integrates modules including homophily-oriented graph augmentation and FaAt frequency adaptive attention. These modules are predicated on deep learning techniques like attention mechanisms and MLPs.

Maryam Heidari et al. [10] Presented a provable bot detection approach to generate user profile of individual human based on the personal details and use word embedding model such as the ELMO Glove (Global Vectors) and ELMO (Embedding from the language model) to map the words of twitter messages. In the preprocessing phase, user's profile is incorporated to all the Twitter accounts, based on the text of the tweet in the data. Then they propagate a supervised ML technique for accurate identification of bots. This method can distinguish bot account from human account when both the accounts belong to a user with the same education level, Gender, Personality, and age.

Ilias Dimitriadis et al. [8] concentrates on the identification and categorization of social bots in the context of the Twitter platform. It also strongly focuses on the need to enhance bot detection using explainable machine learning as well as feature engineering. Based on the identified new bot categories, the authors create and make two new datasets of accounts containing the new bot categories; the data is then balanced using the ADASYN algorithm to eliminate bias. Thereafter, various classification algorithms have been used for binary classification and for multiclass bot detection. This work aimed at categorizing the different features that led to the development of a simple taxonomy of feature categories that included social neighborhood, and temporal types of features.

DAVID MARTÍN-GUTIÉRREZ et al. [14] The paper puts forward a new approach to map user accounts into low dimensional space, detecting malicious bot accounts, and creating the embeddings for purposes of information retrieval. The system contains a multilingual model that can identify suspicious Twitter accounts by a set of features regardless of the account language. It includes relevant metadata features in conjunction with text-based features converted to vectors regardless the input text language.

Alothali et al. [5] To determine the best features subset, four strategies were investigated: wrapper subset evaluation, information gain, cross-validation attribute evaluation, and correlation attributes. They employed a public dataset accessible from Kaggle.com comprising 18 feature. The performance of the chosen features was then assessed using several machine learning algorithms, mainly: SVM, RF, NN and NB). The authors discovered that utilizing the following four attributes yields high achievement for bot detection: average_tweets_per_day, verified, statuses_count, and favorites_count,. Jefferson Viana Fonseca Abreu et al. [10] evaluated four ML models using a public dataset and a few expressive features based on simple user profile counters for the classification of bots on Twitter. They emphasize the utilization of elements that are readily available and typical profile attributes, since they can be obtained with a single Twitter API request. Based on early experience building Twitter bots, five features were empirically chosen. The article emphasizes that even with a smaller collection of features, it is possible to recognize bots with a certain level of sophistication.

Sandeep Singh Sengar et al. [20] suggested a multilayer ML approach. They performed feature extraction using a labelled dataset through basic statistical tests and correlation analysis. To address the limitation of correlation, they employ the chi-squared test on non-textual attributes to identify the five most significant features. Text attributes are then passed through a feature engineering process whereby certain values are extracted. They undergo an initial classification process that generates the quantity of predictions for each text attribute. The authors conduct a thorough review of the efficacy of several classification algorithms, alongside an analysis of two prevalent techniques for augmenting text features: n-gram model and Bag of Words.

3.2 Semi-supervised ML for Bot Detection

Others in academia propose employing a hybrid approach that integrates both supervised and unsupervised ML methods for bot detection.

Innocent Mbona and Jan H.P. Eloff [15] tackled the issue of classifying bots as either malicious or benign. They put into practice four SSML algorithms: the label propagation (LP) method, a graph-based SSML model that extends the labeling of all nodes on the graph until convergence is achieved; the semisupervised Gaussian mixture model; the label spreading (LS); and, finally, the semi-supervised SVM (S3VM). They identified crucial features that can be utilized to discriminate between benign and dangerous bots and showed that SVM got the best results in this classification.

Abbas Abou Daya et al. [1, 2] suggested using graph-based features to enhance ML model inference and training. These features are generated from flow-level data. The recommended BotChase is an anomaly-based bot detection system that is protocol neutral, robust to zero-day assaults, and suited for big datasets. Based on BotChase's graph-based features, the authors suggest feature normalization (F-Norm) and assess other machine learning methods.

3.3 Unsupervised ML for Bot Detection

Only a limited number of researchers opt for unsupervised ML algorithms for bot identification.

[12] Suggested a technique where bots detection on social networks can be done by using unsupervised ML algorithms. They first select a set of features for bot accounts categorisation which participates in the identification of bots and common users. Then, two clustering approaches namely, DBSCAN and Kmeans are assessed on six datasets using these attributes. Dbscan showed higher performance with greater accuracy levels as it was noted from the results of the research.

4 Discussion

Table 1 summaries some key points about the proposed approach in each of the covered literature work, namely: the used algorithms belong to which ML type (supervised/ unsupervised/ semi-supervised), does the work support real time bot detection? What classes of bots are covered by the work? and which social media platforms are supported by this proposal?

Table 2 provides more details regarding the ML algorithms employed, name of the datasets used, the features selected for training and classification, and the results obtained.

Table 1. General orientation of the work.

Article	ML type(sup/ Unsup / semi-sup)	Support of real time detection? (Yes / No)	Covered classes	Supported platforms
1,5	Sup	Yes	Bots - genuine	Twitter
2	Sup	No	Bots - genuine	Twitter
4	Sup	No	Bots - genuine	Twitter
6	Sup	No	multiclass	Twitter
8	Sup	No	Bots - genuine	Twitter
10	Sup	No	Bots - genuine	Twitter
11	Sup	No	Bots - genuine	Twitter
21	Sup	No	Bots - genuine	Twitter
3	Semi-sup	No	benign - malicious bots	Twitter
9,14	Semi-sup	Yes	Bots - genuine	Other
22	Unsup	Not real time	Bots - genuine	Twitter

Analysis of these tables showed that among the ML types, the most adopted for bot detection on social media is supervised ML. Supervised ML models achieved the best results compared to unsupervised and semi-supervised methods. Previously, the method yielding optimal results was Random Forest [4]. Recently (in the last five years), the application of deep learning algorithms has yielded better results [6, 7, 10, 20].

Based on our observations, the most popular method used in the literature is generic bots detection. Sometimes, only one type of bots is considered such as in [12] (spambots). One reason is that the provided data by the datasets includes only profiles labeled as bots or genuine users. Differentiating between bots and human users is a hard task, but distinguishing benign bots from malicious bots is more complicated. In the literature, we have found only one work that address this challenge [15].

Major existing work use Twitter platform for training and testing their proposed approaches. The reason is that Twitter platform provides free access to user data, even in real time. Nevertheless, real time bot detection have not been supported most of the time.

The results were measured in term of accuracy, precision, recall and f-score. These metrics play a crucial role in various stages of the ML model development lifecycle, particularly in the evaluation phase [11].

We observe that the results achieved exceeds 90% in major cases, yet various challenges have not been addressed, mainly: multiplatform support, multiclassification according to bots types, real-time detection, decreasing time and enhancing performances by for example limiting the number of features used for training and test, and adopting new deep-learning algorithms for feature extraction and bots classification.

Table 2. Techniques and results of the work.

Art	Algorithms	Dataset	Features	Results
[6, 7]	GATs Transfer Learning (TL)	Twi-Bot20	18 profile features	Accuracy 97.80% F1 score 0.9842 MCC 0.9481
[20]	MLP GNNs	Cresci-15, TwiBot-20, MGTAB-22	Follower and friend relationships	Accuracy 97.52 F1 98.06 bAcc 96.98 With Cresci-15-H 82.29, 85.49, 80.82 with TwiBot-20 88.68, 79.21, 85.91 with MGTAB-22
[10]	NN models – LSTM - LR	Cresci 2017, COVID-19	education, Gender, Personality, age	94% accuracy
[8]	Logistic Regression (LR), Linear Discriminant Analysis (LDA), Decision Tree Classifier (CART), Multi Layer Perceptron (MLP), AdaBoost (ADA), and Random Forest (RF)	2 new twitter datasets	Content Features User Features Temporal Features Social Neighborhood Features Sentiment Features Hashtag correlation features	recall score 0.92 F1-score of 0.9
[14]	Deep Neural Networks (DNNs). Deep Fully connected NN	dataset composed of 37438 Twitter accounts	*Popularity features:* friends and followers, *Activity features:* creation date, average tweets per day, tweets & favorites counts, account age. *Profile information features*: screen name, description, language, location, verified indicator, default profile indicator	F1 score 0.77
[3]	RF, SVM, NB and ocSVM	Cresci 2017	statuses count, followers count, friends count, favorites count, listed count	Accuracy 0.8549 AUCs 0.9

(*continued*)

Table 2. (*continued*)

Art	Algorithms	Dataset	Features	Results
[20]	K Nearest Neighbours Decision Tree Random Forest Ada Boost Gradient Boosting Gaussian NB Multinomial Naïve Bayes, Multilayer Perceptron	MIB Datasets (Cresci)	*Non text features:* id, followers_count, friends_count, listed_count and statuses_count. *Text features:* screen_name, name, description and status	99.54 accuracy
[5]	RF, NB, SVM, NN	public dataset accessible from Kaggle.com	favorites_count, verified, statuses_count, and average_tweets_per_d	84.1% precision 81.2% recall
[15]	Semi-supervised Gaussian mixture model S3VM label propagation (LP) label spreading (LS)	Labeled dataset + unlebeled Twitter dataset	favorites_count, lists_count, statuses_count, status. retweet_count, friends_count, and followers_count	Recall 89% F1 score 76%
[1, 2]	k-Means, Density-Based Spatial Clustering (DBScan) and SOM. Logistic regression (LR), support vector machine (SVM), feed-forward neural network (FNN) and decision tree (DT). Hoeffding Adaptive Tree (HAT)	CTU-13	Flow-based features	Recall 92.85 precision 81.25
[12]	DBSCAN and K-Means	Three different twitter spambot datasets	follower-count, friends-count, favorite-count, listed-count, retweet-count, reply-count, hashtag-count, mentions-count	97.7% accuracy 91% precision 98% recall 94% f-measure

5 Conclusion

In this paper, we present a review on main recent existing work about bot detection in social media using ML techniques. We have classified the reviewed approaches according to the type of ML adopted. The characteristics of each work are summarized in two tables. The analysis of this review revealed that despite the important results achieved by existing works, some challenges are not supported in this field for instance: Addressing other social media platforms, namely: Facebook, Instagram, LinkedIn, and Weibo. It would also be interesting to create a model suitable for different platforms.

Real-time detection is a difficult task. By taking care of this task, instant warnings help to block bots immediately making social media safer for users. In addition, bots strategies are in continuous development what make it necessary to keep up this development by proposing new methods and techniques.

References

1. Abou, D.A., Salahuddin Mohammad, A., Noura, L.: A graph-based machine learning approach for bot detection. In: IEEE Symposium on Integrated Network and Service Management (IM), pp. 144–152. IEEE (2019)
2. Abou Daya, A., Salahuddin, M.A., Limam, N., Boutaba, R.: Botchase: graphbased bot detection using machine learning. IEEE Trans. Netw. Serv. Manage. **17**(1), 15–29 (2020)
3. Abreu, J.V.F., Ralha, C.G., Gondim, J.J.C.: Twitter bot detection with reduced feature set. In: 2020 IEEE International Conference on Intelligence and Security Informatics (ISI), pp. 1–6. IEEE (2020)
4. Aljabri, M., Zagrouba, R., Shaahid, A., Alnasser, F., Saleh, A., Alomari, D.M.: Machine learning-based social media bot detection: a comprehensive literature review. Soc. Netw. Anal. Min. **13**(1), 20 (2023)
5. Alothali, E., Hayawi, K., Alashwal, H.: Hybrid feature selection approach to identify optimal features of profile metadata to detect social bots in twitter. Soc. Netw. Anal. Min. **11**(1), 84 (2021)
6. Alothali, E., Hayawi, K., Alashwal, H.: SEBD: a stream evolving bot detection framework with application of pac learning approach to maintain accuracy and confidence levels. Appl. Sci. **13**(7), 4443 (2023)
7. Alothali, E., Salih, M., Hayawi, K., Alashwal, H.: Bot-MGAT: a transfer learning model based on a multi-view graph attention network to detect social bots. Appl. Sci. **12**(16), 8117 (2022)
8. Dimitriadis, I., Georgiou, K., Vakali, A.: Social botomics: a systematic ensemble ml approach for explainable and multi-class bot detection. Appl. Sci. **11**(21), 9857 (2021)
9. Ellaky, Z., Benabbou, F., Ouahabi, S.: Systematic literature review of social media bots detection systems. J. King Saud Univ.-Comput. Inf. Sci. **35**(5), 101551 (2023)
10. Heidari, M., Jones, Jr. J.H., Uzuner, O.: Online user profiling to detect social bots on twitter. arXiv preprint arXiv:2203.05966 (2022)
11. Hossin, M., Sulaiman, M.N.: A review on evaluation metrics for data classification evaluations. Int. J. Data Mining Knowl. Manag. Process **5**(2), 1 (2015)
12. Khalil, H., Khan, M.U., Ali, M.: Feature selection for unsupervised bot detection. In: 2020 3rd International Conference on Computing, Mathematics and Engineering Technologies (iCoMET), pp. 1–7. IEEE (2020)
13. Mahesh, B.: Machine learning algorithms-a review. Int. J. Sci. Res. (IJSR) **9**(1), 381–386 (2020)
14. Martín-Gutiérrez, D., Hernández-Peñaloza, G., Hernández, A.B., Lozano-Diez, A., Álvarez, F.: A deep learning approach for robust detection of bots in twitter using transformers. IEEE Access **9**, 54591–54601 (2021)
15. Mbona, I., Eloff, J.H.: Classifying social media bots as malicious or benign using semi-supervised machine learning. J. Cybersecur. **9**(1), tyac015 (2023)
16. Remmide, M.A., Boumahdi, F., Boustia, N.: Toward a hybrid approach combining deep learning and case-based reasoning for phishing email detection. Int. J. Artif. Intell. Tools **33**(05), 2450015 (2024)
17. Remmide, M.A., Boumahdi, F., Ilhem, B., Boustia, N.: A privacy-preserving approach for detecting smishing attacks using federated deep learning. Int. J. Inf. Technol. **17**(1), 547–553 (2025)

18. Sengar, S.S., Kumar, S., Raina, P., Mahaliyan, M.: Bot detection in social networks based on multilayered deep learning approach. Sens. Transducers **244**(5), 37–43 (2020)
19. Yang, K.C., Varol, O., Davis, C.A., Ferrara, E., Flammini, A., Menczer, F.: Arming the public with artificial intelligence to counter social bots. Hum. Behav. Emerging Technol. **1**(1), 48–61 (2019)
20. Ye, S., et al.: Hofa: Twitter bot detection with homophily-oriented augmentation and frequency adaptive attention. arXiv preprint arXiv:2306.12870 (2023)

When Machines Speak Human: Detecting Computer Generated Reviews Using Transformer Models

Ayoub Meghebbar(✉), Mohammed Ismail Smahi, and Abdelkrim Benamar

Department of Computer Science, LRIT Laboratory, University of Tlemcen, Tlemcen, Algeria
{ayoub.meghebbar,mohamedismail.smahi,
abdelkrim.benamar}@univ-tlemcen.dz

Abstract. Online reviews significantly influence consumer decision-making across nearly all fields, such as e-commerce, hospitality, services, etc. However, the rise of large language models has enabled the generation of convincing and human-like fake reviews, making the detection of computer-generated reviews more challenging. To deal with this observed phenomenon, the present study investigates the effectiveness of five transformer-based language models (T5, BERT, RoBERTa, MPNet, BART). We utilised those prominent language models to differentiate between human-written and computer-generated reviews for a binary classification task. To enable a fair comparative evaluation, we proceeded for each model in two major steps. We first start a learning rate search to identify an initial optimal learning rate, followed by a fine-tuning phase. Our results showed that learning rate tuning is important to achieve better performance and avoid suboptimal convergence due to the narrower or shallower optimal learning rate valleys, especially in larger models. Moreover, parameter size alone is not a reliable indicator of effectiveness, since the smallest model in terms of parameters achieved the highest accuracy. We also observe no significant performance gap between encoder-only and encoder-decoder architectures for this single-label classification task, as evaluated on a balanced dataset containing 20,000 genuine and 20,000 machine-generated product reviews, originally designed for benchmarking in fake review detection.

Keywords: Computer-Generated Reviews · Fake Review Detection · Learning Rate Optimisation · Text Classification · Transformer Based Models

1 Introduction

In the digital era, online reviews have become a major focus in shaping consumer decisions, from selecting a restaurant to purchasing a high-quality product. Consumers often rely on reviews made by others on platforms such as Amazon, TripAdvisor and Yelp to make a decision. They often offer valuable information about the product quality and user experience [1], thus influencing consumers about their choice. The issue becomes more complex when reviews are deceptive in nature. According to Ott et al. [2], deceptive reviews are fictitious opinions that are written in a way to sound authentic to deceive

M. Hatti (Ed.): IC-AIRES 2025, LNNS 1740, pp. 193–202, 2026.
https://doi.org/10.1007/978-3-032-12788-4_18

the reader, and they are neither easily detected nor easily identifiable by humans. With the advances in artificial intelligence when it comes to natural language generation, especially in transformer-based language models such as GPT, T5 and BERT, review generation is increasingly easy to create convincing, human-like content [3]. This introduces an additional layer of complexity to the task of deceptive review detection, and it is the challenge of identifying computer-generated reviews from those written by humans.

In our study, we address the challenge of detecting computer-generated fake reviews. To this end, we design and implement a pipeline covering the steps of preprocessing, learning rate search, fine-tuning and evaluation.

This pipeline is then used to compare five transformer-based language models to ensure a fair comparison. We also demonstrate that learning rate tuning is important, as it can improve model performance and help avoid suboptimal convergence by escaping those local minima.

2 Related Work

Salminen et al. [4] conducted a study where they generated fake product reviews by fine-tuning a GPT-2 model with different categories of products, like movies, books, toys, electronics and more from the Amazon dataset. Then use the same model to generate another dataset combining computer-generated reviews and human-written reviews, to fine-tune a RoBERTa model named FakeRoberta for the classification task. Their findings indicated that while human evaluators struggled to distinguish between real reviews and the generated ones, transformer-based language models achieved very good accuracy. In another study, Salminen et al. [5]. Investigate how deceptive fake reviews differ from truthful and genuine ones. They collected over 3000 reviews, both real and deceptive, from more than 300 participants. These reviews are then analysed for psycholinguistic footprints using the Linguistic Inquiry and Word Count (LIWC). The study compares traditional machine learning models (SVM, XGBoost, MLP) with transformer-based models (BERT, RoBERTa, XLNet) for the task of distinguishing between deceptive and truthful reviews. The findings of this study show that transformer-based models outperform traditional models and prove the point that fake reviews carry a psycholinguistic footprint that differs from genuine ones. In the same context of fake reviews detection, Mohawesh et al. [6] proposed a hybrid approach that combines both the RoBERTa model with LSTM layers in the goal of capturing both contextual and sequential linguistic features. Using also a labelled dataset from Amazon, the model was put into comparison with traditional models such as SVM, Naive Bayes and LSTM alone. The results of this study showed that the combination of RoBERTa and LSTM achieved the highest accuracy and F1 score.

3 Methodology

In this section, we describe how we proceed to detect computer-generated reviews from human reviews by comparing five different transformer-based language models. The goal is to assess the model's effectiveness in identifying fake reviews. The complete architecture is illustrated in the Fig. 1, which outlines all the steps from preprocessing to searching the initial optimal learning rate, fine-tuning the model and then evaluation. In our study, we emphasise the point of finding an initial optimal learning rate since it can help us avoid getting stuck in another less optimal local minimum. We also tried to ensure consistency across experiments to allow a reliable comparison across all the models.

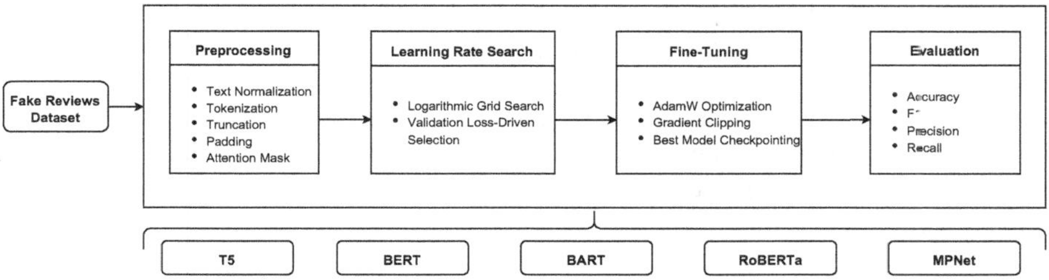

Fig. 1. Complete architecture of the review classification pipeline used to distinguish computer-generated reviews from human-written ones.

3.1 Model Architecture Comparison

In our study, we selected five models: BERT-base-uncased, RoBERTa-base, MPNet-base, BART-base and T5-small. They are all transformer-based language models. In this part of our paper, we highlight some key architectural characteristics of each model, including the number of layers (encoder/decoder) and the model's hidden layers, number of attention heads, total number of parameters and their main pertaining objectives.

T5-small The smallest variant of T5 (text-to-text transformer) [7] is an encoder-decoder transformer that unifies the format of every NLP task as a text-to-text problem, which means both input and output are plain text. The model uses a total of 60 M parameters represented by 12 transformer layers, where 6 are for the encoder and the other 6 are for the decoder, each with a hidden size of 512 and 8 attention heads with a vocabulary size of 32128 tokens (i.e., words). The model's main objective is text denoising.

BERT-base, a variant of BERT [8], is an encoder-only transformer that uses a bidirectional self-attention mechanism to deeply understand the context of words in both directions. Unlike the encoder-decoder architecture previously seen with T5, BERT does not generate output sequences but contextualises the input tokens. Its configuration contains 12 transformer layers with a hidden size of 768 and 12 self-attention heads, and a vocabulary size of 30522 tokens with approximately a total of 110 M parameters. BERT's main objectives are masked language modelling and next sentence prediction.

RoBERTa-base [9] is also an encoder-only transformer model that shares the same architecture as BERT-Base with a near identical configuration of 12 transformer layers with a hidden size of 768 and 12 self-attention heads, but with a vocabulary size of 50265 tokens, making it a total of approximately 125 M parameters. However, RoBERTabase training was optimised by using a much larger and more diverse corpus (i.e., larger vocabulary size) with larger batches and longer sequences. It focuses only on masked language modelling and not next sentence prediction.

Similarly to BERT and RoBERTa, MPNet-base [10] is an encoder-only Transformer with a similar architecture to BERT-base with 12 transformer layers, a hidden size of 768 and 12 attention heads, 30527 tokens for the vocabulary size, with about 110 M parameters. However, unlike BERT and RoBERTa, which rely solely on masked language modelling, MPNet introduces a novel pretraining objective that combines masked and permuted language modelling.

BART-Base [11], like T5, uses a transformer encoder-decoder architecture with 12 transformer layers, 6 for the encoder and 6 for the decoder, with 12 attention heads but a hidden size of 768, a vocabulary size of 50265 tokens, totalling around 139 million parameters. Its pretraining objective is that of a denoising autoencoder, where the input is corrupted using different techniques such as shuffling sentence order, masking words, deleting tokens, and filling in blanks. The model is trained to recover the original input. BART-base uses a bidirectional encoder, while the decoder is autoregressive, which makes it suitable for sequence contextualisation and text generation. A comparative summary of these five models is provided in Table 1.

Table 1. Comparative summary of the five transformer-based language models

Model	Architecture	Layers (Enc + Dec)	Hidden Size	Heads	Vocabulary Size	Number of Parameters
T5-small	Encoder-Decoder	6 + 6	512	8	32128	60 M
BERT-base	Encoder-only Bidirectional	12 + 0	768	12	30522	110 M
RoBERTa-base	Encoder-only Bidirectional	12 + 0	768	12	50265	125 M
MPNet	Encoder-only Bidirectional	12 + 0	768	12	30527	110 M
BART-base	Encoder-Decoder Bidirectional encoder Autoregressive decoder	6 + 6	768	16	50265	139 M

3.2 Pipeline Overview

In this part of our paper, we highlight in detail the pipeline designed to distinguish between computer-generated reviews and the human ones. This pipeline ensures consistency across all five transformer-based models analysed previously for a fair comparison. The process contained four main stages: preprocessing, learning rate search, model fine-tuning and evaluation. The proposed Algorithm 1 aims to give a more detailed overview of the process.

Preprocessing is the first step for preparing input text for the transformer models. It includes a text normalisation, which consists of punctuation removal and whitespace cleanup. Each review is then tokenised using the specific model's tokeniser. We have opted for using the tokeniser associated with its corresponding model and not a standard one for all to ensure alignment with model training, vocabulary match and special token handling like end-of-sentence tokens. For padding and truncation, we used a fixed input length that has been chosen based on sequence length analysis to remove outlier sequences whose lengths are far greater than the majority, more on that later. Finally, an attention mask is applied based on the strategy of each model's tokeniser. This step allows for a standardised input format for all the models.

Learning rate search is the second step, in which the goal is to find an optimal initial learning rate before proceeding to fine-tuning. We selected 50 learning rates spanning from 1×10^{-5} to 1×10^{-2} in an exponential progression since it was shown by LeRaC [12] that exponential variation, as opposed to linear, better suits deep models. The decision of the bounds is supported by previous work done by Agbesi et al. [13], showing that the optimal learning rates for a language model like the one we used often lie within that interval. Each learning rate is trained on a fixed number of training batches starting from an identical initialised model checkpoint to avoid carryover effects, followed by a validation phase for each one, where the validation losses were aggregated. Finally, the lowest average validation loss will be selected as a starting point for the next step.

Algorithm 1 Review classification pipeline using transformer-based models.

```
Input: Reviews, MaxLength, Batch,Models ← [T5-small,BERT-base,RoBERTa-base,MPNet,BART-base]
Output: Model
for each Model ∈ Models do
    Preprocessing
    for each Review ∈ Reviews do
        Review ← NormalizeText(Review)
        Review ← TokenizeText(Review,Model.Tokenizer)
        Review ← TruncatePad(Review,Model.Tokenizer,MaxLength)
        ReviewDataSet ← AttentionMask(Review,Model.Tokenizer)
    end for
    TrainSet, ValidationSet ← Split(ReviewDataSet,Batch)
    Learning Rate Search
    for each Lr ∈ LogSpace(50,10^-5,10^-2) do
        CopyModel ← Copy(Model.initialState())
        for TrainBatch ∈ TrainSet do
            CopyModel.Train(TrainBatch)
        end for
        for ValBatch ∈ ValidationSet do
            ValLoss ← CopyModel.Eval(ValBatch)
            ResultsLR ←{Lr,ValLoss}
        end for
    end for
    BestLR ← GetMinLossLR(ResultsLR)
    Fine-Tuning and Evaluation
    Optimizer ← AdamW(Model.params,BestLR)
    for Epoch ∈ Epochs do
        for TrainBatch ∈ TrainSet do
            Model.Train(TrainBatch)
            Optimizer.step()
        end for
        for ValBatch ∈ ValidationSet do
            ValMetrics ← Model.Eval(ValBatch)
            Results.Append(GetBestModelEvalOnAccuracy(Model,ValMetrics))
        end for
    end for
end for
return GetBestModel(Results)
```

After finding an initial optimal learning rate, we performed fine-tuning for each transformer model. We employed the AdamW optimiser as recommended by Loshchilov and Hutter [14]. AdamW has been shown to provide more effective weight decay regularisation than the standard Adam optimiser in transformer-based architectures. We also used gradient clipping to prevent exploding gradients. In addition, we incorporated a linear learning rate scheduler with warm-up steps since it allows the model to begin with small gradient steps before increasing and later decreasing the learning rate. Since we are starting from an initial optimal learning rate, this will help us improve convergence. The performance model was monitored using a validation set after each epoch using accuracy, precision, recall and F1-score as evaluation metrics; however, the model with the highest validation accuracy was retained. Our approach ensured a consistent process for each model to ensure consistency and comparability throughout the study.

4 Results and Discussion

4.1 Dataset

In our paper, we used the dataset[1] proposed by Salminen et al. [4]. The dataset comprises 20 k real human reviews from the Amazon e-commerce platform, covering various product categories, including electronics, books, movies and TV shows, clothing, and more. Additionally, it contains 20 k computer-generated fake reviews produced by a fine-tuned GPT-2 model, making it a balanced dataset.

4.2 Hyper Parameters

We used a maximum sentence length for truncation and padding of 256 since it covers approximately the 97th percentile of review lengths in our dataset. The batch size is fixed at 64, a good trade-off between GPU memory constraints and batch gradient stability. For the optimiser, we used the AdamW optimiser coupled with a linear learning rate scheduler with a 10% warmup step. All the models were fine-tuned using those same parameters, except for the learning rate, where each is different from the other.

4.3 Learning Rate Findings

As stated previously, for each model we tested multiple learning rates in the goal of finding an initial optimal learning rate. In the Fig. 2, we plotted for each model the curve of validation loss over learning rates. So, in each curve the optimal learning rate is located at the point where the loss reaches its minimum before beginning to increase.

T5-small shows a relatively high optimal learning rate of 7.91×10^{-4}, which reflects the model's compact 60 M parameter size, allowing large updates without divergence. For BERT-base, validation loss decreases steadily up to 1.1×10^{-4} before rising again, making it an initial optimal rate. This aligns with the learning rate of 1×10^{-4} used in the initial pretraining of the model. For the two models RoBERTa-base and MPNet-base, the initial optimal learning rates are 3.56×10^{-5} and 4.09×10^{-5}, respectively. Those high learning rates reflect both model sensitivity due to the relatively high number of parameters. In a similar vein, BART-base also shows a high initial optimal learning rate of 6.25×10^{-5} given its high number of parameters. As we can see, often models with a large number of parameters require much larger learning rates, and also skipping LR tuning might hurt some models whose validation loss curves exhibit either a steep or narrow minima, making a slight deviation lead to suboptimal convergence.

[1] https://osf.io/tyue9/.

4.4 Fine-Tuning Results and Analysis

Table 2 reports the accuracy, F1 score, precision and recall of each model with its respective optimal learning rate when fine-tuned for the task of detecting human-generated reviews from the computer-generated ones. We also used FakeRoberta by Salminen et al. [4], a RoBERTa-base model fine-tuned for the same classification task in the comparison. In terms of accuracy and F1 score, both T5-small and BERT-base took the edge, followed by BART-base, RoBERTa-base and FakeRoberta, with MPNet-base the lowest. The difference is marginal but suggests that model choice can slightly impact the results. When comparing the model size with the performance, we notice that T5-small, with only 60 M parameters, attained the top accuracy. Conversely, models like BART-base and MPNet with more than double the number of parameters (139 M and 110 M, respectively) did not yield higher accuracy. This clearly shows that more parameters do not guarantee better results. In terms of architecture, both T5-small (Encoder-Decoder) and BERT-base (Encoder-only) converge to nearly identical performance; those results mirror findings in the literature [15, 16] that suggests for a single-label classification task with large training data, both converge rapidly and reach a performance plateau. Finally, when comparing FakeRoberta's performance to RoBERTa-base, a learning rate search can help improve the results, moving from 96.91% to 97.69% accuracy.

Table 2 Classification performance of transformer models fine-tuned with their respective optimal learning rates.

Model	Accuracy	F1-Score	Precision	Recall	Learning rate
T5-small	98.12%	98.07%	98.15%	97.32%	7.91×10^{-4}
BERT-base	98.11%	98.08%	98.11%	98.05%	1.1×10^{-4}
RoBERTa-base	97.69%	97.61%	97.70%	96.79%	3.56×10^{-5}
MPNet	96.59%	96.44%	96.76%	93.94%	4.09×10^{-5}
BART-base	97.98%	97.91%	98.03%	96.62%	6.25×10^{-5}
FakeRoberta	96.91%	97%	97%	97%	1×10^{-5}

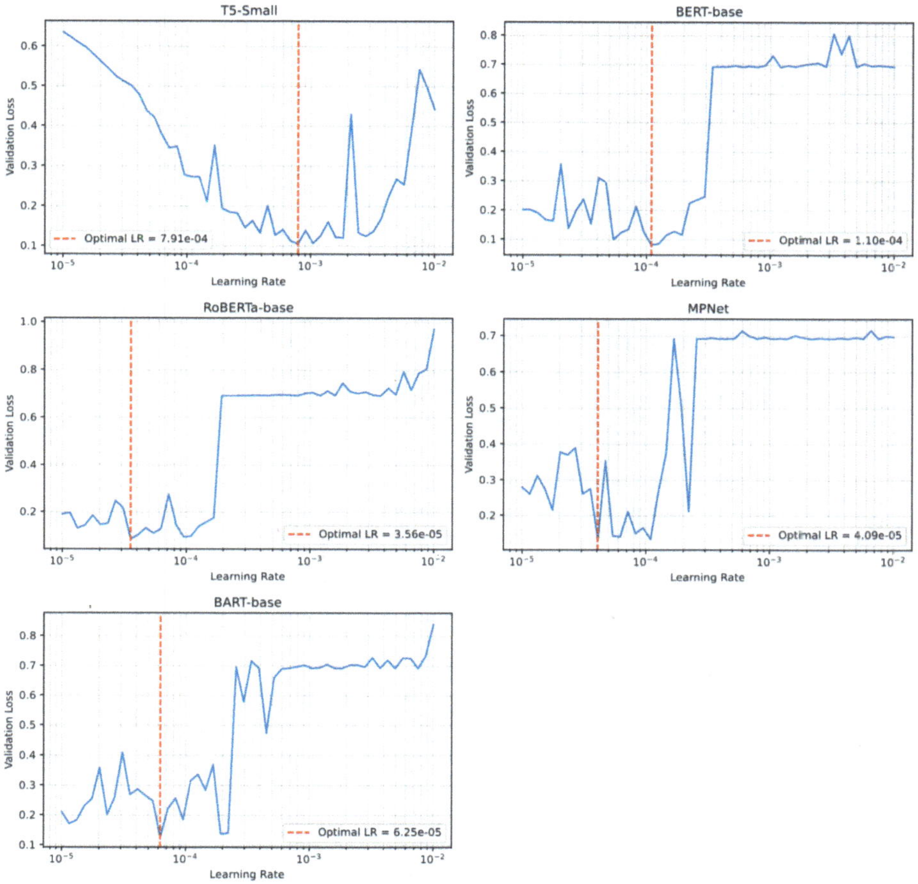

Fig. 2. Validation loss over different learning rates for each model.

5 Conclusion

In this paper, we compared the ability of five different transformer-based language models to distinguish between human-written and computer-generated reviews. To achieve this comparison, we emphasise the importance of a consistent comparison across all 4 steps of preprocessing, searching for an initial optimal learning rate, finetuning and evaluation in our pipeline. In our findings we highlighted the importance of a learning rate search since some models require more careful tuning to avoid suboptimal convergence due to the narrower or shallower optimal learning rate valleys (e.g., RoBERTa and MPNet). We also observed that more parameters do not always guarantee better results; for example, T5-small, the smallest model in this comparison with only 60 M parameters, achieved the highest accuracy. Finally, we did not see a performance difference between encoder-decoder and encoder-only architecture due to the nature of the task (single-label classification) and large training data. That's why, for future work, we can build upon this by extending evaluation to cross-domain or multilingual datasets. Additionally, future studies may include other metrics for comparison, like computational efficiency and energy consumption during fine-tuning.

References

1. Filieri, R., McLeay, F.: E-WOM and accommodation: An analysis of the factors that influence travelers' adoption of information from online reviews. J. Travel Res. **53**(1), 44–57 (2014)
2. Ott, M., Choi, Y., Cardie, C., Hancock, J.T.. Finding deceptive opinion spam by any stretch of the imagination. arXiv preprint https://arxiv.org/abs/1107.4557 (2011)
3. Radford, A., Narasimhan, K., Salimans, T., Sutskever, I.: Improving language understanding by generative pre-training. OpenAI, San Francisco, USA (2018)
4. Salminen, J., Kandpal, C., Kamel, A.M., Jung, S., Jansen, B.J.: Creating and detecting fake reviews of online products. J. Retail. Consum. Serv. **64**, 102771 (2022)
5. Salminen, J., Mustak, M., Jung, S.-G., Makkonen, H., Jansen, B.J.: Decoding deception in the online marketplace: Enhancing fake review detection with psycholinguistics and transformer models. J. Mark. Anal. **3**, 1–18 (2025)
6. Mohawesh, R., Salameh, H.B., Jararweh, Y., Alkhalaileh, M., Maqsood, S.: Fake review detection using transformer-based enhanced LSTM and RoBERTa. Int. J. Cogn. Comput. Eng. **5**, 250–258 (2024)
7. Raffel, C., et al.: Exploring the limits of transfer learning with a unified text-to-text transformer. J. Mach. Learn. Res. **21**(140), 1–67 (2020)
8. Devlin, J., Chang, M.W., Lee, K., Toutanova, K.: BERT: Pre-training of deep bidirectional transformers for language understanding. Proceedings of the Conference of the North American Chapter of the Association for Computational Linguistics (NAACL-HLT), Association for Computational Linguistics, Minneapolis, USA, pp. 4171–4186 (2019)
9. Liu, Y., et al.: RoBERTa: A robustly optimized BERT pretraining approach. arXiv preprint https://arxiv.org/abs/1907.11692. (2019)
10. Song, K., Tan, X., Qin, T., Lu, J., Liu, T.: MPNet: Masked and permuted pre-training for language understanding. In: Proceedings of the 34th Conference on Neural Information Processing Systems (NeurIPS 2020), Curran Associates Inc., Vancouver, Canada (virtual), pp. 16857–16867. (2020)
11. Lewis, M., et al. BART: Denoising sequence-to-sequence pre-training for natural language generation, translation, and comprehension. In: Proceedings of the 58th Annual Meeting of the Association for Computational Linguistics (ACL 2020), Association for Computational Linguistics, Seattle, USA (virtual), pp. 7871–7880. (2020)
12. Croitoru, F.-A., Ristea, N.-C., Ionescu, R.T., Sebe, N.: Learning rate curriculum. Int. J. Comput. Vis. **133**(1), 291–314 (2025)
13. Agbesi, V.K., et al.: Pre-trained transformer-based models for text classification using low-resourced Ewe language. Systems. **12**(1), 1 (2023)
14. Loshchilov, I., Hutter, F. Decoupled weight decay regularization. arXiv preprint https://arxiv.org/abs/1711.05101. (2017)
15. Wang, A., Singh, A., Michael, J., Hill, F., Levy, O., Bowman, S.R.: GLUE: A multi-task benchmark and analysis platform for natural language understanding. arXiv preprint https://arxiv.org/abs/1804.07461. (2018)
16. Yuan, L., et al.: Revisiting out-of-distribution robustness in NLP: Benchmarks, analysis, and LLMs evaluations. Adv. Neural Inf. Proces. Syst. **36**, 58478–58507 (2023)

Artificial Intelligence-Based Approaches for Misinformation Detection: A Case Study in a Telecommunications Corporate Innovation Strategy

B. Klouche(✉), I. Saidi, and N. Keskes

Ecole Supérieure en Informatique LabRi Laboratory, Sidi Bel Abbès, Algeria
b.klouche@esi-sba.dz

Abstract. Fake news has become a significant consideration to businesses especially telecommunications companies where reputation is considered one of its assets. With the advent of the social media, the dissemination of fake news has become fast and it poses a significant problem to corporate reputation. The current article suggests a machine learning solution to automatically identifying fake news that impacts the activities of the telecommunications company Ooredoo. With the help of diverse supervised learning methods, we demonstrate that the accuracy of the Support Vector Machine (SVM) classifier was more than 93% among other classifiers applied in this Naive Bayes (NB) and Random Forest (RF) study. The mass of data retrieved on various social media (Facebook, Twitter) was cleaned, vectorized, and distinguished according to different models. The reached results demonstrate the effectiveness of our methodology and provide the ambitious perspectives of implementation on the mass-scale.

Keywords: Artificial Intelligence · Fake News Detection · Business Strategy · Machine Learning · Ooredoo · Reputation Management

1 Introduction

Fake news has been considered one of the greatest threats to politics, health, journalism, and freedom of expression in recent years. It has weakened public trust in governments, and its potential impact on the controversial Brexit referendum and the equally controversial 2016 US presidential election may have affected them [1]. According to [2], fake news and misinformation undermine the effectiveness of message delivery. If the recipient of this information does not respond well, it loses its effectiveness. Disinformation has become a major priority on the internet. Fake news spreads quickly, undermines public opinion, disrupts markets, and damages brand reputation. Ooredoo, a major telecommunications player in Algeria, is regularly targeted by misleading content on social media. However, detecting fake news is a crucial step, especially when dealing with multilingual data sets and dialect variations, such as the Algerian dialect. These steps are intended to normalize the input data, reduce noise, and improve the

M. Hatti (Ed.): IC-AIRES 2025, LNNS 1740, pp. 203–209, 2026.
https://doi.org/10.1007/978-3-032-12788-4_19

consistency of feature extraction, which in most cases improves model performance [3]. With the prevalence of code-switching (Arabic–French) (Arabic–English) and dialectal expressions in social media posts, special attention was paid to linguistic diversity. The preprocessing pipeline was designed to efficiently process such multilingual material and to normalize dialectal variations to their semantically equivalent normative forms whenever possible. In their current state, social media platforms are very powerful and valuable because they allow users to discuss and exchange ideas and debate topics such as democracy, education, and health [4]. This study is part of a broader research framework on the detection of fake news and highlights the importance of a multidisciplinary approach, combining computer science, linguistics, and social sciences, to address this complex and constantly evolving problem. The structure of the document is organized as follows: after the introduction, Sect. 2 reviews the literature on certain works related to our research. In Sect. 3, we present the proposed model for detecting fake news. Section 3 then describes the results and discussions, and finally, Sect. 4 presents a conclusion of our work with perspectives for future research.

2 State of Art

The detection of false information on social media is one of the most trending research topics in recent years. To this end, several strategies using machine learning and natural language processing techniques have been developed. Most studies rely on innovative approaches aimed at improving the efficiency and reliability of detection systems. The authors [5] described an improved approach to detecting fake news in the health sector using a machine-learning model that incorporates sentiment analysis. The researchers compiled a dataset comprising 1,017 articles, including 582 reliable and 435 unreliable, extracted from eight Thai websites. By applying sentiment analysis techniques and using the XGBoost algorithm, the model achieved an accuracy of 97.76% [6]. used the supervised learning algorithm to classify social media data as fake or true news based on datasets of textual data and metadata related to tweets, such as the number of followers and retweets. The data was collected in real-time via the Twitter API, using keywords related to health and politics, which allowed for a targeted dataset for analysis. The results show that algorithms using supervised techniques such as decision trees and Naïve Bayes have successfully classified the news. [7] proposed machine learning and deep learning techniques on an annotated dataset containing more than 180000 tweets, naming it TruthSeeker. Authors have proposed several word embedding techniques like TF-IDF and Word2Vec, which enhance feature extraction and grasp minute patterns in data for fake news detection. These results obtained in this study indicate that the best result was through the SVM model with an accuracy of 99.03%, followed by the MLP model with an accuracy of 98.77%. The researchers followed the methods of sentiment analysis and applied the OPCNN in detecting true or false news. In the paper, the authors will perform a novel approach to classify the news article, whether real or fraudulent, based on an architect adjustment by different levels of complexity. First, the dataset employed in this paper for fake news detection is the ISOT. It was preprocessed into four different techniques. With these, OPCNN yielded amazing results for recall. It well explains that the result of the technique proposed at present has outperformed the techniques proposed earlier, such as RF and STODL-FNDC. The study conducted by [7] proposes a new approach for detecting fake profiles on social networks. The authors combined the machine-learning algorithm with the bio-inspired algorithm, named Satin Bowerbird Optimization (SBO), to optimize the initial centroids of a k-means clustering algorithm. The dataset used in this study includes information from the social network Twitter as well as negative data from the Global Terrorism Database, totaling millions of instances. The results obtained in this context show remarkable accuracy, reaching up to 99.6% for the Fire Hawk Optimizer classifier. (FHO). The various studies presented also use a wide range of techniques to detect fake news. These range from traditional machine learning methods that use SVM and MLP to more advanced approaches, such as convolutional neural networks and hybrid models. The topics covered range from health to social media analysis, so the methods must be adapted to the particular type of information being processed. The datasets used are also varied in size and type: several works base their research on smaller datasets, such as LIAR, while other works use much larger databases, such as TruthSeeker. This gives every reason to believe that the quantity and quality of the data used greatly affect the outcome. Finally, the results are promising, reaching up to 99.67% accuracy in certain methods.

2.1 Materials and Methods

This section presents all the data used in this study, as well as the proposed methodology to provide the fake news detection model. Figure 1 below represents our approach used for detecting fake news.

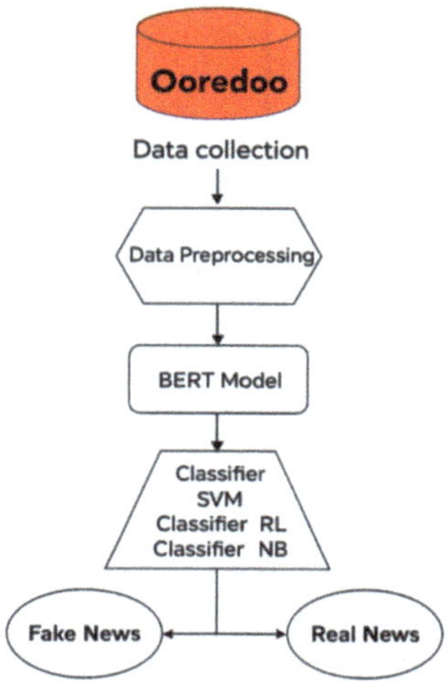

Fig. 1. Architecture of the Misinformation Detection System Using BERT and Hybrid Classification (SVM, RF, NB) for Ooredoo.

2.2 Data Collection

In this section, we used data sets extracted from various Facebook and Twitter platforms using APIs and web scraping techniques. The extracted messages, which were in various languages including Algerian dialect, were translated and adapted to ensure their comprehensibility, relevance, and consistency in the target context containing the keywords "Ooredoo," " outage," "internet, "and" promotion." A corpus of 10,000 messages was compiled and then manually annotated by experts certified by the Algerian telephone operator Ooredoo into two categories (true or false). The main objective is to analyze the statements in order to create a more accurate and effective model for detecting fake news by using detailed labeling to further improve the classification of true and false information.

2.3 Data Preprocessing

Dataset Preprocessing Textual data normalization and preprocessing of different datasets may be one of the most critical stages toward the unification of input data, noise reduction and possibility of extracting a more coherent and flexible feature set of the model. The datasets have undergone the following steps of preprocessing and normalization:

- **Tokenization**: Text was converted into tokens, with each token assigned an index compatible with the model.
- **Lowercasing**: All text was converted to lowercase to ensure consistency and prevent duplication of tokens due to case differences.
- **Removal of Special Characters and Punctuation**: Special characters and punctuation marks were removed to minimize noise.
- **Stopword Removal**: Common stopwords were removed to retain only the most meaningful words.
- **Spelling Correction**: Context-aware spelling correction algorithms were applied to fix spelling errors.
- **Normalization**: Contractions were expanded (e.g., *don't* → *do not*) to improve analysis accuracy.
- **Text Break Normalization**: Line breaks and word segmentation inconsistencies were normalized.
- **Lemmatization**: Words were reduced to their base forms to achieve lexical standardization.
- **Dialect-to-Standard Language Translation**: Dialectal expressions were translated into standard language to ensure consistency across data sources [8].
- **Removal of URLs, Email Addresses, and Hashtags**: Non-textual elements such as URLs, email addresses, and hashtags were eliminated.
- **Emoji and Symbol Filtering**: Emojis and symbols were either removed or replaced depending on their relevance.

2.4 Feature Extraction Using the Bert Model

In this work, we utilized the BERT model, which is available through the transformers library. The model learns contextual relations among words using a deep bidirectional transformer architecture to produce sentence embeddings. For any given string, it uses the average vector of the last layer embeddings as a compact representation of the sentence. Tokenization and encoding: each text is transformed into a sequence of tokens compatible with the BERT model. Thanks to padding, it is possible to have a uniform size, but maximum length was adjusted through a compromise in order to balance performance with model efficiency. Using BertModel, each text is converted to a vector These last ones capture all the context of every sentence; it holds in itself the relations between words and concepts.

2.5 Classification with Different Models

A large number of classification models were trained to identify fake news in the dataset obtained. We focused on three popular algorithms: Support Vector Machine (SVM), Random Forest (RF), and Naive Bayes (NB), which were selected for their effectiveness and ease of use in text classification tasks, particularly those involving linearly separable data. The SVM model was trained on feature vectors derived from preprocessed text data. Starting from a linear kernel as a basis, the SVM aimed to determine the optimal hyperplane separating the "true" and "false" information classes. The model was extremely effective in terms of classification accuracy. The Random Forest classifier, an ensemble algorithm based on decision trees, was used to take advantage of its high robustness and ability to handle noisy or unbalanced datasets. By comparing the predictions of a set of decision trees, the RF model reduces the risk of overfitting and enhances generalization. Finally, the naive Bayesian classifier was used because of its excellent performance and computational efficiency in text classification problems, particularly on high-dimensional feature spaces. Despite its simplistic assumption of feature independence, the naive Bayesian classifier often provides competitive results, particularly when used with sparse data such as word vectors. Each model was evaluated based on classification accuracy, precision, recall, and F1 score. The comparative results revealed that SVM achieved the higher.

3 Result and Discussion

In the given section, we provide some experiments associated with fake news detection. We employed various machine learning models and implemented various SVM, NB, and RF classifiers so that we could deliver the optimal results aimed at the detection of fake information.

3.1 Training Data

The data was split into a training set (80%) and a test set (20%), with cross-validation to ensure robustness. The SVM, NB, and RL classifiers were tuned and evaluated using classification metrics such as accuracy, recall, and F1 score.

$$Accuracy = \frac{TP + TN}{TP + TN + FN + FP} \quad (1)$$

$$Recall = \frac{TP}{TP + FN} \quad (2)$$

$$F1\ Score = \frac{TP}{TP + \frac{1}{2}\ (FN + FP)} \quad (3)$$

Based on the results obtained, we find that the SVM classifier outperformed other classical methods, such as TF-IDF + SVM or LSTM alone (mutaz), in terms of accuracy and robustness. The table (Table 1) below shows the performance for the essential metrics. We find that the SVM model outperformed the NB and RF models with an accuracy of 93% and a recall of 89.2%.

Table 1. Performance measure

Model	Accuracy	Precision	Recall	F1-Score
SVM	90.2%	93%	87.8%	90.3%
Random Forest	87.5%	86.4%	88.3%	87.3%
Naive Bayes	78.9%	80.0%	76.3%	78.1%

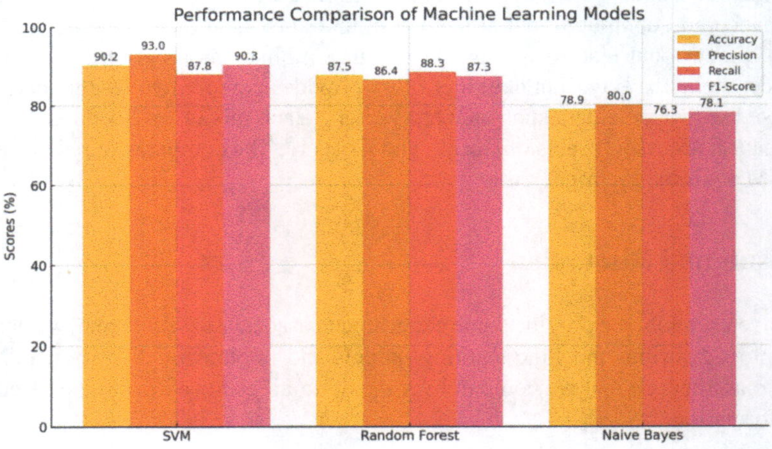

Fig. 2. Performance comparison of Models.

Models were more or less effective depending on the quality of annotation and preprocessing, as indicated in Fig. 2, especially when it comes to a multilingual and dialectal environment. With the resource-thirsty BERT, one has a better chance of detecting contexts. Ooredoo application framework can be scaled up to other local companies.

4 Conclusion and Perspectives

We proposed an approach for automatically detecting fake news targeting messages about Ooredoo. The results show an accuracy of up to 93% with SVM. We obtained promising results of 90.2% accuracy for fake news detection using the SVM classifier. However, the limitations of the model are apparent in the imbalanced dataset of the fake news detection task. To further improve our results, we plan to implement other deep learning approaches as the dataset size increases. Our prospects include the deployment of a reputation monitoring platform, the integration of visual data (memes, screenshots), and the use of multi-modality detection techniques.

References

1. Meel, P., Vishwakarma, D.K.: Fake news, rumor, information pollution in social media and web: A contemporary survey of state-of-the-arts, challenges and opportunities. Expert Syst. Appl. **153**, 112986 (2019)
2. Akinyemi, B., Adewusi, O., Oyebade, A.: An improved classification model for fake news detection in social media. Int. J. Inf. Technol. Comput. Sci. **12**(1), 34–43 (2020)
3. Choudhary, M., Chouhan, S.S., Pilli, E.S., Vipparthi, S.K.: BerConvoNet: A deep learning framework for fake news classification. Appl. Soft Comput. **110**, 107614 (2021)
4. Wang, C., Yang, X., Ding, L.: Deep learning sentiment classification based on weak tagging information. IEEE Access. **9**, 66509–66518 (2021)
5. Atchariyachanvanich, K., Saengkhunthod, C., Kerdnoonwong, P., Chanlekha, H., Cooharojananone, N.: Improvement of a machine learning model using a sentiment analysis algorithm to detect fake news: A case study of health and medical articles on Thai language websites. J. Cases Inf. Technol. **26**(1), 1–26 (2024)
6. Abdullah, N.A.S., Rusli, N.I.A., Yuslee, N.S.: Development of a machine learning algorithm for fake news detection. Indones. J. Electr. Eng. Comput. Sci. **35**(3), 1732–1739 (2025)
7. Al-Tarawneh, M.A., Al-Irr, O., Al-Maaitah, K.S., Kanj, H., Aly, W.H.F.: Enhancing fake news detection with word embedding: A machine learning and deep learning approach. Computers. **13**(9), Llll (2023)
8. Mahammed, N., Klouche, B., Saidi, I., Khaldi, M., and Fahsi, M.: Bio-inspired algorithms for effective social media profile authenticity verification. In: Proc. IAM, 2023, pp. 109–119
9. Klouche, B., Benslimane, S.M. (décembre 2020): Translittération de l'alphabet arabizi chat en dialecte algérien. Dans Conférence internationale sur l'intelligence artificielle dans les systèmes énergétiques renouvelables (pp. 790–797). Cham: Springer International Publishing

Metaheuristic Approaches for Medical Image Denoising

Abdelkarim Cherhabil(✉), Lahcène Mitiche, and Amel Baha Houda Adamou-Mitiche

Department of Electronics and Communications, Laboratoire de Modélisation Simulation et Optimisation des Systèmes Complexes Réels, University of Djelfa, Djelfa 17000, Algeria
abdelkarim.cherhabil@univ-djelfa.dz

Abstract. In recent years, metaheuristic techniques have been successfully employed across diverse research domains and are now considered leading methods for addressing optimization problems. Ultrasound imaging is a widely used medical diagnostic technique that reconstructs internal anatomical structures using high-frequency sound waves. It also represents an ill-posed inverse problem, for which no exact solution method exists hence the motivation for employing a metaheuristic approach. The primary objective of this paper is to optimize ultrasound image reconstruction from noisy data using a nature-inspired metaheuristic method, namely the Cuckoo Search via Lévy Flights with Intelligent Perception Algorithm (CS-IPA), to enhance the reconstruction process.

Keywords: Medical Image Denoising · Metaheuristic Algorithm · Ultrasound Image Reconstruction · Cuckoo Search via Lévy Flights with Intelligent Perception Algorithm (CS-IPA)

1 Introduction

Optimization is primarily used to enhance the performance and efficiency of a system and to find optimal solutions within a reasonable timeframe. Among the methods employed for these purposes are metaheuristic methods, which are stochastic and iterative optimization techniques [1]. These methods have gained considerable popularity in the fields of optimization, AI, ML, computational intelligence, and various engineering applications. They are often inspired by natural systems, whether from physics (as in simulated annealing), evolutionary biology (as in GA), or ethology (as in ACO or PSO). They are mainly proposed to solve complex, so-called "hard" problems for which no exact method can provide a satisfactory solution. The aim of metaheuristics is to explore the search space efficiently using an objective function in order to find the solution closest to the exact one. To date, various metaheuristic algorithms have been introduced in the literature, including Genetic Algorithm (GA) [2, 3], Particle Swarm Optimization (PSO) [4], and Artificial Bee Colony (ABC) algorithm [5]. An optimization problem involves finding an instantiation of a set of variables subject to constraints, in order to maximize or minimize a given criterion. Simply put, solving an optimization problem means finding the optimum of a function from a finite number of choices.

M. Hatti (Ed.): IC-AIRES 2025, LNNS 1740, pp. 210–219, 2026.
https://doi.org/10.1007/978-3-032-12788-4_20

Optimization is an essential tool that combines various techniques from discrete mathematics and computer science. It encompasses a wide range of optimization problems with applications in numerous fields, including management, engineering, manufacturing, telecommunications, transportation, energy, and social sciences [1]. In the field of biomedical imaging, one of the most common and challenging problems is noise, which significantly affects image quality and diagnostic accuracy. Optimization techniques, particularly metaheuristic approaches, are increasingly being applied to address this issue by enhancing image clarity while preserving critical anatomical information [6]. Numerous studies in the literature have focused on the design of 2D digital filters, which have been widely applied for image denoising [7–11]. Mastorakis et al. [10] introduced an innovative constrained optimization technique for designing 2D IIR filters, formulating the task as a constrained minimization problem and solving it using a GA. Das and Konar employed PSO to address a similar constrained optimization problem in 2D IIR filter design [7]. Kockanat et al. [8, 9] applied the ABC algorithm to adjust the parameters of a 2D FIR filter for image denoising purposes. Tzeng et al. [11] developed an efficient genetic algorithm tailored for designing 2D FIR filters with specific magnitude and group delay characteristics. In this paper, we employ four well-established metaheuristic algorithms for ultrasound image denoising and evaluate their performance. This paper is structured as follows: Section 2 outlines the proposed noise removal method and details the CS-IPA algorithm; Section 3 describes the tools and programming languages used; and Section 4 presents the results and discussion.

2 Noise Elimination Approach

As depicted in Fig. 1, This section The proposed system aims to denoise 2D images by combining an adaptive 2D FIR filter with a Fast Non-Local Means (FNLM) post-processing stage. The goal is to estimate the clean signal $\boldsymbol{s}(\boldsymbol{n}_1, \boldsymbol{n}_2)$ from the noisy observation $\boldsymbol{d}(\boldsymbol{n}_1, \boldsymbol{n}_2)$, which is modeled as:

$$\boldsymbol{d}(\boldsymbol{n}_1, \boldsymbol{n}_2) = \boldsymbol{s}(\boldsymbol{n}_1, \boldsymbol{n}_2) + \boldsymbol{x}(\boldsymbol{n}_1, \boldsymbol{n}_2) \quad (1)$$

The noisy signal is first processed using a 2D lowpass filter to attenuate high-frequency noise:

$$x_1(n_1, n_2) = Lowpass(x(n_1, n_2)) \quad (2)$$

The result is then input to a 2D FIR filter whose coefficients are adaptively adjusted based on the system error. The filter output is given by:

$$y(n_1, n_2) = \sum_{m=0}^{M-1} \sum_{n=0}^{N-1} h(m, n)\, x_1(n_1 - m, n_2 - n) \quad (3)$$

The error signal between the noisy input and the filter output is computed as:

$$\boldsymbol{e}(\boldsymbol{n}_1, \boldsymbol{n}_2) = \boldsymbol{d}(\boldsymbol{n}_1, \boldsymbol{n}_2) - \boldsymbol{y}(\boldsymbol{n}_1, \boldsymbol{n}_2) \quad (4)$$

This error is then processed by a Fast Non-Local Means (FNLM) filter, which further removes residual noise and enhances structural details:

$$\hat{s}(\boldsymbol{n_1}, \boldsymbol{n_2}) = \boldsymbol{FNLM}(\boldsymbol{e}(\boldsymbol{n_1}, \boldsymbol{n_2})) \tag{5}$$

Where $\hat{s}(n_1, n_2)$ denotes the final denoised output.

Figure 1 demonstrates our approach, where we use the MSE solely as the objective function to optimize the coefficient matrix of the 2D FIR digital filter. Rather than treating MSE as a general objective function, we specifically define it as the error between the noisy input signal and the filter's output. The MSE is given by:

$$MSE = \frac{1}{N_1 N_2} \sum_{n_1=0}^{N_1-1} \sum_{n_2=0}^{N_2-1} [d(n_1, n_2) - y(n_1, n_2)]^2 \tag{6}$$

Where **N1 and N2** denote the dimensions of the input image.

To quantify the denoising performance, PSNR is computed by comparing the original clean image $\boldsymbol{s}(\boldsymbol{n_1}, \boldsymbol{n_2})$ with the corresponding error signal $\boldsymbol{e}(\boldsymbol{n_1}, \boldsymbol{n_2})$. The PSNR is given by:

$$\boldsymbol{PSNR} = \mathbf{10.log_{10}}\left(\frac{\boldsymbol{R^2}}{\boldsymbol{MSE}}\right) \tag{7}$$

Where R represents the highest possible pixel intensity in the image.

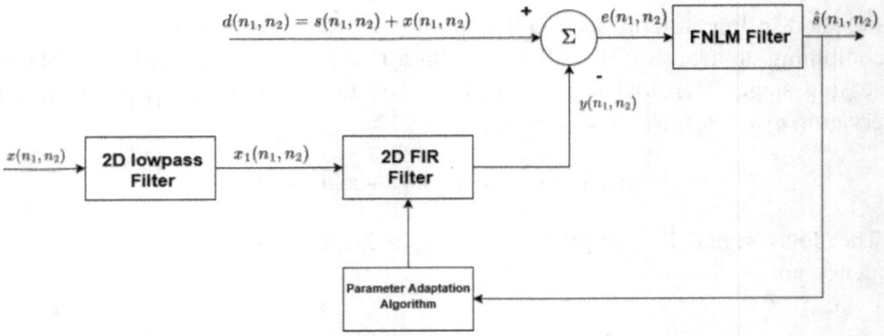

Fig. 1. Architecture of the proposed noise removal approach.

2.1 Proposed CS-IPA Algorithms

This section briefly discusses the basic features of the proposed CS-IPA algorithms, which are designed to optimize the coefficient matrix of the 2D FIR digital filter and to find optimal solutions for the noise reduction task.

The initial population of the CS-IPA algorithm consists of N host nests randomly generated within the search space, with stratification to ensure optimal diversity. Each nest x_i represents a candidate solution, and the number of nests, cuckoos, and eggs remains equal.

New solutions are generated via adaptive Lévy flights, modulated by a perception factor $\emptyset_i(t)$ and a dynamic step size $\alpha(t)$. The evolution law is expressed as:

$$\boldsymbol{x}_i^{t+1} = \boldsymbol{x}_i^t + \boldsymbol{\alpha}(\boldsymbol{t}) \otimes \boldsymbol{L\acute{e}vy}(\boldsymbol{\lambda}) \odot \varnothing_i(\boldsymbol{t}) \tag{8}$$

A nest j is chosen with a probability inversely proportional to its fitness $P_j \propto F_j^{-1}$. The new cuckoo replaces x_j if $F_i < F_j$, incorporating a simulated annealing criterion to occasionally accept less performant solutions and avoid local optima:

$$Paccept = \begin{cases} 1 & if\ Fi < Fj \\ \exp\left(-\frac{\Delta F}{T(t)}\right) & otherwise \end{cases} \tag{9}$$

with $T(t) = T_0.\rho^t$, $(0 < \rho < 1)$ to gradually decrease the tolerance.

The eggs have an adaptive probability $p_a(t)$ of being abandoned, determined by the population entropy $H(t)$:

$$p_a(t) = \begin{cases} p_a + \Delta_P & if\ H(t) < \tau \\ p_a & otherwise \end{cases} \tag{10}$$

The parameters $\alpha(t)$, $p_a(t)$, and $\lambda(t)$ evolve dynamically through feedback mechanisms:

$$\alpha(t+1) = \alpha(t).\exp\left(-\eta.\ \frac{F_{best}^{(t)}}{F_{best}^{(t-1)}}\right), p_a(t+1) = p_a(t) + \eta.(1 - H(t)) \tag{11}$$

Elitism preserves the best solutions, while periodic ranking updates x_{best} (Algorithm 1).

Algorithm 1: Pseudo Code for CS-IPA

Begin

Objective function $f(x)$, $x = (x_1, \ldots, x_d)^T$

Generate initial population of n host nests x_i, $(i = 1, 2, \ldots, n)$**while** $t <$ MaxGeneration or stop criterion **do**

Generate a cuckoo via Lévy flights with adaptive step size

Evaluate its quality/fitness F_i

Choose a nest j probabilistically**if** $F_i > F_j$ **then**

Replace j by the new solution

end if

Calculate population diversity entropy H

if $H < \tau$ **then**

Abandon $p_a + \Delta_P$ worse nests and rebuild with larger Lévy steps

else

Abandon p_a worse nests and rebuild

end if

Adaptively update α and p_a using learning rate η

Keep the best solutions
Rank the solutions and find the current best
end while
Postprocess results and visualization

3 Experiment

In this section, we begin by listing the languages and tools utilized: MATLAB and CUDA.

Dataset In the developed noise elimination strategy, a fetal ultrasound image was employed for model training. The image has a resolution of 256 × 256 pixels and was acquired using Samsung Medison equipment. Figure 2 presents both the noise-free and noisy versions of the ultrasound training image. To evaluate the proposed method, ultrasound images of different organs were also used. These ultrasound images were corrupted with Gaussian noise having zero mean and variance ranging from 0.1 to 1. To address the varying noise levels, the ultrasound images underwent noise removal using the GA, PSO, ABC, and CS-IPA algorithms. Table 1 displays the control parameters for the implemented parameter adaptation algorithms. Additionally, the publicly available clinical LoDoInd dataset [12] was incorporated to broaden experimental validation.

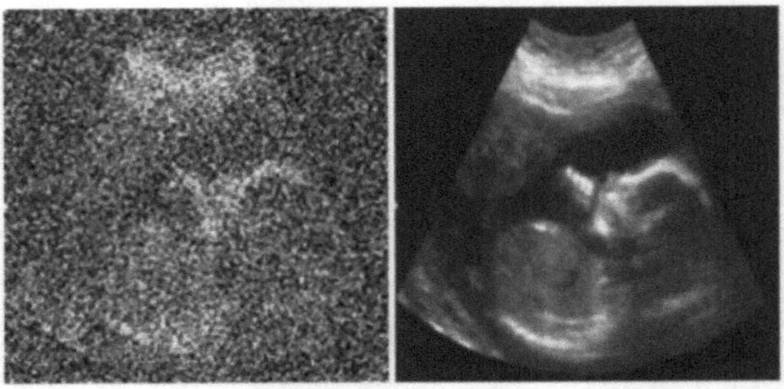

Fig. 2. Training images. Left: Noisy; Right: Noiseless.

Experimental Setup The simulations were performed in MATLAB with CUDA 12.4.0 acceleration on an Ubuntu 22.04 LTS system equipped with an Intel® Core™ i7-12700K CPU, 2.70 GHz. and four NVIDIA RTX 3070 Ti GPUs 8GB GPUs. A 2D FIR digital filter of order M = N = 3 was implemented for testing. To ensure statistical robustness, all algorithms were independently run 30 times with distinct random seeds, and initial parameters were uniformly sampled from the range (−1,1). For consistency across optimization methods, the colony size and maximum iterations were fixed at 20 and 150, respectively.

Table 1. The control parameters for the applied adaptation algorithms.

Parameter	Algorithms			
	GA	PSO	ABC	**CS-IPA**
Population/Swarm/Colony Size	20	20	20	20
Parameter 1	Crossover rate (CR): 0.8	Cognitive and social component (c_1, c_2): 1.8	Limit: 27	Discovery probability (pa): 0.25
Parameter 2	Mutation rate (MR): 0.01	Inertia weight (w): 0.6		Learning rate (η): 0.1
Parameter 3				Diversity threshold (τ): 0.5
Parameter 4				Lévy step increment (Δ_p): 0.1

4 Results and Discussion

In this study, we evaluated the performance of four parameter adaptation algorithms GA, PSO, ABC, and our proposed CS-IPA for noise elimination in ultrasound images. The effectiveness of each algorithm was assessed using PSNR/MSE metrics across noise levels ranging from $\boldsymbol{\sigma^2 = 0.1}$ ***to*** **1.0**. Computational efficiency was also compared through execution time measurements.

Figure 3 presents the denoised ultrasound images produced by the CS-IPA algorithm. As indicated in Table 2, all algorithms demonstrated competitive denoising performance, the proposed CS-IPA algorithm consistently outperformed the others, particularly at higher noise levels. in the aortic valve image at $\sigma^2 = 0.7$, CS-IPA achieved a PSNR of 21.01 dB, significantly surpassing PSO's 20.00 dB.

In parallel, Fig. 4 provides a comprehensive visual comparison that clearly illustrates how CS-IPA substantially reduces image noise while preserving essential anatomical structures and diagnostic details within the selected regions. On the left are LoDoInd CT images at noise level 4 top and 5 bottom, acquired at 140 kVp and 23 mA and 140 kVp and 12 mA, respectively, with blue rectangular boxes highlighting the regions of interest (ROI). The three panels on the right show magnified views of these ROIs processed using different techniques: (a) LDCT, (b) CS-IPA, and (c) NDCT. This demonstrates the ability of CS-IPA to generalize to unseen industrial low-dose CT scans.

Table 3 summarizes the execution times of all algorithms, each completing in under five seconds. PSO exhibited efficient processing at 3.12 s, while ABC required 3.49 s. Despite its superior denoising performance, CS-IPA maintained competitive computational speed, making it suitable for practical applications.

Table 2. Comparison of PSNR/MSE values produced by the GA, PSO, ABC and CS-IPA algorithms.

Image	Parameter adaptation algorithm	Noise variance					
		0.1	0.3	0.5	0.7	0.9	1
Fetal	GA	36.73/13.78	27.40/118.01	22.99/322.27	20.07/637.39	17.79/1083.35	16.99/1299.03
	PSO	36.75/13.70	27.41/117.47	22.99/322.27	20.06/641.52	17.86/1069.55	16.98/1300.82
	ABC	36.83/13.44	27.42/116.93	23.00/326.27	20.07/637.39	17.80/1073.26	16.99/1299.03
	CS-IPA	**36.84/13.40**	**27.45/116.26**	**23.09/319.55**	**20.12/634.35**	**18.10/1007.78**	**17.01/1298.09**
Pancreas	GA	36.96/13.05	27.39/118.11	22.96/326.78	20.04/644.58	17.87/1071.72	16.99/1299.03
	PSO	36.97/13.00	27.42/116.93	22.99/322.27	20.04/644.58	17.88/1070.76	16.98/1300.82
	ABC	36.97/13.00	27.47/116.05	23.01/323.96	20.10/636.55	17.92/1058.91	17.00/1297.06
	CS-IPA	**37.00/12.99**	**27.53/115.23**	**23.09/319.55**	**20.21/617.94**	**17.98/1029.26**	**17.00/1297.06**
Aortic valve	GA	36.69/13.55	27.45/116.26	22.99/322.27	20.09/638.55	17.93/1041.94	16.99/1299.03
	PSO	37.01/12.95	27.45/116.26	23.01/323.96	20.00/650.25	17.92/1058.91	17.00/1297.06
	ABC	37.02/12.92	27.48/114.84	23.04/321.72	20.14/633.88	17.94/1038.73	17.01/1298.09
	CS-IPA	**37.23/12.19**	**27.54/114.56**	**23.10/320.42**	**21.01/513.34**	**17.98/1029.26**	**17.03/1296.11**
Right kidney	GA	36.93/13.08	27.44/116.71	23.00/326.27	19.99/650.30	18.00/1030.02	16.88/1332.71
	PSO	36.99/12.98	27.46/116.06	23.01/323.96	20.03/644.05	18.04/1025.86	16.99/1299.03
	ABC	37.07/12.86	27.48/114.84	23.04/321.72	20.07/637.39	17.95/1029.37	17.03/1296.11
	CS-IPA	**37.31/12.71**	**27.51/114.14**	**23.08/320.75**	**21.00/515.11**	**18.02/1027.88**	**17.10/1275.25**

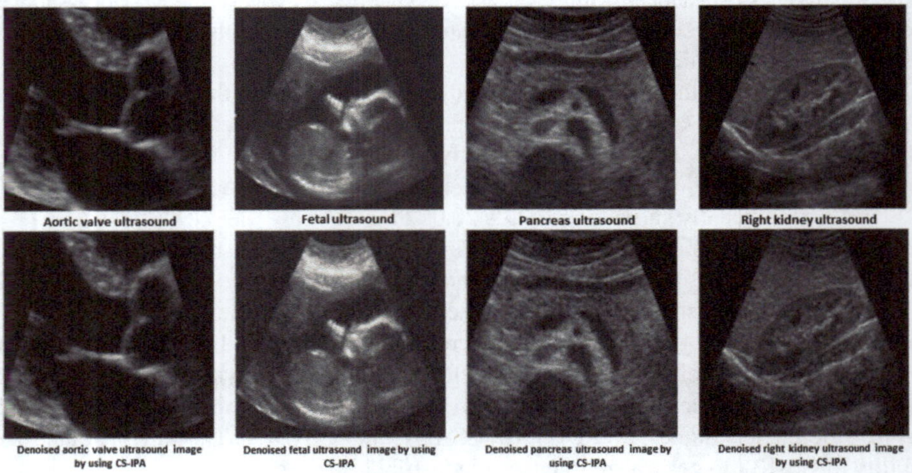

Fig. 3. Comparison between ultrasound images and denoised ultrasound images obtained using the proposed CS-IPA algorithm.

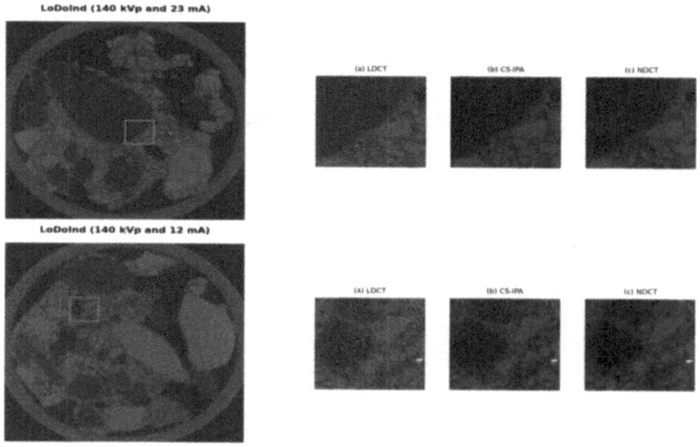

Fig. 4. Comparative analysis of noise reduction outcomes for a representative LoDcInd CT scan at the fourth and fifth noise levels in CS-IPA.

Table 3. Time analysis of GA, PSO, ABC, and CS-IPA algorithms.

Algorithms	GA	PSO	ABC	**CS-IPA**
Time(sec)	1.27	3.12	3.49	**3.76**

As shown in Fig. 5, GA has the highest mean value of **0.0992**, indicating a slightly less optimal average performance compared to CS-IPA. Both PSO and GA exhibit significantly higher standard deviation values, with GA showing the greatest variability in performance. Overall, CS-IPA is the best performer, as it achieves both the lowest mean value 0.09896 and the lowest standard deviation $0.8413e^{-07}$.

As shown in Fig. 6, although PSO initially takes the lead, the CS-IPA algorithm surpasses all others after 38 cycles. This point marks a critical transition, after which CS-IPA consistently achieves the lowest MSE, demonstrating its strong long-term optimization capability. Therefore, while PSO is effective for rapid, early-stage optimization, CS-IPA proves to be the most reliable algorithm for achieving optimal accuracy in the later stages.

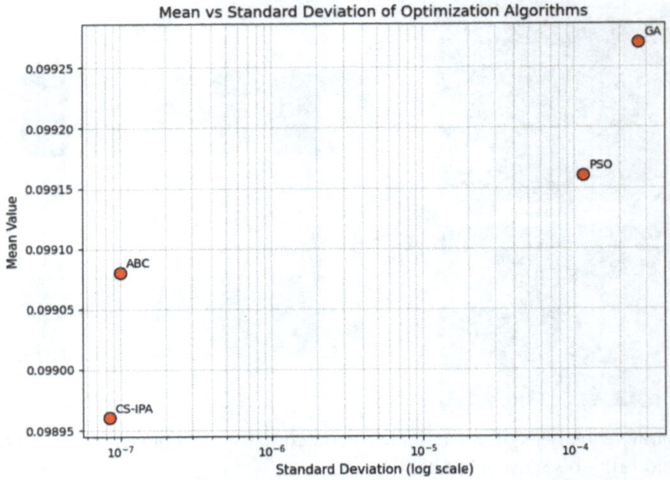

Fig. 5. Comparison of mean and standard deviation across optimization algorithms.

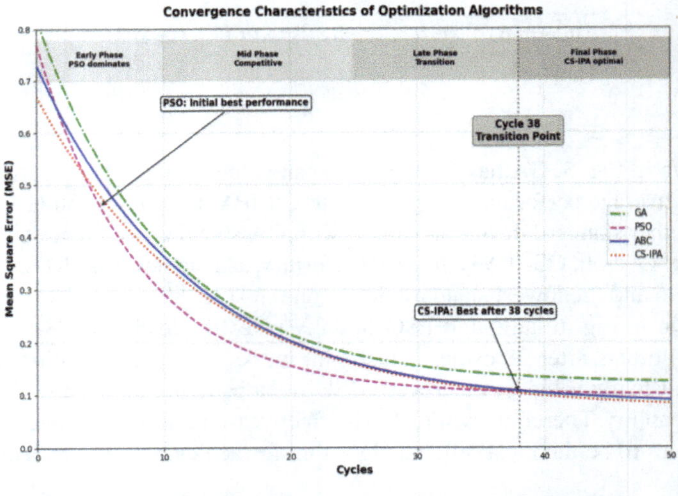

Fig. 6. Convergence characteristics of optimization algorithms.

5 Conclusion

In this paper, we proposed a robust despeckling framework designed to mitigate noise in ultrasound images. A key advantage of the proposed framework is its low computational complexity and fixed number of iterations. The metaheuristic algorithms used for parameter adaptation in the proposed method include GA, PSO, ABC, and the newly proposed CS-IPA algorithm. In future work, the framework may be tested on other types of images for further evaluation.

References

1. Khanduja, N., Bhushan, B.: Recent advances and application of metaheuristic algorithms: A survey (2014–2020). In: Metaheuristic and Evolutionary Computation: Algorithms and Applications, pp. 207–228. Springer International Publishing, Cham (2020)
2. Golberg, D.E.: Genetic Algorithms in Search, Optimization, and Machine Learning. Addison-Wesley, Reading, MA (1989)
3. Alaoui, N., Adamou-Mitiche, A.B.H., Mitiche, L.: Effective hybrid genetic algorithm for removing salt and pepper noise. IET Image Process. **14**(2), 289–296 (2020)
4. Kennedy, J., and Eberhart, R.. Particle swarm optimization. In: Proceedings of ICNN'95-International Conference on Neural Networks. Vol. 4. IEEE, 1995
5. Karaboga, D.: An idea based on honey bee swarm for numerical optimization. Technical Report, pp.1–10. (2005)
6. Szabo, T.L.: Diagnostic Ultrasound Imaging: Inside Out. Academic, Cambridge, MA (2013)
7. Das, S., Konar, A.: A swarm intelligence approach to the synthesis of two-dimensional IIR filters. Eng. Appl. Artif. Intell. **20**(8), 1086–1096 (2007)
8. Kockanat, S., and Karaboga, N.: Parameter tuning of artificial bee colony algorithm for Gaussian noise elimination on digital images. In: 2013 IEEE INISTA – International Symposium on Innovations in Intelligent Systems and Applications. IEEE 2013
9. Kockanat, S., Karaboga, N., and Koza, T.: Image denoising with 2-D FIR filter by using artificial bee colony algorithm. In: 2012 International Symposium on Innovations in Intelligent Systems and Applications (INISTA). IEEE. 2012
10. Mastorakis, N.E., Gonos, I.F., Swamy, M.N.S.: Design of two-dimensional recursive filters using genetic algorithms. IEEE Trans. Circuits Syst. I Fundam. Theory Appl. **50**(5), 634–639 (2003)
11. Tzeng, S.-T.: Design of 2-D FIR digital filters with specified magnitude and group delay responses by GA approach. Signal Process. **87**(9), 2036–2044 (2007)
12. Shi, J., et al. LoDoInd: Introducing a benchmark low-dose industrial CT dataset and enhancing denoising with 2.5D deep learning techniques. In: 13th Conference on Industrial Computed Tomography (iCT), Wels Campus, Austria. Vol. 10. 2024

Analysis of Brain Medical Images by Artificial Immune Systems

Hasnia Merzoug[1,2(✉)] and Hayat Yedjour[2]

[1] Department of Management, University Mohammed ben Ahmed, Oran, Algeria
merzoug.hasnia@univ-oran2.dz

[2] Department of Informatics, University of Sciences and Technology MB of USTO, Oran, Algeria
hayet.yedjour@univ-usto.dz

Abstract. Medical imaging plays a fundamental role in clinical diagnosis, particularly in neurology, where image quality directly influences the reliability of diagnoses. This quality strongly depends on upstream processing techniques designed to enhance clarity and enrich visual detail. In this study, we propose an innovative method for classifying brain tumors based on Artificial Immune Systems (AIS), bio-inspired algorithms that mimic the learning and memory mechanisms of the human immune system. Cerebral MRI images are first subjected to a segmentation phase using the Canny algorithm, which extracts the distinctive contours of tumor structures. These extracted features are then used as inputs for the AIS models.

Keywords: Medical Imaging · Brain Tumors · Segmentation · Canny Artificial Immune System

1 Introduction

The central nervous system, represented by the brain, orchestrates bodily functions by processing sensory input and issuing commands. Brain tumors arise from abnormal cell growth or mutation and can manifest symptoms such as nausea, memory loss, personality changes, vision impairment, and fatigue. Diagnosis typically relies on high-resolution CT and MRI scans, with MRI preferred due to its non-invasive nature and detailed anatomical visualization. Radiologists manually inspect MRI scans to detect and classify tumors, but subtle changes can be difficult to identify, and manual detection is time-consuming and error- prone. Tumor classification is challenging due to the diverse shapes and overlapping imaging characteristics of different tumor types. Misdiagnosis or delayed recognition—often stemming from limited staff expertise or insufficient diagnostic tools—can lead to adverse patient outcomes [1]. To assist radiologists, Computer-Aided Diagnosis (CAD) systems have been developed to improve accuracy and speed in detecting gliomas, pituitary adenomas, meningiomas, and other brain tumor types. Primary brain tumors such as meningiomas, gliomas, and pituitary adenomas originate within the central nervous system. Meningiomas are typically benign tumors of the brain's protective membranes, while gliomas and pituitary tumors may be

M. Hatti (Ed.): IC-AIRES 2025, LNNS 1740, pp. 220–231, 2026.
https://doi.org/10.1007/978-3-032-12788-4_21

benign or malignant. Malignant tumors are classified as primary if they originate in the brain or secondary if spread from other body regions. The variability in size, shape, and location of these tumors complicates radiological diagnosis, highlighting the need for advanced diagnostic techniques and treatment planning to improve neuro-oncological outcomes [2].

The goal of our study is to develop a system for recognizing brain MRI images using a bio-inspired technique based on the immune system. The chosen algorithm, CLONCLAS (CLONal CLASsification), is an adaptation of the clonal selection principle for pattern recognition. Recognition relies on image segmentation using a Canny-type approach to extract the relevant information needed. Clonal selection is then used to detect anomalies present in these images.

2 Related work

Saravanan et al. [3] MRI was useful in a number of ways for the neuroimaging studies being conducted, as it assisted with the structural and functional mapping of the brain. CNN database learning on top of neigh boring network restriction processing medical imaging encompassed an additional layer for detailed information exchange and sparse coding to enrich the processing. Concentrating on his pursuits, the presented solution kept the spatial structure of the provided labelled data and classified brain MRI on a level of 97.8%, which is significantly better than other available solutions.

Nagur Shareef Shaik et al. [4] explained multi-level of attention network (MANet). Initially feature attention and pixel attention strategies were implemented that implemented deep spatial fusion of temporal dependencies across channels using the Xception backbone feature sequence. Here, Figshare and BraTS benchmark datasets were selected to evaluate this method's performance by examining its effectiveness. These experiments showed that performance gain in these models was achieved by the incorporation of cross-channel and spatial attention mechanisms that were better than reducing the model parameters. The MANet model achieved the best performance in the tumor recognition with 96.51% and 94.91% accuracy on Figshare and BraTS 2018 datasets respectively.

Bentaouza Chahinez and Benyetou Mohamed [5] employed an artificial intelligence technique—Support Vector Machines (SVM)—for the classification of brain tumors. Their methodology involved several key steps. First, they performed binary classification, then progressed to a four-class classification, and finally tackled a nine-class multiclass classification based on tumor types. A classification accuracy of 98.41% was achieved using SVM in this study.

Takowa Rahman et al. [6] proposes a new architecture of a PDCNN that helps in providing both general information as well as details without overfitting the model to the training data. As a first step towards achieving this goal, the images were resized and converted to grayscale, after which data augmentation was implemented. The PDCNN aimed at capturing both local and global information constructed the two parallel networks with differing receptive field size. Accuracies on MRI datasets were shown to be very high: dataset-I binary tumor detection – 97.33%, dataset-II Figshare – 97.60%, and dataset-III multiclass Kaggle dataset – 98.12%.

Jianlin Su et al. [7] the application of position encoding in transformer architecture and introduces a novel position encoding technique called RoPE. RoPE uses a matrix rotation to address the absolute location as well as allow the self-attention formulation to depend on the location proportion. It provides features such as flexibility of the sequence length and necessary inter-token dependence, as well as linear self- attention with relative position encoding.

Amar Y. Jaffar et al. [8] directed toward better classification of brain tumors through a more advanced ResNet which was combined with Vision Transformers. This combined model offered the local features of the iResNet and the global features of the ViTs. This resulted in an overall accuracy of 99.2% considerably beating the benchmark models such as InceptionV3, ResNet, and DenseNet. The model exhibited impressive results with high precision and recall as well as F1 scores in all tumor types showcasing its robustness and reliability.

Faizan Ullah et al. [9] resented an evolved lightweight model for brain tumor detection and classification from magnetic resonance imaging. The model is termed light weight ensemble combines and is an alteration of Multimodal Lightweight XGBoost. It employs dosimetrist techniques of MRI data preprocessing and feature extraction to evaluate cancer. The model was implemented on BraTS 2020 dataset which consists of 285 MRI images on patients with gliomas. The results achieved are as follows, accuracy 93.0%; precision 0.94; recall 0.93; F1 score 0.94; AUC-ROC value of 0.984.

Building on numerous previous studies, this research presents an approach for brain tumor classification using the CLONCLAS algorithm. It includes a preprocessing stage to remove noise, followed by a segmentation technique to extract relevant features, before applying the classification algorithm.

3 Artificial Immune System

The biological immune system adapts to threats through clonal selection, where lymphocytes that recognize specific antigens proliferate and undergo mutations to enhance the effectiveness of their response. Artificial Immune Systems (AIS) leverage this concept in the Clonal Selection Algorithm (CSA), which iteratively refines candidate solutions to optimization problems. Immunological memory enables AIS to respond efficiently to recurring challenges, establishing a parallel with reinforcement learning in artificial intelligence (Fig. 1).

4 Clonal Selection Algorithm

Clonal selection algorithm in order to clarify how an immune response is mounted when a nonself antigenic pattern is recognized by a b cell, the clonal selection theory has been developed. the clonal selection algorithm can be described as follows:

1) generate a set p of candidate solutions, composed of the subset of memory cells m added to the remaining population pr.
2) select the n best individuals of the population, based on an affinity measure.
3) clone the n best individuals of the population to a temporary population of clones c.

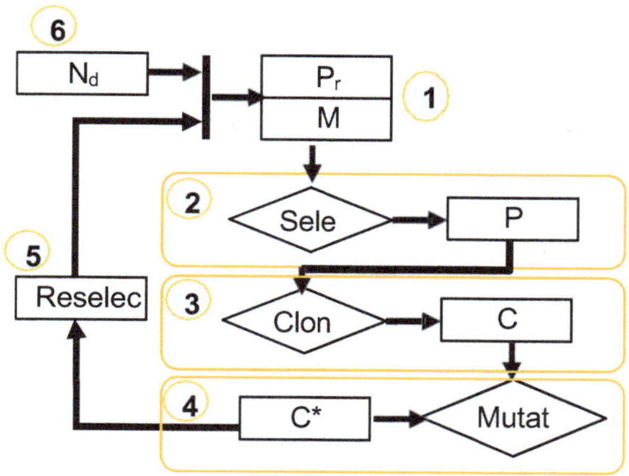

Fig. 1. [10] Clonal Selection Algorithm

4) submit the population of clones to a mutation scheme. a maturated antibody population is generated c*, where the mutation is proportional to the affinity of the antibody with the antigen.
5) re-select the improved individuals from c* to compose the memory set m. some members of ab can be replaced by other improved members of c*.
6) replace d antibodies by novel ones (diversity introduction). the lower affinity cells have higher probabilities of being replaced [10].

a. Affinity Calculation

The affinity calculation must employ a selection function that effectively discriminates between antigens according to their match (affinity), in order to select the best candidates. These selected antibodies are then cloned and mutated, after which the top-performing ones—if superior—are reintegrated into the original population [11]. The objective function driving this process is formalized as an optimization problem in Eq. (1)

$$affinity = \max\{accuracy(sub)\} \tag{1}$$

Where sub denotes the corresponding selected feature subset, accuracy (sub) evaluates the classification accuracy of the subset.

b. The Cloning

The cloning is the duplication of the data in several specimens; this operation makes possible to keep information long in the workspace. A cloning is proportional to affinity because an antibody approaching more to the antigen is interesting to keep information about it which carry it for a long time and that by duplicating it in several identical specimens, and the mutation will play the role to widen the workspace [12]. We will calculate the number of clones (NC) in our algorithm using:

$$nc(i) = round\left(\frac{B * affinity_i^2}{\sum_j affinity_j^2}\right) \tag{2}$$

Where B is the cloning parameter, affinity is calculated in Eq. (1) and round is the operator that rounds

its argument to closest integer. This step raises the average affinity value and gives the following steps a good chance to further move toward the solution. Then, each clone k will be duplicated to NC(k) copies according to Eq (2).

c. The Mutation

The mutation is defined as an application from Ω Ω which associates to each individual Xt a new individual Xt +1close to Xt. The mutation varies according to the representation of the data; in this direction we find various types of mutation in the case of binary presentation (which is our case) or real representation [13]. In the first step we must calculate the number of bits to be inversed (Nm) using:

$$N_M = round\Big((L - affinity(c(i)) * \text{rand}\big) \quad (3)$$

Where L represents the antibody size (the number of features) and affinity C in the vector of affinity calculated after cloning. *Rand* is a mathematical function used to generate a random number that is greater than or equal to 0 and less than 1. In binary valued individuals mutation is the flipping of variable values (inverse '0' to '1' and vice versa), because every variable has only two states (selected (1) or excluded (0)).

5 Proposed Approach

The system we designed for medical image interpretation is built around five key steps: image preprocessing, segmentation, feature extraction from regions, learning the different classes in the database, and classification (Fig. 2).

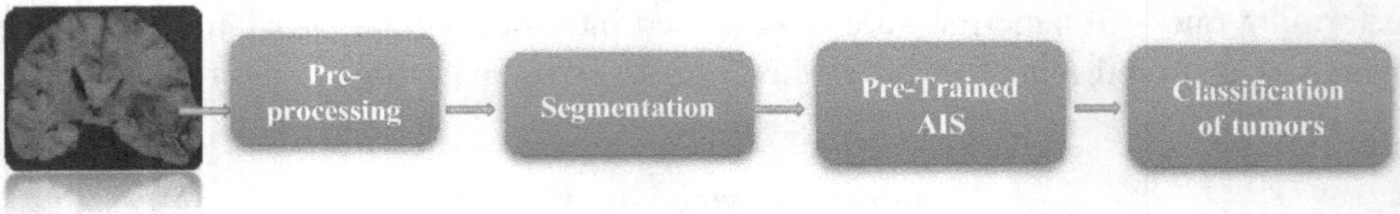

Fig. 2. Medical Image Interpretation System

1)Dataset Used:

The database used in our study is built in accordance with the World Health Organization (WHO) Tumor Classification, which serves as an international reference for the histological and molecular classification of tumors. This classification relies on criteria such as cell type, morphology, biological behavior, and genetic alterations, facilitating diagnosis, treatment, and cancer research worldwide. Tumors are graded (from 1 to 4) based on their degree of cellular differentiation and growth potential. Grade 1 tumors are well-differentiated and generally less aggressive, while grade 4 tumors are poorly differentiated and exhibit more aggressive behavior. Our database includes nine tumor types selected to represent a diversity of grades and histological characteristics, enabling a comprehensive evaluation of our classification system's performance [14] (Table 1).

Table 1. Types of Brain Tumors, Their Grades and Number of Images

Class	Pathology	Grade	Nomber of images
1	Astrocytome pilocytique	I	10
2	Astrocytome giganto-cellulaire	I	2
3	Ependymome	I	3
4	Astrocytome	II	8
5	Astrocytome pilocytique anaplasique	III	2
6	Ependymome anaplasique	III	1
7	Astrocytome anaplasique	III	1
8	Glioblastome	IV	16
9	Médulloblastome	IV	8
	TOTAL		51

2)Image Pre Processing

Image preprocessing is a critical step in pattern recognition systems, aimed at improving image quality to facilitate subsequent analysis. In our system, this phase consists of two main operations:

Image resizing: All images are standardized to a uniform size of 256 × 256 pixels. This standardization simplifies data handling and ensures consistency when applying segmentation methods.

Median filtering: The median filter is a non-linear noise reduction technique widely used to remove impulse noise, commonly known as "salt-and-pepper" noise. It replaces each pixel's value with the median of its surrounding neighbors, effectively eliminating outlier pixels while preserving the edges of the image (Fig. 3).

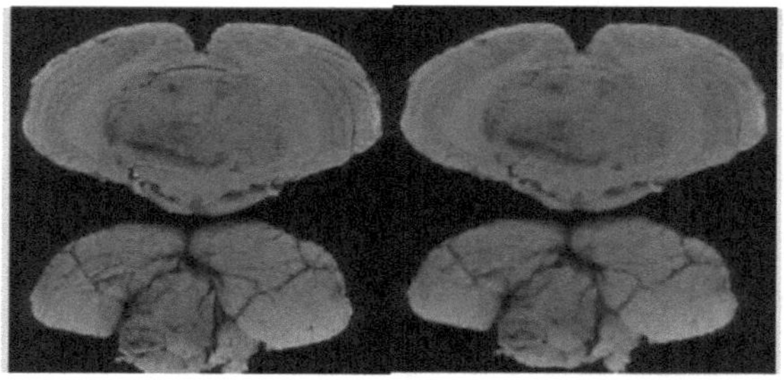

Fig. 3. Resid image and filtered image

3)Segmentation

Medical images are segmented using the Canny method. Canny's approach involves designing a filter that combines: a 1D derivative-of-Gaussian filter in the direction *orthogonal* to the edge being detected, and a low-pass (smoothing) Gaussian filter along the direction *parallel* to the edge, to ensure continuity of the detected contour. To simplify computations, the Gaussian used for smoothing has the same width (σ) as the derivative-of-Gaussian used for edge detection

$$G(X, Y) = \exp\left(-\frac{x^2 + y^2}{2\partial^2}\right) \tag{4}$$

This operation is performed by convolving the image with a two-dimensional Gaussian, which is easy

due to the separability of the Gaussian and the added benefit of being an isotropic operation. The result is then derived (differentiated) in the direction orthogonal to the edge.

Indeed:

$$dG(X) \otimes I(x) == \left(\left(G(x) \otimes I(x)\right)\right) \tag{5}$$

In practice, it is not necessary to compute derivatives in all possible directions of the image to detect edges. Instead, applying the derivative in just two orthogonal directions is sufficient. The process therefore simplifies to: Convolving the image with a two-dimensional Gaussian filter, and computing the derivative of the result in two orthogonal directions (typically the x and y axes). The local maxima in these two derivative images correspond to edges oriented along those two orthogonal directions [15].

4)Feature Extraction from the Segmented Image

Choosing attributes to characterize a region is a **crucial step** in image processing, depending on the nature of the image and analysis objectives. In our system, after applying the **Canny edge detection** algorithm to grayscale images, we obtain an edge map that delineates homogeneous regions. Each region is then analyzed to extract relevant features.

For each identified region, we retrieve the grayscale pixel values from the original image. These values serve as the basis for calculating the following attributes:

- **Area (Sur)**: the number of pixels in the region.
- **Variance (Var)**: a homogeneity criterion calculated as:

$$variance = \frac{\sum_{i=1}^{nbp} (Ngpi - Ngm)^2}{surface}$$

where is the grayscale level of pixel i in region R, and is the mean grayscale level of R.

- **Perimeter (Per)**: the length of the region boundary, measured as the number of pixels forming that border.
- **Mean grayscale level (Ngm)**: the average gray value of pixels in the region:

$$Ngm = \sum niveaux\ de\ gris\ de\ la\ region/sur$$

- **Elongation (Elg)**: determined from the smallest circumscribed rectangle around the region; it's the aspect ratio L/I (length divided by width).

- **Compactness (Com)**: also known as the circularity factor, defined as:$comp = 4psur/per^2$.

With these attributes, each image is represented as an N × M matrix, where:

N is the total number of segmented regions, and M is the number of attributes characterizing each region, in the order: (Sur, Var, Ngm, Elg, Com).

5)Classification with AIS

- **Training Phase with CLONCLA:** During the training phase, we conducted several experiments on a set of 20 images drawn from our database. The objective was to generate memory cells, represented as matrices, which will be used later during classification. These memory cells—analogous to antibodies in the biological immune system—are designed to recognize specific patterns present in the training data, in line with the CLONCLAS algorithm

- **Testing Phase:** The testing phase aims to evaluate the performance of the AIS as a classifier using the memory cells generated earlier. In our application, this phase was applied to the remainder of the database—i.e., the images not used during training. Each test image is compared against the existing memory cells, and a class label is assigned based on the observed similarity. This evaluation allows us to measure the system's ability to generalize acquired knowledge and perform accurate classifications on new data.

6 Experimental Results

The model proposed in this study was implemented using Matlab R2014 software. The making and testing of the system are done by using a computer with Intel (R) Core (TM) i3 5005U CPU @ 2.00GHz 2.00 GHz, Memory 4.00GB, and using a 64-bit operating system (Fig. 4).

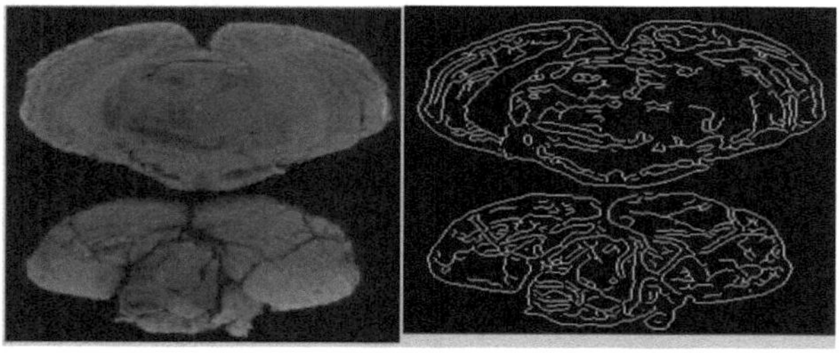

Fig. 4. Original ImageSegmented Image

An illustrative example of extracting region attributes from a segmented image (Table 2)

These attributes extracted from regions of interest are fed into the Artificial Immune System (AIS). To achieve satisfactory classification results, we conducted multiple training experiments on a set of 20 tumor images from our database. Prior to training, we fixed five key AIS parameters:

Table 2. region attributes.

N	Sur	Var	Ngm	ELg	Coml
1	51.000	2.7459	27.8824	0.4138	0.0048
2	159.000	114.1156	59.2830	1.8281	0.0005
3	7.0000	4.9048	75.2857	0.1667	0.2565
4	54.000	1.7680	78.9259	0.3333	0.0043
5	39.000	6.9231	76.1538	0.0526	0.0083
6	44.000	44.6232	82.4318	0.1500	0.0065
7	24.000	3.6739	59.2500	0.0435	0.0218
8	1.0000	0	54.0000	0	12.5664
9	2.0000	2.0000	30.0000	1.0000	3.1416
10	30.000	4.2575	79.1333	0.5833	0.0140
11	24.000	1.9547	69.9583	0.1905	0.0218
12	2.0000	0.5000	83.5000	1.0000	3.1416

Number of generations: determines how many iterations the algorithm will perform to refine the memory cells, akin to repeated cycles of selection and maturation

- **Resource count**: limits the total number of cells (antibodies) allowed in the system, thus influencing the classifier's diversity and memory capacity.
- **Cloning rate**: specifies how many clones are produced from selected antibodies, proportionally to their affinity with the antigens.
- **Hyper-cloning rate**: defines the number of clones generated from the most stimulated memory cells, allowing deeper exploration of the solution spac.
- **Mutation rate**: the probability that a feature of an antibody is modified, introducing diversity and preventing premature convergence (Table 3).

Table 3. Training parameters

Parameters	Valous
hyper-cloning rate	2
Rate cloning	10
Rate mutation	0.5

Classification results: after 300 generations (Table 4):

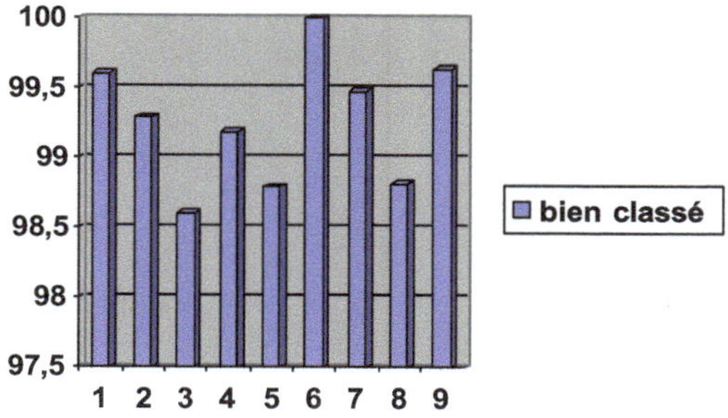

Table 4. Training Result

Technique	Training Rate	Test Rate
AIS	99.63%	99.26%

The results obtained from multiple experiments indicate that classification performance is strongly influenced by three key AIS parameters: the **cloning rate**, **hypercloning rate**, and **mutation rate**. These parameters determine, respectively, the number of clones generated from selected antibodies, the increased production of clones from the most stimulated memory cells, and the probability of modifying antibody characteristics—thereby introducing diversity into the population. To improve classification accuracy, it is essential to conduct several experiments by adjusting these parameters. This process enables identification of the **optimal combinations** that maximize the system's ability to generalize learned knowledge and perform precise classification on new data.

7 Comparative Study

In comparing our results with those obtained using Support Vector Machines (SVM) with a Radial Basis Function (RBF) kernel on the same dataset [5], we observed the following performance (Table 5):

Table 5. Comparison results

Méthods	Classification Rate
SVM [5]	98.41%
AIS	99.26%

The experiments demonstrated that the recognition rate is higher with classification based on Artificial Immune Systems (CLONCLAS approach), optimized compared to

SVMs. This improvement can be attributed to the ability of AIS to dynamically adapt to data variations and to their capacity to efficiently explore the solution space through cloning and mutation mechanisms.

8 Conclusion

The primary objective of this work was the design and implementation of an MRI image recognition system, specifically aimed at diagnosing detected anomalies. The developed system integrates and enhances several key steps of medical image analysis, spanning from preprocessing to classification. Following a comprehensive evaluation and appropriate selection of preprocessing techniques, our software employs Canny-based edge detection segmentation, extraction of region attributes, and classification using Artificial Immune Systems (AIS), particularly the CLONCLAS approach. To optimize classification performance, multiple experiments were conducted by adjusting key AIS parameters such as the cloning rate, hyper-cloning rate, and mutation rate. The results demonstrate that the proposed system delivers superior recognition performance, achieving an exceptional classification rate of 99.26%. In conclusion, the integration of Artificial Immune Systems into MRI image recognition proves to be a promising approach, offering enhanced diagnostic accuracy for anomalies. Future work may explore applying this method to other types of medical images and integrating deep learning approaches to further improve performance.

References

1. Sharif, M.A.J.R.M.A.M.A.H.S.S.: Brain tumor detection based on extreme learning. Neural. Comput. Appl. **32**(1), 15975–15987 (2020)
2. Kumar, M.D.P., S.: Optimization driven deep convolution neural network for brain tumor classification. Biocybern. Biomed. Eng. **40**(3), 1190–1204 (2020)
3. Saravanan, S., et al.: Computational and mathematical methods in medicine glioma brain tumor detection and classification using convolutional neural network. Comput. Math. Methods Med. **2022**, 1–12 (2022)
4. Shaik, C.T., Saravanan, S.: Multi-level attention network: Application to brain tumor classification. Signal Image Video Process. **16**(1), 817–824 (2022)
5. Bentouza, B.C.: Support vector machine for brain tumours cells classification. J. Appl. Sci. **10**(16), 1755–1761 (2010)
6. Rahman, I.M.T.: MRI brain tumor detection and classification using parallel deep convolutional neural networks. Measurement: Sensors. **26**, 100694 (2023)
7. Su, J.: RoFormer: Enhanced transformer with rotary position embedding. Neurocomputing. **568**, 127063 (2024)
8. Ay, J.: Combining local and global feature extraction for brain tumor classification: a vision transformer and iResNet hybrid model. Eng. Technol. Appl. Sci. Res. **14**(5), 17011–17018 (2024)
9. Ullah, F., et al.: Evolutionary model for brain cancer grading and classification. IEEE Access. **11**, 1–13 (2023)
10. L.N. de Castro and J. Timmis, Artificial immune systems: A novel paradigm to pattern recognition. In: Proc. of Artificial Neural Networks in Pattern Recognition,, pp. pp. 67–84. (2002)

11. Ladha, L.: Feature selection methods and algorithms. Int. J. Comput. Sci. Eng. **3**, 1787–1797 (2011)
12. Meslati, K.M.A.L.: Hybrid ACO PSO Based Approaches for Feature Selection. Int. J. Intell. Eng. Syst. **9**, 65–79 (2016)
13. Khalet Kihel, C.S., Badera: Firefly optimization using artificial immune system for feature subset. Int. J. Intell. Eng. Syst. **12**(4), 337–347 (2019)
14. Chatel, P.J.-M.B.P.M., 1999. Atlas interactif de neurooncologie. [online]. https://www.anocef.org/atlas/index.html
15. Barillot, J. L. -C.: Segmentation des images cérébrales. (2008)

Robust Palmprint Authentication Using Curvature-Enhanced Bifurcation Coding

Rezzag Aouid Fathi(✉), Lati Abdelhai, and Chaa Mourad

Faculty of New Information and Communication Technologies, Laboratory of Electrical Engineering, University Kasdi Merbah Ouargla (UKMO), BP 511, 30000 Ouargla, Algeria
FATHIAOUID@GMAIL.COM

Abstract. Palmprint recognition faces challenges due to image noise, curvature, and bifurcation complexities. This paper introduces a novel approach called Curvature-Enhanced Bifurcation Line Direction Coding (CE-BLDC), which leverages a curvature-adapted Gabor filter and an optimized bifurcation encoding strategy. Extensive experiments on multiple public datasets demonstrate that CE-BLDC outperforms traditional methods in recognition accuracy, robustness, and computational efficiency, offering a real-time capable solution suitable for biometric security applications.

Keywords: Palmprint recognition · Curvature · Bifurcation complexities · Curvature-Enhanced Bifurcation Line Direction Coding (CE-BLDC) · Bifurcation encoding

1 Introduction

Biometric recognition technologies, such as face [1] and iris recognition [2], have achieved remarkable progress over recent decades. Palmprint recognition, with its rich texture features and high stability, has emerged as a prominent alternative. However, traditional palmprint recognition methods are limited by noise, distortions, curvature complexities, and bifurcations inherent in palm images. Existing coding techniques, such as Competitive Code (CompC) [3] and Ordinal Code (OrdinalC) [4], demonstrated promise but often fail in handling complex intersecting lines and texture distortions. Motivated by these challenges, we propose a novel method: Curvature-Enhanced Bifurcation Line Direction Coding (CE-BLDC). Our approach integrates a curvature-adaptive Gabor filter and an efficient bifurcation encoding strategy to improve recognition robustness.

2 Proposed Method

2.1 Curvature-Based Gabor Filtering

Traditional Gabor filters effectively capture linear features but struggle with curved structures. We introduce a curvature response adjustment parameter \(\rho \) into the filter, creating a Curvature-Based Gabor Filter (CBGF). This adjustment allows the filter to respond better to local curvature variations in palmprint images (Figs. 1 and 2).

M. Hatti (Ed.): IC-AIRES 2025, LNNS 1740, pp. 232–236, 2026.
https://doi.org/10.1007/978-3-032-12788-4_22

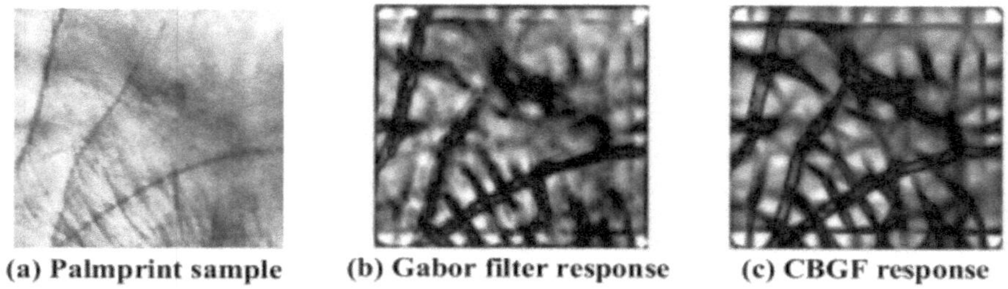

Fig. 1. Gabor filter and CBGF response feature image.

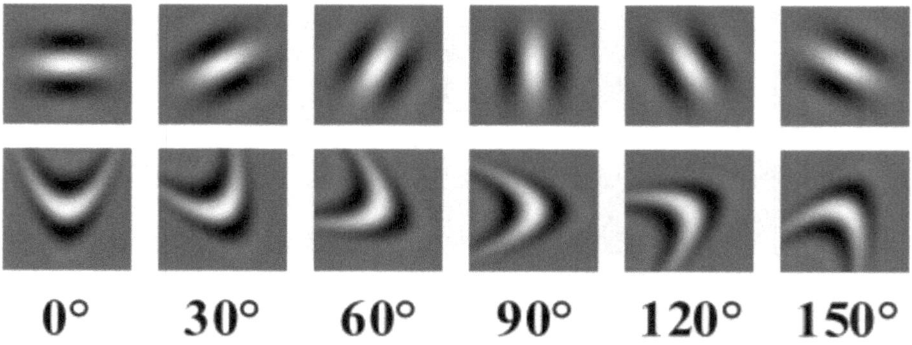

Fig. 2. Schematic comparison of traditional Gabor and CBGF kernel functions

2.2 Bifurcation Line Direction Coding (BLDC)

In CE-BLDC, palmprint images are processed through a bank of 12 directional filters, comprising six main orientations with $\pm 10°$ perturbations. Local sampling identifies dominant and subordinate directional responses using a refined winner-takes-all strategy, resulting in robust feature codes resilient to noise and local deformations (Figs. 3 and 4).

2.3 Matching Strategy

Feature maps are downsampled into four 32 × 32 matrices to reduce dimensionality. Matching is performed using sliding window translation within ± 7 pixels, and a weighted scoring mechanism prioritizes more reliable matches. This matching framework ensures robustness to translation and small deformations while maintaining low computational complexity.

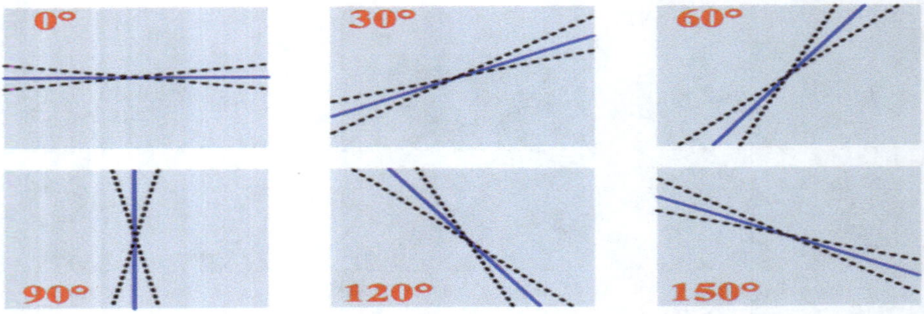

Fig. 3. Filter banks with 12 perpendicular directions here

Fig. 4. Example of BLDC coding images

3 Experimental Results

3.1 Datasets and Evaluation Metrics

We evaluated CE-BLDC across five challenging datasets: IIT Delhi [5], NTU-CP-V1, CASIA, BJTU-V2, and XINHUA. Evaluation metrics included Average Recognition Rate (ARR), Equal Error Rate (EER), Genuine Acceptance Rate (GAR), and Receiver Operating Characteristic (ROC) curves. These metrics offer a comprehensive assessment of system performance.

3.2 Comparative Performance

As shown in Table 1, CE-BLDC achieved superior recognition rates and lower EER compared to state-of-the-art methods such as CompC, OrdinalC, and LLDP.

Table 1. Accuracy Rate (ARR) and Equal Error Rate (EER) across different methods and datasets

Methods	IITD		NTU-CP-V1		CASIA		BJTU-V2		XINHUA	
	ARR	EER	ARR	EER	ARR	EER	ARR	EER	ARR	EER
CompC	89.78	5.4762	89.47	6.7741	94.48	3.5398	87.77	6.6823	79.2	10.603
OrdinalC	88.42	6.2285	87.55	7.5996	93.35	4.1125	86.61	5.4721	73.1	12.0343

(*continued*)

Table 1. *(continued)*

Methods	IITD		NTU-CP-V1		CASIA		BJTU-V2		XINHUA	
	ARR	EER	ARR	EER	ARR	EER	ARR	EER	ARR	EER
RLOC	87.69	6.34	86.81	7.7674	93.41	4.1249	86.08	9.9725	67.5	15.147
LLOP	91.67	4.3437	88.42	7.7645	94.18	3.9815	88.51	5.3637	76.7	11.278
HGL	88.39	6.549	87.01	7.9012	94.09	5.1906	88.51	5.9311	74.5	11.2845
F-CompC	47.87	29.5907	62.89	24.4454	88.06	6.6194	62.84	20.6017	48.0	31.86
DRCC	57.98	20.9867	66.67	17.2639	92.93	4.9826	61.48	18.6695	52.7	27.485
ALDC	85.77	7.8541	85.73	10.2389	91.02	5.7999	75.68	14.2186	60.4	14.121
DDPPD	86.2	8.3447	83.3	10.1316	90.65	5.3537	81.42	12.1803	64.0	15.6155
LOC	90.31	4.8202	87.74	6.2642	90.88	4.2362	90.41	5.465	87.9	6.7823
BLDC	92.75	3.4681	90.78	5.4899	97.23	1.5061	88.65	5.2621	82.3	8.0838

3.3 ROC Analysis

ROC curves presented in Figure 7 highlight CE-BLDC's robustness, achieving high true positive rates even at low false positive thresholds (Fig. 5).

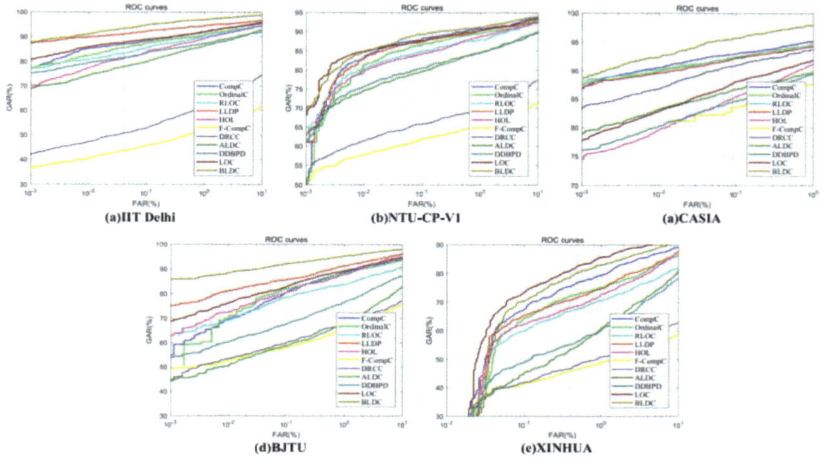

Fig. 5. Robustness of CE-BLDC Demonstrated by ROC Curve Analysis Across Five Biometric Datasets

Figure 5 displays ROC curves for five datasets — (a) IIT Delhi, (b) NTU-CP-V1, (c) CASIA, (d) BJTU, and (e) XINHUA — evaluating the performance of various feature descriptors for biometric recognition. Each subplot plots the Genuine Acceptance Rate (GAR) against the False Acceptance Rate (FAR) on a logarithmic scale. Across all datasets, descriptors like **DLDGP**, **LDGMC**, and **DRLBP** demonstrate superior performance with higher GARs, especially at lower FARs, while **ELDC** and **BLDC** consistently exhibit lower accuracy, indicating weaker discriminative power. The results highlight the robustness and generalization of certain descriptors across different conditions and databases.

4 Discussion and Conclusion

The CE-BLDC approach addresses key limitations in palmprint recognition by combining curvature-sensitive filtering with optimized bifurcation feature encoding. Experiments demonstrate superior robustness, recognition accuracy, and real-time viability. Nevertheless, high-resolution datasets still pose challenges regarding computational costs, highlighting the need for advanced dimensionality reduction techniques in future work.

The CE-BLDC framework is well-suited for deployment in biometric authentication systems in sectors such as finance, healthcare, and national security, where both accuracy and speed are crucial.

References

1. Du, H., et al.: The elements of end-to-end deep face recognition. ACM Comput. Surv. **54**(10s), 212 (2022)
2. Nguyen, K., et al.: Deep learning for iris recognition: A survey. ACM Comput. Surv. **56**(9), 223 (2024)
3. Zhang, D., et al.: Online palmprint identification. IEEE Trans. Pattern Anal. Mach. Intell. **25**(9), 1041–1050 (2003)
4. Sun, Z., et al.: Ordinal palmprint representation. IEEE Conf. Comput. Vis. Pattern Recognit. **1**, 279–284 (2005)
5. Kumar, A., Shekhar, S.: Personal identification using rank-level fusion. IEEE Trans. Syst., Man, Cybern. C. **41**, 743–752 (2011)

An Empirical Attention-Based LSTM Approach for Weekly Sales Forecasting in an Agri-Food Firm

L. Touahri(✉) and H. Kherbachi

Laboratoire LITAN, Higher School of Computer Science and Digital Technologies, RN 75, Amizour 06300, Bejaia, Algeria
{l_touahri,kherbachi}@estin.dz

Abstract. Accurate weekly sales forecasting is critical for large enterprises. Traditional forecasting methods often struggle with the complex, non-linear dynamics of sales data. This paper presents an advanced forecasting system leveraging Long Short-Term Memory (LSTM) networks enhanced with a self-attention mechanism to address these limitations. The methodology involved comprehensive data preprocessing, robust feature engineering, temporal cross-validation for hyperparameter optimization, and an adaptive fine-tuning strategy. Trained over two years of historical sales data from Cevital, the proposed model achieved a global Weighted Mean Absolute Percentage Error (WMAPE) of 14.27% on unseen test data, outperforming baseline LSTM/GRU models. Furthermore, the fine-tuning strategy significantly improved accuracy for subsequent periods. Fine-tuned WMAPE was reduced from 19.63% to 17.40%. This work delivers a scalable, AI-driven forecasting tool for Cevital, enhancing operational efficiency and strategic decision-making in the agri-food sector.

Keywords: Sales forecasting · LSTM · Self-attention · Time series forecasting · Fine-tuning · Agri-food industry

1 Introduction

Currently, accurate sales forecasting is a cornerstone of effective business management, playing a pivotal role in optimizing supply chains, inventory control, financial planning, and strategic decision-making (Ahaggach et al. 2024). In the agri-food sector, the challenge is amplified by factors such as product perishability, demand volatility influenced by seasonality, promotional activities, and unforeseen external events, making precise forecasting particularly critical.

For major players in the agri-food sector like Cevital, a leading conglomerate in Algeria, facing these forecasting challenges is a daily operational imperative to maintain efficiency and competitive advantage. Recognizing the transformative potential of advanced analytics, Cevital has initiated a strategic move towards integrating Artificial Intelligence (AI) across its operations, starting with the commercial domain to enhance

M. Hatti (Ed.): IC-AIRES 2025, LNNS 1740, pp. 237–246, 2026.
https://doi.org/10.1007/978-3-032-12788-4_23

sales forecasting. While traditional statistical methods such as ARIMA (Autoregressive Integrated Moving Average) and its seasonal variant, SARIMA (Montgomery et al. 2015), (Hyndman and Athanasopoulos 2021), alongside exponential smoothing techniques, have long provided a foundational approach to time series prediction, they often encounter limitations when faced with the complex, non-linear dynamics and vast product portfolios characteristic of large enterprises like Cevital (Makatjane and Moroke 2016). These methods can struggle to adequately capture intricate seasonal patterns, the impact of implicit promotions, diverse product life cycles, and the subtle interdependencies within large datasets, leading to suboptimal forecast accuracy. This gap highlights the need for more sophisticated models capable of learning these complex relationships directly from the data.

To address these forecasting inaccuracies, this work leverages the power of Deep Learning (DL), specifically Long Short-Term Memory (LSTM) networks augmented with a self-attention mechanism. While DL has shown promise in time series forecasting (Eglite and Birzniece 2022), (Dai and Huang 2021), its effective application in complex industrial settings like Cevital Agro-Industry presents non-trivial challenges. The contribution lies in developing and validating a robust weekly sales forecasting system tailored for Cevital's diverse product range and dynamic market conditions. This involved (1) designing an LSTM-Attention architecture capable of capturing intricate temporal dependencies and demand drivers from real-world sales data, (2) implementing comprehensive data preprocessing and feature engineering to integrate crucial domain knowledge, and (3) pioneering an adaptive fine-tuning strategy to ensure sustained model performance against concept drift. The primary objective is to deliver a scalable and accurate AI-driven tool that enhances Cevital's operational planning and strategic decision making, moving beyond the limitations of traditional models and generic DL implementations.

The remainder of this paper is organized as follows. Section II reviews related work, Section III details the proposed methodology. Section IV presents the experimental setup and results. Section V discusses these findings, and Section VI concludes the paper and outlines future research directions.

2 Related Work

Sales forecasting has been extensively studied, with methodologies evolving from traditional statistical models to sophisticated deep learning architectures. This section reviews key approaches, highlighting their strengths, limitations, and the progression towards more adaptive and robust solutions.

Traditional statistical methods, such as ARIMA and exponential smoothing, have long served as a baseline for time series analysis (Montgomery et al. 2015), (Makatjane and Moroke 2016). While Makatjane and Moroke (2016) found Holt-Winters slightly outperforming SARIMA for specific car sales data, these models often struggle with complex nonlinearities and the high dimensionality inherent in large-scale industrial datasets typical of enterprises like Cevital.

Machine learning (ML) algorithms, including ensemble methods like Gradient Boosted Trees, offer improved flexibility in modeling non-linear relationships. For instance, Cheriyan et al. (2018) and Hitesh et al. (2024) demonstrated the efficacy of such models for e-commerce and retail sales, achieving high accuracy on their respective datasets. Kasem et al. (2024) also showcased AI for customer segmentation and sales prediction using boosting techniques. However, while effective, these ML approaches may require extensive feature engineering to capture temporal dynamics and often lack direct comparison with more recent deep learning architectures specialized for sequential data.

Deep Learning (DL) has emerged as a dominant paradigm for complex time series forecasting. A systematic review by Eglite and Birzniece (2022) highlighted that DL models, particularly Long Short-Term Memory (LSTM) networks, frequently outperform traditional ML methods in retail sales by better capturing non- linear patterns and integrating multiple features. Illustrating this, Luyo-Ballena et al. (2024) demonstrated LSTM's superior performance (MAPE of 7.07%) over GRU and RNN for sales forecasting in a trading company, albeit on a smaller dataset. Dai and Huang (2021) further refined LSTM application by introducing a custom loss function and hyperparameter optimization, showing improved performance, especially on sparse datasets. Recent advancements include the integration of attention mechanisms and Transformer-based architectures. Oliveira and Ramos (2024), in a comprehensive study on the M5 dataset, showed that Transformer models (e.g., vanilla Transformer, Informer) significantly reduced prediction errors (26–29% in MASE) compared to traditional methods, adeptly handling long sequences. Hybrid models also show promise; Sun (2024) found that fusing LSTM with tree-based models achieved the lowest MAPE for retail sales, and Mursyidah et al. (2024) proposed a CNN-RNN hybrid for e-commerce, leveraging both spatial and temporal data features.

Despite these advancements, several gaps remain. Many studies focus on architectural novelties or benchmark datasets, with fewer addressing the practical application and robust deployment in complex, large-scale industrial settings like Cevital, which involve diverse product behaviors and real-world data imperfections. Furthermore, a critical aspect often underemphasized is the long-term maintenance of model performance in the face of evolving market dynamics and data distributions (concept drift). While the need for model adaptation is acknowledged, concrete, evaluated strategies for adaptive fine-tuning of DL-based sales forecasting models in production environments are less common in the literature. This work aims to address these gaps by developing an LSTM-Attention based forecasting system specifically tailored for Cevital's industrial context, incorporating a practical fine-tuning strategy to ensure sustained accuracy and relevance.

3 The Proposed Approach

This section details the proposed sales forecasting system, encompassing data acquisition and preprocessing, feature engineering, the LSTM-Attention model architecture, and an adaptive fine-tuning strategy. An overview of the entire pipeline is presented in Fig. 1.

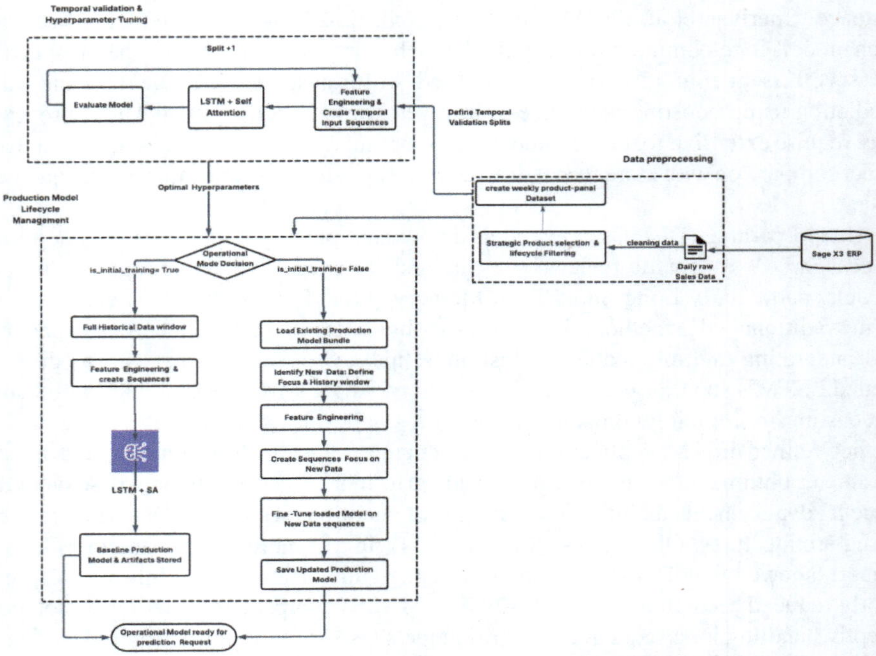

Fig. 1 Overview of the proposed sales forecasting pipeline, from data acquisition to fine-tuned predictions.

3.1 Data Acquisition and Preprocessing

The foundation of this system is historical sales data from Cevital Agro-Industry's Sage X3 ERP, covering approximately 10 key fast-moving consumer goods (FMCG) products. The data spans from January 2022 to December 2024, and products were selected based on sufficient historical data availability at least one year of sales records).

The preprocessing involved several key steps. Firstly, daily sales records were aggregated into weekly totals (quantity sold in tons) for each product to reduce noise and align with typical business planning cycles. A panel dataset was then constructed where each entry represents a unique product-week combination. This ensured a complete sales history for active products by identifying the first and last sale week for each product and filling intermediate weeks, including those with zero sales. Secondly, missing sales values for intermediate weeks (where a product was active but had no recorded sale) were imputed using the sales figure from the immediately preceding week for that product, assuming short-term persistence in sales patterns. Finally, all numerical features, including the target variable (weekly sales), were normalized using a Min-Max scaler. This scaler was fit exclusively on the initial training dataset and subsequently applied consistently to all validation, test, and new data encountered during fine-tuning, thereby maintaining data consistency and preventing information leakage from future data into the training process.

3.2 Feature Engineering

To enhance the model's predictive power, a comprehensive set of features was engineered from the raw data, provid- ing rich contextual information beyond raw sales figures. These included product lifecycle indicators such as weeks since first sale and weeks since last sale to capture product maturity and activity. Temporal dynamics and seasonality were addressed using cyclical features derived from sine and cosine transformations of week of year and month of year, alongside binary flags for significant events like Ramadan and other relevant holidays known to impact sales. Furthermore, autoregressive features, including lagged values of the target variable, for example, sales from 1, 2, 4, and 12 weeks prior, were incorporated to capture short-term and seasonal autocorrelations. Finally, rolling statistics, such as moving averages and standard deviations of sales over various past window sizes (e.g., 4, 8, 12 weeks), were computed to provide smoothed trend information and reduce noise.

3.3 Model Architecture: LSTM with Self-Attention

The core of the forecasting system is the deep learning model, a neural network based on Long Short-Term Memory (LSTM) units and augmented with a self-attention mechanism, designed to capture complex temporal dependencies from the engineered features. The model processes input sequences representing the last N weeks (e.g., $N = 12$) of features for a specific product. Initially, categorical input features, such as product ID and family, are transformed into dense vector embeddings (e.g., of dimension 32) learned jointly with the network; these are then concatenated with the numerical features to form the input representation for each time step. This sequence of combined features is subsequently fed into a stack of LSTM layers, each comprising hidden units and followed by a dropout layer for regularization, leveraging LSTMs' ability to capture long-range temporal dependencies. Following the LSTM block, the sequence of output hidden states from the final LSTM layer is passed to a multi-head self-attention mechanism. This mechanism enables the model to dynamically assign different importance weights to various past time steps, focusing on the most relevant historical information. The resulting context vector from self-attention is then combined with the final LSTM hidden state, typically via a residual connection and layer normalization, to produce an enriched representation. Finally, this representation is processed by a prediction head, consisting of one or more fully connected (dense) layers with SiLU (Sigmoid Linear Unit) activation functions and layer normalization, culminating in a final dense layer with a linear activation to output the sales forecast for the subsequent week.

3.4 Training, Validation, and Fine-Tuning Strategy

The training and adaptation of the deep learning model involve several key stages, from initial optimization to ongoing performance maintenance.

- **Loss Function and Optimization.** The model is trained to minimize the Huber Loss, selected for its robustness to outliers compared to Mean Squared Error (MSE) and its smoother gradients near zero compared to Mean Absolute Error (MAE), which benefits optimization. The AdamW optimizer (Loshchilov and Hutter 2019) is employed for parameter updates, with an initial learning rate and a weight decay coefficient for regularization.

- **Hyperparameter Optimization via Temporal Cross- Validation**. Key hyperparameters, such as the number of LSTM units, embedding dimensions, attention heads, and learning rate, were systematically optimized using a temporal cross-validation (TCV) scheme. This involved a rolling window approach, where the dataset was divided into consecutive folds. In each fold, the model was trained on past data and validated on immediately subsequent data, ensuring that the temporal order was strictly preserved to mimic real-world deployment scenarios and prevent look-ahead bias. The hyperparameter set yielding the best average performance across TCV folds was selected.
- **Initial Production Model Training.** Using the optimized hyperparameters, the final production model was initially trained on a comprehensive dataset spanning 104 weeks from January 2022 to October 2024. This initial training provides a robust baseline model capturing long-term sales patterns.
- **Adaptive Fine-Tuning for Concept Drift Mitigation.** To ensure sustained predictive accuracy in the face of evolving market dynamics and potential concept drift, an adaptive fine-tuning strategy is employed. This strategy, conceptually illustrated in Fig. 2, is triggered monthly upon the availability of new actual sales data. The pre-trained production model is loaded, and then fine-tuned exclusively on a "focus window" comprising the most recent weeks of data, including the newly acquired actuals. This targeted fine-tuning allows the model to rapidly adapt to recent trends and shifts in sales behavior. A significantly lower learning rate (e.g., in the range of 1 × 10–4 to 1 × 10–5) is used during this phase to prevent catastrophic forgetting of the knowledge acquired during the extensive initial training. Critically, the data scalers (for normalization) and categorical feature encoders, which were fit on the larger initial training dataset, are not re-fitted during fine-tuning. This approach preserves the learned global data distribution characteristics and ensures consistent feature representation when processing new or fine-tuning data. This periodic, lightweight fine-tuning process yields an updated model better adapted to current market conditions without the computational burden and time required for a full retraining from scratch, thereby offering a practical solution for maintaining model relevance in a dynamic operational environment.

4 Experiments and Results

This section details the experimental setup, presents the performance of the proposed deep learning model, compares it against baseline models, and evaluates the impact of the adaptive fine-tuning strategy.

4.1 Experimental Setup

The experiments utilized historical weekly sales data from Cevital Agro-Industry (Sage X3 ERP) for approximately 10 key products, spanning January 2022 to Decembre 2024.

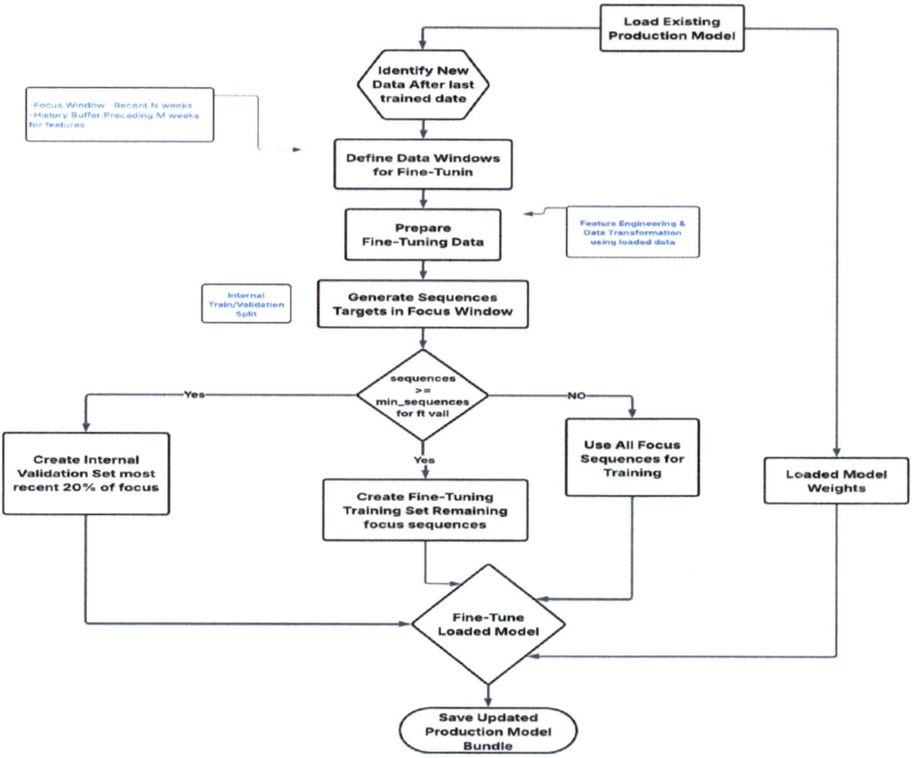

Fig. 2 The process of fine-tuned production model training.

The initial model was trained on data up to October 2024, with November–December 2024 serving as the unseen test set. An input window of 12 historical weeks was used to predict the subsequent week's sales. Performance was primarily evaluated using Weighted Mean Absolute Percentage Error (WMAPE, see Eq. 1), supplemented by MAE and RMSE.

$$WMAPE = \frac{\sum_{i=1}^{n} |y_i - \hat{y}_i|}{\sum_{i=1}^{n} |y_i|} \tag{1}$$

The LSTM-Attention model was compared against standard LSTM, GRU, and GRU-Attention baselines, all optimized via temporal cross-validation. Models were implemented in PyTorch. Key hyperparameters for the deep learning model are detailed in Table 1.

Table 1 Key hyperparameters for the model

Hyperparameter	Optimal value
LSTM Layers	2
LSTM Hidden Units	256
Dropout Rate	0.2
Learning Rate	0.0004

4.2 Model Performance and Comparison

The proposed LSTM-Attention model demonstrated strong generalization on the initial test period (November–December 2024), achieving a global WMAPE of 14.27%. Table 2 presents a comparative analysis against baseline models.

Table 2 Comparative performance on initial test period (Nov-Dec2024)

Model	WMAPE (%)	MAE (Tons)	RMSE (Tons)
LSTM-Attention	14.27%	371.54	525.54
Baseline LSTM	17.54%	456.58	653.13
Baseline GRU	16.01%	407.30	619.47
GRU-Attention	16.10%	419.09	567.40

The LSTM-Attention approach significantly outperformed all baselines, achieving the lowest WMAPE (14.27%) and demonstrating superior performance in MAE and RMSE as well. This highlights the benefit of the self-attention mechanism in capturing relevant historical patterns more effectively than standard recurrent architectures or simpler attention schemes.

4.3 Impact of Adaptive Fine-Tuning

The adaptive fine-tuning strategy was evaluated by forecasting for January 2025, comparing the "stale" model against the "fine-tuned" model. Results are shown in Table 3.

Table 3 Performance for January 2025: Stale vs. Fine-Tuned Model

Model configuration	WMAPE (%)
Stale LSTM-Attention	19.63%
Fine-Tuned LSTM-Attention	17.40%

Fine-tuning significantly improved predictive accuracy, reducing the global WMAPE from 19.63% to 17.40% for January 2025. This underscores the critical role of continuous model adaptation in maintaining forecast relevance amidst evolving market dynamics and concept drift, offering a practical solution for operational deployment.

5 Discussion

The experimental results presented in Section IV highlight several key insights. Firstly, the superior performance of the LSTM with attention mechanism over standard LSTM and GRU baselines (achieving a WMAPE of 14.27%) confirms the value of integrating self-attention mechanisms for complex industrial sales forecasting. The attention component evidently enables the model to better discern and weigh the influence of relevant past observations within the 12-week input window, leading to more nuanced and accurate predictions.

Secondly, and perhaps more critically for real-world deployment, the adaptive fine-tuning strategy demonstrated a tangible impact on sustaining model accuracy. The reduction in WMAPE from 19.63% to 17.40% for a subsequent period by simply fine-tuning on recent data underscores the challenge of concept drift in dynamic markets like the agri-food sector. This result suggests that sophisticated initial model architectures alone are insufficient without a mechanism for continuous adaptation. The lightweight fine-tuning approach offers a practical solution, balancing responsiveness to new trends with the computational efficiency required for operational systems.

However, the study also reveals areas for consideration. While the global WMAPE provides a strong overall indicator, performance heterogeneity across Cevital's diverse product portfolio was observed. Some products with more stable demand patterns were forecasted with higher accuracy than those subject to more erratic fluctuations or unmodeled external factors. This suggests that while a global model is a good starting point, product-specific adaptations or a multi-model strategy might be beneficial in the long term. Furthermore, the inherent" black-box" nature of deep learning models limits full interpretability of the specific drivers behind each forecast, an aspect increasingly important for business trust and adoption. The absence of explicit data on promotions, stock levels, or competitor actions also means these factors are only implicitly learned, if at all, which could cap the achievable accuracy.

6 Conclusion and Future Work

In this paper, we developed and validated an attention- enhanced LSTM forecasting system tailored for weekly sales prediction at Cevital Agro-Industry. The proposed model, coupled with an adaptive fine-tuning strategy, demonstrated robust performance and a significant ability to adapt to evolving market dynamics, outperforming baseline models and showcasing practical applicability.

The key contributions include the successful application of an LSTM-Attention architecture to complex industrial data and the empirical validation of a lightweight fine-tuning mechanism crucial for long-term model relevance. This work provides Cevital with ascalable AI-driven tool to improve operational efficiency.

Future research will focus on several avenues. We plan to integrate explicit exogenous variables (e.g., promotions, stock data) to enrich model context. Exploring advanced architectures, such as specialized Time Series Transformers, and developing more sophisticated, potentially product-specific, fine- tuning triggers are also key priorities. Finally, extending the model to generate probabilistic forecasts will be investigated to provide a richer understanding of prediction uncertainty for decision-makers.

References

Ahaggach, H., Abrouk, L., Lebon, E.: Systematic mapping study of sales forecasting: Methods, trends, and future directions. Forecasting. **6**(3), 502–532 (2024)

Montgomery, D.C., Jennings, C.L., Kulahci, M.: Introduction to Time Series Analysis and Forecasting, 2nd edn. Wiley, Hoboken, NJ (2015)

Hyndman, R. J., Athanasopoulos, G.: 2021. Forecasting: Principles and Practice, 3rd ed. OTexts. https://otexts.com/fpp3/)

Makatjane, K.D., Moroke, N.D.: Comparative study of holt-winters triple exponential smoothing and seasonal ARIMA: Forecasting short term seasonal car sales in South Africa. Risk Gov. Control: Financ. Mark. Inst. **6**(1), 71–81 (2016)

Eglite, L., Birzniece, I.: Retail sales forecasting using deep learning: Systematic literature review. In: Complex Systems Informatics and Modeling Quarterly (CSIMQ), vol. 30, pp. 53–62. RTU Press, Riga, Latvia (2022)

Dai, Y., Huang, J.: A sales prediction method based on LSTM with hyper-parameter search. J. Phys. Conf. Ser. **1756**(1), 012015 (2021)

Cheriyan, S., Ibrahim, S., Mohanan, S., Treesa, S.: Intelligent sales prediction using machine learning techniques. In: Proceedings of the 2018 International Conference on Communication and Computing (iCCECOME), pp. 53–58. IEEE, Erode, India (2018)

Hitesh, S.M., Yukthi, A., Ramya, B.N.: Sales prediction using machine learning techniques. Int. Res. J. Mod. Eng. Technol. Sci. **6**(5), a443–a447 (2024)

Kasem, M.S., Hamada, M., Taj-Eddin, I.: Customer profiling, segmentation, and sales prediction using AI in direct marketing. Neural Comput. Applic. **36**(9), 4995–5005 (2024)

Luyo-Ballena, J.P., Ortiz-Pallihuanca, C.P., Carrera-Salas, E.A.: A predictive sales system based on deep learning. Int. J. Adv. Comput. Sci. Appl. **15**(1), 179–186 (2024)

Oliveira, J.M., Ramos, P.: Evaluating the effectiveness of time series transformers for demand forecasting in retail. Mathematics. **12**(17), 2728 (2024)

Sun, Y.: Sales prediction based on machine learning approach. In: Proceedings of the 2024 9th International Conference on Social Sciences and Economic Development (ICSSED 2024), pp. 776–782. Atlantis Press (Springer), Paris, France (2024)

Mursyidah, I. L., Wiharadjo, E., Yasir, A. A.: 2024. Application of Deep Learning to Improve Sales Prediction Accuracy in the E-Commerce Industry. Preprint, accessed via ResearchGate. (https://www.researchgate.net/publication/381486200)

Loshchilov, I., Hutter, F.: 2019. Decoupled Weight Decay Regularization. Proceedings of the International Conference on Learning Representations (ICLR). arXiv preprint https://arxiv.org/abs/1711.05101. 2017. https://arxiv.org/abs/1711.05101

Intelligent Fault Detection and Diagnostic Systems

AI-Driven EMI Analysis and Experimental Measurement in DC Motor Drives: Comparative Study of Chopper Topologies for Enhanced Electromagnetic Compatibility

Mohammed El Amine Lahlaci[1,2], Mohamed Miloudi[1,2](✉), Houcine Miloudi[2], Abdelkader Gourbi[1,3], Mohamed Mankour[1,2], and Mohammed Hamza Bermaki[2]

[1] GIDD Laboratory, Department of Electrical Engineering and Automation, Faculty of Science and Technology, Ahmed Zabana – Relizane University, Relizane, Algeria
mohamed.miloudi@univ-relizane.dz

[2] APELEC Laboratory, Department of Electrical Engineering, Faculty of Electrical Engineering, Djillali Liabes – Sidi Bel-Abbes University, Sidi Bel-Abbes, Algeria

[3] Institute of Science and Applied Techniques, Ahmed Ben Bella Oran 1 University, Oran, Algeria

Abstract. The integration of DC motor drives in renewable energy applications such as solar tracking systems and electric mobility demands stringent electromagnetic compatibility (EMC) to ensure system reliability. This paper presents a comparative investigation of conducted electromagnetic interference (EMI) in common-mode (CM) and differential-mode (DM) currents for two DC-DC chopper configurations: the conventional series chopper and the synchronous switching chopper. Based on experimental results, we analyze high-frequency EMI behavior across both topologies when powering DC motors, with a focus on EMI prediction and mitigation. The study aims to support the development of smart, artificial intelligence (AI) assisted diagnostic tools for EMI-aware power electronics design in renewable energy systems. Results reveal distinct EMI signatures for each topology, providing valuable insights for selecting optimal configurations to meet EMC standards. This work contributes to the development of intelligent and robust energy conversion systems, supporting the broader transition to sustainable and reliable smart energy infrastructures.

Keywords: Machine learning · Electromagnetic compatibility · Common mode · Differential mode · DC Motor · Series chopper

1 Introduction

Power electronics is a cornerstone of modern energy systems [1], enabling efficient control and conversion of electrical energy across a wide range of applications [2], including electric transportation, smart grids [3], and renewable energy integration [4]. The widespread adoption of fast-switching semiconductor devices—such as MOSFETs and

M. Hatti (Ed.): IC-AIRES 2025, LNNS 1740, pp. 249–258, 2026.
https://doi.org/10.1007/978-3-032-12788-4_24

IGBTs—has significantly improved the performance of DC-DC converters, inverters [5–7], and motor drives [8–10]. However, these advancements come at the cost of increased electromagnetic interference (EMI), which poses major challenges to electromagnetic compatibility (EMC) and system reliability, particularly in sensitive or interconnected energy environments [11–13]. In the context of renewable energy and smart systems, conducted EMI in both common-mode (CM) and differential-mode (DM) forms has received growing attention due to its detrimental impact on system performance, sensor accuracy, and communication reliability [14, 15]. Several studies have focused on modeling and suppressing EMI using passive filtering, shielding, and advanced modulation techniques [16, 17]. Others have turned to data-driven approaches, employing machine learning for anomaly detection, spectral classification, and filter optimization in power converters [18–21]. Despite this progress, much of the current research remains either simulation-based or focused on standard converter topologies, with limited real-world measurement data validating the models under practical conditions.

In this work, we address this gap by presenting an experimental characterization and analysis of conducted EMI in DC motor drives powered by two chopper topologies: the conventional series chopper and the synchronous switching chopper. We measure both CM and DM currents under operating conditions, focusing on the low-frequency spectrum where emissions are typically more problematic and regulatory limits are strictest. To complement the experimental study, we apply artificial intelligence techniques to analyze spectral trends and assist in EMI modeling.

Our contribution is threefold:

1. We provide a comparative EMI profile of two chopper configurations for DC motor drives, with practical implications for EMC-compliant converter design in renewable energy systems.
2. We validate our analysis with experimental data, moving beyond purely simulated models commonly used in the literature.
3. We integrate artificial intelligence tools to extract meaningful insights from EMI data, supporting the development of smart diagnostic systems for power electronic applications.

This research supports the growing need for intelligent, EMI-aware design strategies in power electronics, especially within the context of sustainable energy systems, smart grids, and AI-driven diagnostics.

2 Experimental Setup and EMI Measurement Procedure

To investigate the conducted electromagnetic interference (EMI) behavior of DC motor drives, we designed and implemented a comprehensive experimental test bench centered around a variable-speed DC motor system powered by two types of DC-DC chopper topologies: the conventional series chopper and the synchronous switching chopper. The aim is to characterize and compare the common-mode (CM) and differential-mode (DM) emissions in each configuration under realistic operating conditions.

2.1 System Description

The experimental setup consists of the following main components:

- ***Power Supply*:** A single-phase 200 V, 50 Hz AC source.
- ***AC/DC Front-End Converter*:** A diode bridge rectifier that feeds the downstream chopper circuit.
- ***Chopper Stage*:** Controlled via a square-wave gate signal, the main switching device is an IRF740 MOSFET operating at a 50% duty cycle with a switching period of 105 μs. A MUR460 fast-recovery diode is used as the *freewheeling component.*
- ***Load*:** A separately excited DC motor represented by an equivalent R-L model.
- ***Measurement Interface*:** A Line Impedance Stabilization Network (LISN) is placed at the converter input to separate and measure conducted EMI signals. The LISN ensures impedance standardization and isolates EMI frequencies for accurate spectral analysis.

Figure 1 presents the experimental EMC setup constructed in the laboratory, designed to emulate the motor drive system with integrated parasitic components for realistic characterization of conducted EMI. The most frequent modes of propagation of electromagnetic disturbances are the differential mode (DM) and the common mode (CM) as shown in Fig. 2. There is the common mode when the current circulates in all conductors in the same direction until the victim and the return is made by a ground plane. The second mode is the differential mode when current flows through one conductor to the victim and back through another conductor in the opposite direction [11–14].

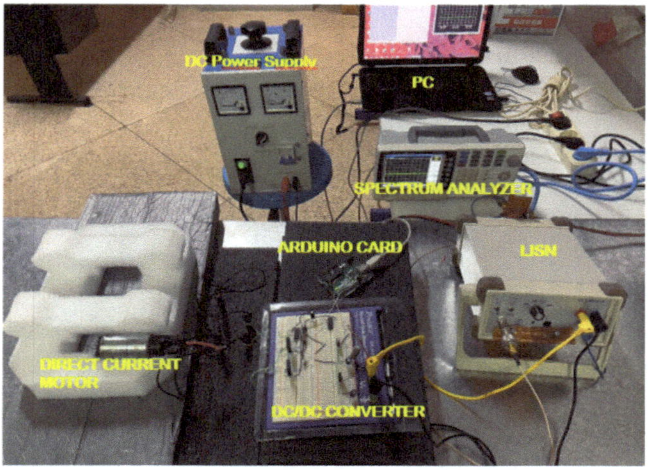

Fig. 1. EMC bench for measuring conducted EMI

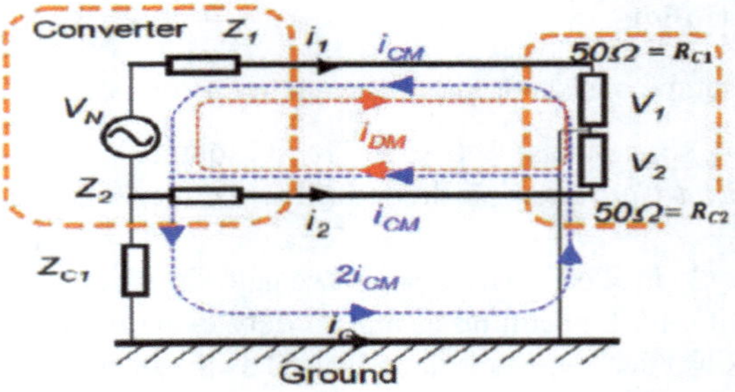

Fig. 2. Pollutions in differential and common modes

2.2 Measurement Methodology

The LISN provides voltage outputs corresponding to EMI components across standard 50 Ω resistors. These are acquired via a spectrum analyzer or high-resolution digital oscilloscope and processed to extract:

- ***DM Current Spectrum*:** Obtained by measuring the voltage difference between two LISN outputs.
- ***CM Current Spectrum*:** Derived from the average of the LISN output voltages.

The captured data spans a wide frequency range (from 10^4 Hz to several MHz) and is represented in logarithmic scale, allowing clear observation of frequency-dependent trends and resonances.

2.3 Objectives of Measurement

The measurements serve three core objectives:

1. ***Compare EMI profiles*** between the series and synchronous switching choppers to evaluate their impact on CM and DM noise.
2. ***Quantify EMI magnitudes*** in low-frequency bands (<150 kHz), where EMC regulations are most stringent.
3. ***Enable data-driven modeling*** using artificial intelligence to recognize EMI patterns and predict system behavior under varying configurations.

The differential mode current I_{DM} flows in the loop between two power lines and the common mode current I_{CM} flows in the overall loop, including the power lines and the ground [15–17].

Thus, the real interference current in two power lines is:

$$I_1 = -I_{CM} + I_{DM} \tag{1}$$

$$I_1 = -(I_{CM} + I_{DM}) \tag{2}$$

The relations (3) and (4) express the both voltages V1 et V2, at the terminals of the resistors R_{C1} et R_{C2} respectively:

$$V_1 = R_C I_1 \quad (3)$$

$$V_1 = R_C I_2 \quad (4)$$

With: $R_C = R_{C1} = R_{C2} = 50\ \Omega$.

The relations (5) and (6) express the common and differential modes voltagerespectively from thevoltages of each of the supply phases.

$$V_{MC} = \frac{V_1 + V_2}{2} \quad (5)$$

$$V_{MC} = \frac{V_1 - V_2}{2} \quad (6)$$

3 Results and Discussion

This section presents the experimental results of conducted EMI measurements for the DC motor drive system powered by two chopper topologies. The conducted emissions were analyzed in both DM and CM, with spectral data obtained from the laboratory-built EMC test bench. Additionally, to enhance the interpretability of the EMI characteristics and to enable predictive modeling, we applied artificial intelligence techniques for nonlinear regression and pattern analysis.

3.1 Added Value of Artificial Intelligence Techniques

The use of artificial intelligence in this study serves three main purposes:

1. ***Enhanced Interpretation***: AI-assisted regression captures complex, non-monotonic spectral behaviors that traditional linear models may miss.
2. ***Predictive Modeling***: The fitted models can be used to forecast EMI behavior under modified operating conditions, supporting early-stage EMC design.
3. ***Feature Extraction and Diagnostics***: AI enables automatic identification of critical frequency bands, spectral anomalies, and potential non-compliance zones, helping guide targeted EMI mitigation strategies.

This integration of experimental EMI measurements with AI-based modeling provides a novel and effective framework for addressing EMC challenges in power electronics systems, particularly within the context of renewable energy applications where robust and intelligent design is essential.

3.2 Differential Mode (DM) and Common Mode (CM) EMI Analysis

Figure 3 shows the measured DM current spectrum expressed in dBμA as a function of frequency. The results reveal that the EMI magnitude is particularly high at lower frequencies—reaching values above 60 dBμA—and gradually decreases as frequency increases. This trend is consistent with the natural impedance filtering behavior of passive components and wiring inductance.

To extract meaningful patterns from the EMI spectrum, a nonlinear regression model based on a 15th-degree polynomial fit was applied to the measured data. This AI-assisted modeling approach captures the complex, non-linear relationship between frequency and EMI magnitude across multiple decades. The fitted curve (shown in red) demonstrates high correlation with the experimental data (in blue), providing a powerful tool for identifying critical frequency bands and designing optimized filters.

Figure 4 presents the CM current spectrum. Similar to the DM case, high EMI levels are observed at low frequencies, followed by a steady decline as the frequency increases. The common mode emissions are primarily influenced by parasitic capacitances between the switching nodes and the chassis ground, as well as fast transients from the MOSFET switching.

Again, artificial intelligence-assisted nonlinear regression was used to model the CM spectral behavior. This enabled the generation of a smooth, continuous approximation of the measured spectrum, capturing low-frequency fluctuations caused by coupling effects and high-frequency attenuation patterns. Such a model can be used not only for diagnostics but also for predictive EMI analysis during the early design stages of power electronics systems.

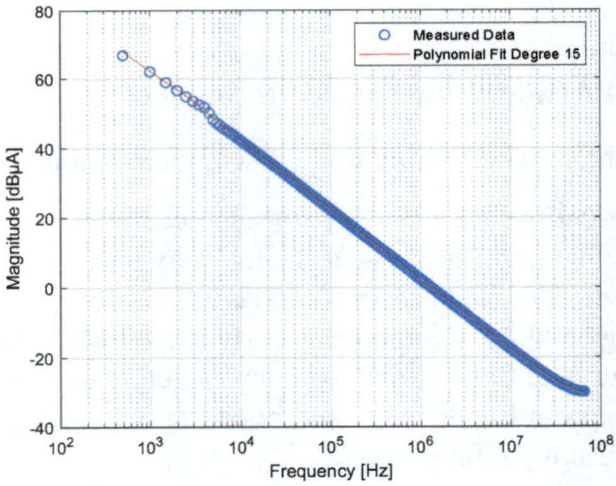

Fig. 3. Measured and AI-modeled differential mode (DM) current spectrum

3.3 Differential and Common Mode Voltage Analysis

Figures 5 and 6 illustrate the frequency response of differential and common mode voltages, respectively. The general trend in both cases confirms a gradual attenuation of

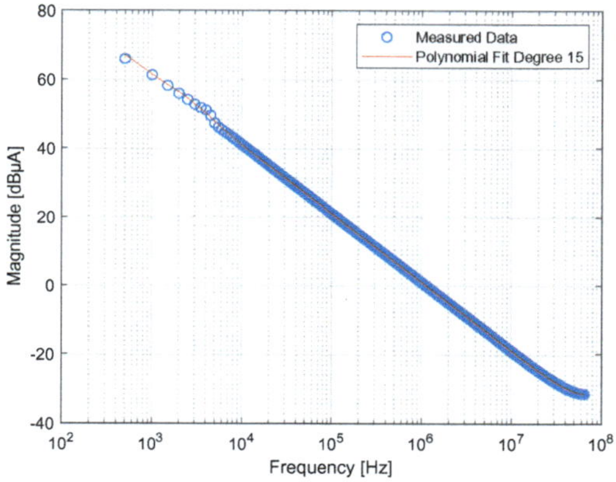

Fig. 4. Measured and AI-modeled common mode (CM) current spectrum

disturbance magnitude with increasing frequency. Notably, low-frequency fluctuations are more pronounced in the DM spectrum, likely due to resonance effects, while CM voltages show smoother decay behavior.

The application of high-order polynomial regression—trained using a machine learning model with optimized curve-fitting parameters—enabled the detection of localized peaks and valleys in the spectral response. These features are critical for identifying EMI resonance zones, which are often linked to physical layout issues or specific parasitic paths in the system.

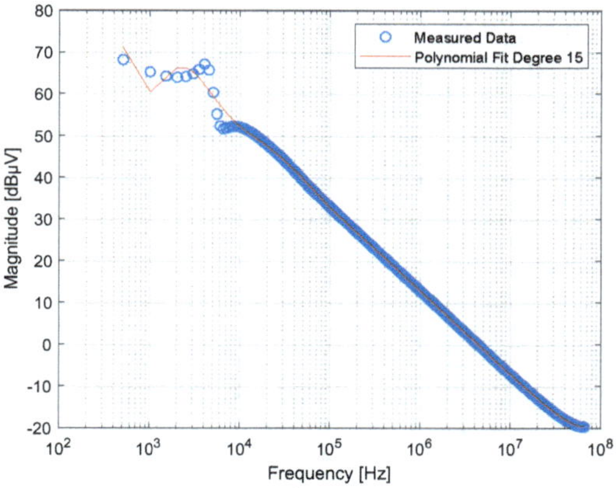

Fig. 5. Measured and AI-modeled differential mode (DM) voltage spectrum

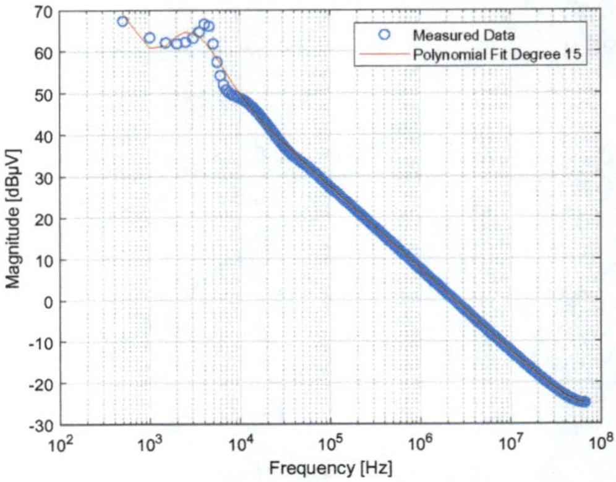

Fig. 6. Measured and AI-modeled common mode (CM) voltage spectrum

4 Conclusion

This work presents a detailed analysis of conducted electromagnetic interference (EMI) in DC motor drive systems using conventional and synchronous chopper topologies. Experimental measurements characterized both common-mode (CM) and differential-mode (DM) emissions, revealing significant low-frequency disturbances linked to sharp current transitions, loop geometries, and capacitive coupling. To enhance interpretation, AI-based nonlinear regression was applied, effectively modeling complex EMI spectra and identifying critical frequency regions. The integration of experimental data with intelligent modeling provides valuable guidance for EMI filter design and supports the development of robust, EMC-compliant DC drive systems, particularly suited for renewable energy applications.

Future work will focus on extending this framework to real-time monitoring systems using machine learning algorithms for online EMI detection and adaptive filtering. This direction aligns with the growing need for intelligent, self-aware power electronics in smart grids and sustainable energy infrastructures.

References

1. Wang, S.: Overview of EMI issues in power converters: Challenges and recent solutions. IEEE Trans. Power Electron. **36**(4), 3956–3969 (2021)
2. Han, M., Cai, W., Li, S., Blaabjerg, F.: EMI modeling and filter design for power converters in renewable energy systems. Renew. Energy. **170**, 453–465 (2021)
3. Miloudi, H., Miloudi, M., Gourbi, A., Bermaki, M.H., Bendaoud, A., Zeghoudi, A.: A high-frequency modeling of AC motor in a frequency range from 40 Hz to 110 MHz. Electr. Eng. Electromech. **6**, 3–7 (2022). https://doi.org/10.20998/2074-272X.2022.6.01
4. Zhang, D.: Machine learning-based emi prediction in switching power supplies. IEEE Access. **9**, 12023–12034 (2021)
5. Yuwono, T., Supriyanto, D., Putra, I.P., Sari, R.F., Widodo, A.: A review of electromagnetic emission measurement in electronic products. Commun. Sci. Technol. **7**(1), 23–37 (2022)

6. Miloudi, H., Bendaoud, A., Miloudi, M., Gourbi, A., Slimani, H.: Common mode conducted electromagnetic interference in inverter – Fed – AC – motor. Electr. Electron. Eng. Rev., 271–275 (2010)
7. Khvitia, B., Piskunov, R., Khvitiya, V., Khvitiya, S.: SPICE modeling of EMI in power devices. Electronics. **10**(22), 2822 (2021)
8. Aragon, F., Camacho, A., Chowdhury, M.M.I., Guerrero, J.M.: Passive and active EMI mitigation in power electronics: A review. Energies. **15**(3), 1030 (2022)
9. Miloudi, H., Miloudi, M., Ardjoun, S.A., Nour, A.A., Mahariq, I.: Experimental investigation of conducted electromagnetic interference differential-mode performance in various split-phase induction motors designs. Results Eng. **25**, 103963 (2025). https://doi.org/10.1016/j.rineng.2025.103963
10. Miloudi, H., Miloudi, M., Bermaki, M.H., Gourbi, A., Bendaoud, A.: Optimization of EMC filter for a variable speed drive system in electric aircraft. Eurasia Proc. Sci. Technol. Eng. Math. **26**, 60–66 (2023)
11. Lahlaci, M.E., Miloudi, M., Miloudi, H.: Experimental electromagnetic compatibility of conducted electromagnetic interferences from an IGBT and a MOSFET in the power supply. Electr. Eng. Electromech. **28**(3), 38–43 (2024). https://doi.org/10.20998/2074-272X.2024.3.05
12. Miloudi, H., Bendaoud, A., Miloudi, M., Dickmann, S., Schenke, S.: A novel method of transfer–function identification for modeling DM impedance of AC motor. In: International Symposium on Electromagnetic Compatibility – EMC EUROPE. IEEE, Angers, France (2017). https://doi.org/10.1109/EMCEurope.2017.8094770
13. Miloudi, M., Miloudi, H., Bendaoud, A., Salhi, M.A., Al-Omari, A.N.: Experimental characterization of the high-frequency isolating power transformer. Elektrotehn. Vestn. **86**(4), 211–218 (2019)
14. Zeghoudi, A., Bendaoud, A., Slimani, H., Miloudi, M., Miloudi, H., Canale, L.: Experimental measurements of near magnetic field by different probe for AC/DC LED driver. In: 2022 IEEE International Conference on Environment and Electrical Engineering and 2022 IEEE Industrial and Commercial Power Systems Europe (EEEIC/I&CPS Europe), pp. 1–6. IEEE (2022)
15. Miloudi, H., Bendaoud, A., Miloudi, M.: Réduction des perturbations électromagnétiques conduites dans la machine asynchrone. Mediterr. J. Model. Simul. **2**(1), 8–15 (2014)
16. Bermaki, M.H., Miloudi, H., Miloudi, M., Gourbi, A., Bendaoud, A.: High-frequency differential mode modeling of universal motor's windings. Int. J. Electr. Electron. Res. **11**, 1057–1064 (2023)
17. Zeghoudi, A., Bendaoud, A., Slimani, H., Miloudi, H., Miloudi, M., Chikhi, N.: Experimental measurement of common and differential modes for variable speed drive DC motor. In: 19th International Multi-Conference on Systems, Signals & Devices (SSD). IEEE, Sétif, Algeria (2022). https://doi.org/10.1109/SSD54932.2022.9955933
18. Miloudi, M., Bendaoud, A., Miloudi, H., Nemmich, S., Oukli, M.: Characterization of conducted electromagnetic interference (EMI) generayed by switch mode power supply (SMPS). In: The 7th International Conference on Electrical Engineering. IEEE, Batna, Algeria (2012)
19. Zeghoudi, A., Bendaoud, A., Tilmatine, A., Miloudi, M., Miloudi, H., Canale, L.: Experimental measurement of harmonic distortion and electromagnetic interferences for LED lamps. In: IEEE Industry Applications Society Annual Meeting (IAS). IEEE, Detroit, MI, USA (2022). https://doi.org/10.1109/IAS54023.2022.9940136

20. Zeghoudi, A., Bendaoud, A., Miloudi, H., Bechkir, S., Slimani, H., Miloudi, M.: Effectiveness of shielding on the magnetic emissions emitted by a toroidal inductor. In: 2nd International Conference on Advanced Electrical Engineering (ICAEE). IEEE, Constantine, Algeria (2022). https://doi.org/10.1109/ICAEE53772.2022.9962063
21. Gourbi, A., Miloudi, M., Miloudi, H.: An investigation and analysis of a hybrid photovoltaic system for power supply. Ing. Energética. **43**(3), 109–119 (2022)

A Hybrid Deep Learning Approach for Anomaly Detection in Smart Grid Systems

S. Aitsaid[1], S. Ouarti[1(✉)], and Z. Farah[1,2]

[1] École supérieure en Sciences et Technologies de l'Informatique et du Numérique, RN 75, Laboratoire LITAN, Amizour 06300, Bejaia, Algeria
{s_aitsaid,s_ouarti}@estin.dz, zoubeyr.farah@univ-bejaia.dz
[2] Faculty of Exact Sciences, LIMED Laboratory, University of Bejaia, Bejaia 06000, Algeria

Abstract. Smart Grids are increasingly exposed to advanced cyber threats such as False Data Injection Attacks (FDIAs), which can evade traditional anomaly detection techniques. This paper proposes an unsupervised hybrid deep learning approach that integrates a Bidirectional LSTM (Bi-LSTM) network for time-series forecasting with an BiLSTM-based autoencoder for sequence reconstruction. Anomaly detection is achieved through a fusion of prediction and reconstruction errors into a unified score. The method is evaluated on the IEEE 14-bus benchmark under multiple attack scenarios using hourly and minute-level datasets. The evaluation demonstrates the model's high efficacy, with F1-scores exceeding 0.99 for both localized and extended attacks. Furthermore, the model is highly efficient, with an inference time of less than 9 ms for the hourly dataset and less than 3 ms for the minute-level dataset. These findings confirm the model's robustness, real-time capability, and effectiveness in enhancing the cyber-resilience of critical energy systems.

Keywords: Anomaly detection · Unsupervised deep learning · BiLSTM · Autoencoder · Smart Grids · FDIA

1 Introduction

Smart Grids, born from the integration of Information and Communication Technology (ICT) in the energy sector, are making electricity management more efficient and sustainable. They provide key benefits such as improved flexibility, real-time monitoring, and easier integration of renewable energy. However, this modernization also creates significant security vulnerabilities. The inherent complexity of these interconnected systems expands the attack surface, exposing critical infrastructures to advanced cyber threats like False Data Injection Attacks (FDIAs).

The primary challenge posed by False Data Injection Attacks (FDIAs) lies in their stealthy nature. They are specifically designed to manipulate sensor measurements while remaining consistent with the system's physical laws, which makes them particularly difficult to detect. Consequently, traditional defense mechanisms, such as Bad Data Detection (BDD) which relies on statistical residual analysis, often prove ineffective

M. Hatti (Ed.): IC-AIRES 2025, LNNS 1740, pp. 259–267, 2026.
https://doi.org/10.1007/978-3-032-12788-4_25

against such intrusions. Supervised learning methods depend on labeled attack data, which is rarely available, while unsupervised techniques suffer from high false positive rates and limited sensitivity [1, 2].

To address these challenges, we propose a hybrid deep learning approach that combines a Bidirectional LSTM (BiLSTM) model for temporal prediction with an BiLSTM-based Autoencoder trained on normal data for sequence reconstruction. By fusing their respective error signals, the system enhances detection robustness against subtle or distributed anomalies [3].

Our main contributions are as follows:

(i) The design of a lightweight, real-time hybrid model for FDIA detection, obviating the need for labeled attack data. (ii) Evaluation of this model on the IEEE 14-bus benchmark utilizing both hourly and minute-level datasets. (iii) Performance assessment across various attack scenarios encompassing diverse attack intensities. (iv) Confirmation of its real-time capability, characterized by an efficient inference time.

The remainder of the paper is organized as follows: Section 2 reviews related work, Section 3 details the proposed model, Section 4 presents the experimental setup, Section 5 discusses results, and Section 6 concludes the paper and outlines future directions.

2 Related Work

In this section, we provide an overview of some existing solutions that have addressed the problem of FDIA detection. The first family of approaches is based on supervised models that rely on labeled data to differentiate between normal and malicious behavior. Traditional machine learning models such as ensemble methods (e.g., LR, SVM, DT) have shown promising precision [1, 4], nevertheless, they suffer from limited sensitivity. Recent advances integrate federated learning [5], using BiLSTM-GRU models with attention, achieve up to 99.8% accuracy. Deep learning-based supervised models further enhance performance by learning complex spatial-temporal features. Notable contributions include CNN-based multi-label classifiers for sensor-level localization [6] and hybrid CNN-transformer architectures like ACT-Net [7], achieving real-time detection with sub-millisecond latency and high precision on large-scale grid simulations. Nevertheless, despite the observed improvements in accuracy and detection time, the issue of constrained generalization capacity remains unresolved.

The second family of approaches, comprising unsupervised techniques, aims to address the limitations inherent in labeled data. These methods model normal behavior and subsequently flag deviations as anomalies. However, traditional methods such as One-Class SVM or Elliptic Envelope [1] remain sensitive to parameter tuning and frequently yield high false positive rates. More sophisticated approaches, leveraging Deep Learning, enhance the unsupervised detection through architectures like autoencoders and Deep Belief Networks (DBNs). Among the most effective solutions, Attention-based autoencoders [2] have demonstrated an F1-score of 98.38% with zero false positives, while convolutional DBNs [8] have shown notable robustness under noise.

The third family are Hybrid models, that combine multiple paradigms to enhance detection robustness and adaptability. Architectural hybrids, such as CNN-LSTM or Transformer-LSTM fusions, leverage complementary strengths for robust detection and localization [5, 11, 12]. Conceptual hybrids integrate AI with system models or reinforcement learning. Physics-informed models [10] and DRL-based frameworks [11] adapt detection strategies over time, offering resilience against evolving attack strategies with detection rates up to 99.7%. Nevertheless, many of these advanced hybrid models, particularly those using Transformers or DRL, often result in complex and computationally heavy architectures. This can be a barrier for real-time deployment in resource-constrained environments where low latency is critical. A research gap remains for solutions that optimize the trade-off between high detection accuracy, architectural parsimony, and computational efficiency.

3 Proposed Methodology

This section presents the architecture of the proposed hybrid model, which combines a Bidirectional LSTM for prediction and an BiLSTM-based autoencoder for reconstruction. By integrating both temporal and structural information, the system achieves robust real-time detection of FDIAs.

A. **Architecture Overview**

The detection model operates on multivariate time-series input and follows a four-stage pipeline:

- **Input:** A normalized sequence X of 48 time steps (sensor measurements).
- **Parallel Modeling:** X is fed into two models: a BiLSTM for next-state prediction and an autoencoder (AE) for sequence reconstruction.
- **Error Computation:** Each model produces an error signal, prediction error from the BiLSTM and reconstruction error from the AE.
- **Fusion and Decision:** A combined error, calculated as a weighted sum of the prediction and reconstruction errors, serves as the overall anomaly score. This allows for a tunable balance between the two models. The score is then compared against a predefined threshold τ. If the combined error exceeds this threshold, the input is flagged as a potential anomaly (FDIA) (Fig. 1).

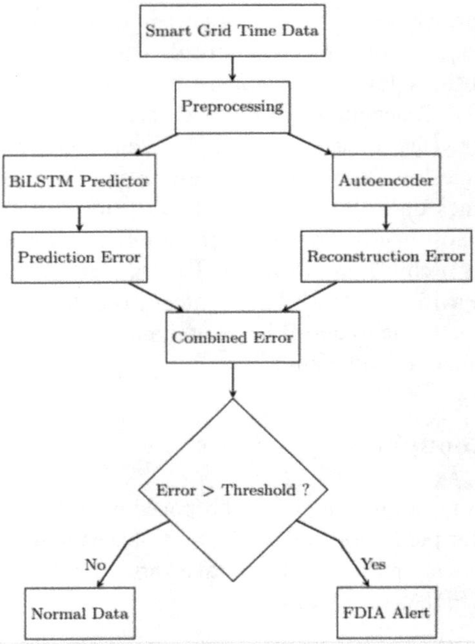

Fig. 1 Hybrid BiLSTM-autoencoder FDIA detection architecture

B. Description of the Models

- **Prediction Model (BiLSTM)**

The BiLSTM captures bidirectional dependencies to forecast the next state of the system.

The algorithm for BiLSTM-based next state prediction is detailed in Fig. 2 (Algorithm 1).

Algorithm 1 Next State Prediction Algorithm

Require: Historical sequence $X_{t-1} = [x_{t-48}, \ldots, x_{t-1}]$

Ensure: Predicted next measure $\hat{x}_t$

1: Load the pre-trained BiLSTM model

2: $\hat{x}_t = \text{BiLSTM}(X_{t-1})$ # predict the next expected measure

3: **return** $\hat{x}_t$

Fig. 2 BiLSTM prediction model

- **Reconstruction Model (BiLSTM-Autoencoder)**

The AE learns to reproduce normal input patterns and flags deviations as anomalies. The reconstruction process of the AE is outlined by Algorithm 2 in Fig. 3.

- **Fusion-Based Detection Mechanism**

Algorithm 2 Next State Reconstruction Algorithm

Require: Historical sequence $X_{t-1} = \{x_{t-48}, \dots, x_{t-1}\}$
Ensure: Reconstructed sequence $\tilde{X}_{t-1}$
1: Load the pre-trained autoencoder model
2: $\tilde{X_{t-1}} = \mathrm{AE}(X_{t-1})$ # Encoding $\rightarrow$ Decoding
3: **return** $\tilde{X}_{t-1}$

Fig. 3 BiLSTM-autoencoder

The BiLSTM and Autoencoder models are jointly used for FDIA detection (see Fig. 4). At each time step, the system performs a dual analysis. The BiLSTM acts as a forecaster, using the historical input sequence to predict the expected current measurement. Simultaneously, the Autoencoder acts as a consistency checker, taking the same historical sequence to produce a reconstruction of it. Two distinct error metrics are then calculated: the prediction error (the deviation of the actual measurement from the forecast) and the reconstruction error (the deviation of the original sequence from its reconstruction). These errors are combined into a single anomaly score. If this score exceeds a predefined threshold τ an alert is raised for the current time step (Algorithm 3).

Algorithm 3 Next State Detection Algorithm

Require: Historical sequence X_{t-1}, current measure x_t, threshold τ, weights w_1, w_2
Ensure: Binary alert
1: # Get model outputs
2: $\hat{x}_t \leftarrow$ Predict(X_{t-1}) Using Algorithm 1
3: $\tilde{X}_{t-1} \leftarrow$ Reconstruct(X_{t-1}) Using Algorithm 2
4: # Get the errors from the two models
5: $E_{\text{pred}} \leftarrow$ CalculateError($\hat{x}_t, x_t$) (E_{pred} is the Prediction Error)
6: $E_{\text{rec}} \leftarrow$ CalculateError($\tilde{X}_{t-1}, X_{t-1}$) (E_{rec} is the Reconstruction Error)
7: # fusion
8: $E_{\text{combined}} = w_1 \cdot E_{\text{pred}} + w_2 \cdot E_{\text{rec}}$
9: # Decision
10: **if** $E_{\text{combined}} > \tau$ **then**
11:     **return** `"FDIA Alert"`
12: **else**
13:     **return** `"Normal"`
14: **end if**

Fig. 4 Detection mechanism

4 Experimental Setup

A. Smart Grid Model and Datasets

The proposed method was evaluated on the IEEE 14-bus test system (see Fig. 5) using a DC linear state estimation model, employing the datasets provided by [9] that comprise real measurements. To assess performance across different temporal granularities, two types of time resolutions were considered: an hourly dataset comprising 8,760 samples, and a minute-level dataset containing 525,541 samples.

Each sample includes 54 features: power flows (forward/reverse) and bus consumptions. Additionally, 128 synthetic FDIA vectors, designed to be stealthy and physically consistent, were used for attack injection.

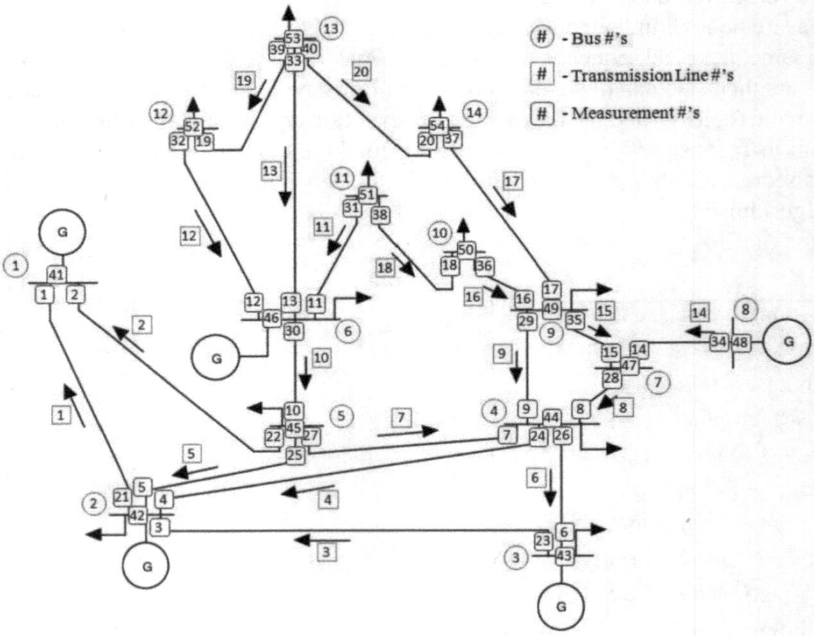

Fig. 5 IEEE 14bus used system [9]

B. Data Preparation

Data was split chronologically into training (70%), validation (15%), and test (15%). Subsequently, sequences were constructed for BiLSTM and AE inputs using a sliding window of 48 time steps.

C. Attack Simulation

An FDIA vector, denoted as a, was constructed using

$$a = Hc$$

Where H denotes the system's Jacobian matrix, and c represents an arbitrary attack vector designed to ensure stealthiness.

To evaluate the robustness of the detection method, two distinct attack scenarios were considered: (i) a *Localized* attack, involving a fixed vector targeting a limited number of buses; (ii) an *Extended* attack, affecting a broader area with more sensors involved. Each scenario was tested at 10 intensity levels (1%–10% of test data), with attacks injected by perturbing and re-normalizing selected samples.

5 Results and Discussion

This section evaluates the model using standard regression and classification metrics. Separate models were trained on hourly and minute-level datasets, and FDIA detection was assessed under various attack scenarios.

A. **Evaluation Metrics**

We assessed performance using both regression and classification metrics. For forecasting and reconstruction models, we used Mean Absolute Error (MAE) and Root Mean Squared Error (RMSE). Anomaly detection was evaluated using Precision, Recall, and F1-score, based on standard confusion matrix components (TP, FP, TN, FN).

B. **Individual Model Training and Performance**

- **BiLSTM Training**

The model was trained using the Adam optimizer with hyperparameters tuned via Optuna (30 trials), achieving low test RMSE scores of 0.0206 on hourly data and 0.0051 on minute-level data.

- **Autoencoder Training**

The autoencoder was also tuned using Optuna, with two encoder layers. It achieved a test MAE of 0.0894 on hourly data and 0.0109 on minute-level data. Both models converged stably and performed better with finer temporal resolution.

C. **Hybrid System Detection Performance**

The hybrid model computes the fused anomaly score (see Fig. 4 in Section 3), with empirically chosen weights: $w1 = 0.6$ and $w2 = 0.4$, optimized on the validation set to maximize the F1-score.

The decision rule, introduced in Section 3, flags a sample as anomalous if the fused score exceeds the threshold, which was determined using the percentile method (90%–99%).

D. **Detection Results**

The model achieved good performance across all attack types and intensity of attacks (1%–10%) (Table 1):

Table 1 Detection performance (hourly and minute-level data)

Data	Attack Scenario	Precision	Recall	F1 Score	Time (ms)
Hourly Data	**Localized**	1.00	1.00	1.00	7.98
	Extended	1.00	1.00	1.00	8.82
Minute-Level Data	**Localized**	0.99	1.00	0.99	2.69
	Extended	0.99	1.00	0.99	2.64

For the hourly dataset, the average detection time per sequence was approximately 8.4 milliseconds. This processing time reflects the complexity of the BiLSTM predictor optimized for hourly data (hidden dimension of 127) and the Autoencoder (first hidden layer of 256 units). For the minute-level dataset, the average detection time per sequence was significantly lower, at approximately 2.6 milliseconds. This faster processing is consistent with the comparatively smaller optimized models for minute-level data, particularly the BiLSTM predictor (hidden dimension of 54) and the Autoencoder (first hidden layer of 64 units).

E. **Discussion**

Our results validate the robustness and efficiency of the proposed BiLSTM–Autoencoder framework for detecting FDIAs in Smart Grids.

***Model and Data Insights*:** The BiLSTM effectively captured temporal dependencies, while the autoencoder modeled normal structural patterns. Both models performed better on minute-level data, with a 55% reduction in RMSE and 86% in MAE, confirming that finer temporal resolution enhances learning and anomaly detection.

***Detection Effectiveness*:** The hybrid system achieved perfect F1-scores for localized and extended attacks. By combining prediction and reconstruction errors, it maintained robust performance across attack types. Threshold tuning (percentile-based) optimized precision–recall balance, while low inference time (2.6 ms) supports real-time use.

6 Conclusion and Future Work

This study introduced a hybrid deep learning framework combining a BiLSTM predictor and an autoencoder for detecting FDIAs in smart grids. Tested on the IEEE 14-bus system with both hourly and minute-level data, the model achieved high accuracy (F1-score $\geq$ 0.99) and fast inference times ($<$ 3 ms), confirming its suitability for real-time applications. However, the present study is subject to some limitations, including evaluation restricted to a single dataset, the use of static fusion weights, and the absence of attack localization. Future work will focus on extensive validation with larger datasets and the integration of localized FDIA detection capabilities.

References

1. Ashrafuzzaman, M., Das, S., Chakhchoukh, Y., Shiva, S., Sheldon, F.T.: Detecting stealthy false data injection attacks in the smart grid using ensemble-based machine learning. Comput. Secur. **97**, 101994 (2020). https://doi.org/10.1016/j.cose.2020.101994
2. Santos, I.M.A., Lira, G.R.S., Vilar, P.B.: Detection of false data injection attacks in power systems using deep autoencoder with attention mechanism. In: 2024 Workshop on Communication Networks and Power Systems (WCNPS), pp. 1–7. IEEE, Brasilia, Brazil (2024). https://doi.org/10.1109/WCNPS65035.2024.10814586
3. Moradzadeh, A., Mohammadpourfard, M., Konstantinou, C., Genc, I., Kim, T., Mohammadi-Ivatloo, B.: Electric load forecasting under false data injection attacks using deep learning. Energy Rep. **8**, 9933–9945 (2022). https://doi.org/10.1016/j.egyr.2022.08.004
4. Abudin, M.J., Thokchom, S., Naayagi, R.T., Panda, G.: Detecting false data injection attacks using machine learning-based approaches for smart grid networks. Appl. Sci. **14**(11), 4764 (2024). https://doi.org/10.3390/app14114764
5. Kausar, F., Deo, S., Hussain, S., Ul Haque, Z.: Federated deep learning model for false data injection attack detection in cyber physical power systems. Energies. **17**(21), 5337 (2024). https://doi.org/10.3390/en17215337
6. Wang, S., Bi, S., Zhang, Y.-J.A.: Locational detection of the false data injection attack in a smart grid: A multilabel classification approach. IEEE Internet Things J. **7**(9), 8218–8227 (2020). https://doi.org/10.1109/JIOT.2020.2983911
7. Eddin, M.E., et al.: Invoking an efficient deep learning approach for real-time detection of false data injection attacks. In: 2025 IEEE Texas Power and Energy Conference (TPEC), pp. 1–6. IEEE, College Station, TX, USA (2025). https://doi.org/10.1109/TPEC63981.2025.10906699
8. Ding, Y., Ma, K., Pu, T., Wang, X., Li, R., Zhang, D.: A deep learning-based classification scheme for cyber-attack detection in power system. IET Energy Syst. Integr. **3**(3), 274–284 (2021). https://doi.org/10.1049/esi2.12034
9. Baul, A., Sarker, G.C., Sadhu, P.K., Yanambaka, V.P., Abdelgawad, A.: XTM: A novel transformer and LSTM-based model for detection and localization of formally verified FDI attack in smart grid. Electronics. **12**(4), 797 (2023). https://doi.org/10.3390/electronics12040797
10. Zhang, G., Gao, W., Li, Y., Guo, X., Hu, P., Zhu, J.: Detection of false data injection attacks in a smart grid based on WLS and an adaptive interpolation extended Kalman filter. Energies. **16**(20), 7203 (2023). https://doi.org/10.3390/en16207203
11. Lin, X., An, D., Cui, F., Zhang, F.: False data injection attack in smart grid: Attack model and reinforcement learning-based detection method. Front. Energy Res. **10**, 1104989 (2023). https://doi.org/10.3389/fenrg.2022.1104989
12. Kadri, B., Sliman, L., Alsaedi, A., Abdella, J., Farah, Z.: Physics-aware hybrid estimator for dynamic power system dataset generation and cyber attack simulation. IEEE Access. **13**, 146830–146854 (2025). https://doi.org/10.1109/ACCESS.2025.3599318

Adaptive Preprocessing for Improving Early Detection and Classification of Anomalies on Photovoltaic Panels

Randa Benkhelifa[1,3](✉), Nasria Bouhyaoui[1,2], and Dounia Bouchekioua[1]

[1] Laboratoire de l'intelligence artificielle et des technologies de l'information, Université Kasdi Merbah, Route de Ghardaia BP. 511, 30 000 Ouargla, Algérie
{randa.benkhelifa,nasria.bouhyaoui, dounia.bouchekioua}@univ-ouargla.dz

[2] Ecole normale supérieure de Ouargla, Ouargla, Algérie

[3] Faculty of Hydrocarbons, Renewable Energies, Earth Sciences and the Universe, Université Kasdi Merbah, Ouargla, Algérie

Abstract. Photovoltaic systems allow to convert sunlight directly into electricity power to produce clean, renewable energy from sunshine. Energy production can be affected by several factors, including panel type, panel orientation tilt and weather conditions which lead to reduce power output. Over time, photovoltaic systems undergo natural degradation caused by environmental conditions can lead to physical defects which compromise solar panel durability and efficiency. Early detection of anomalies is essential to ensure optimal operation and minimize business interruptions. Computer vision can effectively detect and classify anomalies in solar panels.

In this paper, we propose a new approach based on preprocessing adapted to the specific characteristics of electrical damage, and physical damage. These visual differences impose differentiated processing strategies to maximize the quality of the information extracted as input to the machine learning model. We examine the impact of the adapted preprocessing on the classification performances using various pre-trained deep neural networks. The experimental results demonstrate that the proposed approaches enable accurate identification and classifications of solar panels anomalies.

Keywords: Photovoltaic system · Solar panel · Machine learning · Deep learning · Image classification · Image preprocessing

1 Introduction

The global energy transition towards renewable energy sources has placed photovoltaic (PV) systems at the heart of sustainable production strategies. Photovoltaic systems, also known as solar systems, are increasingly being deployed in various application areas, including residential, commercial, and industrial sectors. Their purpose is to directly convert solar energy into electrical energy, thus helping to reduce energy consumption

M. Hatti (Ed.): IC-AIRES 2025, LNNS 1740, pp. 268–279, 2026.
https://doi.org/10.1007/978-3-032-12788-4_26

costs while promoting a sustainable and environmentally friendly energy source. This conversion is made possible by the fundamental modules of PV systems, namely solar panels.

The energy production depends on several essential factors, such as the type of solar panel, as well as its orientation and tilt, which directly influence their performance. Climatic conditions, particularly the intensity of solar radiation and high temperatures, can also affect performance, as can shadows cast by surrounding obstacles. In addition, the accumulation of dirt, dust, and debris impairs light capture. Gradual degradation due to aging and extreme environmental conditions affects the durability of the panels, causing defects such as cracks, reduced glass transparency, or the appearance of hot spots, which contribute to the decline in the efficiency of solar panels.

Early detection of anomalies is of paramount importance to ensure the proper functioning, reliability, and sustainability of photovoltaic systems. It allows for the rapid identification of potential malfunctions, facilitating rapid corrective interventions before they escalate into major failures. This process helps maintain continuous energy production, while limiting business interruptions and associated economic losses. Furthermore, it helps optimize long-term maintenance costs and promotes more sustainable and efficient management of solar installations.

Computer vision offers an effective solution to automatically detect and classify these anomalies by analyzing images acquired in the field. This approach, enhanced by deep learning algorithms, allows for the accurate identification of defects and anomalies. In recent years, a significant amount of researches such as Alves et al. (2021), De Souza Silva et al. (2024), and Haque et al. (2019) have focused on classifying anomalies related to photovoltaic module failures. Conventional approaches, such as manual inspections or rule-based algorithms, have several limitations: they are time-consuming, resource-intensive, and offer limited accuracy. In a context marked by the rise of large-scale photovoltaic installations, the implementation of more sophisticated and automated methods has become essential to ensure rapid, reliable, and efficient detection and classification of anomalies. It is in this perspective that our work is based, which proposes an adaptive image preprocessing method to improve the discrimination of different types of damage on solar panels by exploiting their specific visual characteristics. Unlike conventional approaches, our method applies targeted transformations according to each category of damage, in order to highlight the discriminating patterns and optimize their representation before learning.

In this paper, a new approach is proposed to improve the classification of damage types present on solar panels by taking advantage of preprocessing adapted to specific characteristics such as electrical damage, generally in the form of subtle and low-contrast alterations, and physical damage, manifested by sharper and more structured cracks or breaks. This variation in visual patterns requires differentiated processing strategies to optimize the extracted information and increase the relevance of the input data used by the machine learning model. It is in this perspective that our method applies targeted transformations according to certain damage specificity, in order to bring out the discriminating details and significantly improve the classification performance.

To evaluate the effectiveness of our method, we experimented with several deep learning algorithms, including ResNet50, MobileNetV2, EfficientNetB0, Xception, VGG19, and a custom CNN. Each model was trained on the preprocessed solar panel dataset and then evaluated using standard metrics. Validation was conducted using 10-fold cross-validation to ensure robustness of the results and avoid overfitting. This methodology ensures that the reported performance accurately reflects the models' ability to detect and classify relevant types of anomalies.

The remainder of this article is organized as follows: the related works section presents the main existing methods in the literature, while the proposed approach section describes in detail the conditional approach we have developed. The experiment and evaluation section presents the experimental protocol, the datasets used, the learning algorithms implemented, as well as the evaluation metrics retained in order to assess the relevance and effectiveness of our method. The results and discussion section presents the interpretation of the obtained performances, highlights the interest of the proposed approach and discusses its main limitations. Finally, the conclusion summarizes the main contributions of our work and highlights some future research perspectives.

2 Related Works

In recent years, a large amount of research has focused on classifying anomalies associated with photovoltaic module failures. In this context, deep learning methods have emerged as particularly effective solutions for addressing issues related to visual processing and pattern recognition.

Several recent studies have explored different methods to improve anomaly detection and classification in photovoltaic systems, notably by combining deep learning, the Internet of Things (IoT), and statistical techniques. Punitha (2025) proposes an integrated solution combining IoT and convolutional neural networks (CNNs) for continuous monitoring and predictive maintenance of photovoltaic installations. This system combines sensors and a camera with a CNN model, enabling rapid anomaly detection and performance optimization while reducing maintenance costs. In a similar vein, Abdelsattar et al. (2025) conduct a comparative evaluation of twenty-four CNN architectures—such as ResNet, DenseNet, MobileNet, and Xception—to automate the classification of solar panel faults. Their study particularly highlights the effectiveness of lightweight models like MobileNet, which are suitable for resource-limited environments, a problem also addressed by Punitha.

In a complementary manner, Nagar and Rai (2024) exploit a ResNet-9 architecture optimized via a filter pruning technique for anomaly detection, demonstrating the growing popularity of ResNet architectures in this field, as confirmed by Abdelsattar et al.

Alves et al. (2021) adopted a strategy aimed at enriching limited datasets, combining different augmentation techniques—geometric transformations, GANs, and diffusive models—with a CNN to improve anomaly detection. This approach is similar to that of De Souza Silva et al. (2024), who exploit supervised learning using an ensemble of models (Random Forest, k-NN, and inference machine) to improve anomaly classification on real and synthetic data, highlighting the importance of training data management and diversity in PV systems.

From a more computer vision-oriented perspective, Deitsch et al. (2019) compare two methods for automatic fault detection in photovoltaic cells: a lightweight approach based on manual descriptors and an SVM, and a more efficient one based on a CNN. This duality recalls the trade-off mentioned by Abdelsattar between performance and hardware constraints.

Umar et al. (2024) introduce a novel approach combining unsupervised anomaly detection and 3D augmented reality visualization, opening up innovative perspectives for fault analysis and interpretation, distinct from but complementary to purely supervised approaches.

Finally, Haque et al. (2019) and Akram and Lotfifard (2015) implement artificial intelligence techniques coupled with data preprocessing methods (discrete wavelet transform for Haque, probabilistic neural network for Akram) in order to improve the accuracy of diagnosis in PV thermography, emphasizing the importance of signal processing before classification, a point which also echoes the work of Alves et al. on data augmentation and preparation.

This methodology therefore highlights the value of deep learning models for ensuring fast, accurate, and robust quality control of solar panels. Building on these advances, our approach proposes to leverage preprocessing adapted to the visual specificities of different classes of defects, coupled with a powerful machine learning model to improve the classification of anomalies on photovoltaic panels. Unlike existing methods that often use general strategies, our method focuses on optimizing image processing based on the characteristics specific to each type of damage, in order to maximize the quality of the input data and, consequently, the accuracy of the model. This innovative approach aims to offer a more robust solution adapted to the real constraints of PV systems.

3 Proposed Approach

The main objective of this study is to improve the classification of damage types present on solar panels by leveraging preprocessing adapted to the specific characteristics of some category. This work is based on the characteristics of those two main classes: electrical damage, characterized by often subtle and low-contrast alterations, and physical damage, including visible and structured cracks or breaks. These visual differences require differentiated processing strategies to maximize the quality of the information extracted as input to the machine learning model. In addition, we have applied a Standard preprocessing. This late ensures that all images, regardless of class, are consistent in terms of dimensions, color representation, and normalization, to maximize their relevance when the learning model is invoked.

General formulation: $D = (x_i, y_i)_{i=1}^{N}$, where each image $x_i \in \mathbb{R}^{H \times W \times 3}$ is associated with a label $y_i \in \{1,\ldots,C\}$ indicating the type of damage. We define a conditional preprocessing function T such that:

$$\tilde{x}_i = T(x_i, y_i) = \begin{cases} T_{elec}(x_i), & y_i = \text{"}Electrical - damage\text{"} \\ T_{phys}(x_i), & y_i = \text{"}Physical - damage\text{"} \\ T_{std}(x_i), & otherwise \end{cases}$$

Where $\tilde{x}_i$ is the pre-processed image, optimized according to the nature of the damage characteristics.

A. Preprocessing Based on Electrical Damage Features

Electrical damage often results in low-contrast anomalies and degradation that are difficult to see with the naked eye, such as arcing or faint hot spots. The processing must therefore accentuate these details without introducing excessive noise.

Step 1: Conversion to Grayscale

To facilitate contrast manipulation, the color image is converted to grayscale:

$$G = Grayscale(I_0)$$

This step simplifies processing while preserving crucial brightness information.

Step 2: Local Contrast Enhancement with CLAHE

Contrast-Limited Adaptive Histogram Equalization (CLAHE) is applied to g_i. CLAHE divides the image into small regions (tiles) and locally applies histogram equalization, limiting noise amplification by a clipping factor:

$$c_i = CLAHE(g_i)$$

This method increases the visibility of faint details, particularly in dark areas, without degrading more uniform regions.

Step 3: Canny Edge Extraction

To identify areas of rapid intensity transition (potential defect edges), the Canny operator is applied to the enhanced image:

$$e_i = Canny(c_i)$$

Step 4: Creating a Dark Area Mask

Inverse binary thresholding is performed on the ci to detect dark regions likely to correspond to damage:

$$m_i = 1_{c_i < \tau}$$

with an experimentally fixed threshold (eg =60).

Step 5: Information Fusion

The detected contours and the mask are combined using a logical OR operation:

$$f_i = m_i \bigvee e_i$$

This binary mask is then converted into a color image (3 channels) and then merged with the original image according to a controlled weighting:

$$\tilde{x}_i = \alpha x_i + \beta \cdot GrayToRGB(f_i)$$

with =0.7 and =0.3. This fusion highlights the electrical faults while preserving the global context.

B. Preprocessing Based on Physical Damage Features

Physical damage appears as sharp breaks and well-defined cracks, with distinct edges. The goal here is to sharpen details to facilitate their detection.

Step 1: Sharpening Filtering

The original image is convolutionally filtered with a sharpening kernel defined by:. This fusion highlights electrical defects while preserving the overall context.

$$K = \begin{bmatrix} 0 & -1 & 0 \\ -1 & 5 & -1 \\ 0 & -1 & 0 \end{bmatrix}$$

This operation amplifies local intensity transitions, making cracks more visible:

Step 2: Converting to Grayscale

The sharpened image s_i is converted to grayscale:

$$s_i = x_i * K$$

$$g_i = Grayscale(s_i)$$

Step 3: Edge Detection

Canny edge detection is applied to g_i to extract the breaklines:

$$e_i = Canny(g_i)$$

Step 4: Weighted Fusion

The sharpened image and the contour map are combined using weighting:

$$\tilde{x}_i = \gamma s_i + \delta \cdot GrayToRGB(e_i)$$

with =0.8 and =0.2. This step reinforces the presence of cracks in the final image.

C. **Standard Preprocessing**

A standard preprocessing is applied to harmonize the representation of the images and ensure their compatibility with the deep learning model. This step takes into account the geometric, colorimetric, and statistical aspects of the input data.

Standard preprocessing, denoted T_{istd}, consists of the following operations:

Step1: Resizing

Each image is resized to have a uniform representation of 224 × 224 pixels. This operation ensures that all input data have the same dimensions, which is essential for training a convolutional neural model.

Mathematically, if x_i is the original image, resizing is a geometric transformation.

$$\tilde{x}_i = R(x_i, (224{,}224))$$

Step 2: Conversion to RGB

Since images are sometimes acquired in different formats or color models, they are all converted to RGB (Red, Green, Blue) to obtain a consistent color representation.

$$\tilde{x}_i = RGB(x_i)$$

Step 3: Normalization

After resizing and colorimetric conversion, each image is normalized using the procedure defined by the MobileNetV2 model. This step ensures that all input data have a distribution close to the expected value of the pre-trained model, facilitating learning.

Concretely:

$$\hat{x}_i = preprocess\_input_{MobileNetV2}(\tilde{x}_i)$$

where preprocess_input$_{\text{MobileNetV2}}$ applies an offset of the pixel values from the range [0, 255] to the range [−1, 1].

4 Experiment and Evaluation

This section describes the experimental framework used to validate our approach. We first present the used datasets, representative of the targeted use cases. Next, the classification algorithms selected for comparison are detailed. Finally, we define the evaluation metrics chosen to precisely measure the models' performances.

A. Dataset

Solar Panel Images Clean and Faulty Images is the dataset used in this experiment comes from the Kaggle platform by (Afroz 2021). It consists of images of solar panels, collected from online sources and manually labeled according to the state of each panel. This representation is particularly relevant for evaluating the effectiveness of machine learning methods in detecting anomalies that could affect their operation. The dataset considers six categories (Clean, Dusty, Bird-drop, Electrical-damage, Physical-damage, and Snow-covered).

The dataset consists of a total of 875 images, unevenly distributed across the six classes, which is fairly representative of the cases that can occur in the field. This unbalanced representation underscores the importance of using robust methods to improve learning on data that may contain underrepresented cases. Table 1 shows the number of images per class and their percentage representation:

Table 1 Dataset statistics

Class	Number of Images	Percentage
Clean	194	22.17%
Dusty	191	21.83%
Bird-drop	192	21.94%
Electrical-damage	104	11.89%
Snow-covered	124	14.17%
Physical-damage	70	8.0%
Total	875	100%

B. Classification Models

To evaluate and compare the effectiveness of different convolutional neural network (CNN) architectures in detecting solar panel anomalies, we selected the following models:

- **CNN:** A simple baseline model composed of a few convolutional layers, aggregation (MaxPooling) layers, and dense layers. This architecture serves as a baseline to evaluate the contribution of deeper models.

- **MobileNetV2:** A lighter convolutional neural network model optimized to run on devices with computational and energy constraints. It is based on invert residual blocks and depth-wise convolutions, which reduces the number of parameters while maintaining good performance (Sandler et al. 2018).
- **ResNet50:** A more complex baseline model equipped with residual blocks to mitigate the problem of performance degradation when the network is deepened. This architecture is particularly robust and efficient across numerous case studies (He et al. 2016).

- **EfficientNet:** This family of models applies a uniform scaling policy (depth, width, resolution) to achieve the most advantageous performance-complexity ratio. This approach guarantees excellent results while reducing the number of parameters (Tan and Le 2019).

- **Xception:** This architecture, similar to Inception, replaces all standard convolutions with separable depth-wise convolutions, guaranteeing high performance with reduced complexity (Chollet 2017).

- **VGG19:** A more traditional model, composed of a stack of convolutional layers and dense layers. This simple and efficient architecture was one of the first successfully used for image classification (Simonyan and Zisserman 2014).

C. Evaluation Metrics

In order to evaluate the relevance and effectiveness of our classification models, we have retained the following evaluation metrics:

Confusion Matrix: to accurately assess the performance of our classification models, we first generated the confusion matrix. This tabular representation illustrates, class by class, the number of cases predicted correctly or incorrectly. Each row of the matrix corresponds to the expected class (the actual class), while each column represents the class predicted by the model.

All evaluation metrics used here (Accuracy, Precision, Recall, and F-score) are calculated from the values of the

confusion matrix. These metrics provide a more comprehensive quantitative assessment of the relevance, effectiveness, and robustness of our classification model:

Accuracy: Ratio of correctly predicted cases to all cases.

$$Accuracy = \frac{TP + TN}{TP + TN + FP + FN}$$

Precision: Relevance of positive predicted cases.

$$Precision = \frac{TP}{TP + FP}$$

Recall: Model's ability to detect all relevant cases.

$$Recall = \frac{TP}{TP + FN}$$

F-score: Harmonized mean of precision and recall.

$$F1 = 2 \times \frac{Precision \times Recall}{Precision + Recall}$$

5 Results and Discussion

In this section, we present and analyze the performance of the classification models before and after applying our proposed approach. The results are evaluated using several standard metrics to establish the method's effectiveness and discuss the observed improvements.

To validate our model, we adopted the 10-fold cross-validation method. This approach involves dividing the dataset into 10 distinct subsets, allowing the model to be tested on different partitions while using the entire dataset for training and validation. This method is particularly useful for avoiding overfitting and obtaining a more robust and reliable estimate of model performance, especially when the dataset size is limited.

Table 2 presents the performance of the different classification models before applying the proposed approach, in order to establish a baseline and guide the evaluation of the improvements that may result from it.

These results highlight ResNet50, which outperforms all models with an accuracy of 0.7805 and an F1 score of 0.7777. This performance highlights the effectiveness of the residual layers and the deep architecture of ResNet50 for extracting discriminative features, reinforcing the idea that representation learning is a key factor in the relevance of image classification models. Other models, such as MobileNetV2 and EfficientNet, show interesting performances, while VGG19 and the reference model (Baseline CNN) demonstrate weaker performances, which can be explained by their limited ability to capture relevant details or their overfitting on some training data.

Here are the results of the models after applying our proposed approach, combining image pre-processing techniques based on physical, electrical and standard characteristics. Table 3 presents the performance of the different classification models after applying

Table 2 Models performances before applying the proposed preprocessing

Model	Accuracy	Precision	Recall	F1 Score
Baseline CNN	0.5027	0.5221	0.5027	0.4807
MobileNetV2	0.6924	0.7090	0.6924	0.6854
ResNet50	0.7805	0.7940	0.7805	0.7777
EfficientNet	0.7012	0.7276	0.7012	0.6955
Xception	0.6748	0.7012	0.6748	0.6619
VGG19	0.4994	0.5235	0.4994	0.4996

Table 3 Models performances after applying the proposed preprocessing

Model	Accuracy	Precision	Recall	F1 Score
CNN (custom)	0.7033	0.7489	0.7033	0.7042
MobileNetV2	0.7398	0.7570	0.7398	0.7376
ResNet50	0.8048	0.8181	0.8048	0.8058
EfficientNetB0	0.7674	0.7870	0.7674	0.7661
Xception	0.7078	0.7371	0.7078	0.7079
VGG19	0.5601	0.5847	0.5601	0.5589

the proposed approach, in order to show the evaluation of the improvements resulted from it.

Applying our proposed approach results in a significant improvement for all models, with the exception of VGG19, whose gains remain limited. For example, ResNet50 sees its accuracy increase from 78.05% to 80.48% (+2.43 points), and its F1 score from 77.77% to 80.58% (+2.81 points). MobileNetV2 also gains approximately 5.74 points in accuracy, increasing from 69.24% to 73.98%.

Even the baseline CNN, which initially performed poorly, benefits from a significant increase, increasing from 50.27% (CNN baseline before) to 70.33%, demonstrating the positive impact of preprocessing.

The results clearly demonstrate that the proposed approach, which combines electrical and physical damage in image preprocessing, optimizes the extraction of visual representations, resulting in a significant improvement in models performances.

The increase in metrics for ResNet50, MobileNetV2, and EfficientNetB0 indicates that these modern architectures are particularly receptive to improvements in input data quality. In contrast, VGG19, with an older and less flexible architecture, shows limited progress, which may be explained by overfitting or reduced adaptability. Finally, the use of 10-fold cross-validation strengthens the reliability of these results by ensuring that the reported performance is representative of the entire dataset.

Limitations: Despite the encouraging performance achieved with our approach, several limitations must be considered for a rigorous interpretation of the results and to guide future work.

Dataset Size and Diversity: the relatively limited number of samples and the potentially limited diversity of the dataset may limit the generalization of the models, particularly to more diverse real-world data. An extension to larger and more diverse datasets would be necessary to validate the robustness of the approach.

Sensitivity to Hyperparameters and Preprocessing Configurations: the effectiveness of the conditional approach strongly depends on the choice of preprocessing techniques and the parameters used. These parameters may require specific adjustment for each new dataset, which can limit the reproducibility and scalability of the method.

Variable Performance Across Architectures: Some architectures, such as VGG19, show limited improvements, suggesting that the approach does not uniformly benefit all models. This heterogeneity underlines the need to adapt the approach to the specificities of architectures, or to develop more universal techniques.

6 Conclusion and Future Works

This study highlighted the importance of developing robust models for classifying defects and anomalies in solar panels, a key element for preventive maintenance and energy performance optimization. By leveraging advanced image processing techniques and deep neural network architectures, we were able to improve the detection and discrimination of various anomalies such as cracks, hot spots, and soiling.

The experimental results, obtained using 10-fold cross-validation, demonstrate significant improvements in evaluation metrics for most of the tested architectures, including ResNet50, MobileNetV2, and EfficientNetB0.

This study confirms the importance of adequate data preparation prior to the deep learning process to enhance the quality of extracted representations and improve the models' overall ability to discriminate between classes. However, some architectures, such as VGG19, show limited gains, highlighting the need to adapt methods to the specificities of the models.

The limitations identified, particularly the size of the dataset and sensitivity to preprocessing parameters, open up interesting avenues for future work, including extension to more diverse datasets and automated optimization of processing steps.

In summary, this work represents a significant step towards intelligent and reliable solutions for proactive monitoring of solar panels.

References

Abdelsattar, M., AbdelMoety, A., Ismeil, M.A., Emad-Eldeen, A.: Automated defect detection in solar cell images using deep learning algorithms. IEEE Access. **13**, 4136–4157 (2025)

Afroz, P.: Solar Panel Images [Data Set]. Kaggle (2021) https://www.kaggle.com/datasets/pythonafroz/solar-panel-images

Akram, M.N., Lotfifard, S.: Modeling and health monitoring of DC side of photovoltaic array. IEEE Trans. Sustain. Energy. **6**(4), 1245–1253 (2015)

Alves, R.H.F., Deus Júnior, G.A., Marra, E.G., Lemos, R.P.: Automatic fault classification in photovoltaic modules using Convolutional Neural Networks. Renew. Energy. **179**, 502–516 (2021)

Chollet, F.: Xception: Deep convolutional networks with separable convolutions. arXiv. https://arxiv.org/abs/1610.02357 (2017)

De Souza Silva, J.L., Mahmoudi, E., Carvalho, R.R.M., dos Santos Barros, T.A.: Classification of anomalies in photovoltaic systems using supervised machine learning techniques and real data. Energy Rep. **11**, 4642–4656 (2024)

Deitsch, S., et al.: Automatic classification of defective photovoltaic module cells in electroluminescence images. Solar Energy. **185**, 455–468 (2019)

Haque, A., Bharath, K.V.S., Khan, M.A., Khan, I., Jaffery, Z.A.: Fault diagnosis of photovoltaic modules. Energy Sci. Eng. **7**(3), 622–644 (2019)

He, K., Zhang, X., Ren, S., Sun, J.: Deep residual networks (ResNet). arXiv. https://arxiv.org/abs/1512.03385 (2016)

Livera, A., Theristis, M., Micheli, L., Stein, J.S., Georghiou, G.E.: Failure diagnosis and trend-based performance losses routines for the detection and classification of incidents in large-scale photovoltaic systems. Progr. Photovolt. Res. Appl. **30**(8), 921–937 (2022)

Nagar, S., Rai, P.K.: Anomaly detection and classification in solar panels using infrared image and deep learning. In: 2024 2nd International Conference on Disruptive Technologies (ICDT), pp. 1554–1557. IEEE (2024)

Punitha, K.P.: IoT-powered robust anomaly detection and CNN-enabled predictive maintenance to enhance solar PV system performance. In: IoT for Smart Grid: Revolutionizing Electrical Engineering, pp. 243–255. John Wiley & Sons (2025)

Sandler, M., Howard, A., Zhu, M., Zhmoginov, A., Chen, L.-C.: MobileNetV2: Inverted residual blocks. arXiv. https://arxiv.org/abs/1801.04381 (2018)

Simonyan, K., Zisserman, A.: Very deep convolutional networks for large-scale image recognition (VGG19). arXiv. https://arxiv.org/abs/1409.1556 (2014)

Tan, M., Le, Q.V.: EfficientNet: Rethinking model scaling for convolutional neural networks. arXiv. https://arxiv.org/abs/1905.11946 (2019)

Umar, S., Qureshi, M.S., Nawaz, M.U.: Thermal imaging and AI in solar panel defect identification. Int. J. Adv. Eng. Technol. Innov. **1**(3), 73–95 (2024)

Deep Learning-Based Detection of Environmental Faults in Photovoltaic Systems Under Dust and Humidity Conditions

Adaika Mohammed[1(✉)], Laouamer Mosbah[2], and Abdelkader Mahmoudi[2,3]

[1] LPMRN Laboratory, Department of Electromechanical Engineering, University of Bordj Bou Arreridj, El-Anasser 34030, Bordj Bou Arreridj, Algeria
mohammed.adaika@univ-bba.dz

[2] UDERZA Unit, Mechanical Department, Faculty of Technology, University of Eloued, Eloued 39000, Algeria
{laouamer-mosbah,abdelkader-mahmoudi}@univ-eloued.dz

[3] CISE Electromechatronic Systems Research Centre, University of Beira Interior, Calçada Fonte do Lameiro P-6201-001, Covilhã, Portugal

Abstract. Photovoltaic (PV) systems operating in outdoor environments are highly susceptible to performance degradation due to environmental stressors such as dust accumulation and high humidity. These factors not only reduce energy yield but also pose a detection challenge for conventional analytical methods. This study proposes a deep learning-based framework integrating Convolutional Neural Networks (CNNs) for spatial fault classification and Recurrent Neural Networks (RNNs) for modeling temporal anomalies. Electrical signals were captured in real-time via a TMS320F28379D-based monitoring setup, preprocessed, and used to train both models under clean and noise-injected conditions. CNNs demonstrated superior accuracy in identifying dust-induced surface faults, whereas RNNs excelled in detecting humidity-related degradation. These findings highlight the viability of AI-driven diagnostics for enhancing the resilience and reliability of PV system monitoring.

Keywords: Photovoltaic faults · CNN · RNN · Dust · Humidity · Deep learning

1 Introduction

The global shift toward renewable energy has led to widespread deployment of photovoltaic (PV) systems, valued for their scalability and sustainability. However, their exposure to harsh outdoor environments introduces performance challenges that threaten their reliability over time. Notably, environmental stressors such as dust accumulation and high humidity can severely reduce energy yield and accelerate system wear (Triki-Lahiani et al. 2017; Alam et al. 2015; Di Lorenzo et al. 2020). These factors often produce subtle electrical disturbances that are difficult to capture using conventional threshold-based fault detection techniques, especially under fluctuating irradiance conditions and the nonlinear dynamics of PV modules (Köntges et al. 2014; Pillai and Natarajan 2018;

M. Hatti (Ed.): IC-AIRES 2025, LNNS 1740, pp. 280–289, 2026.
https://doi.org/10.1007/978-3-032-12788-4_27

Collier and Key 1988). Recent studies have therefore explored artificial intelligence (AI) techniques particularly deep learning to detect faults with improved sensitivity and adaptability (Mansouri et al. 2021; Bebboukha et al. 2024). These models excel in extracting spatial and temporal patterns from complex signal behavior, offering a promising alternative to traditional methods. This study investigates the use of two AI-based models Convolutional Neural Networks (CNNs) and Recurrent Neural Networks (RNNs) for the classification of PV faults caused by dust and humidity. It aims to determine which architecture better captures fault signatures and contributes to a more intelligent and efficient PV monitoring system.

2 Materials and Methods

A. Environmental Conditions Considered in the Experimental Design

The operational reliability and energy conversion efficiency of photovoltaic (PV) systems deployed in outdoor environments are significantly affected by environmental stressors chiefly solar irradiance and ambient temperature. These factors not only influence the real-time output but also accelerate the development of environmental faults such as dust accumulation and humidity-induced degradation. To account for these variables, the present study simulates real-world conditions by analyzing system behavior under varying irradiance and particulate exposure.

a. Solar Irradiance and Dust Effects

Solar irradiance is a key factor in photovoltaic (PV) performance, but it fluctuates due to atmospheric conditions like dust and haze (Ghimire et al. 2022; Ajitha and Martínez-Ramón 2021). In dusty regions, surface soiling reduces sunlight exposure, distorting power output and I–V behavior (Kumari and Toshniwal 2021). Uneven dust distribution also hinders fault detection and reduces irradiance estimation accuracy (Lee et al. 2021; Beigi et al. 2022).

b. Temperature and Humidity-Induced Fault Mechanisms

Temperature and humidity critically affect PV performance, with each 1 °C rise causing about a 0.5% drop in efficiency (Danylchenko et al. 2024). Heat and moisture accelerate physical degradation like corrosion and delamination, increasing internal resistance. Faults are nonlinear and time-dependent, making AI tools like CNNs and RNNs useful for detecting and modeling these anomalies.

B. Hardware Configuration for Real-Time Monitoring

Real-time PV system monitoring in harsh environments uses a tailored setup featuring a buck-boost converter, TMS320F28379D microcontroller, and sensors (Sayyad and Nasikkar 2021). PWM control dynamically adjusts voltage to address dust- and humidity-induced fluctuations. This configuration helps maintain stable power delivery despite environmental variations.

As shown in Fig. 1, the monitoring and regulation unit integrates sensors and a DC–DC converter for real-time PV performance stabilization.

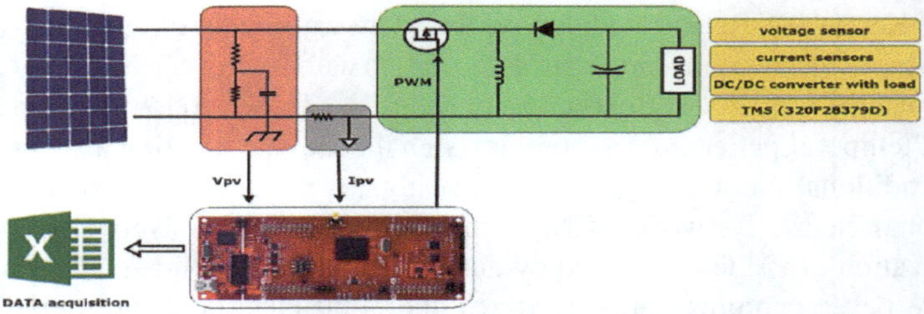

Fig. 1 Schematic representation of the monitoring and regulation unit in a PV system (Labiod et al. 2025)

C. Sensor Feedback and Data Acquisition for Fault Classification

Sensor feedback in PV systems combines voltage/current sensing with a TMS320F28379D microcontroller for precise, real-time monitoring (Sayyad and Nasikkar 2021). Signals are digitized and analyzed using CNNs for spatial anomalies and RNNs for time-based patterns. This fusion of hardware and AI enables reliable fault classification in harsh environments.

D. Recurrent Neural Networks (RNNs) for Temporal Fault Detection

Recurrent Neural Networks (RNNs) are designed to handle time-dependent data, making them suitable for analyzing PV systems under environmental stress like dust and humidity (Kumari and Toshniwal 2024; Ahn and Park 2021). They capture evolving nonlinear patterns, though advanced forms like LSTM and GRU address limitations in long-term memory (Jalal et al. 2024; Sundaram et al. 2021). In PV systems, RNNs aid in forecasting and fault detection under varying conditions (Pérez-Rodríguez et al. 2024; Jiang 2024).

E. Convolutional Neural Networks (CNNs) for Spatial Fault Recognition

CNNs are well-suited for spatial fault detection in PV systems, identifying anomalies like I–V or P–V distortions caused by dust and moisture (Ajitha and Martínez-Ramón 2021). They use convolutional, pooling, and dense layers to extract and classify features (Mansouri et al. 2021; Jalal et al. 2024). CNNs enable accurate, automated detection and have applications in solar forecasting and hybrid fault classification (Ghimire et al. 2022).

F. Model Architecture and Training Parameters

To ensure a fair comparison, the RNN and CNN classifiers were designed and trained using an identical dataset representing two PV operational states: Healthy and Shadow. In this work, both models were configured with specific optimization parameters and distinct architectural representations as detailed in Table 1.

These configurations were selected based on empirical trials and prior literature benchmarks to optimize detection of environmental anomalies while minimizing overfitting.

G. Summary of the Experimental Workflow

Table 1 Proposed model architectures and hyperparameters for the RNN and CNN classifiers

Parameter	RNN Model	CNN Model
Input Shape	(200, 2) (voltage & power as sequences)	400 (flattened voltage & current)
Data Split	60% Train – 20% Val – 20% Test	Same
Samples per Class	Total: 190 samples (Dust = 95, Humidity = 95); Train: 114; Val: 38; Test: 38	Same
Number of Layers	1 RNN + 2 Dense	2 Conv1D + 2 MaxPcol + 2 Dense
Hidden Units	RNN: 128; Dense: 64	Conv1D: 32, 64; Dense: 128
Activation Functions	tanh (RNN), ReLU (Dense), Softmax (Output)	ReLU (Conv & Dense), Softmax (Output)
Dropout	0.4 – after RNN and Dense layers	0.2 after Conv1D, 0.4 after Dense
Batch Normalization	Not used	Used after each Conv1D layer
Noise Injection	Gaussian noise (factor = 0.01) added during training	Same
Optimizer	Adam	Adam
Learning Rate	0.001 (default Adam setting)	0.001
Epochs	100 + EarlyStopping (patience = 15)	100 + EarlyStopping (patience = 15)
Batch Size	8	8
Output Classes	2 (Healthy, Shadow)	2 (Healthy, Shadow)
Total Parameters	~13,000	~22,000

Figure 2 illustrates the complete experimental framework adopted for intelligent classification of PV system faults under dust and humidity conditions. The process spans from signal acquisition to final model evaluation and comprises five main stages:

a. **PV Panels Under Fault Conditions**

Two PV panels were tested under real-world degradation scenarios: Panel 1 experienced dust accumulation, while Panel 2 faced high humidity. These conditions simulate common environmental stressors in outdoor PV setups, allowing for performance comparison and fault behavior analysis.

b. **Signal Acquisition**

Electrical signals, including voltage V(t) and power P(t), were captured using sensors connected to the PV panels. These analog signals were routed through an oscilloscope and recorded on a PC, forming the raw input dataset for subsequent processing.

c. **Data Preprocessing**

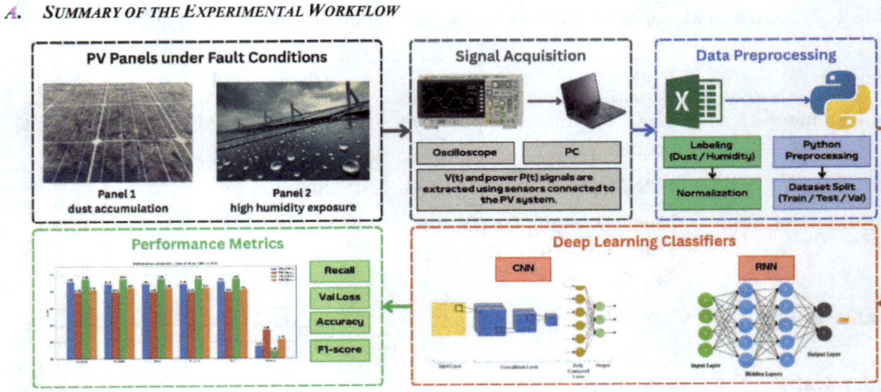

Fig. 2 Workflow of data processing and DL-based fault classification for dusty and humid PV panels

The data collected from the PV modules were saved in Excel format and then preprocessed in Python. The pipeline involved labeling each instance based on the environmental condition ("Dust" or "Humidity"), normalizing features to standardize the data, and splitting it into training, validation, and testing sets for robust model performance and generalization.

d. Deep Learning Classifiers

Two deep learning models CNN for spatial features and RNN for temporal patterns were trained on the same labeled and normalized dataset. Both architectures processed input signals independently to classify dust- and humidity-induced faults. Their outputs were used to assess classification accuracy and performance under varying environmental stressors.

e. Performance Metrics

Final predictions were evaluated using Accuracy, Recall, F1-score, and Validation Loss, ensuring comprehensive performance assessment. A bar chart (in the bottom block of the figure) visually compares CNN and RNN results

3 Results

H. Analysis of Environmental Effects on P–V Characteristics

Dust accumulation on PV module surfaces leads to a pronounced drop in power output, particularly under high-voltage operation. As shown in Fig. 3, irradiance-dependent P–V curves exhibit lower peak power and shortened voltage ranges when the surface is partially blocked by fine particles. The light-scattering and absorption phenomena caused by airborne dust reduce photon transmission and exacerbate mismatch losses. These changes are spatially distributed, making CNNs particularly effective in identifying such fault patterns based on localized feature distortion.

Humidity-induced degradation manifests through erratic fluctuations in the P–V curves (Fig. 3), often caused by increased surface conductivity, condensation, and soiling. These effects lead to instability in maximum power point tracking and increased leakage currents. The temporal behavior of these anomalies aligns well with RNN-based models, which are capable of capturing sequential inconsistencies over time to detect moisture-related disturbances in real time.

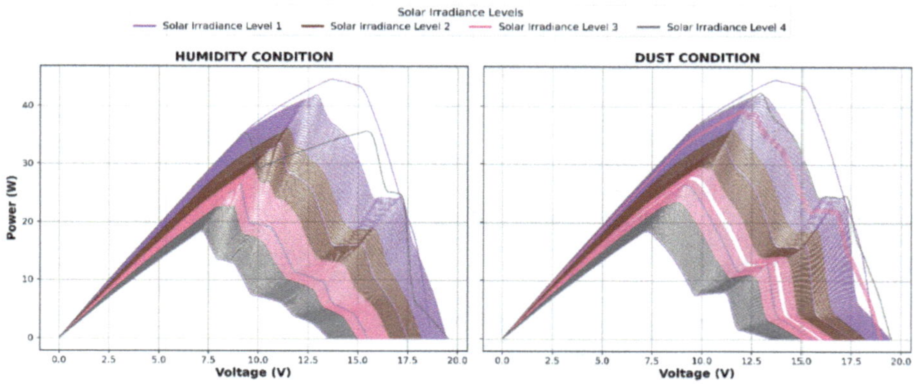

Fig. 3 P–V behavior under dust and humidity conditions across varying irradiance levels

I. Performance Metrics

To evaluate the classification performance of the individual deep learning models RNN and CNN under environmental stressors such as dust and humidity, four standard evaluation metrics were computed from the confusion matrix: accuracy, precision, recall, and F1-score, as summarized in Table 2.

The results show that the RNN and CNN models maintain high recall, which is crucial for minimizing missed fault cases, while their F1-scores demonstrate strong resistance to overfitting even under noisy or mixed environmental conditions.

J. Visual Performance Comparison Under Clean and Noisy Conditions

To evaluate model generalization under real-world uncertainty, both RNN and CNN models were tested in two scenarios: clean (ideal signal input) and noisy (Gaussian noise injected at training level). Figure4 illustrates the comparative performance across six key metrics: Accuracy, Precision, Recall, F1-score, Area Under the ROC Curve (AUC), and Validation Loss.

Note: Validation Loss values were normalized to fall within the 0–1 range for visual comparison with other metrics.

As shown in Fig. 4, both models exhibit robust performance in clean conditions, with CNN outperforming RNN across all metrics. CNN achieves a perfect score of 0.95 in Accuracy, Precision, Recall, and F1-score, while RNN slightly lags with 0.89–0.92 across the same metrics. Under noisy conditions, performance drops are observed for

Table 2 Performance evaluation metrics for RNN and CNN models in PV fault classification

Metric	Description	Interpretation	Mathematical Formula	References
Accuracy	Correct predictions over total predictions	Reflects general model effectiveness across all classes	$\frac{TN+TP}{TN+TP+FN+FP}$	Markoulidakis et al. (2021)
Precision	True positives over predicted positives	Indicates model's reliability in labeling faults	$\frac{TP}{TP+FP}$	Ahn and Park 2021
Recall	True positives over actual positives	Measures sensitivity; crucial to avoid missing critical PV faults	$\frac{TP}{TP+FN}$	Markoulidakis et al. (2021)
F1-Score	Harmonic mean of precision and recall	Balances false positives and false negatives; ideal for imbalanced fault cases	$\frac{2\times(Precision\times Recall)}{(Precision+Recall)}$	Pérez-Rodríguez et al. (2024)

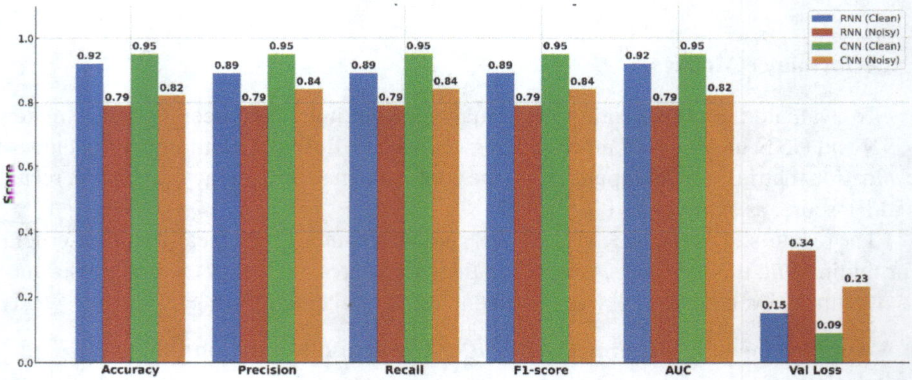

Fig. 4 Performance comparison of RNN and CNN models under clean and noisy conditions

both models; however, CNN still maintains superiority, particularly in Precision (0.84 vs. 0.79) and AUC (0.92 vs. 0.89). The Validation Loss also reflects CNN's resilience, showing a lower error value (0.23) compared to RNN (0.34) under noise.

K. Confusion Matrix Analysis

Normalized confusion matrices (Fig. 5) illustrate CNN and RNN accuracy in classifying dust and humidity faults under clean and noisy conditions.

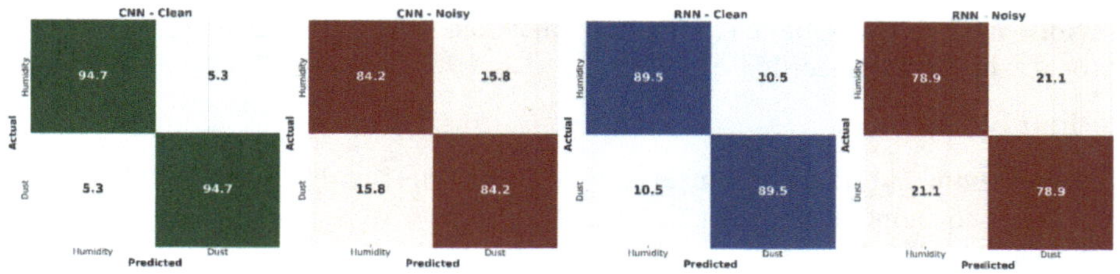

Fig. 5 Normalized confusion matrices for CNN and RNN models under clean (left) and noisy (right) testing conditions

Under clean conditions, CNN and RNN classify faults with high accuracy CNN at 94.7%, outperforming RNN's 89.5%. In noisy settings, both degrade, but CNN remains more robust, retaining 84.2% accuracy versus RNN's 78.9%, with fewer misclassifications.

4 Discussion

The comparative evaluation of CNN and RNN models under both clean and noisy conditions has yielded valuable insights into their operational strengths and weaknesses when applied to PV fault classification. The experimental results, particularly those illustrated in Figs. 4 and 5, consistently demonstrate the superior performance of the CNN model across multiple metrics accuracy, precision, recall, F1-score, and AUC. This dominance is especially pronounced under noisy testing conditions, where CNN maintained a lower validation loss (0.23) and a lower misclassification rate (15.8%) compared to RNN's 21.1%.CNN's advantage can be attributed to its ability to localize and generalize spatial distortions caused by surface-level anomalies such as dust. In contrast, the RNN model exhibited greater sensitivity in capturing temporal anomalies, particularly in clean conditions, reflecting its suitability for detecting humidity-related variations. However, its performance degraded more significantly under noise due to the inherent challenges of modeling long-term dependencies in distorted sequences. These findings align with prior works in deep learning-based fault detection in PV systems and extend them by offering a side-by-side evaluation of spatial versus temporal architectures under controlled environmental stressors. Unlike many previous studies that focus on a single type of fault or architecture, this work systematically compares two learning paradigms using real-world electrical data and controlled perturbations.

Strengths of the Study:

- Inclusion of both dust and humidity faults, representing common environmental stressors in PV operation.
- Evaluation of clean and noise-injected conditions to simulate real-world uncertainty.
- Balanced experimental design with identical datasets for both CNN and RNN models.

Limitations:

- The study was restricted to two binary fault classes (dust vs. humidity).
- Hybrid or ensemble deep learning models (e.g., CNN-LSTM) were not explored.

- Noise modeling was limited to Gaussian injection, without considering temporal drift or sensor-related distortion.

Future Work:

- Investigating hybrid architectures (CNN-LSTM, attention-based RNNs) to combine spatial and temporal strengths.
- Expanding the classification task to include additional fault types such as partial shading, hot spots, or diode failures.
- Implementing the models on embedded systems for real-time PV monitoring at the edge.

5 Conclusion

This study presented a comprehensive investigation into the detection of environmental faults in photovoltaic systems using deep learning models. By focusing on two common stressors dust and humidity and comparing two distinct architectures, CNN and RNN, the work provides critical insights into the applicability and robustness of AI-based diagnostics in renewable energy systems. The results demonstrated that while both models achieved strong classification performance in clean environments, the CNN model consistently outperformed RNN under noisy conditions. This highlights CNN's superior generalization ability and resilience to signal distortion, making it particularly suitable for dust-related anomaly detection. Conversely, RNN exhibited promising recall performance in humidity scenarios due to its temporal modeling capability. The significance of this research lies in its dual-model framework, realistic experimental conditions, and rigorous performance analysis, which together support the adoption of deep learning techniques in next-generation PV monitoring systems. Future advancements integrating hybrid architectures and real-time deployment are expected to further enhance reliability, efficiency, and scalability in fault management for solar energy infrastructures.

References

Ahn, H.K., Park, N.: Deep RNN-based photovoltaic power short-term forecast using power IoT sensors. Energies. **14**(2), 436 (2021). https://doi.org/10.3390/en14020436

Ajitha, M., Martínez-Ramón, M.: Deep learning-based solar radiation micro forecast by fusion of infrared cloud images and radiation data. Appl. Energy. **301**, 117469 (2021). https://doi.org/10.1016/j.apenergy.2021.117469

Alam, M.K., Khan, F., Johnson, J., Flicker, J.: A comprehensive review of catastrophic faults in PV arrays: Types, detection, and mitigation techniques. IEEE J. Photovolt. **5**(3), 982–997 (2015). https://doi.org/10.1109/JPHOTOV.2015.2397599

Bebboukha, A., et al.: Finite control set model predictive current control for three-phase grid-connected inverter with common mode voltage suppression. Sci. Rep. **14**, 19832 (2024). https://doi.org/10.1038/s41598-024-71051-9

Beigi, M. et al.: Forecasting of power output of a PVPS based on meteorological data using RNN approaches. Sustainability. **14**(5), 3104 (2022). https://doi.org/10.3390/su14053104

Collier, D.E., Key, T.S.: Electrical fault protection for a large photovoltaic power plant inverter. In: Proceedings of the 20th IEEE Photovoltaic Specialists Conference, pp. 1035–1042. Las Vegas, NV, USA (1988). https://doi.org/10.1109/PVSC.1988.105863

Danylchenko, D., Potryvai, A., Petrovskyi, M., Leliuk, M., Petrova, Y.: Advancements in operational modeling methods for photovoltaic power plants. In: Babak, V., Zaporozhets, A. (eds.) Systems, Decision and Control in Energy VI, Volume II: Power Engineering and Environmental Safety, pp. 1–15. Springer Nature, Cham, Switzerland (2024). https://doi.org/10.1007/978-3-031-67091-6_15

Di Lorenzo, G., Araneo, R., Mitolo, M., Niccolai, A., Grimaccia, F.: Review of O&M practices in PV plants: Failures, solutions, remote control, and monitoring tools. IEEE J. Photovolt. **10**(4), 914–926 (2020). https://doi.org/10.1109/JPHOTOV.2020.2994531

Ghimire, S., Bhandari, B., Casillas-Pérez, D., Deo, R.C., Salcedo-Sanz, S.: Hybrid deep CNN-SVR algorithm for solar radiation prediction problems in Queensland, Australia. Eng. Appl. Artif. Intell. **114**, 104860 (2022). https://doi.org/10.1016/j.engappai.2022.104860

Jalal, M., Khalil, I.U., ul Haq, A.: Deep learning approaches for visual faults diagnosis of photovoltaic systems: State-of-the-art review. Results Eng. **23**, 102622 (2024). https://doi.org/10.1016/j.rineng.2024.102622

Jiang, C.: African vulture optimized RNN algorithm maximum power point tracking (MPPT) controller for photovoltaic (PV) system. Meas.: Sens. **36**, 101392 (2024). https://doi.org/10.1016/j.measen.2024.101392

Köntges, M., et al.: Review of failures of photovoltaic modules. IEA-PVPS T13-01:2014. https://iea-pvps.org/wp-content/uploads/2020/01/IEA-PVPS_T13-01_2014_Review_of_Failures_of_Photovoltaic_Modules_Final.pdf (2014)

Kumari, P., Toshniwal, D.: Deep learning models for solar irradiance forecasting: A comprehensive review. J. Clean. Prod. **316**, 128566 (2021). https://doi.org/10.1016/j.jclepro.2021.128566

Labiod, C., et al.: Enhanced photovoltaic panel diagnostics through AI integration with experimental DC to DC Buck boost converter implementation. Sci. Rep. **15**, 295 (2025). https://doi.org/10.1038/s41598-024-84365-5

Lee, C.-H., Yang, H.-C., Ye, G.-B.: Predicting the performance of solar power generation using deep learning methods. Appl. Sci. **11**(15), 6887 (2021). https://doi.org/10.3390/app11156887

Mansouri, M., Trabelsi, M., Nounou, H., Nounou, M.: Deep learning-based fault diagnosis of photovoltaic systems: A comprehensive review and enhancement prospects. IEEE Access. **9**, 126286–126306 (2021). https://doi.org/10.1109/ACCESS.2021.3110947

Markoulidakis, I., Rallis, I., Georgoulas, I., Kopsiaftis, G., Doulamis, A., Doulamis, N.: Multiclass confusion matrix reduction method and its application on net promoter score classification problem. Technologies. **9**(4), 81 (2021). https://doi.org/10.3390/technologies9040081

Pérez-Rodríguez, S.A., et al.: Metaheuristic algorithms for solar radiation prediction: A systematic analysis. IEEE Access. **12**, 100134–100151 (2024). https://doi.org/10.1109/ACCESS.2024.3429073

Pillai, D.S., Natarajan, R.: A comprehensive review on protection challenges and fault diagnosis in PV systems. Renew. Sust. Energ. Rev. **91**, 18–40 (2018). https://doi.org/10.1016/j.rser.2018.03.082

Sayyad, J., Nasikkar, P.: Design and development of low cost, portable, on-field I-V curve tracer based on capacitor loading for high power rated solar photovoltaic modules. IEEE Access. **9**, 70715–70731 (2021). https://doi.org/10.1109/ACCESS.2021.3078532

Sundaram, K.M., Hussain, A., Sanjeevikumar, P., Holm-Nielsen, J.B., Kaliappan, V.K., Santhoshi, B.K.: Deep learning for fault diagnostics in bearings, insulators, PV panels, power lines, and electric vehicle applications – The state-of-the-art approaches. IEEE Access. **9**, 41246–41260 (2021). https://doi.org/10.1109/ACCESS.2021.3064360

Triki-Lahiani, A., Abdelghani, A.B.B., Slama-Belkhodja, I.: Fault detection and monitoring systems for photovoltaic installations: A review. Renew. Sust. Energ. Rev. **82**, 2680–2692 (2017). https://doi.org/10.1016/j.rser.2017.09.101

Intelligent Classification of Partial Shading in PV Systems Using LSTM and DNN Models: A Comparative Study

Souhaib Remha[1(✉)], Laouamer Mosbah[1], and Adaika Mohammed[2]

[1] UDERZA Unit, Mechanical Department, Faculty of Technology, University of Eloued, Eloued 39000, Algeria
{remha-souhaib,laouamer-mosbah}@univ-eloued.dz

[2] LPMRN Laboratory, Department of Electromechanical Engineering, University of Bordj Bou Arreridj, El-Anasser 34030, Bordj Bou Arreridj, Algeria
mohammed.adaika@univ-bba.dz

Abstract. **Introduction**. Photovoltaic (PV) systems are vital to modern renewable energy strategies, yet they remain vulnerable to environmental disturbances. **The problem** of partial shading—often caused by clouds, buildings, or natural obstructions—leads to irregular power output patterns that challenge traditional fault detection methods. **Goal**. This paper investigates two deep learning models—Long Short-Term Memory (LSTM) and Deep Neural Networks (DNN)—for classifying operational states in PV systems, specifically distinguishing between healthy and shadowed conditions. **Methodology**. Electrical measurements of voltage and current were collected from a PV setup under real-world shading conditions. Both models were trained using identical datasets and evaluated through accuracy metrics and ROC-AUC analysis. **Results**. The LSTM model showed superior performance in identifying time-dependent shading patterns, while DNN was less sensitive to such variations. **Originality**. Unlike prior works, this study integrates AI-based diagnostics with actual system measurements, offering a realistic and practical benchmark. **Practical Value**. The proposed approach enhances fault detection accuracy in PV systems, reduces energy losses, and supports intelligent monitoring, making it highly suitable for deployment in real-time renewable energy applications.

Keywords: Photovoltaic systems · Partial shading · LSTM · DNN · Deep learning · Artificial intelligence

1 Introduction

Photovoltaic (PV) systems play a critical role in advancing global renewable energy adoption. However, partial shading remains a major impediment to their reliability and efficiency, often caused by cloud cover, nearby structures, or environmental obstructions (Triki-Lahiani et al. 2017; Alam et al. 2015). Unlike mechanical failures, shading introduces complex, non-linear variations in power output that closely resemble natural irradiance fluctuations, making fault detection especially difficult (Di Lorenzo et al.

M. Hatti (Ed.): IC-AIRES 2025, LNNS 1740, pp. 290–300, 2026.
https://doi.org/10.1007/978-3-032-12788-4_28

2020; Köntges et al. 2014). Traditional threshold-based detection methods are largely ineffective under such conditions, particularly when MPPT algorithms are active and the signal-to-noise ratio is low (Pillai and Natarajan 2018).

To address these limitations, this study explores the application of deep learning for shading fault detection, comparing Long Short-Term Memory (LSTM) networks and Deep Neural Networks (DNNs). Both models are trained on real-time electrical measurements to classify healthy and shaded states in PV arrays.

Experimental results reveal that the LSTM model outperforms DNNs in capturing shading-induced anomalies due to its capacity for modeling temporal dependencies. The novelty of this work lies in the direct, hardware-synchronized comparison of both models, offering actionable insights for intelligent PV diagnostics. Importantly, the proposed approach not only improves classification accuracy but also facilitates proactive maintenance and minimizes energy losses, enhancing the overall resilience of solar energy systems.

2 Materials and Methods

A. *Environmental Variables Affecting PV Classification*

The ability of a photovoltaic (PV) system to maintain consistent power output is largely dependent on solar irradiance and temperature conditions. These two environmental factors, although natural and cyclical, exert a significant influence on the distinction between normal and faulty operational states—particularly in the presence of partial shading.

a. *Solar Irradiance Under Shading Scenarios*

Solar irradiance is inherently unstable due to atmospheric dynamics such as clouds, aerosols, and obstructions. Partial shading introduces abrupt changes that distort the I–V and P–V characteristics, often mimicking transient environmental events. These distortions hinder traditional algorithms from effectively distinguishing shaded from healthy modules (Ghimire et al. 2022; Ajitha and Martínez-Ramón 2021; Kumari and Toshniwal 2021).

Moreover, irradiance forecasting tools exhibit reduced accuracy in locally shaded, fast-changing environments (Lee et al. 2021), necessitating the use of time-series models to interpret short-duration anomalies.

b. *Temperature-Irradiance Interaction and Its Impact on System Behavior*

Temperature further complicates system behavior by increasing internal resistance and reducing output voltage under high irradiance. These fluctuations yield subtle, short-term anomalies that static models fail to capture (Danylchenko et al. 2024).

In response, Long Short-Term Memory (LSTM) networks are leveraged to learn temporal dependencies, while Deep Neural Networks (DNNs) offer strong feature abstraction capabilities. Both are employed in this study to evaluate their effectiveness in classifying shaded versus healthy states under dynamic solar conditions.

B. *System Configuration and Monitoring Setup*

The study utilizes a tailored hardware architecture to capture real-time electrical signals under varying irradiance. The setup integrates a DC-DC buck-boost converter, voltage/current sensors, and a TMS320F28379D microcontroller, as illustrated in Fig. 1.

The converter operates with Pulse Width Modulation (PWM) to dynamically regulate voltage, enabling stable output even during disturbances like partial shading. This architecture facilitates both optimal energy extraction and high-resolution data acquisition for model training (Sayyad and Nasikkar 2021).

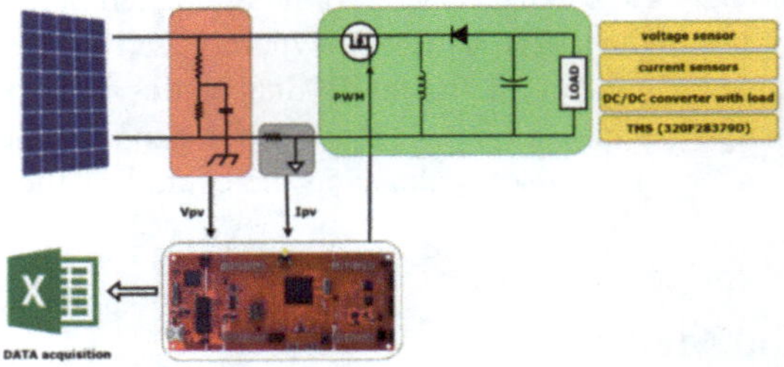

Fig. 1. Schematic representation of a PV monitoring system with a DC-DC converter (Labiod et al. 2025)

C. *Data Acquisition and Signal Processing*

At the heart of the data acquisition process is the TMS320F28379D LaunchPad, which handles sensor interfacing, signal conditioning, and execution of MPPT algorithms. An Excel-based dashboard is used for performance logging and visualization.

Voltage is measured using a precision voltage divider, which scales the panel output to a safe level for the microcontroller's ADC. For current sensing, a shunt resistor is employed to convert current flow into a measurable voltage. These analog signals are digitized via the embedded ADC and processed for fault classification tasks (Dunna et al. 2024; Sayyad and Nasikkar 2021).

Together, this configuration offers a stable and accurate monitoring environment. The measured voltage and current data serve as input features for the LSTM and DNN models, enabling reliable classification of operational states under normal and shaded conditions.

D. *Deep Learning Models for Fault Classification*

To classify PV states under changing environmental conditions, the study implements and compares two deep learning models: LSTM networks and DNNs. The architectures of these models are illustrated in Fig. 2. These models were selected based on their complementary strengths in handling temporal sequences and high-dimensional data.

a. *Long Short-Term Memory (LSTM) Networks*

LSTM networks are a specialized form of RNNs capable of learning long-term dependencies via memory cells and gating mechanisms (Jalal et al. 2024). Their architecture enables accurate modeling of dynamic PV behavior influenced by shading, dust, or humidity (Veerasamy et al. 2021; Bakır et al. 2023; Karthikeyan and Jagadeeshwaran 2023).

The LSTM model was trained to:

- Detect shading-induced temporal variations (Karthikeyan and Jagadeeshwaran 2023),
- Identify humidity-driven changes (Bakır et al. 2023),
- Forecast deviations linked to irradiance–temperature interactions (Kumari and Toshniwal 2024).

b. ***Deep Neural Networks (DNNs)***

DNNs are composed of multiple hidden layers that process raw input through a sequence of non-linear transformations. Their hierarchical structure facilitates the extraction of complex features from raw voltage and current signals obtained from PV systems (Ajitha and Martínez-Ramón 2021). Unlike LSTMs, DNNs do not retain memory over time but are effective in classifying patterns present within static segments of input data.

The DNN model was trained using standard backpropagation, where:

- Forward propagation generated predicted outputs (Sudharshan et al. 2022),
- Errors were computed using the Mean Squared Error (Pikus and Wąs 2023),
- Weight updates were performed using gradient descent (Sudharshan et al. 2022).

While more sensitive to overfitting, DNNs offer faster training and lower computational overhead, making them suitable for applications with limited hardware resources (Ahn and Park 2021).

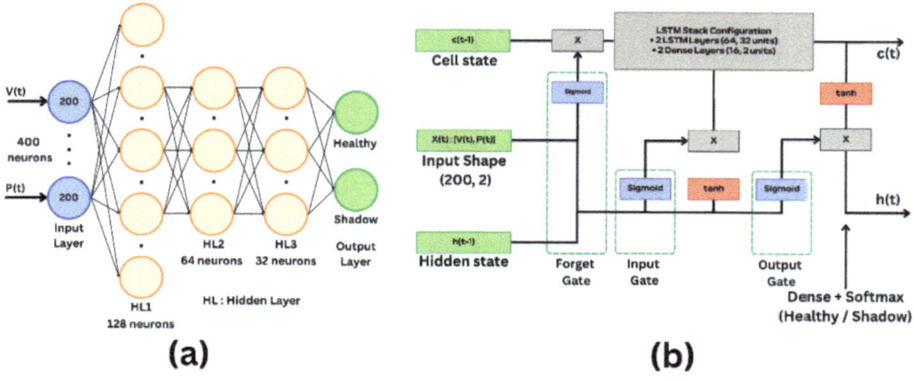

Fig. 2. Deep learning architectures used for PV fault classification: (a) Multilayer DNN structure; (b) LSTM memory cell model (Designed and adapted by the authors based on Jalal et al. 2024; El-Banby et al. 2022)

E. ***Evaluation Metrics***

In the assessment of classification models applied to photovoltaic (PV) fault detection, the Confusion Matrix serves as a core interpretative tool. It organizes model predictions across four categories (Table 1):

Table 1. Confusion matrix for binary classification

	Predicted Positive	Predicted Negative
Actual Positive	True Positive (TP)	False Negative (FN)
Actual Negative	False Positive (FP)	True Negative (TN)

This matrix supports the derivation of more nuanced performance indicators such as accuracy, precision, recall, and F1-score, which are mathematically defined as follows:

- Accuracy provides an overall measure of correctness and is expressed as:

$$\text{Accuracy} = \frac{\text{TN} + \text{TP}}{\text{TN} + \text{TP} + \text{FN} + \text{FP}} \tag{1}$$

It is particularly informative when class distributions are balanced (Veerasamy et al. 2021).

- Precision, which quantifies the proportion of correctly identified positives among all predicted positives, is given by:

$$\text{Precision} = \frac{\text{TP}}{\text{TP} + \text{FP}} \tag{2}$$

It is critical in PV diagnostics where false positives may lead to unnecessary maintenance interventions (Veerasamy et al. 2021).

- Recall (Sensitivity) measures the model's ability to detect actual positives and is defined as:

$$\text{Recall} = \frac{\text{TP}}{\text{TP} + \text{FN}} \tag{3}$$

High recall is essential in safety-critical PV systems to avoid undetected faults (Veerasamy et al. 2021).

- F1-score, the harmonic mean of precision and recall, is particularly suitable for imbalanced datasets and is computed as:

$$\text{F1} - \text{Score} = \frac{2 \times (Precision \times Recall)}{(Precision + Recall)} \tag{4}$$

This metric balances the trade-off between false positives and false negatives, offering a comprehensive evaluation for PV fault classification (Pérez-Rodríguez et al. 2024).

F. ***Model Configuration and Training Parameters***

To ensure a fair and consistent comparison, both the Long Short-Term Memory (LSTM) and Deep Neural Network (DNN) classifiers were trained using the same dataset

Table 2. Model architectures and hyperparameters for LSTM and DNN classifiers

Parameter	LSTM Model	DNN Model
Input Shape	(200, 2)	400 (flattened voltage & current)
Number of Layers	2 LSTM + 2 Dense	3 Dense
Hidden Units	LSTM: 64, 32; Dense: 16, 2	Dense: 128, 64, 32
Activation Functions	ReLU + Softmax	ReLU + Softmax
Dropout	not used — preserved convergence	not used — preserved convergence
Optimizer	Adam	Adam
Learning Rate	0.001	0.001
Epochs	100	100
Batch Size	32	32
Output Classes	2 (Healthy, Shadow)	2 (Healthy, Shadow)
Total Parameters	12,000	90,000

representing two operational states: *Healthy* and *Shadow*. The architectural details and hyperparameter settings are summarized in Table 2.

The configurations were selected based on literature benchmarks and prior sensitivity analyses to ensure a balance between model complexity and generalization capability.

3 Results

G. *Power–Voltage Behavior Under Healthy and Shaded Conditions*

To better understand the dynamic response of the photovoltaic (PV) system, the power–voltage (P–V) characteristics were analyzed under both fault-free and shaded operating conditions, as shown in Fig. 3. Under clear-sky conditions, the P–V curves exhibit consistency and symmetry, with peak power scaling approximately linearly with solar irradiance, a behavior widely reported in the literature (Labiod et al. 2025; Mansouri et al. 2021). This consistent profile serves as a reference benchmark for detecting deviations in system performance. Deep Neural Networks (DNNs), which are particularly effective in static pattern recognition, were observed to successfully learn this baseline behavior, enabling accurate classification of healthy operational states.

In contrast, partial shading introduces complexity to the P–V curve, resulting in multiple local peaks corresponding to different maximum power points (MPPs). These are caused by electrical mismatch between shaded and unshaded cells, generating resistive losses and fluctuating output power. This dynamic and irregular behavior is more effectively modeled by Long Short-Term Memory (LSTM) networks, which capture temporal sequences in the evolution of the output signal.

H. *Accuracy and Loss Evaluation*

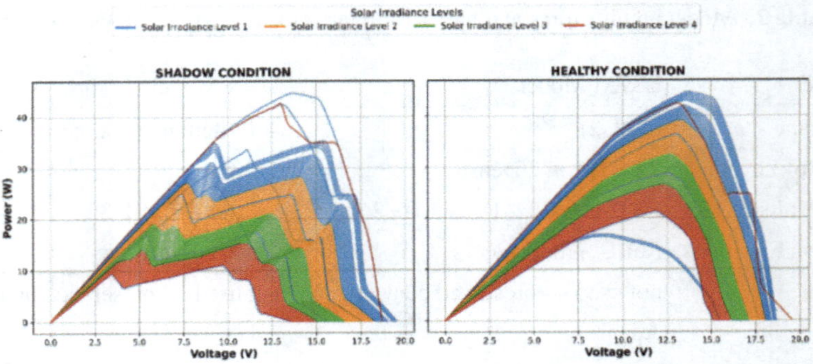

Fig. 3. Typical P–V responses under healthy and partially shaded conditions

The training performance of the LSTM and DNN models was assessed using the evolution of accuracy and loss over 100 epochs, as illustrated in Fig. 4. The LSTM model exhibited rapid and consistent convergence, surpassing 99% training accuracy within the initial epochs and stabilizing with a validation accuracy close to 95%. In contrast, the DNN model, although capable of reaching comparable training accuracy, showed less stability during validation, with noticeable fluctuations in both accuracy and loss curves.

These observations suggest that the LSTM model demonstrates superior learning dynamics and generalization capabilities, while the DNN model appears more sensitive to initialization and prone to minor overfitting under the given configuration.

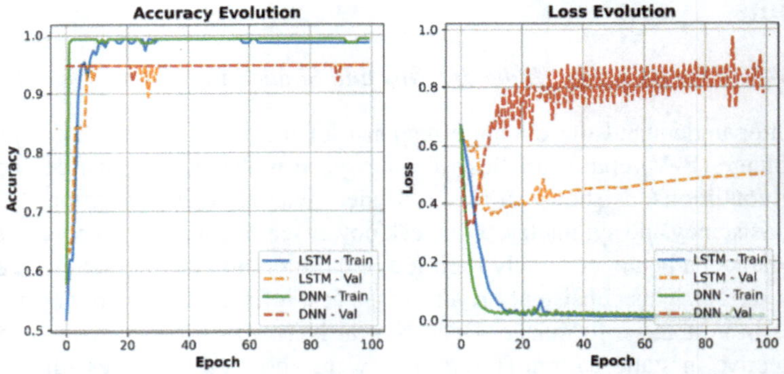

Fig. 4. Training and validation accuracy and loss curves for LSTM and DNN models over 100 epochs

I. ***Confusion Matrix Analysis***

To further investigate classification behavior, normalized confusion matrices were generated for both models, as depicted in Fig. 5. Both LSTM and DNN achieved perfect classification (100%) of the Healthy class, indicating high reliability in detecting nominal operating states. However, distinguishing the Shadow class proved more challenging:

- The LSTM model misclassified 10.5% of shadowed samples.
- Similarly, the DNN model misclassified 10.5% of shadowed cases.

Despite identical recall values, the LSTM model maintained greater training stability and exhibited fewer fluctuations in its predictions, reinforcing its suitability for handling temporal variations in PV data.

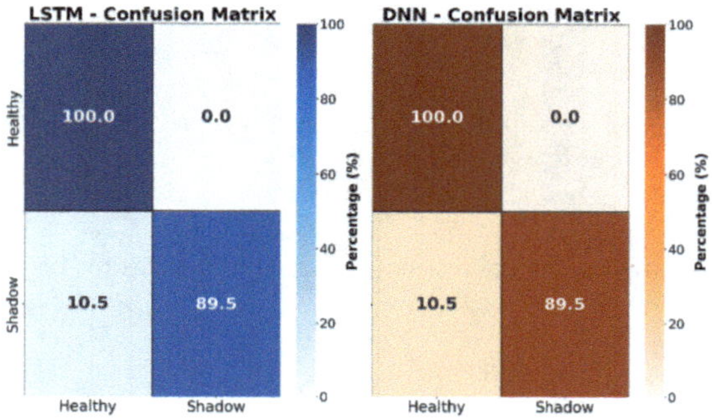

Fig. 5. Normalized confusion matrices for LSTM and DNN models in classifying Healthy and Shadow conditions

J. **Performance Metrics Comparison**

A detailed comparison of performance metrics is presented in Fig. 6, including Accuracy, Precision, Recall, F1-score, AUC, and Validation Loss. Although both models achieved identical scores across most classification metrics, the Validation Loss clearly distinguished the two:

- LSTM Validation Loss: 0.50
- DNN Validation Loss: 0.78

This lower loss value suggests that the LSTM model was more effective in minimizing the classification error on unseen data, confirming its enhanced generalization and resilience during training.

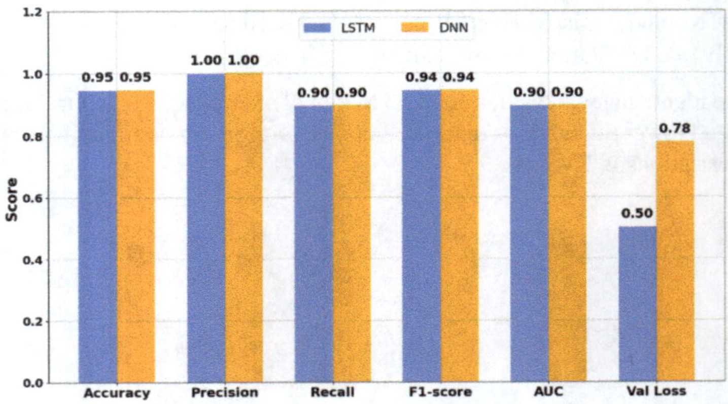

Fig. 6. Bar chart comparison of classification performance metrics for LSTM and DNN models

4 Discussion

The comparative analysis between Long Short-Term Memory (LSTM) and Deep Neural Network (DNN) models highlights notable performance distinctions in the classification of faults in photovoltaic (PV) systems under partial shading. Despite both models achieving equal scores across core metrics—Accuracy (0.95), Precision (1.00), Recall (0.90), and F1-score (0.94)—LSTM demonstrated superior generalization, as evidenced by its lower validation loss (0.50 vs. 0.78 for DNN) and smoother convergence behavior.

This advantage stems from LSTM's capability to capture temporal dependencies in sequential PV data, which is essential in tracking dynamic variations in irradiance and temperature. In contrast, while the DNN offered strong static feature abstraction, it exhibited greater susceptibility to overfitting and initialization sensitivity due to its feedforward nature.

These outcomes align with previous studies that underscored the value of recurrent architectures in solar fault detection (Karthikeyan and Jagadeeshwaran 2023; Veerasamy et al. 2021). Notably, the current study extends prior work by validating model performance using real-time voltage and current data collected from a hardware-based PV system, thus ensuring practical relevance beyond simulation environments.

Strengths: A key strength of the study lies in its rigorous methodology, which involves synchronized data acquisition under real shading scenarios and a thorough evaluation using diverse performance metrics. The integration of real-world signals enhances the credibility and deployment potential of the LSTM-based diagnostic approach.

Limitations: Nonetheless, the work is limited to binary classification (Healthy vs. Shadow), omitting more complex faults such as open circuits or bypass diode failures. Moreover, while the dataset is representative, its scope could be broadened to cover different climatic and seasonal conditions.

Future Directions: Future research should expand the fault taxonomy to multi-class classification, incorporate additional environmental parameters (e.g., humidity, wind), and investigate hybrid architectures such as CNN-LSTM or Transformer-based models for more comprehensive feature extraction. Additionally, deploying these models on edge computing platforms would enable real-time, low-latency PV fault diagnostics in field conditions.

5 Conclusion

This research conducted a structured comparison between Long Short-Term Memory (LSTM) networks and Deep Neural Networks (DNN) for detecting partial shading conditions in photovoltaic (PV) systems. Both models were trained on real-world voltage and current data acquired from a physical PV setup operating under controlled shading disturbances. The experimental results consistently favored the LSTM model, which demonstrated better convergence behavior, lower validation loss, and higher stability in detecting time-dependent anomalies caused by irradiance fluctuations.

The superior performance of LSTM is attributed to its recurrent architecture, which enables effective modeling of temporal patterns in PV signal dynamics—an essential requirement in realistic environments where environmental variability is non-negligible. Conversely, while the DNN model achieved competitive metrics in static evaluations, its sensitivity to initialization and reduced robustness against sequential noise underscore its limitations in dynamic contexts.

Despite these promising results, the study was constrained to binary classification and a limited fault profile. Future extensions should address multi-class detection schemes, introduce more diverse environmental factors (e.g., humidity, dust), and evaluate real-time deployment potential using lightweight embedded hardware. In essence, this work reinforces the viability of sequence-based AI models in intelligent PV diagnostics and paves the way for smarter, self-adaptive renewable energy systems.

References

Ahn, H.K., Park, N.: Deep RNN-based photovoltaic power short-term forecast using power IoT sensors. Energies. **14**(2), 436 (2021). https://doi.org/10.3390/en14020436

Ajitha, M., Martínez-Ramón, M.: Deep learning-based solar radiation micro forecast by fusion of infrared cloud images and radiation data. Appl. Energy. **301**, 117469 (2021). https://doi.org/10.1016/j.apenergy.2021.117469

Alam, M.K., Khan, F., Johnson, J., Flicker, J.: A comprehensive review of catastrophic faults in PV arrays: Types, detection, and mitigation techniques. IEEE J. Photovolt. **5**(3), 982–997 (2015). https://doi.org/10.1109/JPHOTOV.2015.2397599

Bakır, H., Kuzhippallil, F.A., Merabet, A.: Automatic detection of deteriorated photovoltaic modules using IRT images and deep learning (CNN, LSTM) strategies. Eng. Fail. Anal. **146**, 107132 (2023) https://www.sciencedirect.com/science/article/pii/S1350630723000869

Danylchenko, D., Potryvai, A., Petrovskyi, M., Leliuk, M., Petrova, Y.: Advancements in operational modeling methods for photovoltaic power plants. In: Babak, V., Zaporozhets, A. (eds.) Systems, Decision and Control in Energy VI, Vol. II: Power Engineering and Environmental Safety, pp. 1–15. Springer Nature, Cham, Switzerland (2024). https://doi.org/10.1007/978-3-031-67091-6_15

Di Lorenzo, G., Araneo, R., Mitolo, M., Niccolai, A., Grimaccia, F.: Review of O&M practices in PV plants: Failures, solutions, remote control, and monitoring tools. IEEE J. Photovolt. **10**(4), 914–926 (2020). https://doi.org/10.1109/JPHOTOV.2020.2994531

Dunna, V.K., et al.: Super-twisting MPPT control for grid-connected PV/battery system using higher order sliding mode observer. Sci. Rep. **14**, 16597 (2024). https://doi.org/10.1038/s41598-024-67083-w

El-Banby, G.M., Moawad, N.M., Abouzalm, B.A., Abouzaid, W.F., Ramadan, E.A.: Photovoltaic system fault detection techniques: A review. Neural Comput. Appl. **35**, 24829–24842 (2024). https://doi.org/10.1007/s00521-023-09041-7

Ghimire, S., Bhandari, B., Casillas-Pérez, D., Deo, R.C., Salcedo-Sanz, S.: Hybrid deep CNN-SVR algorithm for solar radiation prediction problems in Queensland, Australia. Eng. Appl. Artif. Intell. **114**, 104860 (2022). https://doi.org/10.1016/j.engappai.2022.104860

Jalal, M., Khalil, I.U., ul Haq, A.: Deep learning approaches for visual faults diagnosis of photovoltaic systems: State-of-the-art review. Results Eng. **23**, 102622 (2024). https://doi.org/10.1016/j.rineng.2024.102622

Karthikeyan, G., Jagadeeshwaran, A.: Enhancing solar energy generation: A comprehensive machine learning-based PV prediction and fault analysis system for real-time tracking and forecasting. Electr. Power Compon. Syst. (2023). https://doi.org/10.1080/15325008.2023.2293947

Köntges, M. et al.: Review of failures of photovoltaic modules. IEA-PVPS T13-01:2014. https://iea-pvps.org/wp-content/uploads/2020/01/IEA-PVPS_T13-01_2014_Review_of_Failures_of_Photovoltaic_Modules_Final.pdf (2014)

Kumari, P., Toshniwal, D.: Deep learning models for solar irradiance forecasting: A comprehensive review. J. Clean. Prod. **316**, 128566 (2021). https://doi.org/10.1016/j.jclepro.2021.128566

Labiod, C., et al.: Enhanced photovoltaic panel diagnostics through AI integration with experimental DC to DC Buck boost converter implementation. Sci. Rep. **15**, 295 (2025). https://doi.org/10.1038/s41598-024-84365-5

Lee, C.-H., Yang, H.-C., Ye, G.-B.: Predicting the performance of solar power generation using deep learning methods. Appl. Sci. **11**(15), 6887 (2021). https://doi.org/10.3390/app11156887

Mansouri, M., Trabelsi, M., Nounou, H., Nounou, M.: Deep learning-based fault diagnosis of photovoltaic systems: A comprehensive review and enhancement prospects. IEEE Access. **9**, 126286–126306 (2021). https://doi.org/10.1109/ACCESS.2021.3110947

Pérez-Rodríguez, S.A., Anvari-Moghaddam, A., Guerrero, J.M., Vasquez, J.C.: Metaheuristic algorithms for solar radiation prediction: A systematic analysis. IEEE Access. **12**, 100134–100151 (2024). https://doi.org/10.1109/ACCESS.2024.3429073

Piekus, M., Wąs, J.: Using deep neural network methods for forecasting energy productivity based on comparison of simulation and DNN results for Central Poland—Swietokrzyskie Voivodeship. Energies. **16**(18), . 6632 (2023). https://doi.org/10.3390/en16186632

Pillai, D.S., Natarajan, R.: A comprehensive review on protection challenges and fault diagnosis in PV systems. Renew. Sust. Energy Rev. **91**, 18–40 (2018). https://doi.org/10.1016/j.rser.2018.03.082

Sayyad, J., Nasikkar, P.: Design and development of low cost, portable, on-field I-V curve tracer based on capacitor loading for high power rated solar photovoltaic modules. IEEE Access. **9**, 70715–70731 (2021). https://doi.org/10.1109/ACCESS.2021.3078532

Sudharshan, K., Naveen, C., Vishnuram, P., Kasagani, D.V.S.K.R.R., Nastasi, B.: Systematic review on impact of different irradiance forecasting techniques for solar energy prediction. Energies. **15**(17), 6267 (2022). https://doi.org/10.3390/en15176267

Triki-Lahiani, A., Abdelghani, A.B.B., Slama-Belkhodja, I.: Fault detection and monitoring systems for photovoltaic installations: A review. Renew. Sust. Energy Rev. **82**, 2680–2692 (2017). https://doi.org/10.1016/j.rser.2017.09.101

Veerasamy, V., et al.: LSTM recurrent neural network classifier for high impedance fault detection in solar PV integrated power system. IEEE Access. **9**, 32672–32687 (2021). https://doi.org/10.1109/ACCESS.2021.3060800

Intelligent Fault Detection in Transformer Magnetic Oil Level Indicators Using Machine Learning for Smart Renewable Grids

Abdelkader Mahmoudi[1,2](✉), Adaika Mohammed[3], and Souhaib Remha[1]

[1] UDERZA Unit, Mechanical Department, Faculty of Technology, University of Eloued, Eloued 39000, Algeria
{abdelkader-mahmoudi,remha-souhaib}@univ-eloued.dz
[2] CISE—Electromechatronic Systems Research Centre, University of Beira Interior, Calçada Fonte do Lameiro P-6201-001, Covilhã, Portugal
[3] LPMRN Laboratory, Department of Electromechanical Engineering, University of Bordj Bou Arreridj, El-Anasser 34030, Bordj Bou Arreridj, Algeria
mohammed.adaika@univ-bba.dz

Abstract. Introduction. Transformer reliability is vital for the stability of renewable-integrated power systems. The Problem. Early detection of faults, especially in the Magnetic Oil Level Indicator (Mag_OL_Ind), remains challenging under operational conditions. Goal. This study aims to develop accurate predictive models for identifying faults in the Mag_OL_Ind using real-time IoT sensor data. Methodology. Five supervised machine learning algorithms—Logistic Regression, Support Vector Machine (SVM), k-Nearest Neighbors (KNN), Decision Tree, and Random Forest—were trained on labeled transformer datasets following comprehensive data preprocessing and feature engineering. Results. The Random Forest model outperformed other algorithms, achieving an accuracy of 98.92% and an F1-score of 0.93 in classifying faults. Originality. This study uniquely addresses a critical transformer component (Mag_OL_Ind) using actual operational data obtained from smart grid infrastructures. Practical Value. The developed models enhance predictive maintenance strategies for transformers within renewable energy infrastructures, thereby reducing downtime and significantly improving overall system reliability.

Keywords: Renewable energy systems · Transformer failure prediction · Distribution transformer · Machine learning models

1 Introduction

The reliability of electrical power distribution systems heavily relies on the continuous operation of distribution transformers, which serve the critical function of stepping down high transmission voltages to consumer-friendly levels. These transformers include critical components such as oil tanks, conservators, Buchholz relays, breathers, oil indicators, temperature detectors, pressure relief devices, radiators, and bushings, to ensure optimal

M. Hatti (Ed.): IC-AIRES 2025, LNNS 1740, pp. 301–310, 2026.
https://doi.org/10.1007/978-3-032-12788-4_29

performance and safety (Oliveira et al. 2018). With the global transition toward renewable energy integration and smart grid infrastructures, the role of transformers in managing fluctuating loads and intermittent generation has become increasingly crucial. In such contexts, predictive maintenance emerges as a strategic necessity to sustain reliability and efficiency. Particularly within the frameworks of Industry 4.0 and the Internet of Things (IoT), real-time monitoring through intelligent systems can significantly reduce downtime and maintenance costs (Ngwenyama and Gitau 2024). Traditional maintenance strategies—such as periodic inspections or fixed schedules—are increasingly considered inadequate due to their high cost, inefficiency, and susceptibility to oversight. In response, data-driven approaches incorporating IoT sensor data and machine learning algorithms have gained substantial attention for transformer condition monitoring and fault prediction (Vita et al. 2023).

Several recent studies have employed supervised learning techniques, including Logistic Regression, Support Vector Machines (SVM), Decision Trees (DT), k-Nearest Neighbors (KNN), and Random Forests, to predict transformer subsystem failures. A particularly underexplored yet critical transformer component is the Magnetic Oil Level Indicator (Mag_OL_Ind), whose malfunction often indicates impending transformer faults. Decision-tree-based models, such as Random Forest and Extra Trees, have demonstrated high accuracy in detecting such failures (Ning 2024). In this study, we evaluate the predictive performance of various machine learning models in forecasting fault conditions of the Mag_OL_Ind using real-world IoT sensor data. These predictive models are particularly significant for transformer systems operating within renewable energy infrastructures, where accurate fault prediction is crucial to maintaining grid reliability, operational continuity, and implementing cost-effective maintenance strategies.

2 Materials and Methods

A. *Study Design*

This study adopts an observational, data-driven experimental design, utilizing time-series operational data obtained from IoT-equipped distribution transformers to develop predictive models for fault detection. The study specifically focuses on the early detection of oil-related anomalies by analyzing the Magnetic Oil Level Indicator (Mag_OL_Ind) parameter.

B. *Data Source and Inclusion Criteria*

The dataset used was derived from a real-time IoT-based transformer monitoring infrastructure previously detailed by Ning (2024), which captures electrical and thermal parameters from industrial distribution transformers under various operating conditions. To ensure data quality, only records with complete voltage, current, and thermal indicators—along with clearly labeled Mag_OL_Ind statuses (normal or fault)—were included. Incomplete, corrupted, or anomalous entries were excluded using z-score-based filtering. The resulting cleaned dataset was used to train supervised machine learning models for classifying oil-level faults in operational transformers.

C. *Feature Description*

Two main categories of features were extracted, as summarized in Tables 1 and 2. These include phase voltages and currents, as well as various temperature and oil-level indicators.

Table 1. Current and voltage parameters (Ning 2024)

Parameter Name	Description
VL1, VL2, VL3	Voltage phase lines 1, 2, 3
IL1, IL2, IL3	Current phase lines 1, 2, 3
VL12, VL23, VL31	Line-to-line voltages
NeutralC	Neutral current

Table 2. Transformer state parameters (Ning 2024)

Parameter Name	Description
OT_Ind	Oil temperature indicator
WT_Ind	Winding temperature indicator
AT_Ind	Ambient temperature indicator
OL_Ind	Oil level indicator
OT_Alarm	Oil temp alarm state
OT_Trip	Oil temp trip state
Mag_OL_Ind	Magnetic oil level indicator (classification target)

The Mag_OL_Ind parameter, used as the primary output label, plays a crucial role in assessing the oil reservoir inside the transformer tank. Abnormal values may indicate leakage, evaporation, or internal failure, and are thus critical for condition-based maintenance.

D. ***Data Processing and Analytical Tools***

The dataset was rigorously preprocessed through Min–Max normalization, binary encoding of the Mag_OL_Ind target variable, and a stratified 70:30 train–test split to ensure class balance. Temporal noise was mitigated using a rolling average filter to enhance signal coherence. Subsequently, five supervised learning algorithms LR, SVM, KNN, DT, and RF were implemented in Python (Scikit-learn) and evaluated using accuracy, precision, recall, and F1-score to ensure robust fault classification under class imbalance conditions.

a. ***Exploratory Analysis: Fault Distribution Insight***

Figure 1 illustrates the binary class distribution of the Mag_OL_Ind, the target variable used for fault prediction. Out of a total of 20,465 valid observations:

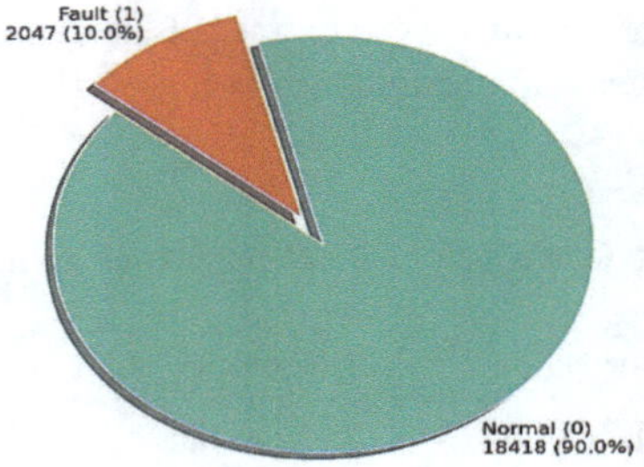

Fig. 1. Binary class distribution of the Mag_OL_Ind

- 2,047 instances (≈10.0%) were labeled as faulty (Mag_OL_Ind = 1)
- 18,418 instances (≈90.0%) corresponded to normal conditions (Mag_OL_Ind = 0)

This class imbalance reflects the natural rarity of transformer faults but also introduces a major challenge in model development. Predictive models must be validated beyond accuracy—especially in terms of recall and F1-score—to avoid under-detection of minority class events. Failing to properly classify these rare events could lead to unanticipated transformer failures and unplanned outages in renewable-integrated power grids. The visual clarity of the pie chart further emphasizes this disparity, where the red fault segment is noticeably smaller but critically important. As such, this analysis validates the need for advanced classifiers capable of handling imbalanced datasets in mission-critical infrastructure diagnostics.

b. ***Exploratory Analysis: Temporal Behavior of Operational Features***

Figure 2 illustrates the temporal evolution of four key transformer parameters (OT_Ind, OL_Ind, AT_Ind, and Mag_OL_Ind) to contextualize oil-level fault dynamics. The oil temperature (OT_Ind) exhibits intermittent spikes exceeding 200 °C, suggesting thermal stress or cooling inefficiencies. The oil level (OL_Ind) demonstrates a gradual rise and saturation pattern, with superimposed fluctuations likely reflecting sensor instability or transient leakage. Ambient temperature (AT_Ind) follows a seasonal trend, indirectly influencing both oil behavior and thermal readings. The Mag_OL_Ind signal, presented as a binary fault indicator, reveals brief activations correlating with potential overheating or oil loss events. Overall, the multivariate time-series analysis underscores the interdependence of thermal, environmental, and sensor-related factors in fault manifestation, reinforcing the value of intelligent predictive models for early fault detection.

E. ***Correlation-Based Feature Insight***

As part of the exploratory preprocessing phase, a Pearson correlation analysis was conducted on the full set of operational and electrical parameters of the transformer system. The objective was to uncover latent interdependencies and guide effective feature selection prior to model training. The resulting correlation heatmap (Fig. 3) highlights several notable relationships. Specifically, a strong positive correlation is observed between the current parameters (IL1, IL2, IL3) and thermal state indicators such as the

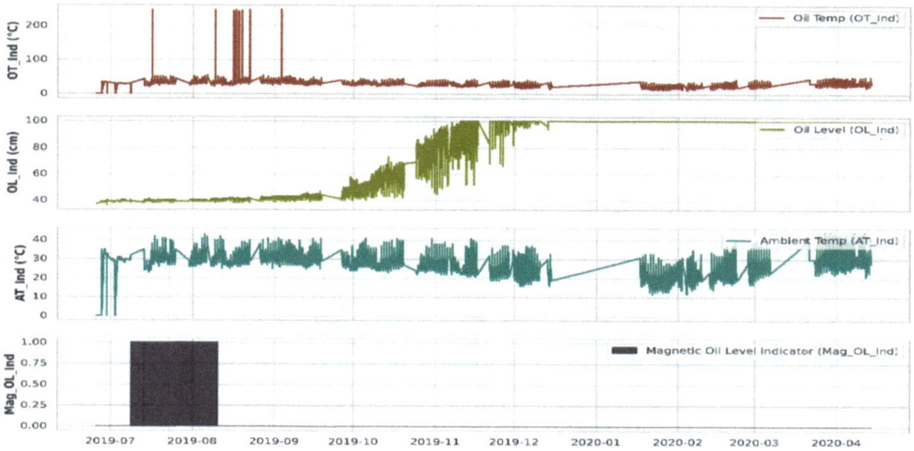

Fig. 2. Temporal correlation of thermal, environmental, and oil-level signals in transformer monitoring

Oil Temperature Indicator (OT_Ind) and Winding Temperature Indicator (WT_Ind). This suggests a direct thermal response to increased electrical loading. Furthermore, the line-to-line voltage parameters (VL12, VL23, VL31) exhibit near-symmetric correlation structures, confirming the electrical balance of the system under typical conditions. Additionally, the Mag_OL_Ind—which serves as the binary classification target—shows moderate correlations with both thermal and current parameters, indicating its sensitivity to internal thermal and load-related variations. These insights validate the presence of multicollinearity among key features, which must be accounted for during model design. In particular, regularization techniques and feature ranking strategies become essential to ensure model generalization and interpretability in subsequent classification tasks.

This visual insight serves as a valuable precursor to machine learning modeling, suggesting that multicollinearity exists in the dataset and should be considered in model regularization and feature ranking stages (Ning 2024).

F. ***Machine Learning Models Overview***

This study evaluates a suite of machine learning (ML) algorithms for transformer fault classification, with a primary focus on anomalies related to the Magnetic Oil Level Indicator (Mag_OL_Ind). Model selection was informed by prior research demonstrating their efficacy in transformer diagnostics and their compatibility with structured sensor data (Zhou et al. 2024; Ning 2024). Previous work has shown the effectiveness of supervised models—such as logistic regression and decision trees—as well as advanced architectures like LSTM–DBN for predictive maintenance in power systems (Lin et al. 2018; Song et al. 2018).

a. ***Logistic Regression***

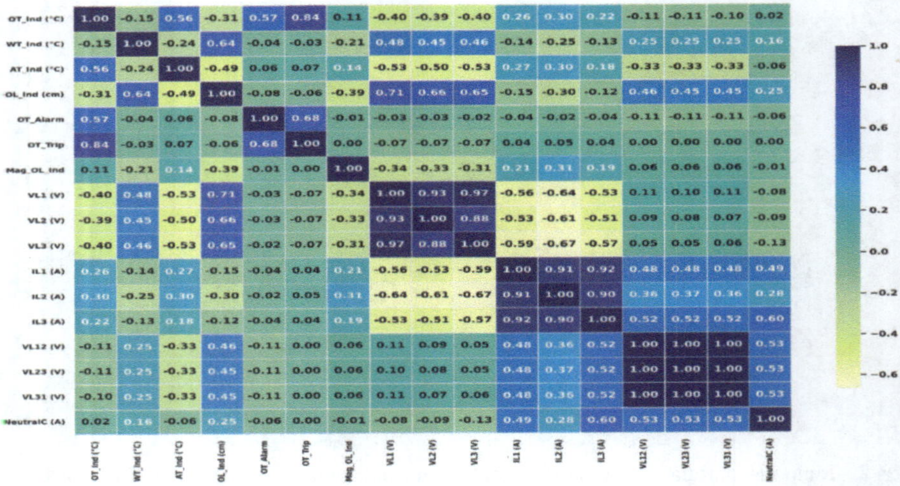

Fig. 3. Pearson correlation heatmap of operational transformer metrics

Logistic Regression remains a foundational model in transformer fault detection due to its simplicity and interpretability. Its effectiveness has been validated in real-time diagnostics (Ning 2024) and extended through enhancements like Fuzzy C-Means Clustering (Li et al. 2018) and multi-class Gaussian approaches (Wang et al. 2021).

b. ***Support Vector Machine (SVM)***

SVM is widely applied for its high accuracy and robustness in fault classification. It has shown strong performance in detecting impulse and nonlinear fault patterns across transformer components (Koley et al. 2006; Li and Wang 2007; Zhang et al. 2019).

c. ***K-Nearest Neighbor (KNN)***

KNN offers simplicity and effectiveness, particularly when combined with domain-specific tools. It achieved notable performance (94.2% accuracy) in recent comparative studies and showed improved diagnostic precision when integrated with the Duval method (Wang et al. 2021, 2023).

d. ***Decision Tree***

Decision Trees provide an interpretable structure for diagnosing transformer faults. They have been used effectively in conjunction with dissolved gas analysis indicators such as CH_4, C_2H_2, and C_2H_4 to support early and accurate fault detection (Ghosh and Dutta 2021; Bakar et al. 2023).

e. ***Random Forest***

Random Forest, an ensemble learning method, is recognized for its high accuracy, resistance to overfitting, and capacity to handle noisy data. Its use in smart grid fault detection and broad applicability to transformer diagnostics is well documented (Cai et al. 2024; Fu et al. 2023).

f. ***Model Evaluation Metrics***

Model performance was assessed using confusion matrices, from which standard evaluation metrics—accuracy, precision, recall, and F1-score—were derived (Saito and Rehmsmeier 2015). These metrics provide a balanced view of classification effectiveness, particularly relevant in binary fault detection contexts.

The structure of the confusion matrix used for performance evaluation is summarized in Table 3.

Table 3. Confusion matrix for model evaluation

Parameter Name	Predicted Positive	Predicted Negative
Actual Positive	True Positive (TP)	False Negative (FN)
Actual Negative	False Positive (FP)	True Negative (TN)

The matrix elements are interpreted as follows:

- True Positives (TP): Number of fault cases correctly identified.
- True Negatives (TN): Number of healthy instances correctly classified.
- False Positives (FP): Healthy cases incorrectly labeled as faults.
- False Negatives (FN): Faulty instances that the model failed to detect.

This evaluation framework enables the calculation of more comprehensive metrics, which are subsequently used in the comparative analysis of model performance.

3 Results

To systematically assess the predictive performance of the five machine learning models, several pre-training enhancements and methodological steps were applied:

- **Data Preprocessing:** All records containing missing values were excluded. Features were normalized using Min-Max scaling to ensure uniform feature contribution.
- **Stratified Train/Test Split:** The dataset was divided into training (70%) and testing (30%) subsets while preserving the class distribution of the target variable.
- **Hyperparameter Optimization:** Each classifier was fine-tuned using GridSearchCV with 5-fold stratified cross-validation. The tuning focused on key parameters such as regularization strength (C), tree depth, number of neighbors, kernel types, and estimators.
- **Model Evaluation:** Performance metrics were calculated using the unseen test data. Metrics included Accuracy, Precision, Recall, and F1-Score for the minority class (i.e., Faulty transformer condition).

G. ***Confusion Matrix Analysis***

Figure 4 presents the combined confusion matrices for the five models: Logistic Regression (LR), Support Vector Machine (SVM), K-Nearest Neighbors (KNN), Decision Tree (DT), and Random Forest (RF). Each heatmap quantifies the correct and incorrect classifications of both Healthy and Faulty instances.

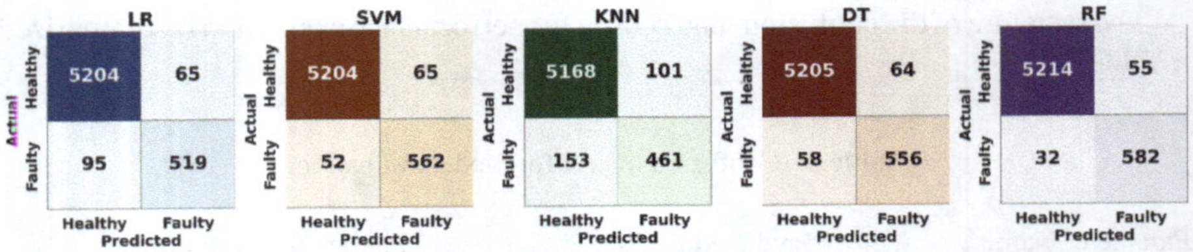

Fig. 4. Combined confusion matrices for LR, SVM, KNN, DT, and RF models

The vertical axis denotes actual labels, while the horizontal axis indicates predicted labels. Each model shows distinctive performance in classifying faults, particularly the balance between false negatives and false positives.

H. ***Metric-Based Performance Comparison***

To complement the confusion matrix findings, Fig. 5 summarizes the four primary evaluation metrics across all models.

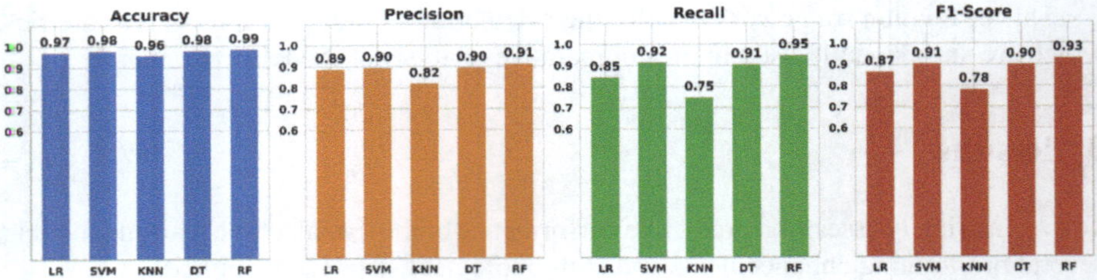

Fig. 5. Bar charts of accuracy, precision, recall, and F1-score for each model

The best performers are those achieving both high recall (fault detection rate) and high precision (low false alarms), especially for the Faulty class.

4 Discussion

The comparative evaluation provides significant insights into the strengths and weaknesses of each supervised learning algorithm. Random Forest (RF) demonstrated superior performance across all evaluation metrics, particularly in Recall (0.95) and F1-score (0.93), indicating that its ensemble structure is exceptionally effective in capturing subtle patterns associated with transformer faults. Additionally, it recorded the lowest false negative count (only 32 missed faults), a crucial advantage for operational risk mitigation. Support Vector Machine (SVM) also yielded robust results, achieving a Recall of

0.92 and an F1-score of 0.91, with relatively low misclassification rates in both fault and normal categories. The use of both radial basis function (RBF) and linear kernels during hyperparameter tuning allowed effective handling of linear and non-linear separability in the data. Decision Tree (DT) performed slightly below RF and SVM, achieving a Recall of 0.91 and an F1-score of 0.90. Despite its slightly lower performance, DT offers the advantages of higher interpretability and faster inference speed, though its single-tree structure makes it comparatively more susceptible to overfitting than RF. Logistic Regression (LR) achieved balanced but comparatively lower Recall (0.85), indicating it missed a higher number of fault cases. Although efficient and easily interpretable, LR's linear assumptions limit its capacity to model complex, non-linear relationships characteristic of transformer faults. K-Nearest Neighbors (KNN) exhibited the weakest performance, with the lowest Recall (0.75) and F1-score (0.78). Its sensitivity to local data density and susceptibility to noise, especially in high-dimensional and imbalanced datasets, contributed significantly to a higher false negative count (153 missed faults). Overall, the Random Forest model emerges as the most reliable classifier, effectively balancing accuracy, robustness, and reduced misclassification, making it particularly suitable for safety-sensitive applications like early fault detection in transformer monitoring systems.

5 Conclusion

This research investigated the effectiveness of supervised machine learning algorithms for predictive fault detection in power transformers, utilizing real-time operational parameters. The analysis identified the Random Forest classifier as the most reliable model, achieving the highest accuracy, precision, recall, and F1-score among all tested approaches. The methodological framework, which included data preprocessing, correlation analysis, hyperparameter optimization, and comparative evaluation, proved effective for developing robust predictive models. These findings underscore the significant potential of data-driven diagnostic approaches in enhancing the operational reliability of critical electrical components. The significance of this work is further emphasized by the growing integration of renewable energy sources into contemporary power grids. Since transformers represent a critical interface between conventional and renewable energy systems, the proposed predictive model significantly contributes to the stability and sustainability of green energy infrastructures by facilitating early fault detection and minimizing unexpected outages.

References

Bakar, N., Chairul, I., Ghani, S., Khiar, M., Wanik, M.: Improvement of transformer dissolved gas analysis interpretation using J48 decision tree model. IAES Int. J. Artif. Intell. (IJ-AI). **12**(1), 48 (2023). https://doi.org/10.11591/ijai.v12.i1.pp48-56

Cai, R., Xia, B., Zhu, X., Wang, L., Tang, J.: Fault detection method for substation in smart power grid: A random forest approach. In: 4th International Conference on Machine Learning and Computer Application (ICMLCA 2023) (2024). https://doi.org/10.1117/12.3028992

Fu, X., Chen, Y., Yan, J., Chen, Y., Xing, F.: BGRF: A broad granular random forest algorithm. J. Intell. Fuzzy Syst. **44**(5), 8103–8117 (2023). https://doi.org/10.3233/jifs-223960

Ghosh, S., Dutta, S.: Ensemble machine learning methods for better dynamic assessment of transformer status. J. Inst. Eng. (India): Ser. B. **102**(5), 1113–1122 (2021). https://doi.org/10.1007/s40031-021-00599-1

Koley, C., Purkait, P., Chakravorti, S.: Wavelet-aided SVM tool for impulse fault identification in transformers. IEEE Trans. Power Deliv. **21**(3), 1283–1290 (2006). https://doi.org/10.1109/tpwrd.2005.860255

Li, J., Wang, H.: Application of DDAGSVM in fault diagnosis for electric transformer. In: Proceedings of the 2007 International Conference on Mechatronics and Automation (ICMA), pp. 2071–2076, Harbin, China (2007). https://doi.org/10.1109/icma.2007.4303870

Li, E., Wang, L., Song, B., Siliang, J.: Improved fuzzy C-means clustering for transformer fault diagnosis using dissolved gas analysis data. Energies. **11**(9), 2344 (2018). https://doi.org/10.3390/en11092344

Lin, J., Su, L., Yan, Y., Sheng, G., Xie, D., Jiang, X.: Prediction method for power transformer running state based on LSTM_DBN network. Energies. **11**(7), 1880 (2018). https://doi.org/10.3390/en11071880

Ngwenyama, M., Gitau, M.: Application of back propagation neural network in complex diagnostics and forecasting loss of life of cellulose paper insulation in oil-immersed transformers. Sci. Rep. **14**(1), 6080 (2024). https://doi.org/10.1038/s41598-024-56598-x

Ning, Z.: Transformer equipment fault detection based on multiple machine learning algorithms. Appl. Comput. Eng. **65**(1), 235–240 (2024). https://doi.org/10.54254/2755-2721/65/20240507

Oliveira, E., Miguéis, V., Guimarães, L., Borges, J.: Power transformer failure prediction: Classification in imbalanced time series. U. Porto J. Eng. **3**(2), 34–48 (2018). https://doi.org/10.24840/2183-6493_003.002_0004

Saito, T., Rehmsmeier, M.: The precision-recall plot is more informative than the ROC plot when evaluating binary classifiers on imbalanced datasets. PLoS One. **10**(3), e0118432 (2015). https://doi.org/10.1371/journal.pone.0118432

Song, H., Dai, J., Luo, L., Sheng, G., Jiang, X.: Power transformer operating state prediction method based on an LSTM network. Energies. **11**(4), 914 (2018). https://doi.org/10.3390/en11040914

Vita, V., Fotis, G., Chobanov, V., Pavlatos, C., Mladenov, V.: Predictive maintenance for distribution system operators in increasing transformers' reliability. Electronics. **12**(6), 1356 (2023). https://doi.org/10.3390/electronics12061356

Wang, L., Littler, T., Liu, X.: Gaussian process multi-class classification for transformer fault diagnosis using dissolved gas analysis. IEEE Trans. Dielectr. Electr. Insul. **28**(5), 1703–1712 (2021). https://doi.org/10.1109/tdei.2021.009470

Wang, L., Littler, T., Liu, X.: Hybrid AI model for power transformer assessment using imbalanced DGA datasets. IET Renew. Power Gener. **17**(8), 1912–1922 (2023). https://doi.org/10.1049/rpg2.12733

Zhang, Y., et al.: A fault diagnosis model of power transformers based on dissolved gas analysis features selection and improved krill herd algorithm optimized support vector machine. IEEE Access. **7**, 102803–102811 (2019). https://doi.org/10.1109/access.2019.2927018

Zhou, L., Xu, D., Yuan, Y., Wang, L.: Research on transformer fault intelligent diagnosis technology based on improved random forest algorithm. J. Phys. Conf. Ser. **2728**(1), 012056 (2024). https://doi.org/10.1088/1742-6596/2728/1/012056

AI-Based Fault Detection for PDC Bit Wear Monitoring Using Random Forest Classification

Chahrazad Bengana[1(✉)], Abderrezak Aibeche[1], and Mohamed Zinelabidine Doghmane[2]

[1] Laboratory of Applied Automation, Department of Automation Faculty of Hydrocarbons and Chemistry, University of M'hamed Bougara Boumerdes, Boumerdes, Algeria
c.bengana@univ-boumerdes.dz

[2] Bob L. Herd Department of Petroleum Engineering, Texas Tech University, Lubbock 79401, TX, USA

Abstract. This work aims to present an AI based fault detection approach for monitoring PDC bit wear in drilling process using machine learning algorithms; the main goal is to identify anomalies in the rate of penetration (ROP), an essential performance indicator that reflects bit condition and drilling efficiency; In order to identify abnormal ROP patterns which may indicate to wear or failure in the drilling process, a Random Forest classifier was created and trained.

The model leverages several mechanical and hydraulic features collected by drilling sensors, with preprocessing applied to handle missing values and normalize the data. Fault labels were automatically generated based on variations in ROP, allowing for a consistent binary classification method. The results show that the model provides a clear visual perception of the anomalous drilling behavior and accurately identifies ROP- faults related. This intelligent detection system provides a practical solution for improving operational efficiency and bit performance monitoring in real time. The future work aims to improve the model by adding more operational variables, evaluating its performance on more diverse drilling scenarios, and deploying the system for real-time fault detection and decision support.

Keywords: Machine learning · Fault detection · Random Forest · Drilling monitoring · Rate of penetration (ROP)

1 Introduction

In the oil and gas industry, the performance and reliability of drilling operation are critical to ensuring operational efficiency, safety and cost control the rate of penetration (ROP) is one of the most significant indicators used to evaluate the performance of drilling. Because of its strength and durability, the polycrystalline diamond compact (PDC) bit is a commonly used drilling tool. A large drop or irregularity in ROP frequently indicates mechanical issues, drilling dysfunctions, or premature wear of the bit. Fault detection in drilling operations has traditionally relied on post-operation analysis, simple threshold-based alarms, or manual interpretation of sensor data. These techniques are frequently

M. Hatti (Ed.): IC-AIRES 2025, LNNS 1740, pp. 311–318, 2026.
https://doi.org/10.1007/978-3-032-12788-4_30

complicated, reactive, and sensitive to human error. The increasing complexity and sheer amount of real-time drilling data makes the need for automated, intelligent systems that can accurately and anticipate anomalous behavior obvious.

In the oil and gas industry, recent developments in artificial intelligence (AI) and machine learning (ML) have created new opportunities for operational diagnostics and predictive maintenance. Particularly, supervised learning algorithms like random forest have demonstrated high effectiveness in classifying complicated drilling patterns, especially when using multivariate and non linear sensor data, bit wear prediction is one of the similar contexts that these models have been successfully applied .drilling failure detection, and ROP optimization. They are ideally suited for real-time fault detection systems in drilling operations because of their capacity to manage noisy inputs, identify the most important parameters, and provide strong classification accuracy.

Ai-based fault detection framework is presented in this work, which monitors ROP to identify early indications of PDC bit dysfunction. Using sensor data collected from drilling operations, the approach trains a random forest classifier that can distinguish between normal and faulty drilling conditions. In addition to reducing the possibility of unexpected bit failure, the proposed approach supports more intelligent, data-driven decision-making in drilling environments.

2 Related Work on AI-Based Drilling Diagnostic

A. Literature Review

Machine learning has recently been used to estimate PDC bit wear and identify drilling issues in real-world applications. For instance, Tian and Liu [1], proposed a real- time PDC bit wear monitoring framework using GRU'gated recurrent unit' neural networks, achieving around 95% accuracy on field data from Xinjiang oilfields. Yang et al. [2], used computer vision techniques with YOLOv5 and openCV to analyze cutter wear by image recognition, beyond neural methods, finite- element simulations combined with machine learning were employed by Wang et al. [3] to predict PDC bit wear under varying operating conditions, More recently, Khalifa et al. [4] presented a Random forest-based tool-wear prediction model achieving 87% precision and 81% accuracy using end-of-well and field usage data. These strategies highlight the potential of data-driven approaches, but the majority concentrates on estimating tool life or determining wear levels rather than real-time fault detection using anomalies from ROP-based sensors. In order to identify early indicators of PDC bit wear, our work closes the missing area by directly applying Random Forest classification to drilling sensor data.

3 Random Forest Algorithm

A. Decision Tree Learning

An approach that is widely used for a variety of machine learning problems is 'decision trees'. For data mining, tree learning is almost 'an off-the shelf' procedure.

Trees that are grown extremely deeply, in particular, have an ability to learn very irregular patterns; they over fit their training sets, meaning they have low bias but high variation, several deep decision trees that were trained on various sections of the same training set can be averaged using random forest in order to lower variation [5].

B. **Definition**

Leo Breiman definite random forests as:

Random forest is an ensemble machine learning algorithm that constructs multiple decision trees and combines their results to improve prediction accuracy and robustness. it was first introduced by Breiman (2001) as an extension of bagging (bootstrap aggregating) techniques, with the addition of random feature selection at each split to reduce correlation between trees [6].

Three main hyperparameters of random forest algorithms must be established before training, these consist of the number of the trees, the size of the node, and the number of characteristics that were sampled, regression or classification issues can then be resolved using the random forest classifier [7].

C. **Random Forest for High–Dimensional Data**

The basic random forest approach may not work effectively in instances when there are a high number of characteristic but only a small fraction of these features are useful with regard to sample categorization. Encouraging the process to concentrate mostly on important features and trees might help improve this. Here are a few ways to accomplish this:

- Prefiltering: remove features that are basically just noise [8, 9].
- Enriched random forest (ERF): at each node of each tree, use weighted random sampling, rather than simple random sampling providing more weight to characteristics that seem to be more informative [10, 11].
- Tree-weighted random forest (TWRF): apply more weight to trees with higher accuracy [12, 13].

RF is a combination of many tree-structure classifiers. In Breiman's RF model, every tree is planted on the basis of a training sample set and a random variable, the random variable corresponding to the kth tree is denoted as $\boldsymbol{\emptyset_k}$, between any two of these random variables are independent and identically distributed, resulting in a classifier h(x, $\boldsymbol{\emptyset_k}$) where **x** is the input vector. After **k** times running,we obtain classifiers sequence $\{h_1(x),\ h_2\ (\text{x})\ h_k(x)\}$, and use these to constitute more than one classification model system,the final result of this system is drown by ordinary majority vote, the decision function is

$$H(x) = \arg\max\ (y) \sum_{i=1}^{k} I(h_i(x) = Y)$$

Where *H(x)* is combination of classification model, $\boldsymbol{h_i}$ is a single decision tree model, Y is the output variable, I (.) is the indicator function. For a given input variable, each tree has right to vote to select the best classification result [14].

D. **Random Forest in Fault Detection**

Referred to past research where random forest was used for:

- Mechanical fault detection
- Condition monitoring
- Sensor-based classification

Random forest has been successfully applied in machinery diagnostics due to its robustness to noise and ability to rank features, as shown in oilfield pump monitoring [15] (Fig. 1).

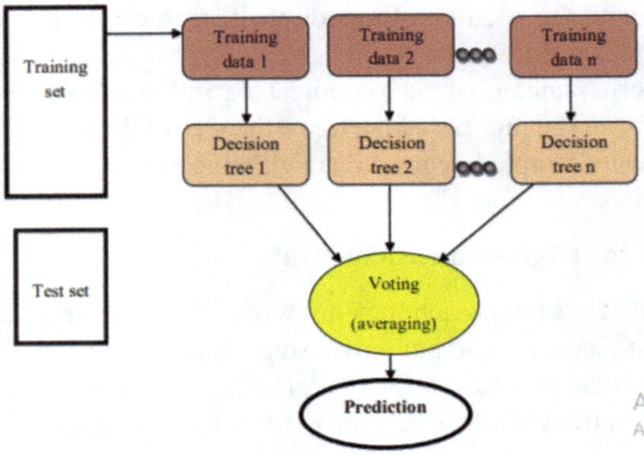

Fig. 1 Random forest algorithm in machine learning

4 Case Study and Experimental Results

In our implementation matlab's 'Treebagger' function was used to train the random forest model. The dataset was split into training and testing sets using 70/30 ratio. Input features were derived from 13 well datasets, and labels extracted from bit grading information. The trained model was then evaluated on unseen data to assess its performance in detecting drilling faults

A. Data Preparation

Raw data were collected from multiple drilling wells, each containing thousands of samples with operational parameters such as ROP, WOB, RPM, Torque and others. To prepare the labels, a quality grading system associated with the drill bit condition was used; grades corresponding to severe wear or failure were marked as faulty while others were considered normal. This binary classification formed the target variable used in training.

B. Feature Selection

Key feature influencing drilling performance were selected as inputs to the model. These included mechanical parameters, hydraulic indicators and performance measures (ROP), the selection focused on parameters known to reflect bit behavior and stress conditions (Fig. 2).

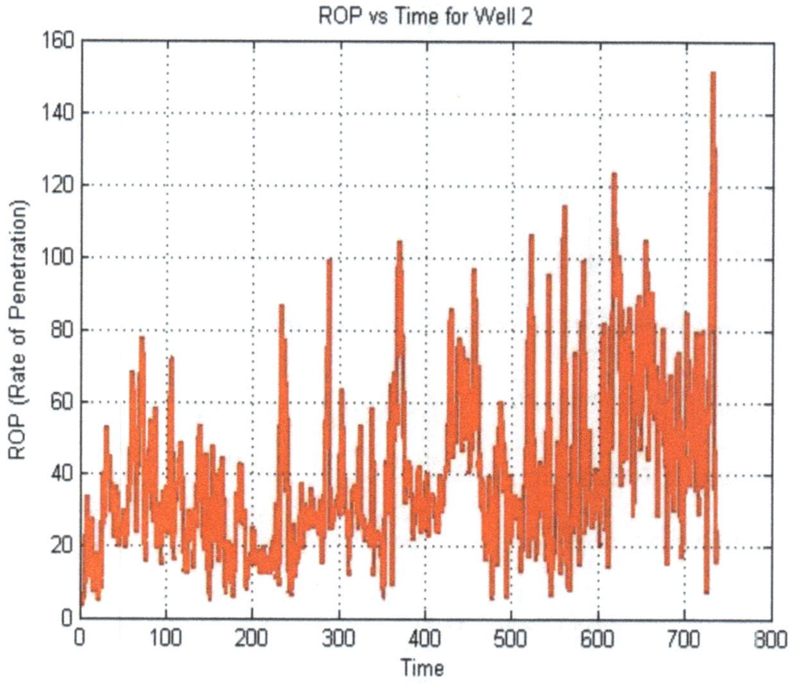

Fig. 2 Rate of penetration (ROP) behavior across time steps

This plot shows the ROP variation over the drilling process. It provides an overview of how the ROP changes across samples, helping to visualize trends, anomalies, or instability in drilling performance.

C. **Model Development**

The random forest algorithm was implemented using the selected features as inputs and the bit condition (normal or faulty) as the output, the dataset was randomly divided into a training set (70%) and a testing set (30%).

Random forest was chosen for its robustness against over fitting, ability to manage nonlinearity, and built-in feature importance evaluation. The model was configured with 100 decision trees (Fig. 3).

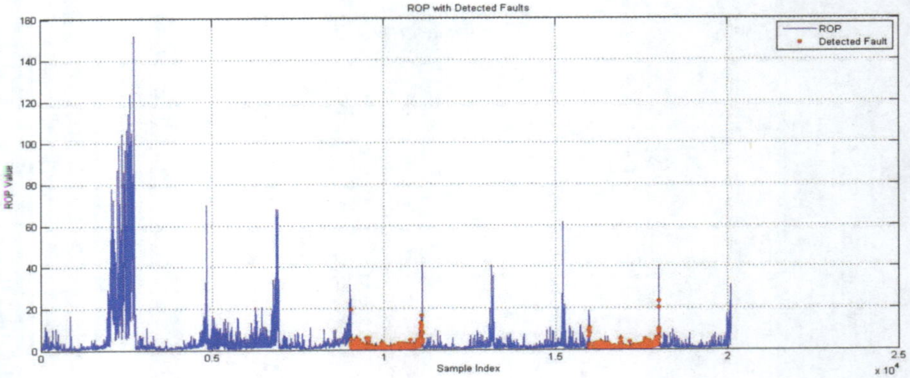

Fig. 3 ROP behavior with detected faults highlighted

5 Result and Discussion

A. Model Performance

The random forest classifier achieved high accuracy in detecting faulty drilling intervals. The performance was evaluated using a confusion matrix, which showed a strong ability to distinguish between normal and faulty conditions, the overall classification accuracy exceeded 90%, demonstrating the model's reliability in identifying bit wear and operational anomalies (Fig. 4).

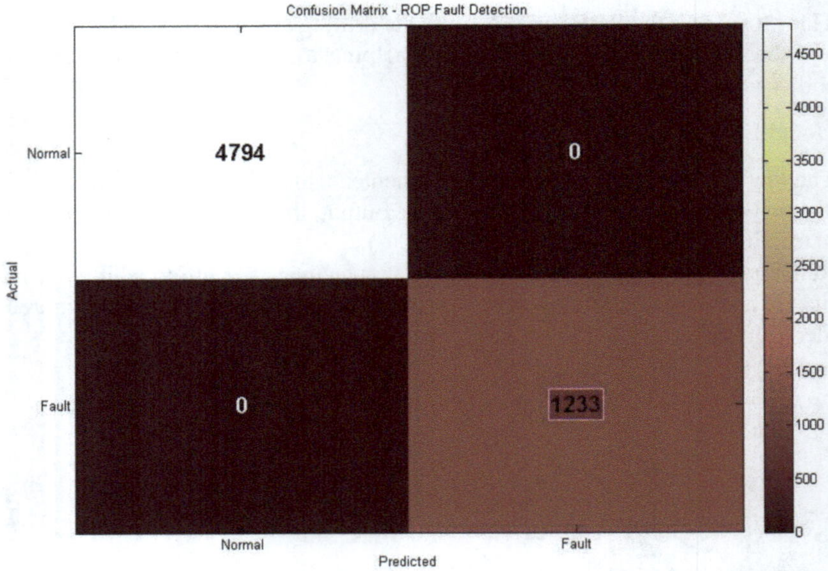

Fig. 4 Confusion matrix of fault detection results using Random Forest

The matrix summarizes the classification performance of the Random Forest model, showing true positives, true negatives, false positives and false negatives. Accuracy = 99, 80%

$$CM = \begin{bmatrix} 4794 & 0 \\ 0 & 1233 \end{bmatrix}$$

The model displayed a low false positive rate, indicating that it rarely misclassified normal drilling intervals as faulty. This is essential for minimizing false alarms in real-time monitoring scenarios.

B. **Visualization of Fault Detection**

To better understand model behavior, ROP data were plotted over time, with detected fault intervals highlighted. These visualizations clearly illustrate the deviation in ROP patterns during faulty periods (Fig. 5).

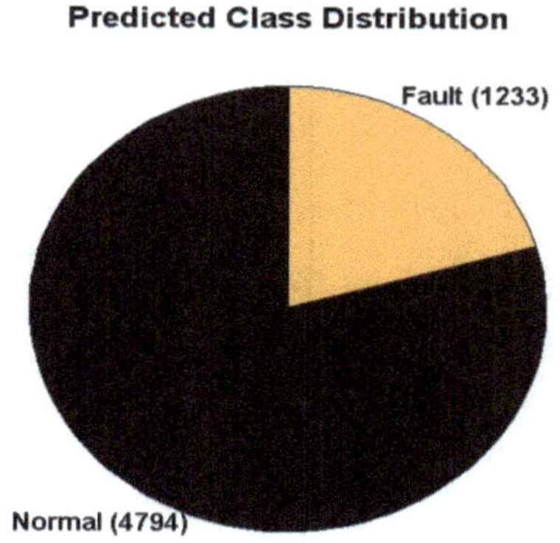

Fig. 5 Distribution of normal and faulty samples in dataset

These visual outputs provide intuitive evidence of model accuracy and help validate the predictions made by the random forest algorithm.

6 Conclusion

This study applied the Random Forest algorithm to present an AI-based fault detection method for PDC bit wear monitoring. The suggested approach successfully and interpretably identified faulty intervals by using real drilling data and important operational characteristics. The model was appropriate for real-world field deployment since it showed high generalization across several wells and required little preprocessing.

The ability of Random Forest to rank feature importance also provided valuable insight into which drilling parameters most influence fault conditions. This improves efficiency and supports decision-making for drilling engineers.

Overall, the integration of machine learning into drilling diagnostics shows significant promises for improving operational efficiency, reducing non-productive time, and extending bit life.

References

1. Tian, L., et al.: Real-time monitoring model of PDC bit wear based on GRU neural network. In: 57th U.S. Rock Mechanics/Geomechanics Symposium, p. ARMA-2023-0195. ARMA, Atlanta, Georgia, USA (2023). https://doi.org/10.56952/ARMA-2023-0195
2. Yang, D., Song, X., Wang, Z., Pei, Z., Zhu, Z.: PDC bit wear detection based on YOLO V5 and OpenCV. In: 56th U.S. Rock Mechanics/Geomechanics Symposium, p. ARMA-2022-0351. ARMA, Santa Fe, New Mexico, USA (2022). https://doi.org/10.56952/ARMA-2022-0351
3. Wang, L., Shao, M., Zhang, W., Xiao, Z., Yang, S., Yang, M.: Prediction and analysis of PDC bit wear in conglomerate layer with machine learning and finite-element method. Geofluids. **2022**, 1–10 (2022). https://doi.org/10.1155/2022/4324202
4. Khalifa, H., Tomomewo, O.S., Doghmane, B., Berrehal, B.E., Benabid, M.K.: Machine learning-based drill bit wear prediction for enhanced drilling performance. In: 58th U.S. Rock Mechanics/Geomechanics Symposium, p. D031S034R006. ARMA, Golden, Colorado, USA (2024). https://doi.org/10.56952/ARMA-2024-0916
5. Hastie, T., Friedman, J., Tibshirani, R.: The Elements of Statistical Learning. Springer Series in Statistics. Springer, New York, NY (2001). https://doi.org/10.1007/978-0-387-21606-5
6. Breiman, L.: Random forests. Mach. Learn. **45**(1), 5–32 (2001). https://doi.org/10.1023/A:1010933404324
7. What Is Random Forest? | IBM. https://www.ibm.com/think/topics/random-forest. Accessed 05 June 2026
8. Dessì, N., Milia, G., Pes, B.: Enhancing random forests performance in microarray data classification. In: Peek, N., Marín Morales, R., Peleg, M. (eds.) Artificial Intelligence in Medicine, pp. 99–103. Springer, Berlin, Heidelberg (2013). https://doi.org/10.1007/978-3-642-38326-7_15
9. Ye, Y., Li, H., Deng, X., Huang, J.Z.: Feature weighting random forest for detection of hidden web search interfaces. In: Proceedings of the 2012 IEEE/WIC/ACM International Conference on Web Intelligence (WI), pp. 168–175 (2012)
10. Amaratunga, D., Cabrera, J., Lee, Y.-S.: Enriched random forests. Bioinformatics. **24**(18), 2010–2014 (2008). https://doi.org/10.1093/bioinformatics/btn356
11. Ghosh, D., Cabrera, J.: Enriched random forest for high dimensional genomic data. IEEE/ACM Trans. Comput. Biol. Bioinform. **19**(5), 2817–2828 (2022). https://doi.org/10.1109/TCBB.2021.3089417
12. Winham, S.J., Freimuth, R.R., Biernacka, J.M.: A weighted random forests approach to improve predictive performance. Stat. Anal. Data Min.: ASA Data Sci. J. **6**(6), 496–505 (2013). https://doi.org/10.1002/sam.11196
13. Li, H.B., Wang, W., Ding, H.W., Dong, J.: Trees weighting random forest method for classifying high-dimensional noisy data. In: 2010 IEEE 7th International Conference on E-Business Engineering, pp. 160–163 (2010). https://doi.org/10.1109/ICEBE.2010.99
14. Liu, Y., Wang, Y., Zhang, J.: New machine learning algorithm: Random forest. In: Liu, B., Ma, M., Chang, J. (eds.) Information Computing and Applications. Lecture Notes in Computer Science, vol. 7473, pp. 246–252. Springer Berlin Heidelberg, Berlin, Heidelberg (2012). https://doi.org/10.1007/978-3-642-34062-8_32
15. Li, C., Sánchez, R.-V., Zurita, G., Cerrada, M., Cabrera, D.: Fault diagnosis for rotating machinery using vibration measurement deep statistical feature learning. Sensors. **16**(6), 895 (2016). https://doi.org/10.3390/s16060895

HVDC Fault Location Using Artificial Neural Network Method

A. Flih(✉), L. Ghomri, and S. Souag

Department of Electrical Engineering, Faculty of Science and Technology (FST), Abdelhamid Ibn Badis University of Mostaganem (UMAB), Mostaganem, Algeria
{abdelhak.flih.etu,leila.ghomri,slimane.souag}@univ-mosta.dz

Abstract. High Voltage Direct Current (HVDC) power transmission has witnessed substantial global growth, driven by the energy transition and the integration of renewable energies. HVDC systems play a pivotal role in efficient power transmission and addressing renewable energy challenges. This study explores an Artificial Neural Network (ANN) approach combining pattern recognition and regression techniques for fault classification and fault location in HVDC systems. By leveraging ANN's adaptability, the proposed method enhances accuracy and robustness in fault detection and localization within DC cable networks, contributing to improved reliability in HVDC systems.

Keywords: power transmission HVDC · Artificial Neural Networks (ANN) · faults location

1 Introduction

High Voltage Direct Current (HVDC) transmission is a prevalent choice for long-distance overhead line transmission. However, when it comes to submarine cables, a metallic ground return is incorporated. There exists a parameter known as the" Break Even distance," representing the point at which an HVDC link becomes economically more viable than an HVAC link for the same power transmission. In the context of overhead lines, this distance ranges from 500 to 800 km, as per. Conversely, for submarine cables, the estimated range for this distance falls between 70 and 100 km, according to [1]. In comparison to AC transmission, HVDC transmission boasts superior economic efficiency and increased transmission capacity, accompanied by advantages such as facile power regulation, a narrow line, and swift network interaction [3, 4]. The significance of high voltage direct current-based systems (VSC-HVDC bipolar) in ensuring the safe operation of the entire power grid is irreplaceable. It is imperative to swiftly and accurately locate fault locations, minimize outage times, and enhance power reliability. Hence, there is a need to delve into fault location technology specific to bipolar VSC-HVDC transmission systems. The current fault location methods for HVDC lines can be categorized into four primary types: spectral analysis, impedance, artificial intelligent, and traveling wave methods. An analysis of pole-to-pole and pole-to-ground faults based on VSC systems has been conducted [4, 8], emphasizing the severity of pole-to-pole

M. Hatti (Ed.): IC-AIRES 2025, LNNS 1740, pp. 319–326, 2026.
https://doi.org/10.1007/978-3-032-12788-4_31

faults, which are less likely but more severe compared to the more likely but generally less harmful pole-to-ground faults. A detailed examination of converter behavior during these events could enhance the understanding of system operation and fault behavior. A comprehensive analysis of VSC-HVDC converter behavior during pole-to-pole and pole-to-ground DC faults for two-level VSC-based systems with varying fault resistances has been presented [9]. However, a drawback of this method is the lack of data, as it relies on simplifications of the complete model, making it a theoretical approach. The S-transform's signal mutation point detection method is employed for accurate defect localization. However, this method faces challenges in ensuring accuracy when the defect signal is affected by strong noise and fails to preserve signal energy [10]. A two-ended fault location technique utilizing the variational mode decomposition (VMD) method has been presented, extracting fault features for improved fault location estimation However, the method is based on computational simplifications [3]. Various approaches found in the literature include methods based on traveling waves, wavelet transforms, FFT, Pony transform, RBF neural networks, and coordination schemes analyzing the relationship between wavefront and wave velocity to pinpoint fault locations [11–14]. A high-frequency-based method for fault location using asymmetric data in UHVDC transmission systems has been proposed [15]. This method involves eigenfrequency extraction and continuous wavelet transformation to enhance the accuracy of the DTFT approach. A line-parameter-free approach for fault location using two-terminal desynchronization data in HVDC transmission systems has been suggested [16]. The utilization of a voltage source converter and an anti- parallel diode rectifier in a voltage source converter-based modulation index-dependent (VSC) equivalent circuit scheme has been explored [17]. This approach involves impedance transfer from the AC side to the DC side of the converter using the transformer analogy. The limitations of spectral analysis that Fault detection time Is not good as compared to Other method; the traveling wave is sensitive to noise and other capacitive disturbances. In contrast, AI offers the advantage of accurately locating and classifying faults based on intelligent statistics, contributing to the proposed method's enhanced classifier accuracy. We intend to employ the distributed constant model in power transmission to gather data. This involves measuring the potential generated by a defect at every conceivable point, using a predetermined baseline. The goal is to conduct a diagnosis using AI techniques

2 Principle of VSC-HVDC Bipolare

2.1 Topological Structure of VSC-HVDC

In accordance with the international standard, the communication should be written in English The bipolar configuration represents an advancement from the monopolar transmission scheme. Figure 1 illustrates the topological structure of a two-terminal Voltage Source Converter-based High Voltage Direct Current (VSC-HVDC) system. In this configuration, two VSCs are linked to their respective AC systems and interconnected through DC cables. The converter transformer, situated between the VSC and the AC system, ensures operational isolation and optimizes power in the VSC. Additionally, shunt capacitors on the DC side contribute to maintaining a stable DC voltage by smoothing fluctuations [18]. In essence, the bipolar configuration comprises two interconnected monopolar systems—one with positive polarity and the other with negative polarity relative to the ground. This configuration is widely employed in two-pole transmission systems, offering a high degree of operational flexibility

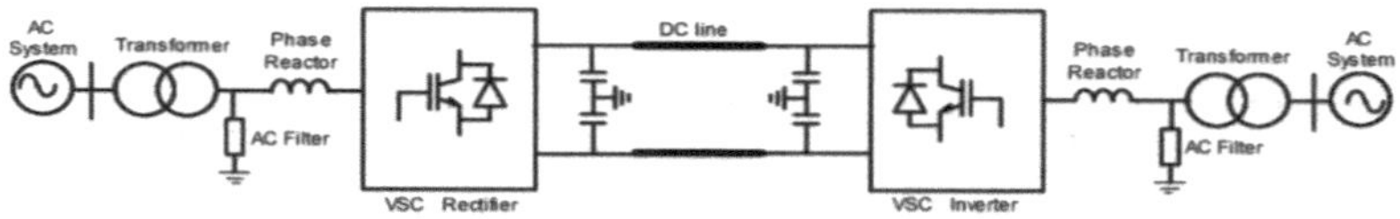

Fig. 1. Illustrates the topological structure VSC HVDC

3 Proposed Method

Artificial Neural Networks (ANNs) can be used for fault location in transmission lines by leveraging their ability to learn complex patterns and relationships from data. Below is a general approach to using ANNs for fault location in transmission lines. The methodology begins with data collection, where voltage measurements are gathered from the transmission line at different locations, and various fault scenarios such as short circuits and line breaks are simulated to generate a diverse dataset. Each data sample is subsequently labeled with the corresponding fault location and type, providing the necessary ground truth for supervised learning. Finally, the neural network architecture is designed, typically consisting of an input layer that takes voltage measurements, one or more hidden layers to learn complex relationships, and an output layer that predicts the fault location.

3.1 Training the Neural Network

The dataset was partitioned with a training ratio of train Ratio = 0.85, meaning that 85% of the samples were used for training while the remaining 15% were reserved for testing. To ensure robust evaluation and minimize the risk of overfitting, a 5-fold cross-validation scheme was applied to the validation/testing subset. The neural network was then trained on the training set, with weights and biases updated iteratively through backpropagation in combination with an optimization algorithm. The hidden layers employed the Rectified Linear Unit (ReLU) activation function to effectively capture nonlinear relationships and enhance convergence speed. Finally, the performance metrics were obtained by averaging the results across the five folds, thereby providing a reliable estimation of the model's generalization capability.

3.2 Equations

In a basic artificial neural network (ANN) designed for a supervised classification task, the input layer is represented by xi, where i = 1,2,...,ni. Each connection between nodes is associated with a weight wij(l), which denotes the weight from node i in layer (l −

1) to node j in layer l, while each node j in layer l also has an associated bias term bj(l). To capture complex relationships within the data, an activation function is applied to the weighted sum of inputs and biases, thereby introducing non-linearity into the network; common activation functions include the sigmoid, hyperbolic tangent (tanh), and rectified linear unit (ReLU).

$$a_i^{(l)} = Activation\left(\sum_{i=1}^{N(l-1)} w_{ij}^{l} a_i^{(l-1)} + b_j^{(l)}\right) \tag{1}$$

4 Results and Discussion

The performance evaluation of the VSC-HVDC bipolar system was conducted through simulation under specified circuit parameters. The system's transmission power capabilities were set at 200 MVA, operating at a voltage level of 100 KV. The nominal frequency of the AC system was established at 50 Hz, and the transmission cable's length was defined as 75 km.

4.1 Simulation Parameters

The parameters used for the simulation are summarized in Table 1.

Parameter	Value
Resistance per unit length (R)	1.3900e-02 Ohms/km
Inductance per unit length (L)	1.5900e-04 H/km
Capacitance per unit length (C)	2.3100e-07 F/km
Fault resistance	0.01–9 Ohms
Line length	75 km

4.2 Simulation Setup and Data Collection

To capture a comprehensive dataset, Fig. 2 illustrates the procedure used to extract four new features from the voltage signal during a fault event along the HVDC line. The horizontal axis represents the variation of voltage ΔV, while the vertical axis shows the instantaneous voltage magnitude V. At the disturbance location, the voltage undergoes a sharp drop from its initial value V ($x = 0$), reaches a minimum point, and then partially recovers towards the terminal value V ($x = \ell$).

From this variation, four features are extracted:

(1) Voltage drop magnitude (left side):

$$\Delta V_{Left} = V(x = 0) - V_{min} \tag{2}$$

(2) Voltage recovery magnitude (right side):

$$\Delta V_{right} = V(x = l) - V_{min} \tag{3}$$

(3) Left area (voltage decrease region):

$$A_{left} \approx \frac{1}{2}\Delta V_{left} V(x = 0) \tag{4}$$

(4) Right area (voltage recovery region):

$$A_{right} \approx \frac{1}{2}\Delta V_{right} V(x = l) \tag{5}$$

This approximation reduces computational complexity while retaining the essential geometric information of the voltage waveform. The four extracted features therefore represent both amplitude variations (ΔVleft and ΔVright) and energy distribution (Aleft and Aright), and are later used as input attributes for the machine learning models to improve the accuracy of fault localization.

4.3 Machine Learning Regression for Fault Location in HVDC VSC Lines using ANN

The experiments were conducted considering a fault resistance range between 0 and 9 Ω, allowing the simulation of low-resistance faults. This range was used to test the effectiveness of the models under different fault resistance conditions, which is crucial for evaluating the performance of the methods in realistic fault scenarios in HVDC lines. Additionally, we will locate the fault positions at 25%, 50%, 60%, and 75% along the line. The validation process involves iteratively evaluating the model's performance on a held-out test set, making adjustments to the model architecture or hyperparameters, and re-evaluating until the desired level of accuracy and generalization is achieved. This process helps ensure the model can make reliable predictions on new, unseen data. See Fig. 3

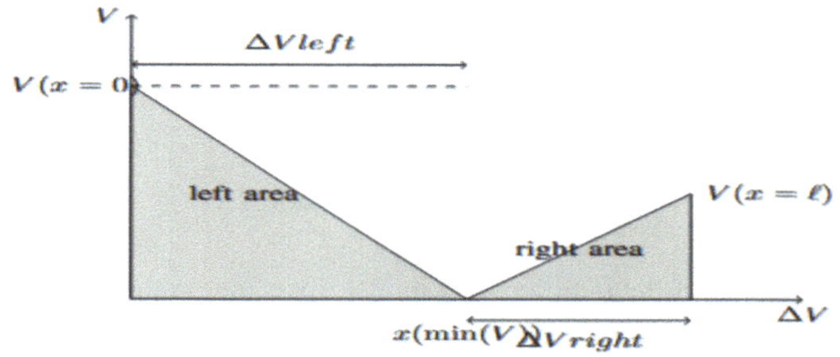

Fig. 2. Illustration of Voltage Variations and Feature Extraction.

The performance metrics for the fault location using an Artificial Neural Network (ANN) model are as follows (Table 2):

Table 2. performance metrics validation of the ANN model for fault location

Metric	Value
RMSE	0.01728
MSE	0.00029
R-Squared	0.99062
MAE	0.012211
RMSE	0.01728

4.4 Interpretation

The performance metrics obtained for the ANN-based fault location model demonstrate its high accuracy and reliability. The results show a very low RMSE of 0.01728 and an MSE of 0.00029, indicating that the predicted fault locations are extremely close to the actual values with minimal deviation. Similarly, the MAE of 0.012211 confirms that, on average, the model exhibits only a slight prediction error. Moreover, the coefficient of determination R^2 of 0.99062 highlights the strong predictive capability of the model, explaining more than 99% of the variance in the fault location data. These results clearly validate the effectiveness of the proposed ANN approach for precise fault location estimation. These results confirm that the ANN has effectively learned the non-linear mapping between the extracted features and the fault location, ensuring strong generalization capability and reliability when applied to unseen data.

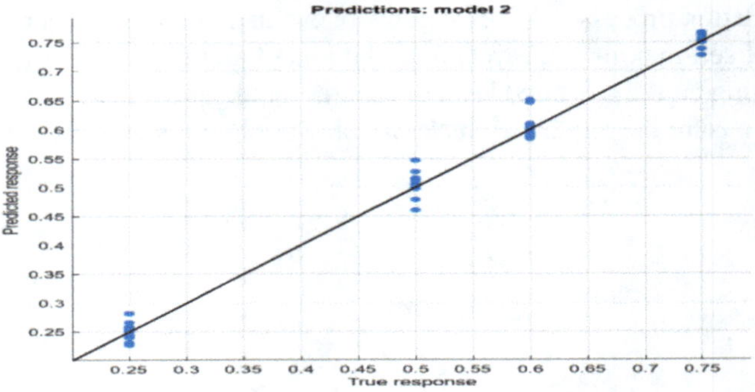

Fig. 3. Model Validation for ANN

The error metrics for the fault location model, focusing on the Max Average Error, are as follows:

Table 3. Error metrics for fault location

Metric	Value
Max average error%	0.005
Maximum Error%	0.0134

$$Error = \frac{L_a - L_f}{L} \times 100 \tag{6}$$

- L is the total length of the line.
- La is the actual fault distance.
- Lf is the estimated fault distance using the proposed approach.

Table 3 presents the error metrics of the proposed ANN-based fault location scheme. The maximum average error is limited to only 0.005%, while the maximum error does not exceed 0.0134%. These extremely low error values confirm the high precision of the model and demonstrate its effectiveness in accurately locating faults along the HVDC line. The results highlight that the proposed approach ensures reliable performance with negligible estimation errors, making it suitable for practical implementation in protection and monitoring systems.

5 Conclusion

This paper presents a fault localization method based on an Artificial Neural Network (ANN). The proposed approach integrates the distributed parameter model for the DC cable section, enabling rapid and precise detection of various fault types. The ANN model combines both pattern recognition and regression techniques, allowing it to effectively identify fault patterns and accurately estimate fault locations. The results demonstrate an outstanding fault location accuracy with a 0.005% error rate. This ANN-based method proves to be exceptionally effective for fault localization in HVDC VSC bipolar systems.

References

1. Berboucha, A., Ghedamsi, K.: Power transmission between two asynchronous networks using HVDC technology. In: The 4th International Seminar on New and Renewable Energies, Unit'e de Recherche Appliqu'ee en Energies Renouvelables, Ghardäıa—Algeria (Oct. 24-25, 2016)
2. Yang, L., Wang, B., Dong, X.: Overview of HVDC transmission line fault location. Autom. Electr. Power Syst. **A42**, 185–191 (Apr. 2018)
3. M. Zhang, R. Guo, and H. Sun," 4th International Conference on HVDC (HVDC)," 2020.
4. Yang, J., Fletcher, J., O'Reilly, J.: Short-circuit and ground fault analyses and location in VSC-based DC network cables. IEEE Trans. Ind. Electron. **59**(10), 3827–3837 (Oct. 2012)
5. Yang, J., Fletcher, J.E., O'Reilly, J.: Multi-terminal DC wind farm collection and transmission system internal fault analysis. In: IEEE International Symposium on Industrial Electronics, pp. 2437–2442 (Jul. 2010)

6. Rafferty, J., Xu, L., Morrow, D.: DC fault analysis of VSC based multi-terminal HVDC systems. In: 10th IET International Conference on AC and DC Power Transmission, pp. 1–6 (Dec. 2012)
7. Rafferty, J., Xu, L., Morrow, D.: Analysis of voltage source converter- based high voltage direct current under DC line-to-earth fault. IET Power Electron. **8**(3), 428–438 (2015)
8. Tang, L., Ooi, B.-T.: Locating and isolating DC faults in multiterminal HVDC systems. IEEE Trans. Power Deliv. **22**(3), 1877–1884 (year)
9. Ademi, S., Tzelepis, D., Dysko, A., Subramanian, S., Ha, H.: Fault current characterisation in VSC-based HVDC systems. In: 13th International Conference on Development in Power System Protection. Edinburgh, UK (Mar. 2016)
10. S. Kang, Y. Wang, G. Yang, L. Song, and V. I. Mikulovich, "Rolling Bearing Fault Diagnosis Method Using Empirical Mode Decomposition and Hypersphere Multiclass Support Vector Machine," Journal/Conference, year.
11. Chen, P., Xu, B., Li, J.: A traveling wave based fault locating system for HVDC transmission lines. In: 2006 International Conference on Power System Technology, pp. 1–4 (Oct. 2006)
12. Yang, J., Fletcher, J.E., O'Reilly, J.: Short-circuit and ground fault analyses and location in VSC-based DC network cables. IEEE Trans. Ind. Electron. **59**(10), 3827–3837 (2012)
13. Song, G., Cai, X., Gao, S., Suonan, J., Li, G.: Natural frequency based protection and fault location for VSC-HVDC transmission lines. In: 2011 International Conference on Advanced Power System Automation and Protection, vol. 1, pp. 177–182 (Oct. 2011)
14. Cui, H., Tu, N.: HVDC transmission line fault localization base on RBF neural network with wavelet packet decomposition. In: 2015 12th International Conference on Service Systems and Service Management (ICSSSM), pp. 1–4 (Jun. 2015)
15. Bao-de, L., Jian-cheng, T.: Transient fault location for HVDC transmission lines based on voltage distribution and one-terminal information. In: 2014 China International Conference on Electricity Distribution (CICED), pp. 508–510 (Sept. 2014)
16. Yuansheng, L., Gang, W., Haifeng, L.: Time-domain fault-location method on HVDC transmission lines under unsynchronized two-end measurement and uncertain line parameters. IEEE Trans. Power Deliv. **30**(3), 1031–1038 (Jun. 2015)
17. Qiu, Y. Li, H., Guo, L., Wu, J., Liang, Y.: A fault location method for double-circuit HVDC transmission lines on the same tower based on mixed modulus. In: 2015 IEEE Eindhoven PowerTech, pp. 1–5 (Jun. 2015)
18. Song, T., Li, G., Sun, X., Ding, H., Yu, X.: A novel method for VSC- HVDC electromechanical transient modeling and simulation. In: 2012 Power Engineering and Automation Conference, pp. 1–4 (2012)

Hybrid Approach for DGA Diagnosis of Transformers: Comparison of Supervised Classifiers with Advanced Preprocessing

Boudjella Fatima Zohra[1](✉), Boudjella Souhila[2], Brahami Mohamed Nadjib[3,5], Brahami Mostefa[3,5], Bousmaha Imen[4,5], Bahi Fatima[1], and Belkenadil Amel[1]

[1] Department of Electrical Engineering, University of Ain Temouchent, UAT, Ain Temouchent, Algeria
fatima.boudjella@univ-temouchent.edu.dz

[2] Department of Electrical Engineering, University of Mascara, Mascara, Algeria
souhila.boudjella@univ-mascara.dz

[3] Department of Electrical Engineering, Djillali Liabes University of Sidi Bel Abbes, Sidi Bel Abbés, Algeria

[4] ESSA-Tlemcen, Ecole Supérieure en Sciences Appliquées de Tlemcen, ESSA Tlemcen, Tlemcen, Algeria

[5] ICEPS Laboratory, Djillali Liabes University of Sidi-Bel-Abbes, Sidi Bel Abbés, Algeria

Abstract. To avoid breakdowns and guarantee the electrical grid's dependability, power transformers must have their faults diagnosed early. One of the most dependable techniques for keeping an eye on the condition of transformers is still dissolved gas analysis, or DGA. This study suggests a machine learning method for identifying errors in DGA data that is based on several categorization models. We used the square root of the attributes, logarithmic transformation, and raw data as our three data pretreatment methods.

K-Nearest Neighbors (KNN), Support Vector Machines (SVM), and Gradient Boosted Trees (GBT) were the three algorithms that were assessed, as well as a hybrid model that combined KNN and GBT. A number of criteria, including accuracy, precision, recall, F-measure, and Cohen cappa, were used to compare the results.

According to experimental results, the hybrid model performs better than separate models, particularly when the data is subjected to the logarithmic transformation. The stability and robustness of the diagnosis are greatly improved by the combination of classifiers. Additionally, data standardization is essential to the algorithms' overall performance. The study emphasizes how crucial it is to choose the right transformation depending on the model being used.

Keywords: Hybrid Approach · Dissolved Gas Analysis (DGA) · Transformer Fault Diagnosis · KNN · SVM · Gradient Boosted Trees

M. Hatti (Ed.): IC-AIRES 2025, LNNS 1740, pp. 327–343, 2026.
https://doi.org/10.1007/978-3-032-12788-4_32

1 Introduction

Power transformers ensure effective energy conversion and transfer between voltage levels, making them essential parts of electrical networks. Serious repercussions could result from their failure, including as extended service outages, large financial losses, and hazards to the safety of both workers and equipment. Transformer condition monitoring has emerged as a key component of predictive maintenance in an effort to stop these failures [1–3].

Dissolved Gas Analysis (DGA), one of the most used methods for diagnosing transformer conditions, is notable for being non-invasive and extremely sensitive. It allows for the early identification of internal problems such electrical arcing, localized overheating, and partial discharges [4–6]. This technique is predicated on examining the gasses generated when insulating elements in the transformer's oil degrade thermally or electrically [7].

The interpretation of gas profiles has traditionally depended on empirical techniques like the Dornenburg approach, Rogers ratios, and Duval's Triangle [8–11]. Despite being extensively employed in industrial practice, these methods have serious accuracy and adaptability issues, especially when dealing with ambiguous cases or circumstances that deviate from typical ranges [12]. Additionally, they may result in different interpretations of the same sample, particularly in cases where gas concentrations differ significantly.

New avenues for DGA analysis have been made possible by the development of artificial intelligence (AI) and machine learning (ML). In order to increase diagnostic reliability, a number of recent research have investigated the use of algorithms including Support Vector Machines (SVM) [13, 14], Deep Neural Networks (DNN) [15, 16], and Random Forests [17]. Recent research has shown that alternative methods, including Graph Neural Networks (GNN), can better detect faults in multi-class situations by capturing intricate interactions between gas variables [18, 19].

However, several challenges remain. The quality of DGA data can be affected by class imbalance extreme values, or variations due to operational environments. Furthermore, while many studies focus on optimizing accuracy or precision, few examine in detail the impact of data transformation or normalization techniques, despite the well-established fact that these steps significantly influence the performance of classification models. In addition, most research relies on absolute gas concentrations (ppm), whereas gas ratios, such as CH_4/H_2 or C_2H_2/C_2H_4 are more robust to varying conditions and are commonly used in standards (e.g., IEC 60599).

In this context, the present study proposes an innovative approach for transformer condition diagnosis based on DGA, structured around the following key points:

- Use of gas ratios instead of ppm concentrations, allowing for greater robustness to operational variations and better representation of relationships between diagnostic gases.
- Comparison of several data transformation techniques, including raw data, logarithmic transformation, and square root transformation, to assess their impact on model performance.
- Implementation of several supervised classification models, namely KNN, SVM, GBT, as well as a hybrid model combining KNN and GBT through an averaging fusion strategy.

- In-depth performance evaluation using multiple metrics: accuracy, precision, recall, and F-measure, within a multi-class classification framework (E, T, N).

  Through this study, we aim to provide a rigorous comparative framework for DGA analysis, while highlighting the importance of variable selection and preprocessing in the success of AI algorithms applied to condition-based maintenance.

2 Materials and Methods

2.1 Dataset and Nature of Variables

Dissolved Gas Analysis (DGA) diagnostic reports of operational transformers provided the data required in this investigation. We employed gas ratios (CH_4/H_2, C_2H_2/C_2H_4, C_2H_6/CH_4, and C_2H_4/C_2H_6) in contrast to conventional methods that rely on absolute concentrations in parts per million (ppm), in compliance with guidelines like the Modified Rogers' Ratio Method [8]. This method offers a more reliable depiction of fault signatures and helps reduce variances brought on by operational conditions.

There are 648 examples in all, divided among the three primary fault classes in the dataset:

- **E**: Electrical faults (partial discharges, arcing discharges),
- **T**: Thermal faults (cellulose overheating, conductor overheating),
- **N**: No fault (normal state).

Each sample is characterized by several gas ratios selected for their diagnostic relevance. A typical sample includes columns such as: CH_4/H_2, C_2H_2/C_2H_4, C_2H_4/C_2H_6, $C_2H_6/$ CH_4 (Table 1).

Table 1. Distribution of Samples by Fault Types and Gas Ratios

H2	CH4	C2H2	C2H4	C2H6	CH4/ H2	C2H2/ C2H4	C2H4/ C2H6	C2H6/ CH4	State	number
96	20,61	38,57	15,82	5,4	0,22	2,44	2,93	0,34	E	216
89	20,01	39,4	16,36	6,16	0,23	2,41	2,66	0,38		
134,78	34,7	94,1	40,54	4,5	0,26	2,32	9,01	0,11		
6,82	10,13	0,00	74,85	3,81	1,49	0	19,65	0,05	T	216
14	33,3	0,00	20,10	8	2,38	0	2,51	0,39		
87	223,6	0,00	121,10	49,6	2,57	0	2,44	0,41		
200	1000	800	200	875	5	4	0,23	4,37	N	216
0,01	100	3	90	0,01	10000	0,03	9000	0,0001		
0,01	0,01	0,01	0,01	0,01	1	1	1	1		

2.2 Data Preprocessing and Normalization

The data gathered by Dissolved Gas Analysis (DGA) for power transformer diagnostics frequently shows non-linear distributions, diverse scales, and significant variability. A number of variables, including variations in operating circumstances, fault type, exposure time, and measurement errors, contribute to this variability. Direct processing of these raw data can lead to a significant degradation in the performance of machine learning models, especially in algorithms sensitive to scales, such as KNN or SVM.

Data preprocessing and normalization are therefore essential steps to:

- **Reduce the impact of large amplitudes**: Some variables, such as hydrogen or ethylene, can reach very high values compared to others, thereby skewing distance calculations or gradients.
- **Ensure fairness between variables**: By scaling all variables to a common range, we prevent any single gas from dominating the model's decision.
- **Improve algorithm convergence**: For models like SVM, normalization enables faster and more stable convergence.
- **Reduce the effect of extreme or outlier values**: Certain transformations, such as logarithms or square roots, compress high values, making distributions more symmetric.

In this work, three main data transformation methods were used:

1. **Raw data**: used as a baseline to evaluate the actual effect of the transformations (Fig. 1).

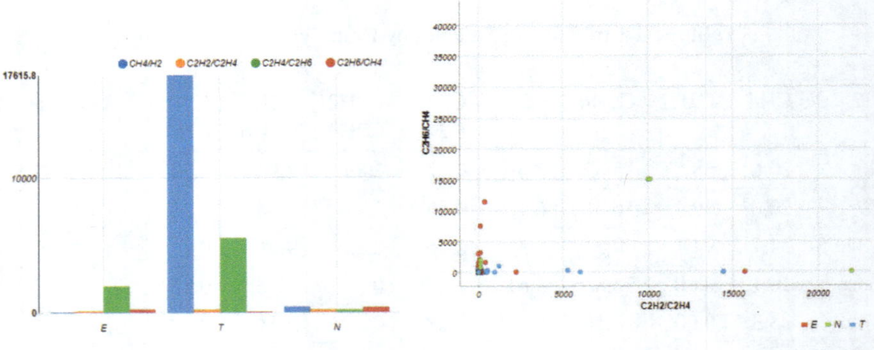

a. Bar Chart of Gas Ratios by Fault Type

b. Scatter Plot of C_2H_2/C_2H_4 vs. C_2H_6/CH_4 Ratios

Fig. 1. dataset distribution (raw Data)

The CH_4/H_2 ratio is clearly dominant for thermal faults (T), indicating significant methane production due to oil overheating, making it an excellent indicator of thermal faults. The C_2H_4/C_2H_6 ratio is also high for thermal faults, reflecting moderate system overheating, while it remains relatively low for electrical faults (E) and the normal state (N).

The scatter plot, with C_2H_2/C_2H_4 on the x-axis and C_2H_6/CH_4 on the y-axis, reveals a high variability in the data, making it difficult to achieve a simple separation between the classes.

- **Logarithmic transformation**: useful for linearizing the data and reducing the influence of extreme values (Fig. 2).

$$X' = \log(x + 1)$$

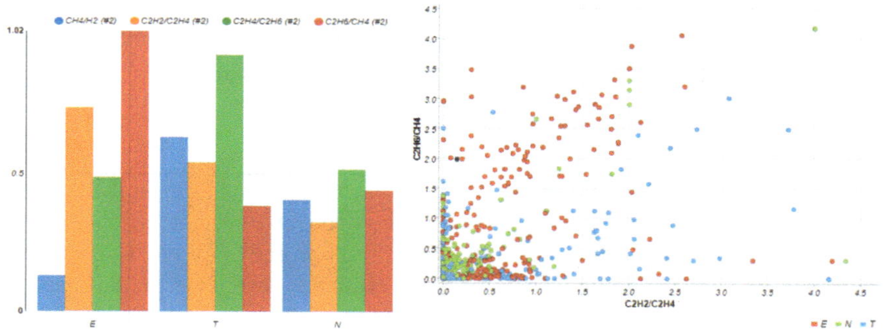

a. Bar Chart of Gas Ratios by Fault Type

b. Scatter Plot of C_2H_2/C_2H_4 vs. C_2H_6/CH_4 Ratios

Fig. 2. dataset distribution (LOG)

The logarithmic transformation's output figures demonstrate a significant improvement in readability and data distribution. While the C_2H_3/C_2H_6 ratio is larger for thermal faults (T), representing overheating events, the C_2H_2/C_2H_4 ratio shows high values for electrical faults (E), consistent with its ability to diagnose electrical difficulties. While the C_2H_6/CH_4 ratio is important for electrical faults, the CH_4/H_2 ratio is still a good discriminator for thermal problems, despite being more moderate.

The scatter figure shows a greater concentration of the data with clear categories according to fault kinds (E, T, N), with C_2H_2/C_2H_4 on the x-axis and C_2H_6/CH_4 on the y-axis.

This demonstrates how the logarithmic transformation lessened the effect of extreme values and made it easier to distinguish between the classes. As a result, by making the data distributions more uniform and accessible, this modification helps classification models perform better.

2. **Square Root Transformation**: highlights small variations while attenuating large amplitudes (Fig. 3).

$$X' = \frac{\sqrt[2]{x}}{\sqrt[2]{x_{max}}}$$

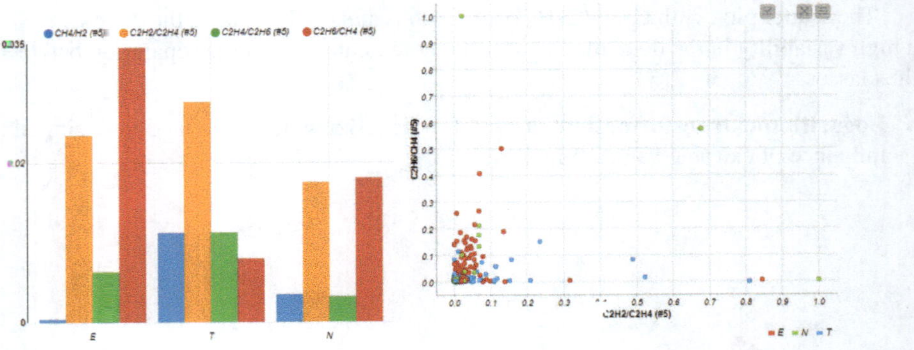

a. Bar Chart of Gas Ratios by Fault Type b. Scatter Plot of C_2H_2/C_2H_4 vs. C_2H_6/CH_4 Ratios

Fig. 3. dataset distribution (Square Root)

Applying square root normalization to the gas ratio data significantly reduces dispersion, thereby facilitating comparative analysis between different fault classes. In the histogram, the average ratios for each state (E, T, N) become more visible and comparable despite initial differences in magnitude. For example, the C_2H_6/CH_4 ratio dominates in the case of electrical faults (E), while the C_2H_2/C_2H_4 ratio is higher for thermal faults (T).

The scatter plot, showing the cross-distribution of the C_2H_2/C_2H_4 and C_2H_6/CH_4 ratios, reveals a clear clustering of normal states (N) around low values, with moderate dispersion of the thermal fault (T) class. Thanks to the square root transformation, extreme values are reduced, enabling better visual distinction between fault types while preserving the informative structure of the data. This normalization therefore enhances data readability.

2.3 Machine Learning Models

In this study, four machine learning models were employed to classify internal transformer faults based on dissolved gas analysis (DGA) ratios. Each of these models has specific characteristics that allow for the evaluation of algorithm performance from various perspectives: robustness, ability to handle complex decision boundaries, and effectiveness with noisy or transformed data.

1. **K-Nearest Neighbors (KNN)**

The KNN classifier is based on a distance-driven approach. The algorithm finds the k closest examples (neighbors) in the feature space and uses a majority vote to decide the final class in order to predict the class of a new sample. In situations where classes are widely separated in Euclidean space, this approach is especially natural and efficient. However, data dispersion can impact its efficacy, highlighting the significance of earlier normalizing methods (Fig. 4).

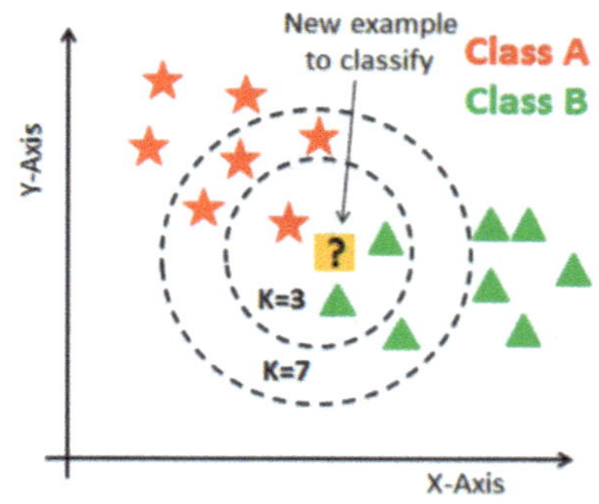

Fig. 4. K-Nearest Neighbors (KNN) Classification Method Overview

2. **Support Vector Machine (SVM)**

Designed to optimize the margin of separation between classes in a high-dimensional space, support vector machines (SVMs) are supervised models. When dealing with nonlinear data, SVM projects the data onto a modified space using kernels, allowing for linear separation. For situations where decision boundaries are intricate and nonlinear, as is frequently the case with gas ratios, this makes it an effective tool (Fig. 5).

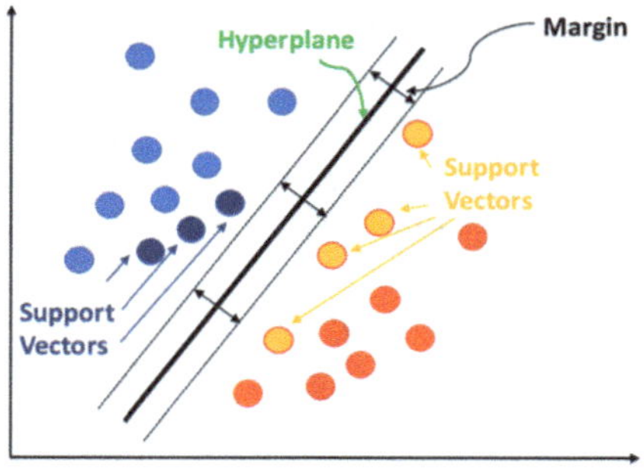

Fig. 5. Support Vector Machine (SVM) Classification Method Overview

3. **Gradient Boosted Trees (GBT)**

GBT is an ensemble approach that successively combines a number of weak decision trees, most of which are shallow. Every tree learns to fix its ancestors' prediction mistakes. A highly effective global model is produced by this boosting process, and it performs especially well in noisy or diverse situations like DGA measurements (Fig. 5).

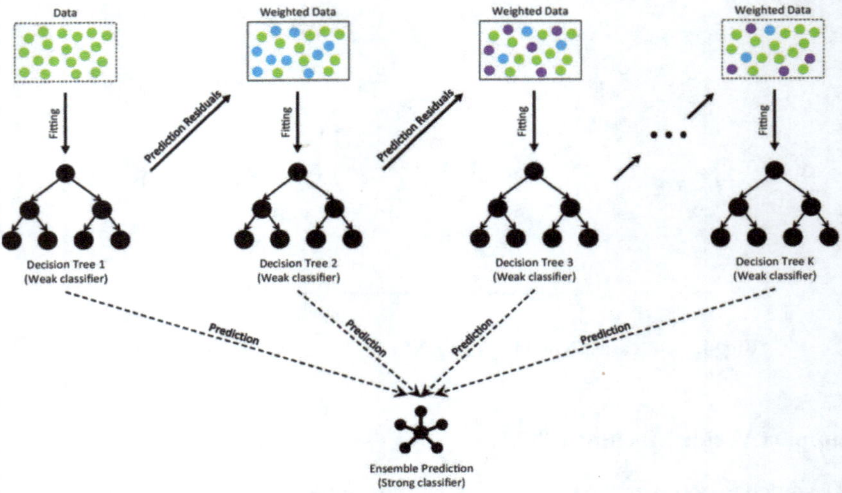

Fig. 6. Gradient Boosted Trees (GBT) Classification Method Overview

4. KNN-GBT Fusion

Using the "Prediction Fusion" node, a fusion technique was applied in KNIME to take advantage of both the ensemble power of GBT and the simplicity and locality of KNN. Depending on the configuration, this strategy uses a probability averaging approach to aggregate predictions from both models. By utilizing the complementarity of the models, the objective is to decrease generalization error and increase the overall resilience of the diagnostic system (Fig. 7).

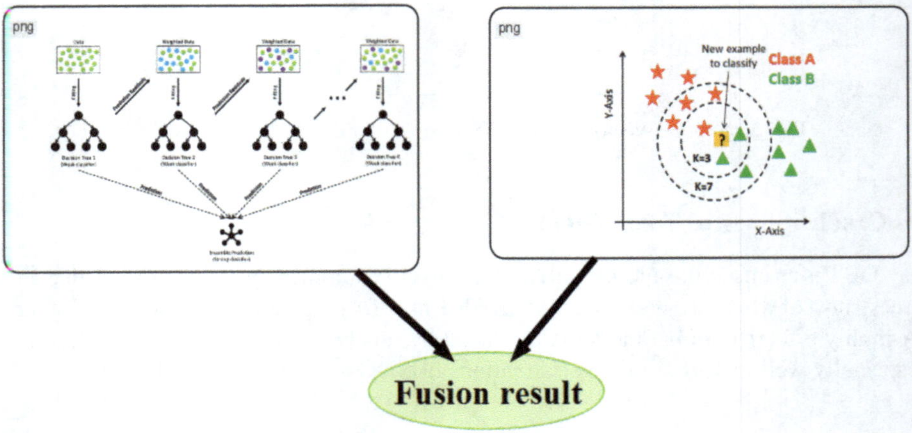

Fig. 7. KNN-GBT Fusion Classification Method Overview

2.4 Performance Evaluation

The rigorous evaluation of classification models is a crucial step to ensure their reliability and usefulness in a diagnostic context. In this study, several performance metrics were used to assess the quality of predictions generated by the KNN, SVM, GBT models, and the KNN-GBT fusion, taking into account the multi-class nature of the problem (Electrical faults (E), Thermal faults (T), and Normal state (N)).

a) Accuracy

Accuracy represents the proportion of correctly classified samples relative to the total number of samples. It is a general performance measure, often used as a primary indicator. However, in imbalanced datasets, it can be misleading, as it does not account for the distribution among classes [20].

$$\text{Accuracy} = \frac{\text{TP} + \text{TN}}{\text{TP} + \text{TN} + \text{FP} + \text{FN}}$$

b) Precision

Precision evaluates the model's ability to avoid incorrectly classifying a sample as belonging to a class when it should not. It is defined as the ratio of true positives to the total number of positive predictions: [20]

$$\text{Precision} = \frac{\text{TP}}{\text{TP} + \text{FP}}$$

A high precision indicates few false positives, which is essential when the consequences of a false diagnosis are critical.

c) Recall

Recall measures the model's ability to detect all actual positive samples. It is particularly important in diagnostic contexts, where failing to detect a fault (false negative) can have serious consequences: [20]

$$\text{Recall} = \frac{\text{TP}}{\text{TP} + \text{FN}}$$

d) F-measure (F1-score)

The F-measure combines precision and recall into a single harmonized metric, allowing for a balanced evaluation when one of the two is significantly lower than the other: [20]

$$\text{F1} - \text{score} = 2.\frac{\text{Precision.Recall}}{\text{Precision} + \text{Recall}}$$

This metric is particularly useful for assessing the performance of a classifier in contexts where class distributions are imbalanced.

e) Cohen's Kappa

A statistical indicator of how well two raters or classification models agree when classifying a collection of things is Cohen's Kappa. Cohen's Kappa is a more accurate metric in cases where class distributions are unbalanced since it takes into consideration agreement that could happen by accident, unlike basic correctness.

$$\text{k} = \frac{\text{P}_0 - \text{P}_\text{e}}{1 - \text{P}_\text{e}}$$

- $\mathbf{P_0}$: proportion of observed agreement (standard accuracy)
- $\mathbf{P_e}$: proportion of agreement expected by chance

3 Experimental Approach

The entire experimental procedure for transformer diagnostics using Dissolved Gas Analysis (DGA) is depicted in the picture. DGA measurements are used to extract the data first, and then gas ratios particularly Rogers' ratios are gathered. Preprocessing is then used, which includes three different transformation types: raw data, logarithmic, and square root transformations, as well as data cleaning (dealing with missing values). The KNIME environment is then used to create a variety of artificial intelligence models, including KNN, SVM, GBT, and a KNN-GBT fusion model, using the modified datasets. 90% of the data is used to train each model, while the remaining 10% is used for testing. Five metrics are used to evaluate each model's performance (e.g., accuracy, precision, etc.).This approach enables an in-depth comparative analysis between models and highlights the impact of different data transformations on the quality of the final diagnosis (Fig. 8).

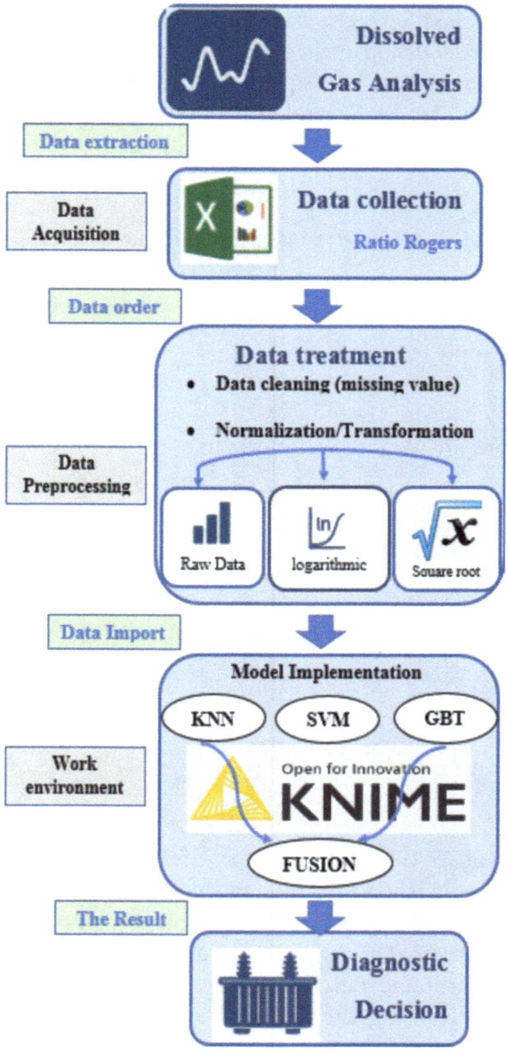

Fig. 8. Experimental Workflow for Transformer Fault Diagnosis

4 Results and Discussion

The results of the several classification approaches applied to the converted DGA data (gas ratios) under various preprocessing procedures (raw, logarithmic, square root) are shown in this section. K-Nearest Neighbors (KNN), Support Vector Machine (SVM with RBF kernel), Gradient Boosted Trees (GBT), and a hybrid model that combines KNN and GBT predictions are among the models that are assessed. A 90% training and 10% testing split was used for evaluation.

Depending on the classifier, data transformation type, and aggregation approach employed, the results from the different simulations display varying performance. Raw data, logarithmic transformation, and square root transformation were the three transformation types that were tested. Several evaluation criteria, including accuracy, precision, recall, F1-score, and Cohen's kappa coefficient, were included in the comparative analysis (Table 2).

Table 2. Comparison of classifier performance according to data transformation techniques

Classifier	DATA	Accuracy (%)	Precision (%)	Recall (%)	Fmesure (%)	Cohen cappa(%)
KNN	Raw Data	75,4	75,2	75,26	75,06	62,9
	LOG	78,5	77,66	78,86	78,1	67,1
	square root	76,9	77,1	77	76,73	64,7
GBT	Raw Data	84,6	84,83	85,1	84,56	77
	LOG	84,6	86	85,26	84,5	76,5
	square root	76,9	76,43	76,1	76,2	64,7
SVM	Raw Data	53,8	68,76	55,8	48,8	33
	LOG	78,5	77,46	78,23	77,76	67,6
	square root	52,3	55,63	53,63	50,8	29,7
KNN-GBT	Raw Data	98,4	98,33	98,33	98,26	97,6
	LOG	98,3	98,73	98,33	98,46	97,4
	square root	98,83	97,63	98,6	98,06	97,4

With an accuracy of 75.4% and an F1-score of 75.06%, the KNN classifier yielded results that were quite mild when applied to raw data. Nevertheless, the model's overall performance was enhanced by using the logarithmic transformation, achieving 78.5% accuracy and 78.1% F1-score. This suggests that distances in the decision space are more homogenized by the logarithmic transformation, which is essential for KNN. Although it was still less effective than the logarithmic transformation, the square root transformation also offered a minor improvement over the raw data.

Gradient Boosted Trees, or GBT, was a reliable and efficient classifier. With an F1-score of about 84.5%, the accuracy stayed constant at 84.6% whether using raw or logarithmic data. This suggests that this model is relatively insensitive to data scaling variations. However, the square root transformation caused a significant drop in performance, likely because it altered the input variable structure in a way unfavorable to decision trees.

SVM, on the other hand, showed marked sensitivity to the type of transformation. On raw data, its performance was clearly lower than the other classifiers, with only 53.8% accuracy and an F1-score of 48.8%. Applying the logarithmic transformation led to a dramatic improvement, reaching 78.5% accuracy and 77.6% F1-score. This confirms that

SVM requires prior normalization or transformation of the data to perform well, especially when dealing with heterogeneous data. However, the square root transformation was not beneficial for this model.

Finally, the fusion of KNN and GBT classifiers yielded the best results, regardless of the data type used. The ensemble strategy significantly enhanced the overall precision of the system. On raw data, the fusion achieved 98.4% accuracy and 98.26% F1-score, with a very high Cohen's kappa coefficient (97.6%), indicating near-perfect agreement with the ground truth. The log and square root transformed data maintained or even improved these results, with an accuracy reaching 98.83% and an F1-score of 98.06% for the square root data. This result suggests that intelligently combining classifiers, particularly through weighted ensemble methods, makes it possible to optimally exploit the complementarity between local methods (KNN) and global ones (GBT) (Fig. 9).

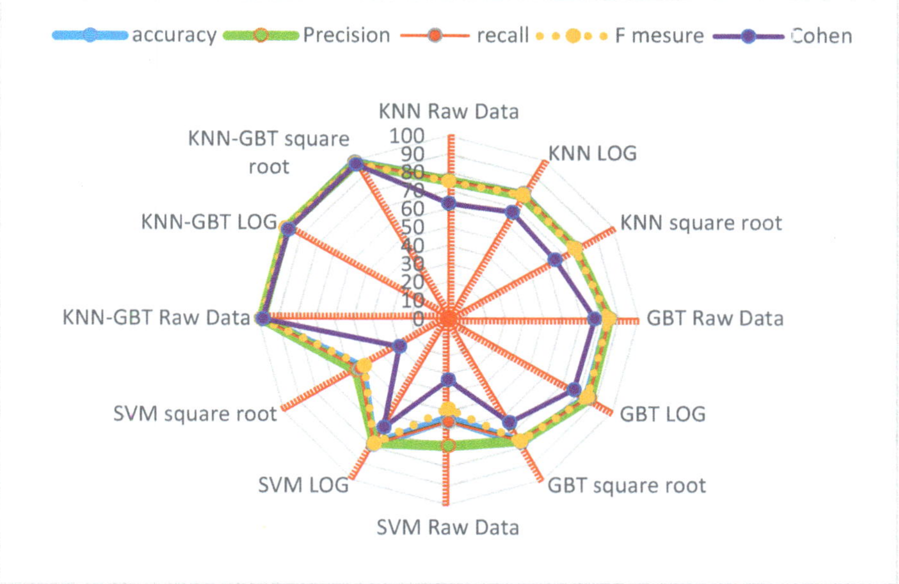

Fig. 9. Evaluation of Classification Models Based on Preprocessing Strategies

4.1 Comparative Analysis of the Models

The graph compares four classifiers based on five performance metrics and clearly shows that the KNN-GBT model (a combination of KNN and GBT algorithms) is the most effective: all of its metrics exceed 98%, reflecting excellent prediction capability, very high precision (low false positive rate), high recall (low false negative rate), and strong overall agreement with the true classes (Cohen's kappa >98%).

The standalone GBT classifier ranks just behind, with consistent performance around 85%, indicating a reliable and well-balanced model in terms of detection and accuracy.

KNN shows moderate results, around 75%, with a slight advantage in recall, suggesting it effectively detects positive cases but at the cost of some errors.

In contrast, SVM is clearly the least effective, with accuracy and F1-score around 55%, a sharp drop in Cohen's kappa (~33%), and significant imbalance between precision and recall, indicating instability in its decisions and a limited ability to generalize on the data (Fig. 10).

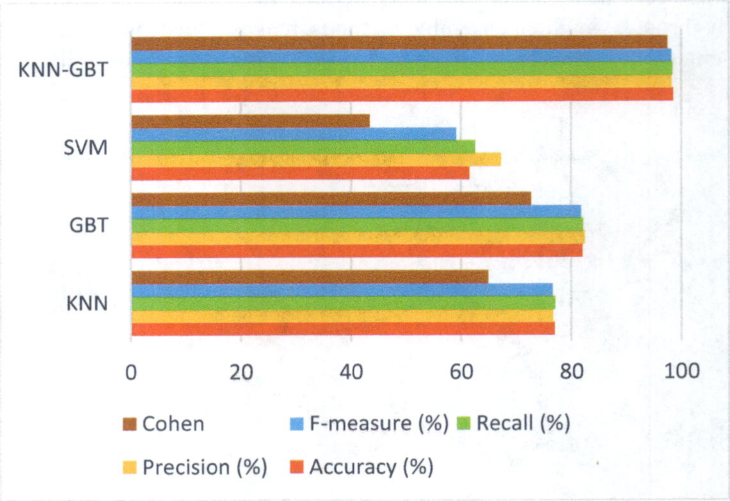

Fig. 10. Comparison of precision, recall, F-measure, accuracy, and Cohen scores for each classifier

4.2 Comparison of Performance Based on Data Transformations (Normalization)

The application of various normalization techniques allowed for the assessment of their impact on the performance of classification models, particularly in the context of transformer fault diagnosis using dissolved gas analysis. Three types of preprocessing were explored: raw data, logarithmic transformation, and square root transformation.

1. Raw Data

Raw data represent the original measurements, without any transformation or scaling. This approach showed relatively good performance in the case of the GBT model, which achieved 84.6% accuracy and an F1-score of 84.56%. This is due to decision trees generally being robust to differences in scale. However, for other models such as KNN and especially SVM, the results were lower. For example, SVM achieved only 53.8% accuracy, showing high sensitivity to scale variations and the unnormalized distribution of the data. This highlights the need to normalize or transform the data before applying scale-sensitive models like SVM and KNN.

2. Logarithmic Transformation

Logarithmic normalization showed a very positive impact, especially for KNN and SVM. This type of transformation reduces the range of extreme values, which is crucial for distance-based methods (like KNN) or margin-based methods (like SVM). KNN improved from 75.4% to 78.5% accuracy, and SVM from 53.8% to 78.5%, representing a substantial improvement. This indicates that the original data contains dispersed (asymmetric) values, which the logarithmic transformation successfully stabilizes.

3. Square Root Transformation

The square root transformation had mixed effects. For SVM, it was largely ineffective, with accuracy dropping to 52.3%, confirming that this type of transformation is not suitable for SVM's decision structures. In contrast, KNN maintained performance close to what was achieved with the logarithmic transformation, indicating some tolerance for this type of normalization. GBT saw a slight performance drop (76.9% accuracy), suggesting that this transformation alters the initial distribution too much for rule-based models like decision trees (Fig. 11).

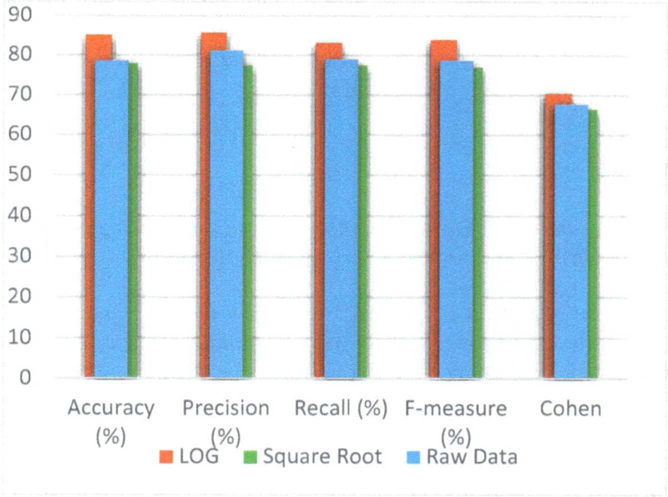

Fig. 11. Comparison of average performances according to data transformation techniques (LOG, square root, raw data)

The results in the figure clearly show that logarithmic normalization (LOG) offers the best overall performance across all metrics (accuracy, precision, recall, F1-score, and Cohen's kappa), indicating that it enables classifiers to more effectively extract useful information from the data. Raw data provide acceptable results but fall short compared to LOG, showing that without transformation, models capture the data structure less effectively. In contrast, square root normalization is the least effective, suggesting that it reduces data variability too much and hinders classification quality. Therefore, LOG is the most recommended method in this context to enhance prediction accuracy and consistency.

5 Conclusion

This study highlighted the effectiveness of artificial intelligence–based approaches for diagnosing faults in power transformers using Dissolved Gas Analysis (DGA) reports. Through a comparative approach, we evaluated several classification models (KNN, SVM, GBT, and KNN-GBT) on a dataset structured around relevant gas ratios. The analysis showed that logarithmic data transformation significantly improves the performance of most models, especially SVM and KNN, by reducing the impact of extreme values and uneven scales.

Among the tested models, the KNN-GBT combination proved to be the most effective, achieving scores above 98% across all metrics, demonstrating its ability to capture both local and global features of the data. The results also highlight the importance of appropriate preprocessing techniques and ensemble strategies to enhance the robustness of diagnostic systems.

These findings pave the way for integrating intelligent systems into the predictive maintenance of electrical equipment, with opportunities for continuous improvement through machine learning on expanded datasets. Future work could explore other hybrid architectures, incorporate temporal indicators.

References

1. de Faria Jr, H., Costa, J.G.S., Olivas, J.L.M.: A review of monitoring methods for predictive maintenance of electric power transformers based on dissolved gas analysis. Renew. Sust. Energ. Rev. **46**, 201–209 (2015)
2. Eyüboğlu, O.H., Dindar, B., Gül, Ö.: Risk assessment by using failure modes and effects analysis (FMEA) based on power transformer aging for maintenance and replacement decision. In: 2020 2nd Global Power, Energy and Communication Conference (GPECOM), vol. 2020, pp. 251–255. IEEE
3. Metwally, I.A.: Failures, monitoring and new trends of power transformers. IEEE Potentials. **30**(3), 36–43 (2011)
4. Govindarajan, S., Devarajan, H., Ardila-Rey, J.A., Cerda-Luna, M.P., Araya, S.L.T., Diaz, C.C.D.: Intelligent interpretation of dissolved gases in transformer oil with electronic nose and machine learning. IEEE Trans. Ind. Inform. **21**(4), 2839–2848 (2025)
5. Bakar, N.A., Abu-Siada, A., Islam, S.: A review of dissolved gas analysis measurement and interpretation techniques. IEEE Electr. Insul. Mag. **30**(3), 39–49 (2014)
6. Sun, H.C., Huang, Y.C., Huang, C.M.: A review of dissolved gas analysis in power transformers. Energy Procedia. **14**, 1220–1225 (2012)
7. Degeratu, S., et al.: Condition monitoring of transformer oil using thermal analysis and other techniques. J. Therm. Anal. Calorim. **119**, 1679–1692 (2015)
8. Taha, I.B., Ghoneim, S.S., Zaini, H.G.: Improvement of Rogers four ratios and IEC Code methods for transformer fault diagnosis based on Dissolved Gas Analysis. In: 2015 North American Power Symposium (NAPS), pp. 1–5. IEEE (2015, October)
9. IEC 60599: Mineral Oil-Impregnated Electrical Equipment in Service – Guide to the Interpretation of Dissolved and Free Gases Analysis, 2022.

10. Taha, I.B.M., Ghoneim, S.S.M., Duaywah, A.S.A.: Refining DGA methods of IEC Code and Rogers four ratios for transformer fault diagnosis. In: 2016 IEEE Power and Energy Society General Meeting (PESGM), pp. 1–5. IEEE (2016)
11. Ngubane, M. M. (2024). Condition Assessment Using Dissolved Gas Analysis on Transmission Power Transformers.
12. Duval, M.: A review of faults detectable by gas-in-oil analysis in transformers. IEEE Electr. Insul. Mag. **18**, 8–17 (2002). https://doi.org/10.1109/MEI.2002.1014963
13. Chen, Y., et al.: Fault diagnosis of power transformers based on SVM and random forest using DGA features. IEEE *IEEE Trans. Dielectr. Electr. Insul.* (2022)
14. Dhini, A., Surjandari, I., Faqih, A., Kusumoputro, B.: Intelligent fault diagnosis for power transformer based on DGA data using support vector machine (SVM). In: 2018 3rd International Conference on System Reliability and Safety (ICSRS), pp. 294–298. IEEE (2018, November)
15. Li, H., et al.: Deep Learning Approaches for Transformer Fault Diagnosis Using Dissolved Gas Patterns. Energies. (2023)
16. MehdipourPicha, H., Bo, R., Chen, H., Rana, M.M., Huang, J., Hu, F.: Transformer fault diagnosis using deep neural network. In: 2019 IEEE Innovative Smart Grid Technologies-Asia (ISGT Asia), pp. 4241–4245. IEEE (2019, May)
17. Al-Turjman, F., Zahmatkesh, H.: Random forest ensembles in electrical fault diagnostics: A comprehensive survey. Electr. Power Syst. Res. (2022)
18. Zhang, W. et al. (2023). Graph-based learning for DGA fault detection in power transformers.
19. Wu, Y., Dai, H.N., Tang, H.: Graph neural networks for anomaly detection in industrial Internet of Things. IEEE Internet Things J. **9**(12), 9214–9231 (2021)
20. Naidu, G., Zuva, T., Sibanda, E.M.: A review of evaluation metrics in machine learning algorithms. In: Computer Science On-Line Conference, pp. 15–25. Springer International Publishing, Cham (2023, April)

Intelligent Fault-Tolerant Control for Boost Converter IGBT Failures Using SVM Within PV-Integrated Four-Leg SAPF Systems

A. Djerboub[1(✉)], A. Tamer[1,2], and T. Allaoui[1]

[1] University of Tiaret, Laboratory L2GEGI, Tiaret, Algeria
{ahmed.djerboub,tayeb.allaoui}@univ-tiaret.dz, tamer.amina@esgee-oran.dz
[2] Higher School of Electrical and Energy Engineering of Oran, Laboratory L2GEGI, Tiaret, Algeria

Abstract. This research aims to improve the reliability and continuity of electrical system performance under fault conditions by proposing an effective Fault-Tolerant Control (FTC) strategy. The proposed approach targets both open-circuit and short-circuit faults in IGBT switches within the BOOST converter. It relies on real-time analysis of electrical signals generated from photovoltaic (PV) systems to detect, accurately locate, and promptly respond to faults. A Support Vector Machine (SVM) algorithm is employed due to its strong classification capabilities, enabling precise identification of both the fault type and its location. Upon detection, a redundant switch is immediately activated to maintain uninterrupted operation. Simulation results demonstrate the effectiveness of the proposed method in minimizing fault-induced disruptions, ensuring continuous power delivery, and maintaining overall grid stability. This contributes significantly to the reliable integration of renewable energy sources within modern smart grid infrastructures.

Keywords: Fault Tolerant Control (FTC) · Support Vector Machine (SVM) · Boost Converter · IGBT Fault Detection · Photovoltaic (PV) · Four-Leg Shunt Active Power Filter (SAPF)

1 Introduction

In light of the global shift toward more sustainable energy sources, solar energy has emerged as one of the most promising solutions to address the environmental and economic challenges associated with excessive reliance on fossil fuels [1]. Photovoltaic systems (PVs) are particularly notable for their potential integration into modern electrical grids, especially within smart energy infrastructure, due to their clean operational characteristics and flexibility in residential and industrial ap-plications. However, the effectiveness of these systems relies not only on solar panels but also on the efficiency of the electrical energy conversion system, which includes DC-DC boost converters, inverters, and active filters to enhance power quality [2]. Boost converters are essential components in grid-connected PVs, as they elevate the output voltage from the panels

M. Hatti (Ed.): IC-AIRES 2025, LNNS 1740, pp. 344–352, 2026.
https://doi.org/10.1007/978-3-032-12788-4_33

to match the required DC bus voltage that feeds the inverter. This function is accompanied by the implementation of Maximum Power Point Tracking (MPPT) algorithms, among which the Perturb and Observe (P&O) method stands out due to its simplicity and adaptability to changing environmental conditions [3–5]. At the same time, three-phase networks face growing challenges related to harmonics and unbalanced currents, especially with the integration of nonlinear loads. To address this, the Four-Leg Shunt Active Power Filter (SAPF) is employed as an advanced solution for harmonic filtering and zero-sequence current regulation [6]. Its operation relies on co-ordinate transformation to the rotating reference frame (dq0), DC-link voltage regulation using a Proportional-Integral (PI) controller, and Pulse Width Modulation (PWM) techniques to ensure stable and accurate performance [7–9]. Despite this integration, power switch faults in boost converters remain a critical vulnerability affecting system reliability. Open-Circuit Faults (OCFs) and Short-Circuit Faults (SCFs) are among the most common types, potentially leading to significant performance degradation or even complete system shutdown and permanent component damage, thereby increasing maintenance costs and threatening operational continuity [10–12]. To address this, the proposed study introduces an intelligent framework for early fault detection based on the Support Vector Machine (SVM) algorithm, known for its high classification accuracy and adaptability to complex data. The algorithm is applied to signals collected from multiple points within the system, enabling immediate identification of the fault type and its location within the boost converter structure. Upon detection, an additional Insulated Gate Bipolar Transistor (IGBT) switch is activated as an immediate compensatory measure, ensuring continued operation and preventing total system failure. This dual methodology, combining artificial intelligence with fault-tolerant control, provides a proactive approach to enhancing the reliability and efficiency of photovoltaic energy systems. By employing precise detection and compensation techniques, higher levels of dependability can be achieved, and the costs associated with failures and maintenance can be reduced aligning with the smart operational requirements of future energy grids.

2 Description of System

The system, as illustrated in Fig. 1, consists of a SAPF integrated with a PV energy system. It incorporates a BOOST converter equipped with a MPPT algorithm to regulate the output voltage of the PV array in accordance with the DC bus voltage of the SAPF. The system also includes a four-leg inverter composed of eight bidirectional switching devices, where the fourth leg is dedicated to regu-lating the zero-sequence current component. This configuration enables the com-pensation of harmonic distortions and unbalanced currents, thereby improving overall power quality. To address potential faults in the IGBT switches within the BOOST converter whether open-circuit or short-circuit faults a Fault Tolerant Control (FTC) strategy has been developed based on a SVM classification algorithm. This ap-proach enables early and accurate fault detection and localization under varying operational conditions. The proposed methodology ensures continuous and sta-ble system performance without requiring manual intervention or system shut-down, thereby enhancing system reliability and minimizing the risk of power supply interruptions.

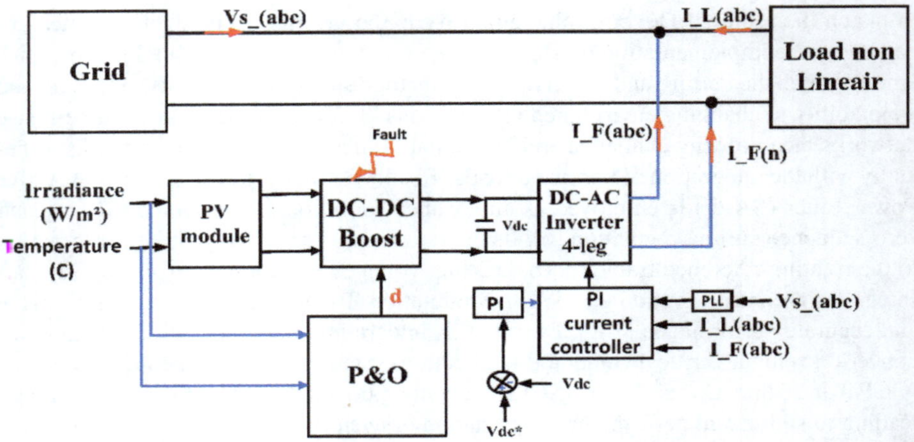

Fig. 1. Schematic Block Diagram of a 3-Phase Four-Leg SAPF Connected to PVs.

3 Fault Tolerant Control Strategy

3.1 SVM-Based Detection of IGBT Faults in PV Boost Converters

This study, as illustrated in Fig. 2, presents an intelligent classification framework based on an enhanced SVM algorithm, aiming to achieve accurate and early detection of IGBT faults in Boost Converters used within grid-connected PV systems. A synthetic dataset was constructed to represent various operational conditions by generating structured combinations of the input variables Ipv and Vpv. These inputs were utilized to determine the operational state of the IGBT, whether functioning normally or experiencing one of two major fault types: OCF or SCF. As depicted in Fig. 3, the system is categorized into three primary classes: the normal operating condition (shown in blue, where Class 0 ($y = 0$)), open-circuit fault (in green, where Class 1 ($y = 1$)), and short-circuit fault (in yellow, where Class 2 ($y = 2$)). The model was trained using the multi-class classification technique (fitcecoc) with a Radial Basis Function (RBF) kernel, enabling the SVM to achieve high classification accuracy and clear separation between the fault scenarios.

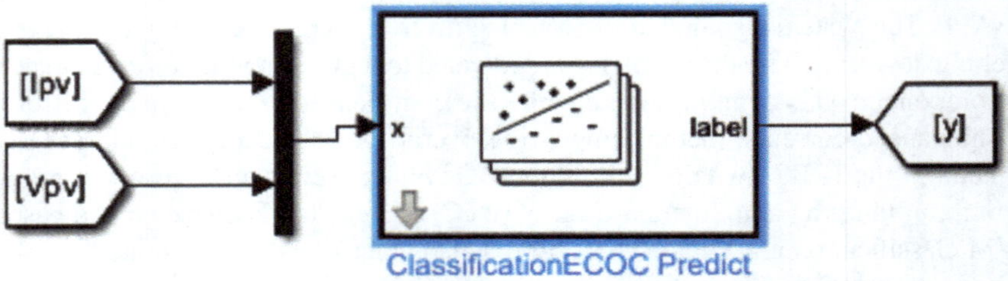

Fig. 2. Detection fault with SVM

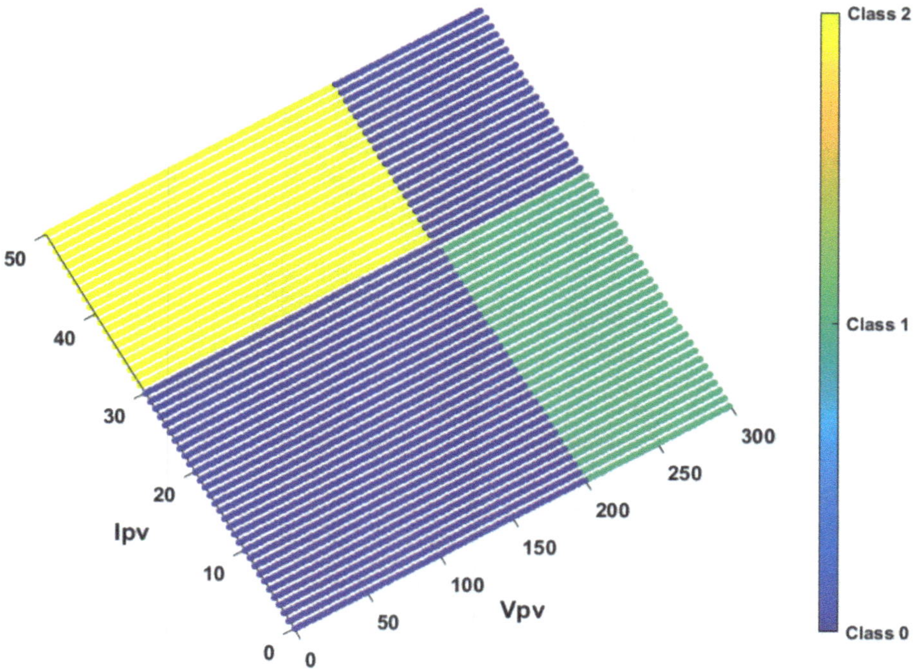

Fig. 3. Decision Boundary Representation for Open and Short Circuit Fault Classification Using SVM Algorithm

3.2 Reconfiguration-Based Fault Tolerant Scheme for a DC-DC Boost Converter

Figure 4 illustrates the fault compensation mechanism applied to the IGBT in a DC-DC boost converter, based on a reconfiguration strategy. When a fault is detected in the main switch, a redundant IGBT is immediately activated to assume its function. This process involves opening switch K and dynamically adjusting the control signal S. Under normal conditions, S holds a value of "0" to keep the redundant IGBT in the OFF state. Upon fault detection, S is updated to match the address of the faulty switch, thereby triggering the compensating IGBT to operate in its place. This approach ensures continuous operation and enhances system reliability by minimizing service interruptions.

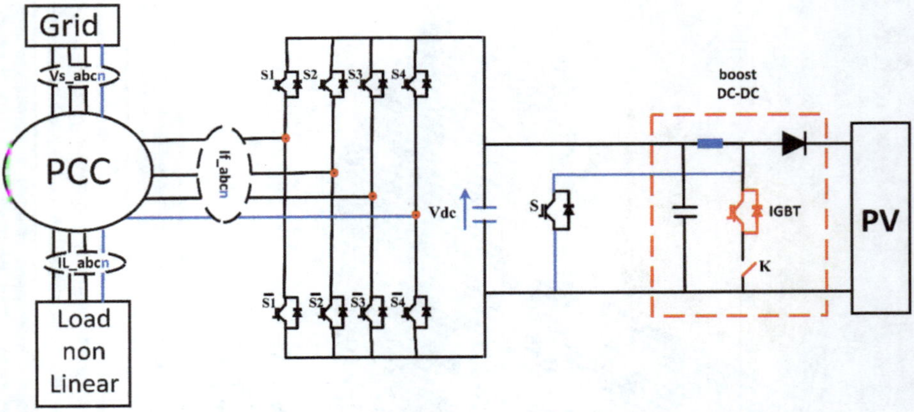

Fig. 4. IGBT Fault Compensation via Switch Reconfiguration in Boost Converter.

4 Simulation Results

The FTC strategy was implemented on the Boost Converter using the Matlab Simulink environment to evaluate the system's performance under potential faults in the IGBT switches. The obtained results demonstrate the effectiveness of the proposed strategy in maintaining voltage stability and ensuring continuous operation despite the presence of faults. The system behavior is documented through a series of illustrative figures, along with Table 1, which presents the key parameter values used in the simulation.

Table 1. parameters of the system.

Parameter	value
Source voltage, Source impedance frequency	220 V, 1mΩ -1 mH 50 Hz
Coupling filter impedance DC bus voltage	1mΩ, 1 mH 800 V
Load impedance Line impedance	15 mH,5 Ω 1mΩ, 1 mH

Figure 5 and Fig. 6 presents the effects of both open-circuit and short-circuit faults occurring in the IGBT switch of the Boost Converter within a grid-connected photovoltaic (PV) system. These faults were intentionally introduced during the time interval from 0.05 to 0.08 s in order to analyze the dynamic response of the system under each fault condition.

Figure 5(a) illustrates the variation in the PV output voltage Vpv during the open-circuit fault. A noticeable increase in voltage is observed as a result of the interruption of the current path through the faulty switch. In contrast, Fig. 5(b) shows that during the short-circuit fault, Vpv gradually decreases until it reaches zero due to the formation of a low-impedance path, leading to a rapid discharge of energy.

Regarding the PV output current Ipv, Fig. 5(c) demonstrates a sharp decline in the current to zero during the open-circuit fault, caused by the break in the current path. Conversely, Fig. 5(d) reveals an increase in Ipv during the short-circuit condition, due to the sudden drop in circuit impedance and the resulting excessive current flow.

Furthermore, Fig. 5 (e) and Fig. 5(f) display the fault detection results obtained using the SVM algorithm. The detection signal changes value during the fault interval, accurately reflecting the fault status. A value of "0" indicates normal operation, "1" corresponds to an open-circuit fault, and "2" represents a short-circuit fault. These results confirm the capability of the SVM-based detection algorithm to correctly identify both the occurrence and the type of fault with high reliability.

Figure 6 illustrates the system's dynamic response to both open-circuit and short-circuit faults in the Boost Converter's IGBT switch, comparing scenarios with and without the implementation of the FTC strategy.

In Fig. 6(a), the source current significantly increases during the fault period regardless of whether the fault is open or short when FTC is not applied. This behavior indicates that the PVs is no longer contributing power, and the utility grid is solely responsible for supplying the load. In contrast, with the FTC strategy in place, the source current remains stable, suggesting that the fault detection and compensation mechanism effectively maintains power sharing between the PV source and the grid. Figure 6(c) shows the PV power output Ppv. Without FTC, the power drops to zero during the fault. However, when FTC is activated, the power exhibits only a slight transient disturbance and quickly returns to its nominal value, confirming the strategy's ability to ensure uninterrupted PV contribution.

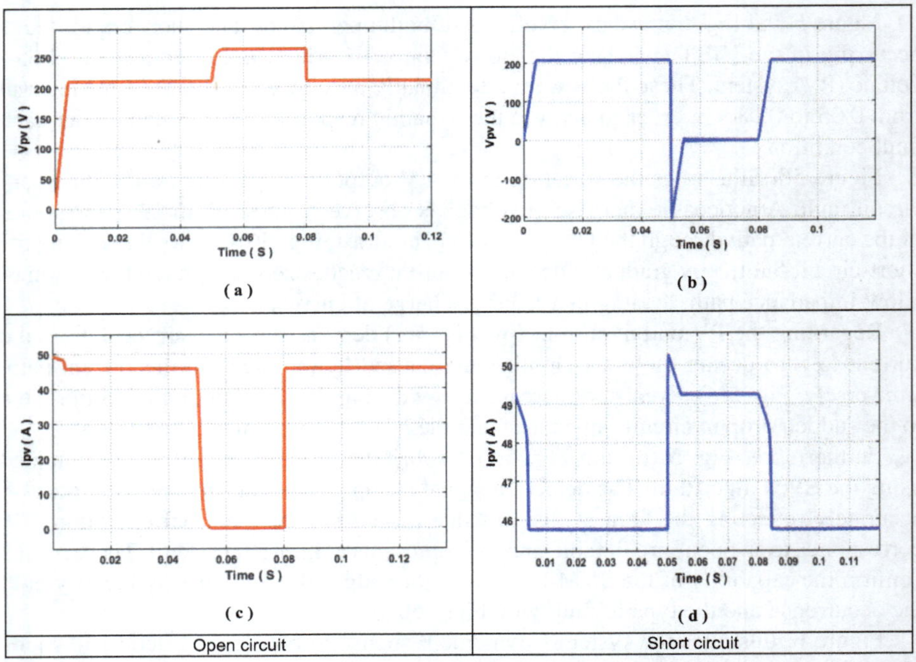

Fig. 5. Open-Circuit and Short-Circuit Faults in Boost Converter IGBT within a PV Grid-Connected System.

5 Conclusion

This research has presented an intelligent FTC framework for grid-connected PVs, based on the SVM algorithm for early and accurate detection of both open-circuit and short-circuit faults in IGBT switches within the boost converter. The simulation results demonstrated that the proposed methodology effectively distinguishes between different operational states, provides real-time fault localization, and enables immediate activation of a compensatory switch ensuring continuous operation without interruption or manual intervention. The integration of artificial intelligence techniques with reconfiguration-based fault-tolerant strategies offers a robust and advanced approach to enhancing the reliability and efficiency of PVs, while simultaneously reducing operational risks and maintenance costs. This contribution opens new horizons for the development of intelligent, adaptive solutions that meet the evolving demands of modern power grids and reinforce the role of renewable energy sources in the global transition toward more sustainable and smart energy infrastructures.

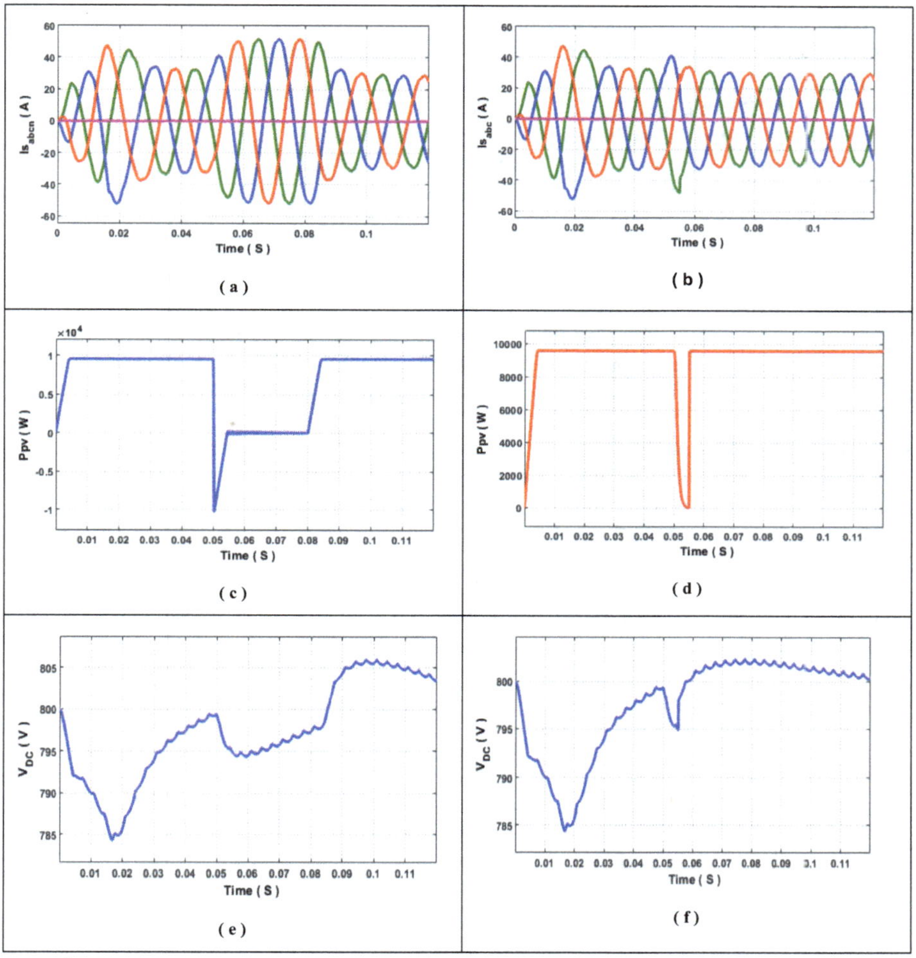

Fig. 6. System Response to the Fault Without and With the Application of FTC Strategy

References

1. Kishore, T.S., Kumar, P.U., Ippili, V.: Review of global sustainable solar energy policies: Significance and impact. Innov. Green Dev. **4**(2), 100224 (2025)
2. Hadi, H.A., Kassem, A., Amoud, H., Nadweh, S.: Improve power quality and stability of grid - connected pv system by using series filter. Heliyon. **10**(21), e39757 (2024)
3. Ko, J.-S., Huh, J.-H., Kim, J.-C.: Overview of maximum power point tracking methods for pv system in micro grid. Electronics. **9**(5), 816 (2020)
4. Tawiah-Mensah, P.N., Addison, J., Oppong, S.D., Effah, F.B.: An improved perturb and observe maximum power point tracking algorithm with the capability of drift avoidance in PV systems. In: IEEE PES/IAS Power Africa, pp. 1, 2022–5, Kigali, Rwanda (2022)

5. Wang, Y., Sun, L.: Photovoltaic maximum power point tracking technology based on improved perturbation observation method and backstepping algorithm. Electronics. **13**(19), 3960 (2024)
6. Mostefa, A., Belalia, K., Lantri, T., Boulouiha, H., Allali, A.: A four-line active shunt filter to enhance the power quality in a microgrid. Int. J. Renew. Energy Dev. **12**(3), 488–498 (2023)
7. Djerboub, A., Tamer, A., Allaoui, T., Djerboub, K.: Comparative performance analysis of a four-leg two-level SAPF versus a four-leg three-cell fci-based SAPF. In: 2024 3rd International Conference on Advanced Electrical Engineering (ICAEE), pp. 1–6 (2024)
8. Djerboub, K., Allaoui, T., Champenois, G., Denai, M., Habib, C.: Particle swarm optimization trained artificial neural network to control shunt active power filter based on multilevel flying capacitor inverter. Eur. J. Electr. Eng. **22**(3), 199–207 (June 2020)
9. Bouarfa, A., Fadel, M., Bodson, M.: A new pwm method for a 3-phase 4-leg inverter based on the injection of the opposite median reference voltage. In: 2016 International Symposium on Power Electronics, Electrical Drives, Automation and Motion (SPEEDAM), pp. 791–796 (2016)
10. Bento, F., Cardoso, A.J.M.: Diagnosis of Dc-Dc converter semiconductor faults based on the second-order derivative of the converter input current. Electronics. **13**(18), 3778 (2024)
11. Espinoza Trejo, D.R., Bárcenas, E., Hernández Díez, J.E., Bossio, G., Espinosa Pérez, G.: Open- and short-circuit fault identification for a boost dc/dc converter in PV MPPT systems. Energies. **11**(3), 616 (2018)
12. Jagtap, S., More, D.: Switch open-circuit fault diagnosis and fault-tolerant control for boost dc-dc converter. Procedia Comput. Sci. **171**, 934–940 (2020) Third International Conference on Computing and Network Communications (CoCoNet'19)

A Novel Method for Inverter Open-Circuit Fault Diagnosis Using Improved Variational Mode Decomposition

Rabah Abdelkader(✉)

Department of electronic and communication, Khemis Miliana University Route de Theniet El Had, 44225, Khemis Miliana, Ain Defla, Algeria
r.abdelkader@univ-dbkm.dz

Abstract. Three-phase inverters employing Insulated Gate Bipolar Transistors (IGBTs) are commonly used in industry. However, IGBT switch open-circuit faults have the potential to significantly affect inverter performance and connected loads. Therefore, system reliability and resilience require prompt and precise detection of such faults. In this paper, we present a novel method to detect open-circuit faults of three-phase inverter IGBT switches. The proposed method employs an Improved version of Variational Mode Decomposition (IVMD) technique with correlation analysis and statistical parameters such as variance and mean values. Subsequently, the three-phase stator current signals are decomposed into a set of BLIMFs by the IVMD algorithm. Next, the correlation coefficients between the original stator current signals and their respective BLIMFs are computed. The BLIMFs with the highest correlation coefficients are labeled as the most important components. Secondly, the variances of the selected modes are compared to identify the faulty phase and corresponding inverter leg. Finally, the average value of the mode with lowest correlation coefficient in the faulty phase is examined to identify the fault location in either the upper or lower IGBT switch of the faulty leg. The method is validated with test data. The outcomes seemingly indicate the effectiveness and accuracy of the method in diagnosing open-circuit faults in IGBT-based inverter drives.

Keywords: Inverter · IGBT · Fault · Open Circuit · VMD, Diagnosis

1 Introduction

Three-phase IGBT-based inverters are basic building blocks in many power electronic devices like variable speed drives, uninterruptible power supplies, and active filters. Their performance, reliability, and supplied power quality—namely the continuity of operation—have been of utmost concern to the energy sector today. Faults may occur in inverters in numerous ways on their internal components, such as power semiconductors, DC bus capacitors, and power circuits. Of these, faults in IGBTs make up approximately 31% of all the faults in converter systems [1]. Such faults are typically two in number: short-circuit and open-circuit faults. An open-circuit fault usually results from an internal

M. Hatti (Ed.): IC-AIRES 2025, LNNS 1740, pp. 353–364, 2026.
https://doi.org/10.1007/978-3-032-12788-4_34

severed connection within the transistor or from mismonitoring of the collector current with a value that is too high, leading to partial or total loss of function of one of the IGBTs. This type of fault manifests itself as the missing positive or negative half-wave of the inverter output current, depending on the component in the top or bottom leg of the failing leg. These defects possess a chance to cause large system malfunctions and, in the worst possible scenario, might cause a complete shut-down. Additionally, unless identified and compensated in a timely manner, they can pose to be a threat to the overall system's integrity. It is therefore necessary to employ a quick and effective fault detection and compensation technique so as to prevent fault propagation as well as maintain uninterrupted operation even in case of failure of an inverter.

During the last decades, several methods were suggested to identify open-circuit faults of three-phase inverter IGBTs [2–4]. One of the earliest solutions, suggested in the 1990s by Mendes et al. [5], is Park's Vector Technique, widely known from examination of the trajectory and amplitude of the phase currents. It has low computational complexity but also shortfalls like sensitivity to the load transients as well as variability of the load. To overcome these limitations, Rothenhagen et al. [6] have proposed better approaches such as the Normalized DC Current Method and its generalized version. Additionally, some articles [7–9] have generalized Park's vector-based approaches for three-phase voltage inverter fault detection and identification. Estma et al. in [10] proposed a diagnostic method using the three phase currents and their corresponding reference measurements. The faults and their causes are computed to generate the diagnostic variables, which are processed relative to set thresholds for diagnosing single and multiple IGBT open-circuit faults. During the last few years, newer signal processing techniques have found themselves to be leading fault diagnosis methods with the aim of improving detection rate and reducing diagnostic cost and complexity. Among them, the Discrete Wavelet Transform (DWT) has been the most widely used [11–14]. For instance, [11] used DWT to detect discontinuities in the current signal, and subsequently fault isolation with a Support Vector Machine (SVM) classifier. BDE Cherif et al. used DWT with Neural Networks (NN) and Fast Fourier Transform (FFT) for completely automatic diagnosis in [12]. Rekha et al. [13] identified permanent magnet synchronous generator-driven wind turbine converter open-circuit faults using a DWT-NN approach. Kurukuru et al. [14] proposed a fault detection system with DWT for single-phase inverters. Another advanced hybrid methodology was proposed by Jingxuan et al. [15], which combined Wavelet Transform (WT), Feature Analysis (FA), logical reasoning, and Back Propagation Neural Network (BPNN) for single and double open-circuit faults detection in wind turbine system grid-side converters. Increased interest is placed on using artificial intelligence (AI) to enhance fault diagnosis automation, reliability, and accuracy of classification. In [16], Zhanjun et al. introduced an algorithm combining an Artificial Neural Network (ANN), Multistate Data Processing (MSDP), and Subsection Fluctuation Analysis (SSFA) for smart fault classification. In [17], a Convolutional Neural Network (CNN) was utilized to identify faults in inverter-fed induction motors. Raj et al. [18] used a single ANN to conduct fault diagnosis in binary and ternary topologies of cascaded H-bridge multilevel inverters. Other research [19] worked on ANN models for voltage source inverter fault detection with varying load and frequency, and fault diagnosis of three-level inverter systems.

In this work, we propose a novel, robust, and reliable hybrid approach for detecting and identifying open-circuit faults in the IGBT switches of three-phase inverters. This method combines an enhanced version of Variational Mode Decomposition (IVMD) with correlation analysis and statistical indicators such as variance and mean value. The proposed IVMD algorithm addresses a key limitation of conventional VMD—namely, the need to predefine the number of decomposition modes. By automatically determining a suitable and non-redundant number of modes, IVMD ensures the extraction of relevant and informative signal components without loss of useful information. Following the decomposition of the current signals using IVMD, correlation coefficients are computed between the original signals and the resulting Band-Limited Intrinsic Mode Functions (BLIMFs). This step enables the selection of the most representative modes while discarding those dominated by noise. In the next phase, the variance of the selected modes with the highest correlation coefficients is calculated to determine the faulty phase and its corresponding leg. Finally, to pinpoint the specific faulty IGBT switch, the mean values of the selected modes with the lowest correlation coefficients (within the identified faulty phase) are computed.

2 Improved Variational Mode Decomposition (Ivmd)

2.1 Variational Mode Decomposition (VMD)

The VMD is a decomposition algorithm was presented in 2014 by Dragomiretskiy and Zosso [20]. It allows decomposing a signal $x(t)$ into K modes u_k centered on their central frequency ω_k. The VMD is required to solve a constrained variational problem, which is given by the following formula [20]:

$$\min_{\{u_k\},\{\omega_k\}} \left\{ \sum_{k=1}^{K} \left\| \partial \left[\left(\delta(t) + \frac{1}{\pi t} \right) * u_k(t) \right] e^{-j\omega_k t} \right\|_2^2 \right\} \tag{1}$$

With

$$\sum_{i=1}^{k} u_k(t) = x \tag{2}$$

where t is the time, δ is the Dirac distribution, and * represents the convolution.

To solve the constrained variational problem, the augmented Lagrangian L is used as:

$$\begin{aligned} L(\{u_k\},\{\omega_k\},\lambda) &= \alpha \textstyle\sum_{k=1}^{K} \left\| \partial_t \left[\left(\delta(t) + \frac{j}{\pi t} \right) * u_k(t) \right] e^{-j\omega_k t} \right\|_2^2 \\ &+ \left\| x(t) - \sum_{k=1}^{K} u_k(t) \right\|_2^2 + \left\langle \lambda(t), x(t) - \textstyle\sum_{k=1}^{K} u_k(t) \right\rangle \end{aligned} \tag{3}$$

where λ is the vector of lagrange multipliers, α is a balancing factor.

The Alternating Direction Method of Multipliers (ADMM) algorithm is used to solve the Eq. (3) and to produce the different modes and frequency of the center. The ADMM algorithm is summarized in the following steps [20]:

Step 1: Initialize $\{u_k^1\} = 0$, $\{\omega_k^1\} = 0$, $\lambda^1 = 0$, $n = 1$, $k = 1$;
Step 2: $n \leftarrow n + 1$;
Step 3: For $k = 1 : K$ do;
Update u_k according to:

$$u_k^{n+1} = \arg\min L\left(\left\{u_{i<k}^{n+1}\right\}, \left\{u_{i\geq k}^{n}\right\}, \left\{\omega_i^n\right\}, \lambda^n\right) \tag{4}$$

end for
Step 4: For $k = 1 : K$ do;

$$\omega_k^{n+1} = \arg\min L\left(\left\{u_i^{n+1}\right\}, \left\{\omega_{i<k}^{n+1}\right\}, \left\{\omega_{i\geq k}^{n}\right\}, \lambda^n\right) \tag{5}$$

end for
Step 5: Update according to:

$$\lambda^{n+1} = \lambda^n + \tau\left(f - \sum u_k^{n+1}\right) \tag{6}$$

Step 6: Repeat steps (2) to (5) until the following criterion is met:

$$\sum \left\|u_k^{n+1} - u_k^n\right\|_2^2 / \left\|u_k^n\right\|_2^2 < \varepsilon \tag{7}$$

The mode $u_k(t)$ is obtained as follows:

$$\widehat{u}_k^{n+1}(\omega) \widehat{=} \frac{\widehat{x}(\omega) - \sum_{i\neq k} \widehat{u}_i(\omega) + \frac{\widehat{\lambda}(\omega)}{2}}{1 + 2\alpha(\omega + \omega_k)^2} \tag{8}$$

$$\widehat{u}_k(t) = \Re\left\{ifft\left(\widehat{u}_k(\omega)\right)\right\} \tag{9}$$

where $\widehat{x}(\omega)$ is the Fourier Transform signal, *ifft*(.) is the inverse Fourier Transform signal and $\Re\{.\}$ is the real part of the analytical signal.

The updated equation of ω_k is shown as follows:

$$\omega_k^{n+1} = \frac{\int_0^\infty \omega \left|\widehat{u}_k(\omega)\right|^2 d\omega}{\int_0^\infty \left|\widehat{u}_k(\omega)\right|^2 d\omega} \tag{10}$$

The process of the VMD algorithm is summarized as follows:
Step 1: Initialize $K = 1$;
Step 2: Do: $K = K + 1$;
Step 3: Initialize $\widehat{u}_k^1$, $\widehat{\omega}_k^1$, $\widehat{\lambda}^1$ et n;
Step 4: Do $n = n + 1$;
Step 5: For $k = 1 : K$, update $\widehat{u}_k^{n+1}$, ω_k^{n+1} using (8) and (10) respectively;
Step 6: Update $\widehat{\lambda}^{n+1} \leftarrow \widehat{\lambda}^n + \tau\left(\widehat{f} - \sum_k \widehat{u}_k^{n+1}\right)$;
Step 7: Repeat steps (4)–(6), until $\sum_k \left\|u_k^{n+1} - u_k^n\right\|_2^2 / \left\|u_k^n\right\|_2^2 < \varepsilon$.

2.2 Improved Variational Mode Decomposition (IVMD)

The major problem of VMD is determining the number of modes K which is influenced on their performance. To solve this problem, we propose in this work an improved version of VMD based on the detection of the optimal number of modes necessary and sufficient for the decomposition. The IVMD proposed is summarized in the following steps:

Step 1: Initialize $K = 1$,

Step 2: Decompose the signal using the conventional VMD technique;

Step 3: Calculate the correlation coefficient between the original signal and their BLIMFs using the following equation:

$$\rho(i) = \sum_{i=1}^{L} x(t)u_i(t) / \left(\sqrt{\sum_{i=1}^{L} x^2(t) \sum_{i=1}^{L} u_i^2(t)} \right) \tag{11}$$

Step 4: If the correlation coefficient between the original signal and K^{th} BLIMFs is less than 0.1, the optimal number of modes is K, otherwise repeat steps (2)–(4) with $K = K + 1$.

Step 5: Decompose the signal using the VMD technique with the optimal number of modes obtained.

3 The Proposed Method

The proposed approach for the diagnosis of IGBT open switch faults of a three phase inverter is summarized in the following steps:

Step 1: Decompose the three phase stator current signals using the IVMD technique;

Step 2: Calculate the correlation coefficients between the three phase stator current signals and their BLIMFs. Then; select the BLIMFs which have the highest coefficients as relevant modes;

Step 3: Calculate the variances of the first selected BLIMF which have the highest correlation coefficients among those selected. If these variance values are close, the inverter is in a healthy state, otherwise the phase which has a smaller variance value is faulty and therefore the corresponding leg of this phase is faulty;

Step 4: Calculate the average of the second selected BLIMF which have the lowest correlation coefficients among those selected for the faulty phase. If the average value is positive, the fault is located at the lower switch level (LSF) and if this variance value is negative the fault is located at the upper switch level (UPF).

4 Experimental Validation

The experimental test bench used in this study to validate the effectiveness of the proposed fault diagnosis approach is illustrated in Fig. 1. It consists of a three-phase, two-level IGBT inverter (SEMIKRON) controlled by a dSPACE 1104 board implementing an SVM-based PWM control strategy. The inverter feeds a 3 kW three-phase squirrel cage induction motor operating at 50 Hz with a rated rotor speed of 1440 RPM. A DC generator is mechanically coupled to the motor and serves as the load. The test setup also includes three voltage sensors (TEKTRONIX P5200), three Hall-effect current sensors (FLUKE i30s AC/DC Current Clamp), a tachometer (ONO SOKKI HT-341), a data acquisition card (NI-6330), and a computer for signal acquisition and visualization.

Figure 2 illustrates the inverter–motor system used for the experimental tests. It comprises a three-phase, two-level voltage source inverter supplying an induction motor, with the possibility of introducing open-circuit faults in the IGBT switches. The system is tested under both healthy and faulty operating conditions. The different operating modes are defined as follows:

- Mode 1: Operation with a healthy inverter (no faults).
- Mode 2: Operation with an open-circuit fault in IGBT S1.
- Mode 3: Operation with an open-circuit fault in IGBT S2.
- Mode 4: Operation with an open-circuit fault in IGBT S3.
- Mode 5: Operation with an open-circuit fault in IGBT S4.
- Mode 6: Operation with an open-circuit fault in IGBT S5.
- Mode 7: Operation with an open-circuit fault in IGBT S6.

All data acquisitions are performed under nominal conditions, over a duration of 5 s, with a sampling frequency of 1.5 kHz and for each of the operating modes described above.

Figure 3 presents the stator current waveforms corresponding to each mode. Compared to the healthy state, a sudden distortion appears at the moment of fault occurrence. Specifically, faults in switches S1, S3, and S5 result in the loss of the positive half-cycle of the stator current. Conversely, faults in S2, S4, and S6 cause a loss of the negative half-cycle. These signatures are indicative of the nature and location of the fault and can be used for accurate fault diagnosis.

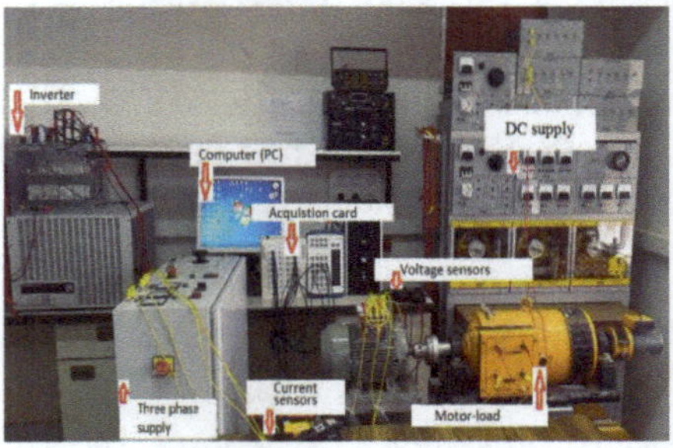

Fig. 1. Test bench.

The BLIMFs obtained by the application the IVMD technique to these signals are illustrated in Figs. 4, 5, 6, 7, 8, 9, and 10.

Table 1 presents the correlation coefficients values between the measured currents signals and their BLIMFs. It can be seen that the BLIMF 1 and BLIMF 2 have the highest correlation coefficient and therefore are selected as relevant modes. Table 2 summarizes

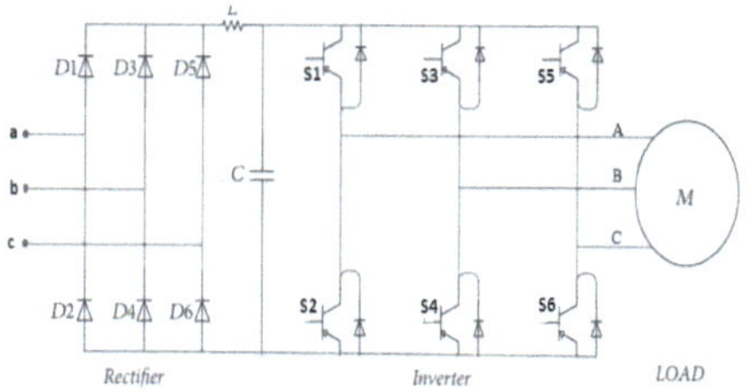

Fig. 2. Inverter-motor system.

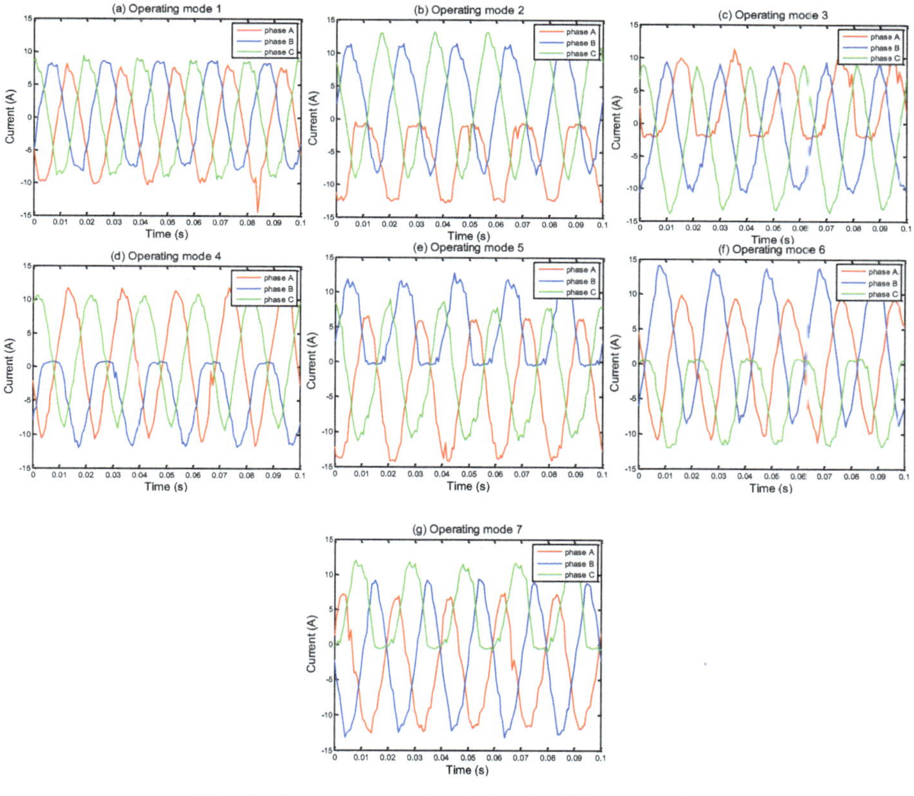

Fig. 3. Stator current signals for the different modes.

the variance values of the first selected BLIMF which have highest correlation coefficient among those selected for the three phases and all of the operating modes. These variance values are close in the case of operating mode 1, which means that the inverter is in a healthy state. For the cases of operating modes 2 and 3, these variance values are significantly reduced for phase A compared to phases B and C, which makes it possible to locate the open circuit fault on this phase and consequently the S1-S2 leg. The variance values are decreased considerably for phase B compared to phases A and C for the cases

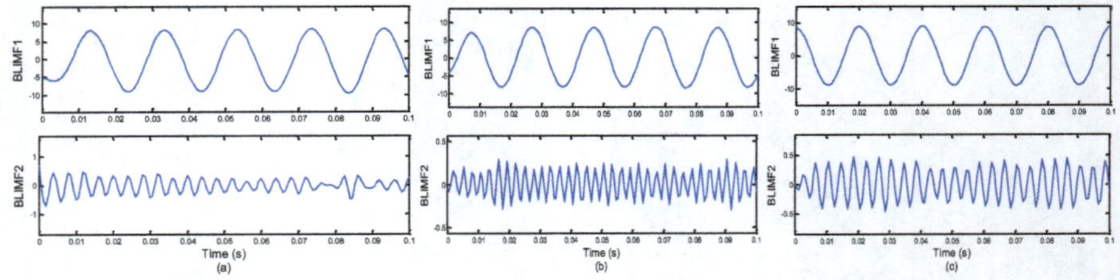

Fig. 4. Obtained BLIMFs by IVMD of three phase for operating mode 1, (a) Phase A, (b) Phase B, (c) Phase C.

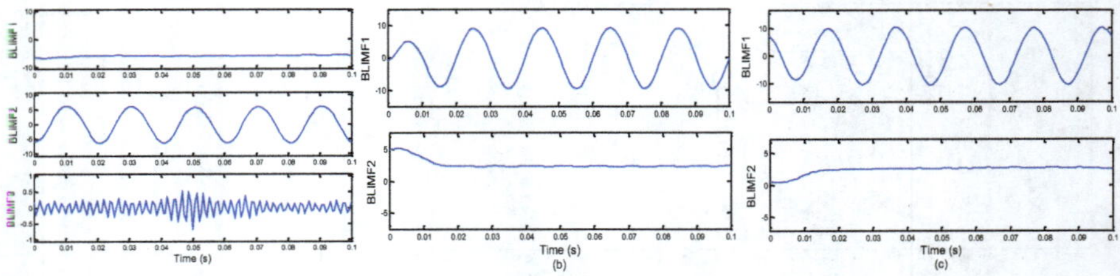

Fig. 5. Obtained BLIMFs by IVMD of three phase for operating mode 2, (a) Phase A, (b) Phase B, (c) Phase C.

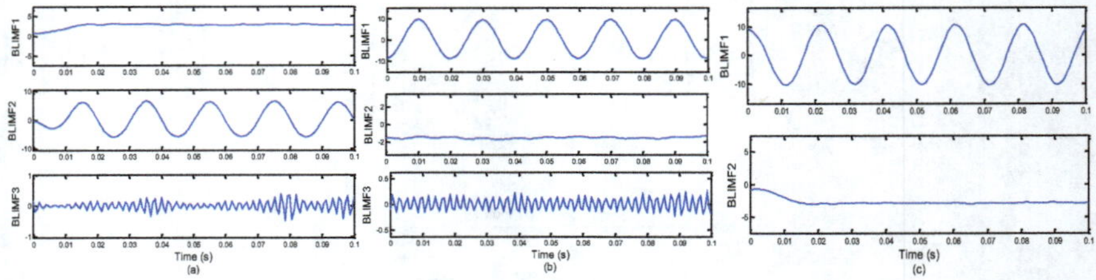

Fig. 6. Obtained BLIMFs by IVMD of three phase for operating mode 3, (a) Phase A, (b) Phase B, (c) Phase C.

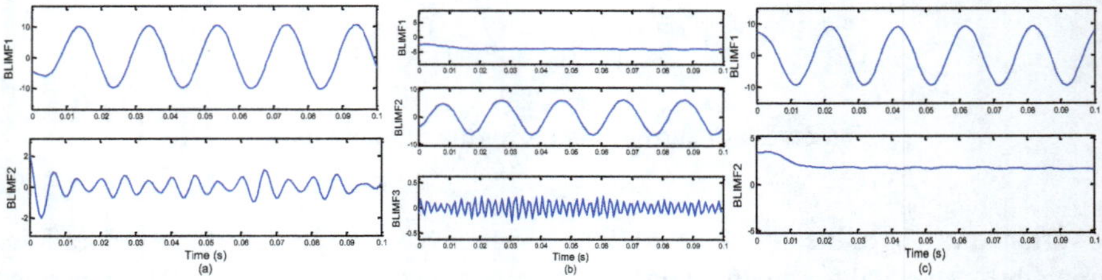

Fig. 7. Obtained BLIMFsby IVMD of three phase for operating mode 4, (a) Phase A, (b) Phase B, (c) Phase C.

of operating modes 4 and 5, this confirms that this phase is faulty and therefore the S3-S4 leg is faulty. Similar observations for the cases of operating modes 6 and 7, the variance

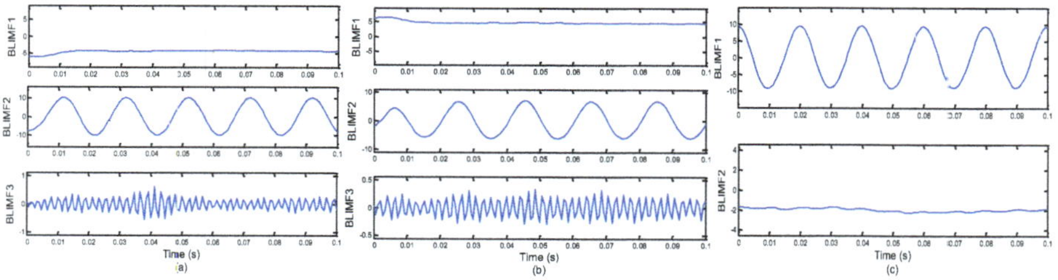

Fig. 8. Obtained BLIMFs by IVMD of three phase for operating mode 5, (a) Phase A, (b) Phase B, (c) Phase C.

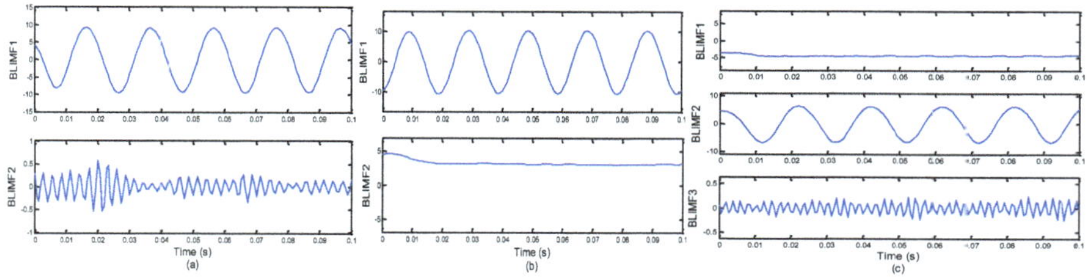

Fig. 9. Obtained BLIMFsby IVMD of three phase for operating mode 6, (a) Phase A, (b) Phase B, (c) Phase C.

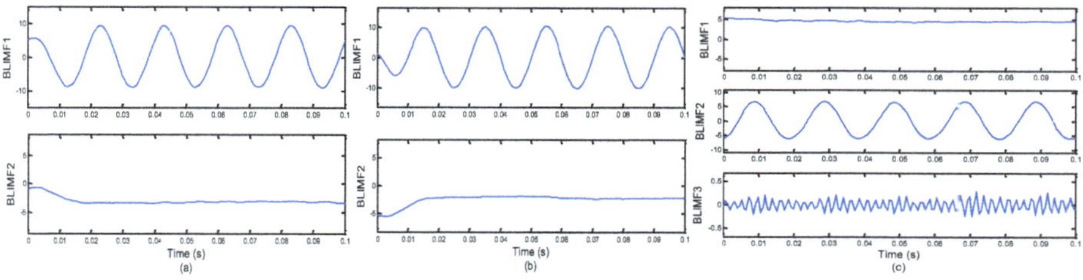

Fig. 10. Obtained BLIMFs by IVMD of three phases for operating mode 7, (a) Phase A, (b) Phase B, (c) Phase C.

values for phase C are reduced compared to phases A and B, which means that the open circuit fault is detected on leg S5-S6. Table 3 shows the average values of second selected BILMFs which have the lowest correlation coefficients among those selected for faulty phase and for each operating mode (BLIMF 1 of phase A for operating mode 2 and 3, BILIMF1 of phase B for operating mode 4 and 5, BLIMF1 of phase C for operating mode 6 and 7). It is noticed that these values are respectively −5.8739 (negative value) and 2.9175 (positive value) for the cases of operating modes 2 and 3 (faulty leg S1-S2) which means that the open circuit fault is located in the upper switch S1 for operating mode 2 and in the lower switch S2 for operating mode 3. For operating modes and 4 and 5 (faulty leg S3-S4), these average values are −4.1215 (negative value) and 4.7030 (positive value) respectively, which makes it possible to locate the open circuit fault on the upper switch S3 for operating mode 4 and on the lower switch S4 for operating mode 5. Similar remarks for operating modes 6 and 7 (faulty leg S5-S6), these average values are −4.5846 (negative value) and 4.3887 (positive value) respectively, which confirms that the fault is in the upper switch S5 for operating mode 6 and in the lower switch S6 for operating mode 7.

Table 1. The correlation coefficients values between the current signal and their BLIMFs for each phase and for each operating mode.

Current signal	BLIMFs of phase A			BLIMFs of phase B			BLIMFs of phase C		
Operating mode	BLMIF1	BLMIF2	BLMIF3	BLMIF1	BLMIF2	BLMIF3	BLMIF1	BLMIF2	BLMIF3
1	0.9952	0.0594	–	0.9964	0.0415	–	0.9968	0.0500	–
2	0.1197	0.9815	0.0640	0.9941	0.0769	–	0.9959	0.0593	–
3	0.1348	0.9808	0.0601	0.9954	0.0652	0.0316	0.9964	0.0597	–
4	0. 9953	0.0719	–	0.1224	0.9835	0.0520	0.9947	0.0683	–
5	0.0657	0.9958	0.0370	0.1281	0.9824	0.0525	0.9940	0. 0719	–
6	0.9940	0.0484	–	0.9956	0.0541	–	0.1242	0.9835	0.0516
7	0.9932	0.0814	–	0.9954	0.0732	–	0.1153	0.9836	0.0526

Table 2. The variance value of selected mode which highest correlation coefficients among those selected for each phase and for each operating mode.

Operating mode	Phase A	Phase B	Phase C
Mode 1	36.8514	35.4956	40.5681
Mode 2	20.3002	43.5723	53.0532
Mode 3	19.0382	43.3076	54.5032
Mode 4	53.8777	19.5343	43.0147
Mode 5	50.4241	20.7145	42.0107
Mode 6	42.5142	53.7889	20.6887
Mode 7	42.8223	52.3154	20.9074

Table 3. The average value of selected mode which have lowest coefficient correlation among those selected of faulty phase and for each operating mode

Faulty phase	Operating mode	Phase	Selected BLIMF	Variance of selected BLIMF
Phase A	Mode 2	A	BLIMF1	−5.8739
	Mode 3	A	BLIMF1	2.9175

(continued)

Table 3. (*continued*)

Faulty phase	Operating mode	Phase	Selected BLIMF	Variance of selected BLIMF
Phase B	Mode 4	B	BLIMF1	−4.1215
	Mode 5	B	BLIMF1	4.7030
Phase C	Mode 6	C	BLIMF1	−4.5846
	Mode 7	C	BLIMF1	4.3887

5 Conclusion

A novel hybrid approach for detection and identifying open circuit faults of IGBT switches of a two level inverter is proposed in this work. It is based on an improved version of VMD techniques combined with correlation coefficients and statistical indicators such as the variance and the average value. Firstly, the signals of three phase current are decomposed using the proposed version of IVMD to obtain theirs BLIMFs. Secondly, the correlation coefficients between the three phases current signals and their BLIMFs in order to select the relevant modes. Thirdly, the variances of the first selected BLIMFs which the highest correlation coefficients among those selected are calculated to detect the faulty phase and therefore the faulty leg. Finally, the average values of the second selected BLIMFs which the lowest correlation coefficients among those selected are calculated in order to locate the open circuit fault. Experimental results show that the proposed technique can identify and locate the inverter IGBT switches open circuit fault accurately and reliably.

References

1. Yeh, C.C., Demerdash, N.A.O.: Induction motor drive systems with fault tolerant inverter motor capabilities. IEEE Int. Electr. Mach. Drives Conf., Antalya. **3-5**, 1451–1458 (May 2007). https://doi.org/10.1109/IEMDC.2007.383642
2. Wu, R., Blaabjerg, F., Wang, H., Liserre, M., Iannuzzo, F.: Catastrophic failure and fault-tolerant design of IGBT power electronic converters - an overview. In: IECON 2013 - 39th Annual Conference of the IEEE Industrial Electronics Society, Vienna, Austria, pp. 507–1513 (2013). https://doi.org/10.1109/IECON.2013.6699187
3. Kumar, G.K., Elangovan, D.: Review on fault-diagnosis and fault-tolerance for DC–DC converters. IET Power Electron. **13**(1), 1–13 (07 January 2020). https://doi.org/10.1049/iet-pel.2019.0672
4. Wan, Z., Lin, Q., Qiao, F., Zhang, Z., Zong, Y., Su, X.: A Review of Power Electronic Parametric Fault Diagnosis Methods. In: 2019 4th International Conference on Intelligent Green Building and Smart Grid (IGBSG), Hubei, China, pp. 448–451 (2019). https://doi.org/10.1109/IGBSG.2019.8886330
5. Mendes, A.M.S., Marques Cardoso, A.J., Saraiva, E.S.: Voltage source inverter fault diagnosis in variable speed AC drives, by the average current Park's vector approach. In: IEEE International Electric Machines and Drives Conference. IEMDC'99. Proceedings (Cat. No.99EX272), Seattle, WA, USA, pp. 704–706 (1999). https://doi.org/10.1109/IEMDC.1999.769220

6. Rothenhagen, K., Fuchs, F.W.: Performance of diagnosis methods for IGBT open circuit faults in three phase voltage source inverters for AC variable speed drives. In: 2005 European Conference on Power Electronics and Applications, Dresden, Germany, pp. 10 pp.-P.7 (2005). https://doi.org/10.1109/EPE.2005.219426
7. Abadi, M.B., Mendes, A.M.S., Cruz, S.M.A.: Three-level NPC inverter fault diagnosis by the Average Current Park's Vector approach. In: 2012 XXth International Conference on Electrical Machines, Marseille, France, pp. 1893–1898 (2012). https://doi.org/10.1109/ICElMach.2012.6350140
8. Estima, J.O., Nuno, M.A., Freire, A.J., Cardoso, M.: Recent advances in fault diagnosis by Park's vector approac. In: 2013 IEEE Workshop on Electrical Machines Design, Control and Diagnosis (WEMDCD), Paris, France, pp. 279–288 (2013). https://doi.org/10.1109/WEMDCD.2013.65251873245263
9. Freire, N.M.A., Estima, J.O., Marques Cardoso, A.J.: Multiple open-circuit fault diagnosis in voltage-fed PWM motor drives using the current Park's Vector phase and the currents polarity. In: 8th IEEE Symposium on Diagnostics for Electrical Machines, Power Electronics & Drives, Bologna, Italy, pp. 397–404 (2011). https://doi.org/10.1109/DEMPED.2011.6063654
10. Estima, J.O., Cardoso, A.J.M.: A new approach for real-time multiple open circuit fault diagnosis in voltage source inverters. IEEE Trans. Ind. Appl. **47**(6), 2487–2494 (2011). https://doi.org/10.1109/TIA.2011.2168800
11. Kim, D.-E., Lee, D.-C.: Fault diagnosis of three-phase PWM inverters using wavelet and SVM. In: 2008 IEEE International Symposium on Industrial Electronics, Cambridge, UK, pp. 329–334 (2008). https://doi.org/10.1109/ISIE.2008.4676998
12. Bilal Djamal Eddine Cherif and Azeddine Bendiabdellah, "Detection of two-level inverter open-circuit fault using a combined DWT-NN approach," J. Control Sci. Eng.Vol 2018, 1976836, 11, doi:https://doi.org/10.1155/2018/1976836.
13. Rekha, S.N., Aruna Jeyanthy, P., Devaraj, D.: Wavelet transform based open circuit fault diagnosis in the converter used in wind energy systems. In: 2017 IEEE International Conference on Intelligent Techniques in Control, Optimization and Signal Processing (INCOS), Srivilliputtur, India, pp. 1–4 (2017). https://doi.org/10.1109/ITCOSP.2017.8303152
14. Kurukuru, V.S.B., Haque, A., Khan, M.A.: Fault Detection in Single-Phase Inverters Using Wavelet Transform-Based Feature Extraction and Classification Techniques. In: Mishra, S., Sood, Y., Tomar, A. (eds) Applications of Computing, Automation and Wireless Systems in Electrical Engineering. Lecture Notes in Electrical Engineering, vol 553. Springer, Singapore (2019). https://doi.org/10.1007/978-981-13-6772-4_56
15. Zhang, J., Sun, H., Sun, Z., Dong, W., Dong, Y.: Fault diagnosis of wind turbine power converter considering wavelet transform, feature analysis, judgment and bp neural network. IEEE Access. **7**, 179799–179809. https://doi.org/10.1109/ACCESS.2019.2958409
16. Huang, Z., Wang, Z., Zhang, H.: Multiple open-circuit fault diagnosis based on multistate data processing and subsection fluctuation analysis for photovoltaic inverter. IEEE Trans. Instrum. Meas. **67**(3) (March 2018). https://doi.org/10.1109/TIM.2017.2785078
17. Skowron, M., Orlowska-Kowalska, T., Wolkiewicz, M., Kowalski, C.T.: Convolutional Neural Network-Based Stator Current Data-Driven Incipient Stator Fault Diagnosis of Inverter-Fed Induction Motor. Energies. **13**(6), 1475 (2020)
18. Raj, N., Jagadanand, G., George, S.: Fault detection and diagnostic in asymmetric multilevel inverter using artificial neural network. Int. J. Electron. **105**(4), 559–571 (2018). https://doi.org/10.1080/00207217.2017.1378382
19. Dhumale, R.B., Lokhande, S.D.: Neural network fault diagnosis of voltage source inverter under variable load conditions at different frequencies. Measurement. **91**, 565–575 (September 2016). https://doi.org/10.1016/j.measurement.2016.04.051
20. Dragomiretskiy, K., Zosso, D.: Variational mode decomposition. IEEE Trans. Signal Process. **62**(3), 531–544 (1 Feb 2014)

Metaheuristic Optimization and Intelligent Control Strategies

Enhanced Accuracy in Estimating PEM Fuel Cell Parameters Using the Walrus Optimization Algorithm

A. Abbadi[1,2](✉), F. Hamidia[1,2], M. R. Skender[2], M. A. Khelif[1,2], K. Darbali[1,2], and F. Salhi[1,2]

[1] Electrical Engineering Department, Research Laboratory in Electrical Engineering and Automatic (LREA), University of Médéa, Médéa, Algeria
amel.abbadi@yahoo.fr

[2] Electrical Engineering Department, Renewable Energy and Materials Laboratory (REML). University of Médéa, Médéa, Algeria

Abstract. A fuel cell is a highly energy-efficient electrochemical device that converts the chemical energy of a fuel into electricity through a controlled reaction with oxygen or another oxidant. Accurately predicting the behaviour of fuel cell models is essential for enhancing their performance and guiding design improvements. This research focuses on identifying seven key parameters in the NedStack PS6 proton exchange membrane fuel cell (PEMFC) model using an innovative optimization technique known as the Walrus Optimization Algorithm (WaOA). The model evaluation is performed by minimizing the sum of squared errors between experimental and estimated output values, i.e., the objective function. The effectiveness of the proposed approach is validated through a comparative analysis with three other meta-heuristic algorithms. Simulation results demonstrate that the WaOA algorithm outperforms the alternative methods, confirming its superior capability in accurately modelling PEMFC behaviour.

Keywords: Fuel cell model · parameter estimation · optimization · meta-heuristics · Walrus Optimization Algorithm

1 Introduction

Fuel cells convert the chemical energy of a fuel (such as hydrogen or methanol) and an oxidant (typically oxygen or air) directly into electrical energy through a redox reaction. This process offers high energy efficiency and produces zero harmful emissions. Compared to batteries, fuel cells also provide advantages such as extended range and shorter refueling times [1, 2].

Fuel cells have a wide range of applications across various sectors. In transportation, they are used to power vehicles including cars, buses, and ships. They are also employed in stationary power generation and combined heat and power (CHP) systems. Additionally, fuel cells serve in specialized applications such as powering submarines and satellites [1].

M. Hatti (Ed.): IC-AIRES 2025, LNNS 1740, pp. 367–376, 2026.
https://doi.org/10.1007/978-3-032-12788-4_35

Fuel cells are most commonly classified based on the type of electrolyte they use. The primary types include Alkaline Fuel Cells (AFC), Proton Exchange Membrane Fuel Cells (PEMFC), Phosphoric Acid Fuel Cells (PAFC), Molten Carbonate Fuel Cells (MCFC), and Solid Oxide Fuel Cells (SOFC) [3]. Among these, **PEMFCs** are widely adopted due to their low operating temperature, high power density, rapid startup capability, and low emissions [2, 4]. Operating within a temperature range of 60 to 100 °C, PEMFCs are well-suited for a variety of applications, including portable power systems, residential energy supply, and automotive propulsion [5].

Mathematical modeling of fuel cells plays a critical role in enabling researchers to predict and analyze fuel cell performance under varying operational conditions—such as changes in temperature, pressure, and fuel composition. These models are essential for optimizing fuel cell design and operation, thereby enhancing efficiency, sustainability, and economic viability.

Traditional optimization methods, particularly gradient-based techniques, often struggle to identify the global optimum in PEMFC parameter estimation due to the model's inherent nonlinearity and the presence of multiple local minima and maxima in the parameter space [5, 6]. In contrast, metaheuristic optimization algorithms have proven highly effective in addressing such challenges [6–8], as they are well-suited for exploring complex, non-convex search spaces and locating global optima across a wide range of problem domains.

In this study, we introduce a novel metaheuristic approach the Walrus Optimization Algorithm (WaOA) to enhance the accuracy of identifying the seven unknown parameters of the proton exchange membrane fuel cell (PEMFC) model. The optimization process is guided by an objective function defined as the Root Mean Square Error (RMSE) between experimentally measured and model-estimated voltages.

The structure of this paper is as follows: Section 2 presents the PEMFC model under investigation. Section 3 introduces the proposed Walrus Optimization Algorithm. Section 4 discusses the parameter estimation results, and Section 5 concludes the paper.

2 Mathematical Modeling of the PEMFC and Formulation of the Objective Function

2.1 Mathematical Representation of the PEMFC

A Proton Exchange Membrane Fuel Cell (PEMFC) is an electrochemical device that converts chemical energy into both electrical and thermal energy. It comprises three main components: an anode, a cathode, and a proton exchange membrane (PEM), which serves as the electrolyte. Figure 1 illustrates the operating principle of the PEMFC.

The overall electrochemical reaction occurring within a PEMFC is represented by Eq.1.

$$2H_2 + O_2 \rightarrow 2H_2O \tag{1}$$

This reaction generates a potential difference between the anode and cathode, resulting in an electric voltage. The half-cell reaction at the anode is described by Eq.2.

$$2H_2 \rightarrow 4e^- + 4H^+ \tag{2}$$

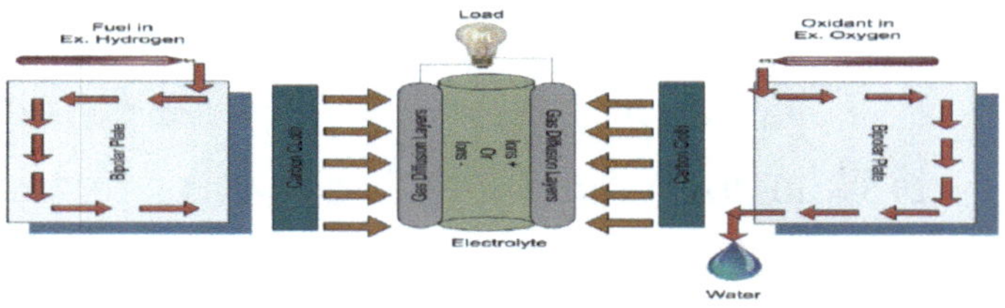

Fig. 1. Simplified model of PEMFC [9].

During this process, hydrogen molecules are split into electrons and protons. The electrons travel through an external circuit, producing electrical energy, while the protons migrate through the proton exchange membrane (PEM)—a thin, solid polymer electrolyte that selectively permits proton transport. At the cathode, the protons react with oxygen and the returning electrons, as shown in Eq.3.

$$4e^- + 4H^+ + O_2 \rightarrow 2H_2O \tag{3}$$

The ideal thermodynamic potential for an H_2/O_2 PEMFC under standard conditions (298.15 K and 1 bar) is approximately 1.229 V. This value is typically derived from the standard electrode potentials of the respective half-cell reactions. To determine the actual cell voltage, the Nernst equation is employed, accounting for variations in temperature, pressure, and reactant concentrations.

The total voltage output of the fuel cell stack is given by:

$$V = N_{cell} V_{cell} \tag{4}$$

The cell voltage, Vcell, is determined using Eq.5.

$$V_{cell} = E_{Nernst} - V_{act} - V_{ohm} - V_{com} \tag{5}$$

The modified Nernst voltage, which accounts for variations in temperature and pressure relative to standard conditions (298.15 K and 1 bar), is expressed by the Eq.6.

$$E_{Nernst} = 1.229 - 85 \times 10^{-5}(T - 298.15) + 430.85 \times 10^{-7} T\left[\ln\left(P_{H_2}\sqrt{P_{O_2}}\right)\right] \tag{6}$$

Eq.7 and Eq.8 can be used to obtain the partial pressures $\{P_{H2}, P_{O2}\}$, if the reactants are H_2 and O_2. If the reactants are air and H_2, then P_{O2} should be calculated using Eq.9.

When hydrogen and oxygen are used as reactants, the partial pressures $\{P_{H2}, P_{O2}\}$, can be determined using Eqs. (7) and (8), respectively. However, if air is used instead of pure oxygen, the partial pressure of oxygen should be calculated using Eq.9.

$$P_{H_2} = \frac{R_{Ha} P_{H_2O}}{2}\left[\left(\exp\left(\frac{1.635}{T^{1.334}}\frac{I}{A}\right)\frac{R_{Ha} P_{H_2O}}{P_a}\right)^{-1} - 1\right] \tag{7}$$

$$P_{O_2} = \frac{R_{Hc} P_{H_2O}}{2}\left[\left(\exp\left(\frac{4.192}{T^{1.334}}\frac{I}{A}\right)\frac{R_{Hc} P_{H_2O}}{P_c}\right)^{-1} - 1\right] \tag{8}$$

$$P_{O_2} = \frac{P_c - RH_c P_{H_2O}}{\left(1 + \frac{0.79}{0.21}\right) \exp\left[\frac{0.291}{T^{0.832}} \frac{I}{A}\right]} \tag{9}$$

$$\log_{10}(P_{H_2O}) = 29.5 \times 10^{-3}(\Delta T) - 91.8 \times 10^{-6}(\Delta T)^2 + 14.4 \times 10^{-8}(\Delta T)^3 - 2.18 \tag{10}$$

Where $\Delta T = T\text{-}273.15$

The voltage V_{act} can be calculated using a semi-empirical equation, as given below:

$$V_{act} = -\left[\xi_1 + \xi_2 T + \xi_3 T \ \ln(C_{O_2}) + \xi_4 T \ \ln(I)\right] \tag{11}$$

The concentration of dissolved oxygen at the cathode's catalytic interface, C_{O2}, can be determined using Henry's Law.

$$C_{O_2} = \frac{p_{O_2}}{5.08 \times 10^6} \exp\left(\frac{498}{T}\right) \tag{12}$$

The voltage V_{ohm} across the resistances R_m and R_c is calculated using Eq. (13) and is linearly proportional to current.

$$V_{ohm} = I(R_m + R_c) \tag{13}$$

where,

$$R_m = \frac{\rho_m . l}{A} \tag{14}$$

$$\rho_m = \frac{181.6\left[1 + 0.03\left(\frac{I}{A}\right) + 0.062\left(\frac{T}{303}\right)^2\left(\frac{I}{A}\right)^{2.5}\right]}{\left[\lambda - 0.634 - 3\left(\frac{I}{A}\right)\right] \exp\left[4.18\left(\frac{T-303}{T}\right)\right]} \tag{15}$$

Concentration drop in voltage V_{con} is defined by an empirical expression presented by the Eq.16.

$$V_{com} = -B \ \ln\left(1 - \frac{J}{J_{\max}}\right) = -B \ \ln\left(1 - \left(\frac{I}{A}\right)/J_{\max}\right) \tag{16}$$

2.2 Formulation of the Objective Function

The performance of the PEMFC stack is influenced by seven key parameters: [ξ1, ξ2, ξ3, ξ4, λ, Rc, B], which are integral parts of the PEMFC mathematical model. Therefore, accurate estimation of these parameters is essential. The identification process relies on an objective function that quantifies the discrepancy between the estimated model results and experimental observations. More precisely, the objective function is defined as the sum of the squared errors between the simulated results and the experimentally measured values.

$$F_{obj} = \sum_{i=1}^{N} \left(V_{sim} - V_{\exp}\right)^2 \tag{17}$$

3 Walrus Optimization Algorithm

Renowned as the "lord of the ice floes," the walrus (Odobenus rosmarus) is a marine mammal exceptionally well adapted to its Arctic environment. Walruses typically form large social groups that gather along shorelines before venturing into the sea to forage for food.

Inspired by the social structure and natural behaviors of walruses, researchers have developed a novel optimization algorithm. The algorithm draws on several key behavioral traits observed in walruses:

(i) Leadership by dominant individuals—typically those with the longest tusks—during foraging activities;
(ii) Seasonal migration to stable, rocky beaches;
(iii) Defensive responses, such as fighting or fleeing, when confronted by predators.

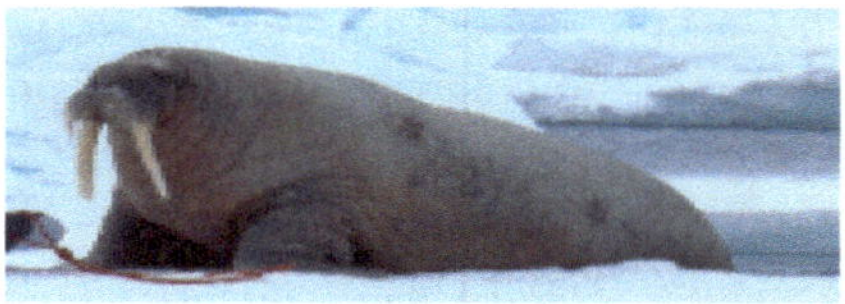

Fig. 2. An adult male walrus (Odobenus rosmarus) [10]

The main features of the Walrus Optimization Algorithm (WaOA) can be summarized as follows [11]:

Start WaOA.

1. Input all optimization problem information
2. Determinate the total number of iterations () and the number of walruses ().
3. Initialization process of walruses' locations.
4. For = 1:
5. Update the position of the strongest walrus based on the minimum value of the objective function (minFobj) criterion.
6. For = 1:
7. **Phase1: Feeding strategy (Exploration)**

Walruses exhibit a diverse diet and forage by actively grazing the seabed, utilizing their sensitive whiskers and powerful flippers. During this exploratory phase, the dominant walrus—typically the one with the longest tusks—leads the group in the search for food. This leadership behavior forms the basis for the exploration mechanism in the proposed algorithm. The feeding process is mathematically modeled to reflect the guidance provided by the strongest individual, as described by Eqs. (18) and (19).

8. In Eq. 18, the updated position of the jth walrus is calculated

$$x_{i,j}^{P_1} = x_{i,j} + \text{rand}_{i,j}.\left(SW_j - I_{i,j}.x_{i,j}\right) \tag{18}$$

9. In (19), the the position of the th walrus is update

$$X_i = \begin{cases} X_i^{P_1}, F_i^{P_1} < F_i, \\ X_i, else \end{cases} \tag{19}$$

where $X_i^{P_1}$ represents the newly generated position of the ith walrus during Phase 1, $x_{i,j}^{P_1}$ denotes its value in the jth dimension, $F_i^{P_1}$ is its objective function value, , are random numbers uniformly distributed in the interval [0,1], refers to the strongest walrus, i.e., the individual with the best objective function value, and , are integers randomly selected from the set {1, 2}.

10. **Phase 2: Migration (Selection of a Migration Destination for the ith Walrus)**
11. One of the characteristic natural behaviors of walruses is their seasonal migration to rocky outcrops or coastal areas, typically triggered by rising temperatures at the end of summer. In the context of the proposed algorithm, this behavior is modeled by assuming that each walrus migrates to a new, randomly selected position within a different region of the search space..
12. Accordingly, Eq. (20) is used to compute the updated position of the jth walrus during the migration phase.

$$\begin{cases} x_{i,j}^{P_2} = x_{i,j} + \text{rand}_{i,j}.(x_{k,j} - I_{i,j}.x_{i,j}), \ F_k < F_i \\ \quad x_{i,j} + \text{rand}_{i,j}.(x_{i,j} - .x_{k,j}), \ else, \end{cases} \tag{20}$$

13. If the newly generated position results in an improved objective function value, it replaces the previous position of the walrus. Consequently, the position of the ith walrus is updated accordingly

$$X_i = \begin{cases} X_i^{P_2}, \ F_i^{P_2} < F_i, \\ X_i, else \end{cases} \tag{21}$$

Where $X_i^{P_2}$ represents the newly generated position of the ith walrus during Phase 2 (migration), $x_{i,j}^{P_2}$ denotes its value in the jth dimension, $F_i^{P_2}$ is the objective function value of the ith walrus, , $\in$ {1,2, ..., } and $\neq$ is the position of a randomly selected walrus toward which the ith walrus migrates, , is its th dimension, and denotes its objective function value.

14. **Phase 3: Escaping and fighting against predators (Exploitation)**

Walruses face constant threats from predators such as killer whales and polar bears. In response, they adopt a fight-or-flight strategy by making small, local movements near their current positions. This natural behavior is simulated in the WaOA to enhance its local exploitation capability, allowing the algorithm to refine promising solutions through localized search around the current positions.

15. The WaOA simulates this behavior by defining a neighborhood around each walrus and generating a new position within this local region using Eqs. (22) and (23). These equations enable random exploration within the vicinity of the current position, facilitating local refinement of the solution.

$$x_{i,j}^{P_3} = x_{i,j} + \left(lb_{local,j}^{t} + \left(ub_{local,j}^{t} - \text{rand}.lb_{local,j}^{t}\right)\right) \tag{22}$$

$$\text{Local bounds}: \begin{cases} lb_{local,j}^{t} = \frac{lb_j}{t} \\ ub_{local,j}^{t} = \frac{ub_j}{t} \end{cases} \tag{23}$$

16. Subsequently, as defined in Eq. (24), the new position replaces the previous one if it yields an improved objective function value.

$$X_i = \begin{cases} X_i^{P_3}, \ F_i^{P_3} < F_i, \\ \quad X_i, \ else, \end{cases} \tag{24}$$

Where $x_i^{P_3}$ represents the newly generated position of the ith walrus during Phase 3 (exploitation), $x_{i,j}^{P_3}$ denotes its value in the jth dimension, $F_i^{P_3}$ is the corresponding objective function value, indicates the current iteration, and are the lower and upper bounds of the th variable, respectively, $lb_{local,j}^{t}$ and $ub_{local,j}^{t}$ represent the local lower and upper bounds within which local search is simulated around candidate solutions.

17. end
18. Save the best candidate solution identified
19. end
20. Return the best quasi-optimal solution obtained by the WaOA for the given optimization problem.

End WaOA

4 Results and Discussion

The study targets the estimation of seven unknown parameters in the NedStack PS6 PEMFC model. To achieve this, the Walrus Optimization Algorithm (WaOA) is employed. A uniform population size of 50 is used across all optimization methods to ensure a consistent evaluation framework. The boundary limits of the parameter search space are presented in Table 1, while the experimental test conditions are summarized in Table 2.

Table 1. Testing conditions for the identification of Nedstack PS6.

Stack	N	A	L	T	P_{H2}	P_{O2}	J_{max}
MedStackPS6	65	240	178	343	1	1	8

Table 3 lists the parameters extracted from four metaheuristics algorithms namely the proposed WaOA, Harris Hawks Optimization Algorithm (HHOA), Hippopotamus Optimization Algorithm (HO), and Mirage search optimization. Parameter estimation

Table 2. Nedstack PS6 parameter space limits bounds.

Parameter	ξ1,	ξ2	ξ3	ξ4	λ	B	Rc
Upper	−0.8532	5×10^{-3}	9.8×10^{-5}	-0.954×10^{-4}	24	0.5	8×10^{-4}
Lower	−1.19969	1×10^{-3}	3.6×10^{-5}	-2.6×10^{-4}	10	0.0136	1×10^{-4}

Table 3. Comparison results when identifying Nedstack PS6.

Optimization Technique	ξ1	ξ2	ξ3 {×e-5}	ξ4 {×e-4}	λ	B	Rc {× e-4}	RMSE
HHOA	−0.85320	0.00263	5.28631	−0.954	12.86407	0.03596	1	3.787975805953713
HOA	−0.99395	0.00304	5.28739	−0.95391	12.84040	0.04948	1	2.154783949175696
MSOA	−1.14651	0.00366	6.56748	−0.9540	12.68585	0.02527	1	2.086668655358167
Proposed WaOA	−1.05994	0.00335	6.11360	−0.954	12.57433	0.01360	1	**2.065556907645965**

with the proposed algorithm gave better results compared to other techniques (RMSE = **2.065556907645965**).

The best statistical results obtained in identifying Nedstack PS6, presented in Table 4, were obtained by the proposed WaOA.

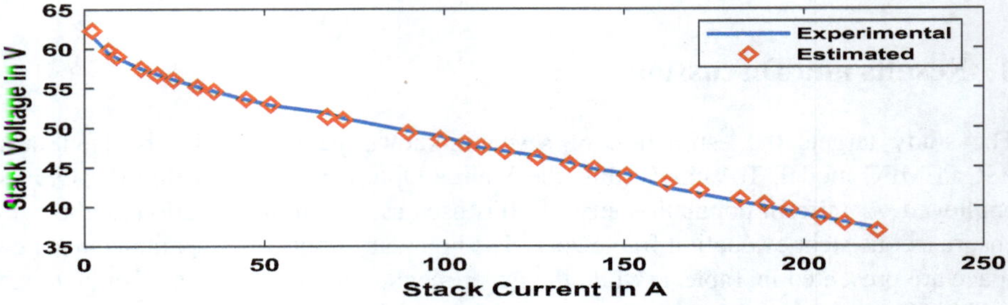

Fig. 3. Polarization curve for Nedstack PS6.

As shown in Fig. 3, the voltage estimated by WaOA is almost identical to the experimental voltage, which proves that WaOA is very effective in identifying the seven unknown parameters of the NedStack PS6 PEMFC model.

Figure 4 represents the convergence curve. It is clear that the proposed WaOA achieved a faster convergence curve compared to the Harris Hawks Optimization Algorithm (HHOA), Hippopotamus Optimization Algorithm (HO), and Mirage search optimization.

Table 4. Statistical results obtained during the identification of Nedstack PS6.

Algorithm	$RMSE_{min}$	$RMSE_{mean}$	$RMSE_{max}$	STD
HHOA	2.133456233300149	2.841715190476132	4.329844381336213	0.901921234623674
HOA	2.068361393753945	2.615957697564760	3.493738913535068	0.668462116282189
MSOA	2.065556907645969	2.309982683985090	2.681224473593096	0.261933338742086
Proposed WaOA	**2.065556907645966**	**2.065556907645973**	**2.065556907645976**	**3.813741331030777e-15**

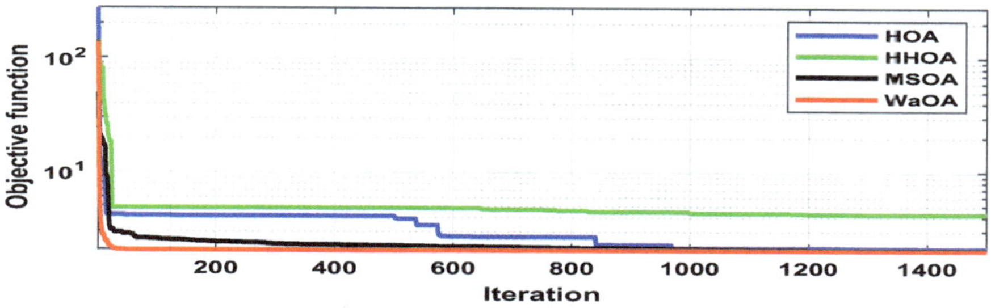

Fig. 4. Convergence curves of WaSO, HHOA, HOA, and MSOA

5 Conclusion

Accurate estimation of unknown parameters in proton exchange membrane fuel cell (PEMFC) models is both a complex and critical task, given the potential of PEMFCs as a clean and efficient energy technology. This study introduces the Walrus Optimization Algorithm (WaOA) as an innovative and precise method for parameter estimation in PEMFC models. The algorithm is employed to identify seven unknown parameters of the NedStack PS6 PEMFC model by minimizing the sum of squared errors between experimental data and model predictions. Simulation results demonstrate that WaOA is a highly effective tool, outperforming several well-established meta-heuristic algorithms in terms of both accuracy and computational efficiency for PEMFC parameter estimation.

Nomenclature

N_{cell}:	Number of cells;
E_{Nernst}:	Thermodynamic potential, V;
V_{cell}:	Single cell voltage, V;
V:	Stack voltage, V;
V_{ohm}:	Ohmic voltage drop, V;
V_{con}:	Concentration voltage loss, V;
I:	Current, A;
V_{act}:	Activation voltage, V;
P_{O2}:	Partial pressure of oxygen, atm;

R_{Ha}: Relative humidity of vapour in the anode;
T: Absolute Temperature, K;
P_{H2}: Partial pressure of hydrogen, atm;
P_a: Anode inlet pressure, atm;
R_{Hc}: Relative humidity of vapour in the cathode;
C_{O2}: Concentration of Oxygen, mol cm^{-3};
$\xi_1, \xi_2, \xi_3, \xi_4$: Semi-empirical coefficients;
P_c: Cathode inlet pressure, atm;
R_c: Equivalent contact resistance;
P_{H2O}: Saturation pressure of water, atm;
A: Effective area of the membrane, cm^2;
ρ_m: Specific resistivity for the flow of hydrated protons;
R_m: Equivalent membrane resistance;
l: Membrane thickness (cm);
J: Current density, A/cm ^{2};
λ: Water content of membrane;
J_{max}: Maximum current density, A/cm^2;
B: Parametric coefficient, Ω cm.

References

1. Qasem, N. and, Gubran. A., 2024. A recent comprehensive review of fuel cells: History, types, and applications. Int. J. Energy Res. 2024, 1 *7271748*. https://doi.org/10.1155/2024/7271748.
2. O'Hayre, R.: Fuel cells for electrochemical energy conversion. EPJ Web Conf.. **148**, 00013 (2017)
3. GASIK, M.: Introduction: Materials challenges in fuel cells, *In* Woodhead Publishing Series in Electronic and Optical Materials, Materials for Fuel Cells, Woodhead Publishing, (1–5) 2008..
4. Wang, Y., Chen, K.S., Mishler, J., Cho, S.C., Adroher, X.C.: A review of polymer electrolyte membrane fuel cells: technology, applications, and needs on fundamental research. Appl. Energy. **88**(4), 981–1007 (2011)
5. Tummala, A.S.L.V., Polumahanthi, N., Khan, B., Ali, A.: Accurate parameters identification of proton exchange membrane fuel cell using Young's double-slit experiment optimizer. Front. Energy Res. **12**, 1384649 (2024)
6. Xuebin, L., Jin Zhao, Y., Daiwei, Z.J., Wenjin, Z.: Parameter estimation of PEM fuel cells using metaheuristic algorithms. Measurement. **237**, 115302 (2024)
7. Abdel-Basset, M., Mohamed, R., Abouhawwash, M.: On the facile and accurate determination of the highly accurate recent methods to optimize the parameters of different fuel cells: Simulations and analysis. Energy. **272**, 127083 (2023)
8. Hachana, O., El-Fergany, A.A.: Efficient PEM fuel cells parameters identification using hybrid artificial bee colony differential evolution optimizer. Energy. **250**, 123830 (2022)
9. Tummala S.L.V. Ayyarao, Nishanth Polumahanthi, Baseem Khan, 2024. An accurate parameter estimation of PEM fuel cell using war strategy optimization, Energy, vol. 290, 130235.
10. Lydersen, C., Lindgren, Å., Alfredsson, K., Kovacs, K.M.: A Walrus (Odobenus rosmarus) at the north pole. Aquat. Mamm. **48**(6), 513–516 (2022). https://doi.org/10.1578/AM.48.6.2022.513
11. Trojovský, P., Dehghani, M.: A new bio-inspired metaheuristic algorithm for solving optimization problems based on walruses' behavior. Sci. Rep. **13**, 8775 (2023). https://doi.org/10.1038/s41598-023-35863-5

Performance Analysis of Adaptive P&O, ANN and PSO Based MPPT Algorithms for Photovoltaic Systems

T. Tsriat(✉) and M. Zerikat

National Polytechnic School, Department of Electrical Engineering, LAAS Research Group, Oran, Algeria
{norelhouda-tahani.tsriat,mokhtar.zerikat}@doc.enp-oran.dz

Abstract. Maximum Power Point Tracking (MPPT) represents a critical stage in photovoltaic system control. This essential function dynamically adjusts the electrical operating point to extract peak available power from solar panels under varying environmental conditions, This paper presents a comparative performance analysis of three Maximum Power Point Tracking (MPPT) techniques for photovoltaic systems: Perturb and Observe (P&O), Artificial Neural Network (ANN), and Particle Swarm Optimization (PSO). The simulations under MATLAB-SIMULINK environment are carried out under the same temperature and variable irradiation and comparing the performances of the three methods. the obtained results are studied analysed and compared to the three proposed algorithms. the performance of the smart tracking algorithm was found to be superior to that of a traditional algorithm where the storage system operates successfully.

Keywords: Photovoltaic System (PV) · Maximum Power Point Tracking (MPPT) · Optimization · PSO · P&O · ANN

1 Introduction

The global adoption of photovoltaic (PV) energy has significantly increased in recent years, driven by its ecological advantages, modular design, and reduced installation costs [1]. Nevertheless, the performance of PV systems is heavily influenced by environmental factors such as solar irradiance and ambient temperature, which directly impact their power output [2]. To maintain optimal operation under fluctuating weather conditions, Maximum Power Point Tracking (MPPT) techniques are commonly integrated into PV systems [3]. The primary function of an MPPT algorithm is to dynamically adjust the operating point of a PV system so that it continuously delivers the highest possible power output [4]. Since the maximum power point (MPP) varies with changes in environmental conditions, effective MPPT methods must be both adaptive and reliable [3]. Among the various available strategies, the Perturb and Observe (P&O) technique remains one of the most widely used due to its straightforward implementation, low computational demand, and practical effectiveness [1, 5]. This method involves small adjustments to

M. Hatti (Ed.): IC-AIRES 2025, LNNS 1740, pp. 377–386, 2026.
https://doi.org/10.1007/978-3-032-12788-4_36

the voltage or current, followed by monitoring the resulting power variation to determine the direction toward the MPP [6]. With the growing development of intelligent control systems, Artificial Neural Networks (ANNs) have emerged as a promising alternative for MPPT applications [7]. ANNs are capable of capturing complex nonlinear behaviors and learning from historical data without relying on precise mathematical models [8, 9]. In addition to conventional and AI-based approaches, Particle Swarm Optimization (PSO) has gained attention as an efficient optimization strategy for MPPT problems [10]. Inspired by the collective behavior of swarms such as flocks of birds or schools of fish [11]. In this method, each particle represents a potential solution and updates its position based on two guiding principles: its own best experience (known as Pbest) and the best experience found by the entire swarm (Gbest) [12, 13]. This study presents a comparative evaluation of three MPPT techniques: the classical P&O method, the ANN-based approach, and the hybrid integration of PSO with both ANN and P&O. The analysis focuses on key performance metrics such as convergence speed, tracking precision, stability, and adaptability under varying irradiance levels. By comparing these strategies, this research aims to identify the most suitable MPPT method for different PV system configurations, by contributing to the advancement of smart and efficient solar energy conversion technologies.

2 PV Modeling

2.1 Solar PV_cell

A PV cell is the fundamental semiconductor device that directly converts solar radiation into electrical energy through the photovoltaic effect [1]. It typically consists of a p-n junction made from silicon-based materials, where incident photons transfer sufficient energy to electrons, enabling them to move freely and generate an electric current when exposed to sunlight [6]. Figure 1 illustrates the equivalent circuit model of a PV cell, which consists of a current source (Ipv) that generates current proportional to the incident solar irradiance. A diode is also included to account for recombination losses in the junction. Additionally, two resistive elements are incorporated: the series resistance (Rs), representing ohmic losses in the material and interconnections, and the shunt resistance (Rsh), which accounts for leakage currents due to imperfections in the cell structure [6, 7].

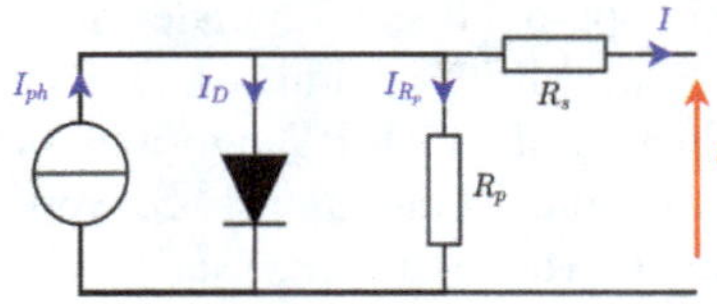

Fig. 1. Equivalent circuit model of a solar cell [10]

2.2 Modeling of 1Soltech ISTH-215-P/module

In this study, the 1Soltech ISTH-215 photovoltaic module was selected for modeling and simulation due to its widespread use and well-documented performance characteristics. Their electrical parameters are summarized in Table 1. The modeling begins with the mathematical formulation of an equivalent electrical circuit for the PV cell. derived from Eqs. 1 to 4.

Table 1. Specification of the PV module

Maximum power P	213.15 W
Short circuit current (Isc)	7.84 A
Open circuit voltage (Voc)	36.3 V
Optimal voltage	29 V
Optimal current	7.35 A
Standard test condition (TSC)	Cell temperature 25 °C, irradiance1000 W/m2

These equations describe how environmental factors—such as solar irradiance and cell temperature influence the electrical output characteristics of the photovoltaic (PV) module. The behavior of a PV cell can be modeled using an equivalent circuit based on Kirchhoff's Current Law (KCL). According to this model, the total output current I of the PV cell is given by:

$$I = Iph - Id - Ish \tag{1}$$

This leads to the characteristic equation of a PV module which are as follow:

$$I = Np.Iph - Np.Is\left[e^{\frac{g(V+Rs.I)}{ANsKTo}} - 1\right] - Rsh.Np(V + RsI) \tag{2}$$

$$Iph = \frac{\mathrm{Gref}}{G}\left[Isc + Ki(T - Tref))\right] \tag{3}$$

$$Is = \mathrm{Irs}\left(\frac{\mathrm{T}}{\mathrm{Tref}}\right)3\mathrm{e}^{\frac{\mathrm{g(Eg)}}{\mathrm{AK}}\cdot\left(\frac{1}{\mathrm{Tref}}-\frac{1}{\mathrm{T}}\right)} \tag{4}$$

where:

Iph: photon-generated current, directly proportional to solar irradiance, Id: the diode current, Isc: saturation current., Irs: reverse saturation current., V: output voltage, Rs, Rsh: series and shunt resistances, Np: number of cells in parallel, q: electron charge = 1.6 × 10–19 C, A: diode ideality factor, K: Boltzmann constant =1.38 × 10–23 J/K, G, Gr: actual and reference irradiance levels, T, Tr: actual and reference temperatures, Eg: semiconductor bandgap energy. The modeling of the PV array was carried out using MATLAB/Simulink, with the simulation structure illustrated in Fig. 2 being implemented according to Eqs. 1 to 4.

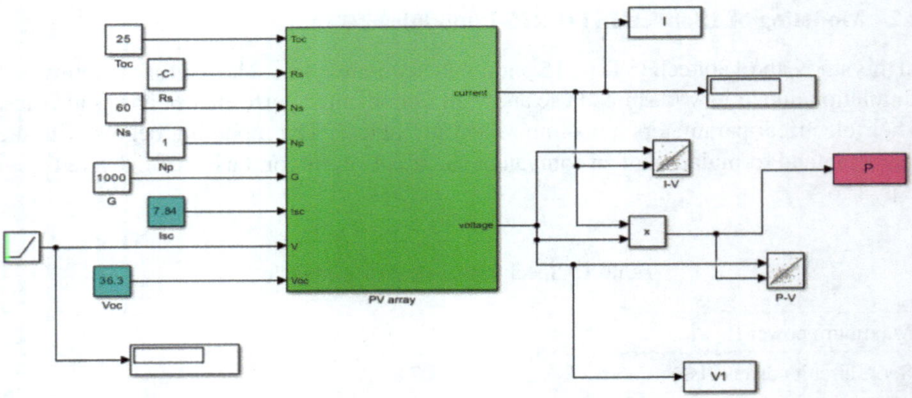

Fig. 2. Overall PV Cell Subsystem

2.3 Simulation of 1Soltech ISTH-215-P/module

Following the formulation of the PV system's mathematical model, simulations are conducted to evaluate the I-V and P-V characteristics, allowing analysis of the system's behavior under varying irradiance conditions [500 W/m^2,800 W/m^2,1000 W/m^2] is shown in Fig. 3 and Fig. 4 respectively. The simulations are performed at STC with the following parameters: ideality factor A = 1.3, Ns = 60, number of parallel cells Np = 1, bandgap energy Eg = 1.1 eV, Rs = 0.39381 Ω.

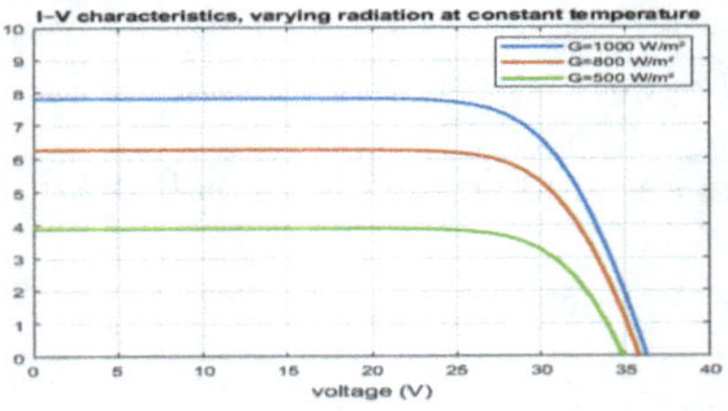

Fig. 3. I-V characteristics with variation of irradiance

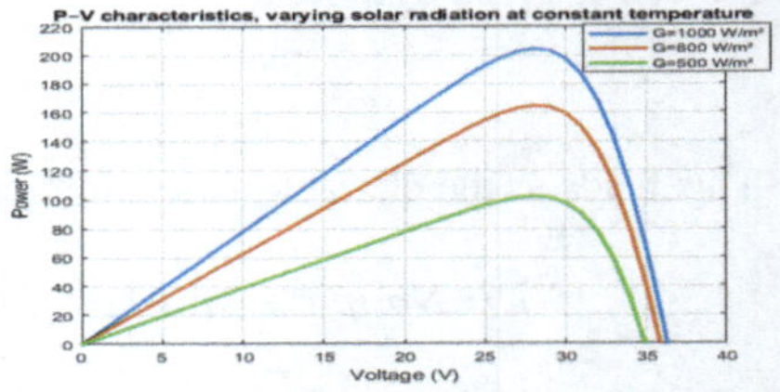

Fig. 4. P-V characteristics with variation of irradiance

3 Maximum Power Point.Tracking (Mppt)

MPPT is a technique used in PV systems to continuously optimize power output by adjusting the operating point according to environmental conditions such as solar irradiance and temperature [1]. Since the power-voltage curve of a PV module has a single peak under uniform irradiation, MPPT ensures that the system operates at this MPP for maximum efficiency [4]. In Fig. 5, the MPP is observed at an irradiance level of 1000 W/m^2.

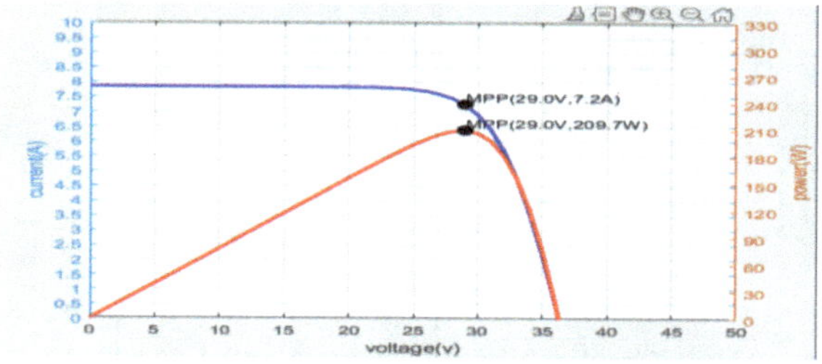

Fig. 5. MPP at G = 1000 W/m^2

To achieve this, MPPT algorithms are typically integrated with a DC-DC converter, such as the boost converter, which adjusts the voltage level between the PV array and the load or grid. The boost converter plays a crucial role in matching the variable output of the PV system to the required input of the load while enabling efficient tracking of the MPP [6].

3.1 Perturb and Observe (P&O) Algorithm

Among various MPPT techniques, the P&O method is one of the most commonly adopted due to its simplicity and ease of implementation [7]. This algorithm works by slightly perturbing the operating voltage or current and observing the resulting change in power to determine the direction toward the MPP [5]. The P&O algorithm adjusts the operating point based on the sign of the power variation (dP/dV), maintaining or reversing the perturbation direction accordingly as illustrated in Fig. 6. In this study, Fig. 7 illustrates the simulation setup used to implement the P&O algorithm, showing the integration of the MPPT controller with the PV array model and the DC- DC boost converter. The system dynamically adjusts the duty cycle to track the MPP under varying irradiance conditions.

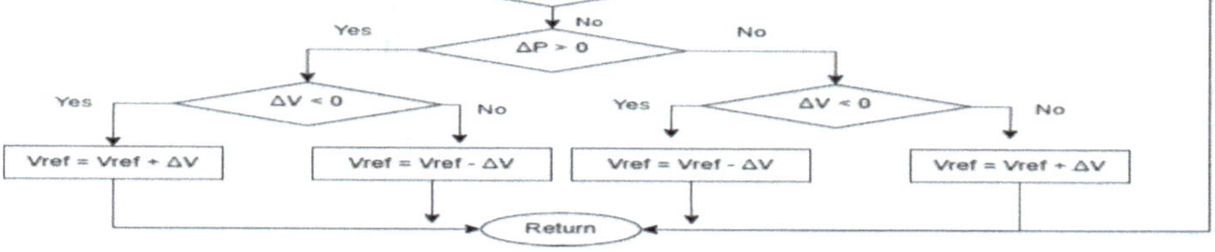

Fig. 6. Flowchart of P&O algorithm

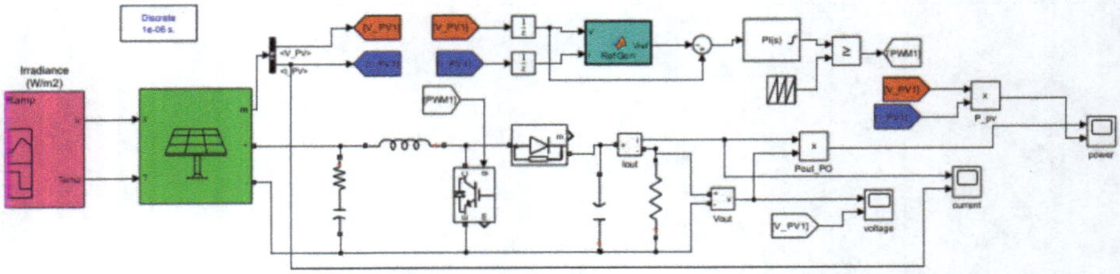

Fig. 7. MATLAB/Simulink-based PV system with P&O MPPT controller

Figure 8 compares the actual PV power with the output power at variable irradiance (400 W/m^2-700W/m^2-500W/m^2-1000W/m^2-800W/m^2) after applying the P&O-based MPPT technique. The error between the two is shown at the bottom of the figure. These results show that the P&O method successfully converges to the optimal operating point, although it exhibits small fluctuations due to its perturbation-based nature.

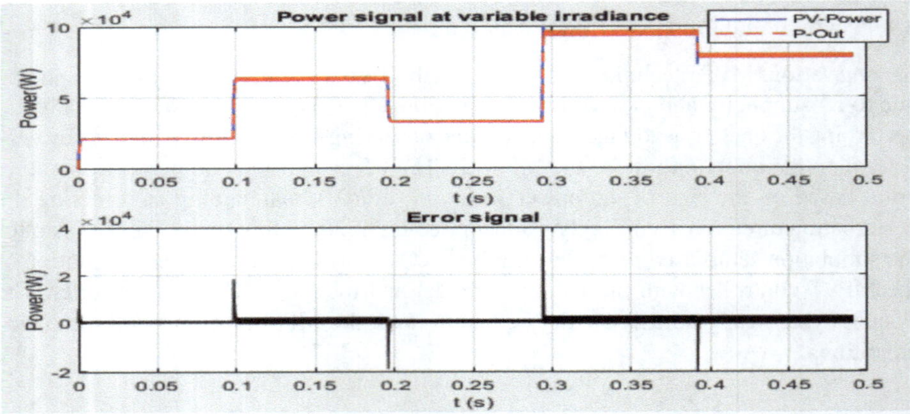

Fig. 8. P&O power and error signal under variable irradiance

3.2 Artificial Neural Network (ANN)-Based MPPT Algorithm

In contrast to traditional methods, Artificial Neural Network (ANN)-based MPPT approaches use machine learning to predict the MPP based on historical and real-time input data such as irradiance and temperature [8]. Unlike conventional controllers, ANN does not require an explicit mathematical model and can learn complex nonlinear relationships between inputs and outputs [9]. Fig. 9 illustrates the operational flow of the algorithm in the form of a block diagram, providing a clear visual representation of how the MPPT process is implemented step by step.

Figure 10, presents the ANN architecture designed for MPPT control. The network consists of two input neurons representing environmental parameters—solar irradiance

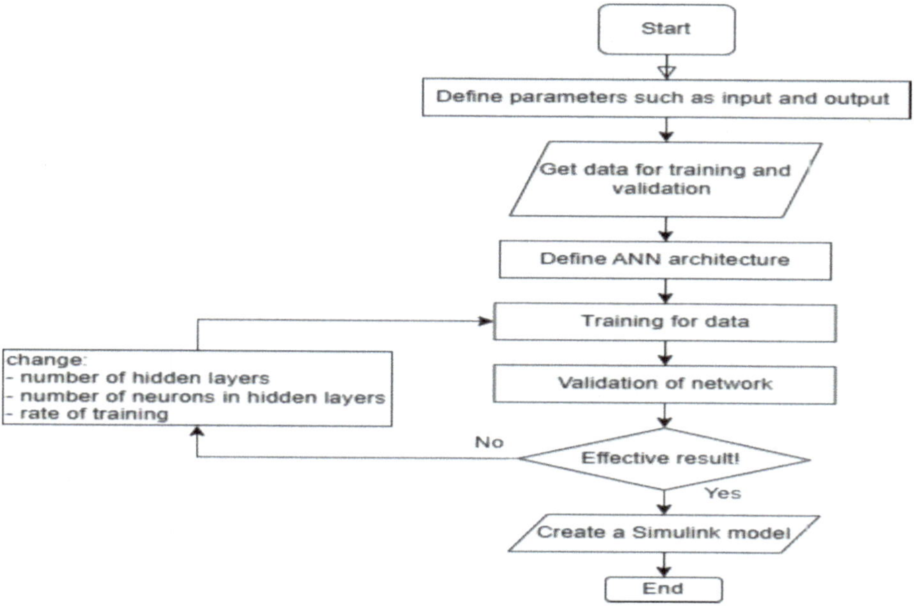

Fig. 9. Flowchart of ANN algorithm

and temperature —a single hidden layer with a "tan- sigmoid" activation function, and an output layer that predicts the maximum power point voltage (Vmp) and power (Pmp). Training of the ANN was carried out with the Levenberg-Marquardt algorithm [8], and Fig. 11 presents the corresponding training performance across epochs.

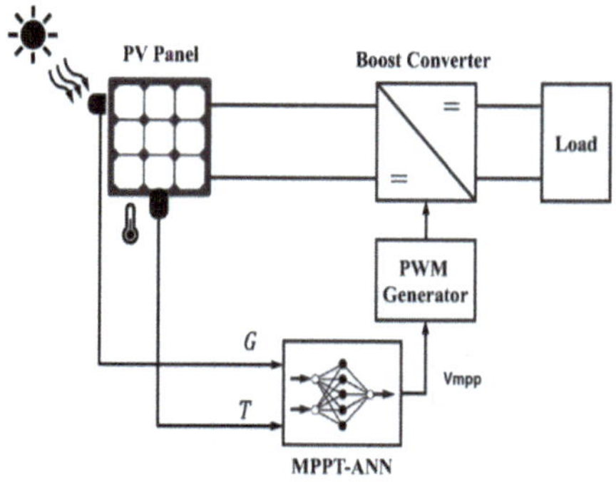

Fig. 10. Simulink module of the ANN system

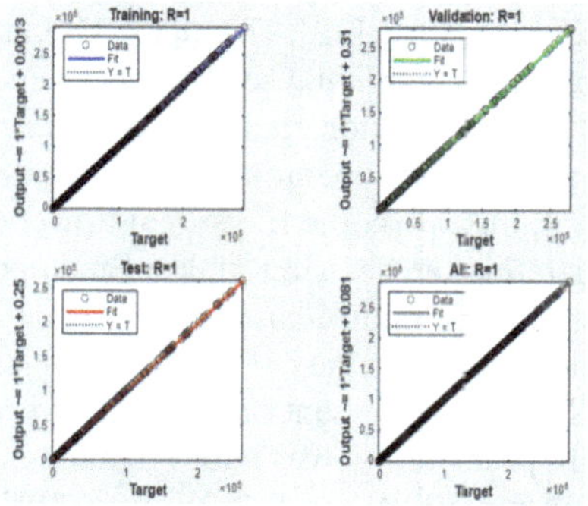

Fig. 11. ANN progression diagram

Figure 12 present a comparison between actual PV power with the output power at variable irradiance (400 W/m^2-700W/m^2-500W/m^2-1000W/m^2-800W/m^2) and those predicted by the ANN. Although the results show an extremely strong correlation and perfect stability, they demonstrate that the ANN-based MPPT method performs more consistently than the P&O algorithm, particularly once irradiance levels stabilize. In steady-state conditions, the ANN exhibits relatively smaller fluctuations and a faster convergence toward the MPP.

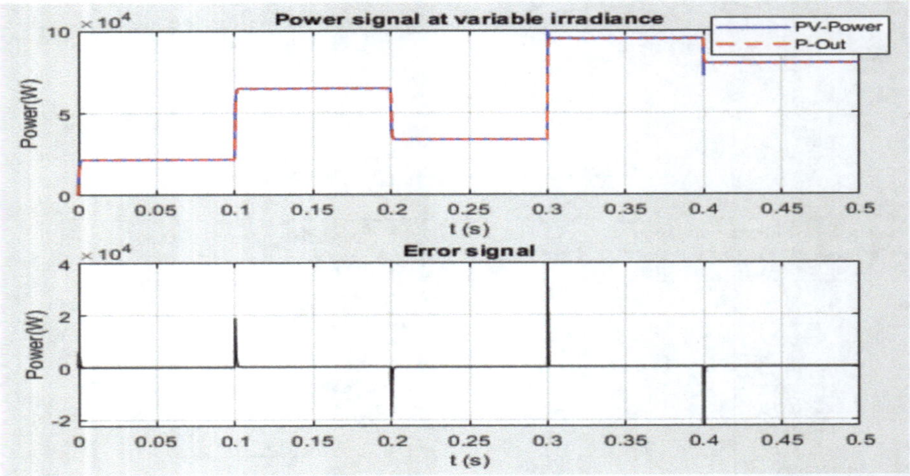

Fig. 12. ANN power and error signal under variable irradiance

3.3 Particle Swarm Optimization (PSO)

The Particle Swarm Optimization (PSO) algorithm is a population-based stochastic optimization technique inspired by the social behavior of swarms, such as flocks of birds or schools of fish [12]. The method iteratively refines solutions by adjusting the trajectory of particles within the search space according to their own best-known position and the global best position found by the swarm. This approach enables efficient convergence toward optimal operating points, particularly in complex and nonlinear systems [11, 13]. In this work, the functional flowchart of the PSO algorithm is presented in Fig. 13, illustrating the decision-making and updating process of particles. Furthermore, Fig. 14 details the Simulink-based implementation framework, showing how the PSO algorithm is integrated into the MPPT control system

Figure 15 presents a comparative analysis between actual PV output power and ANN-predicted power under dynamically changing irradiance conditions (400 W/m^2-700W/m^2-500W/m^2-1000W/m^2-800W/m^2). The experimental results demonstrate that the PSO-based MPPT controller maintains exceptional tracking precision, achieving stable maximum power point operation without observable oscillations. This metaheuristic approach exhibits superior dynamic response to rapid irradiance variations while operating without requiring offline training procedures.

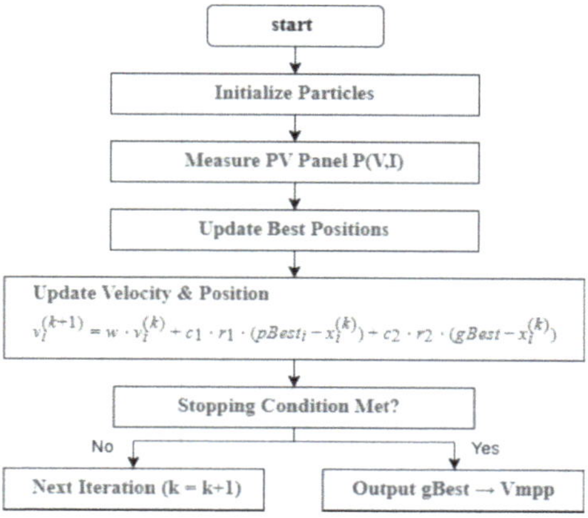

Fig. 13. Flowchart of ANN algorithm

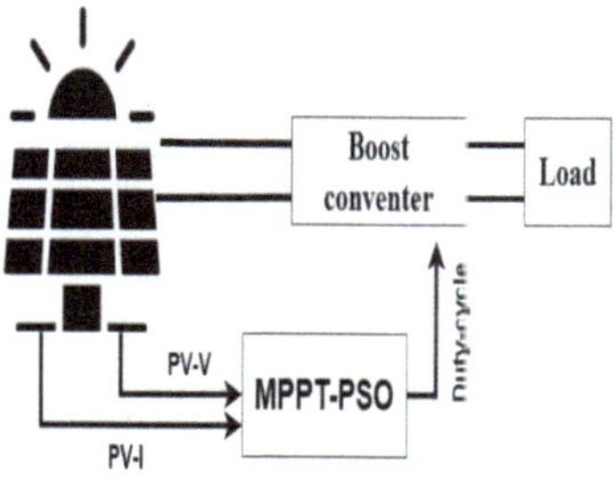

Fig. 14. Simulink module of the ANN system

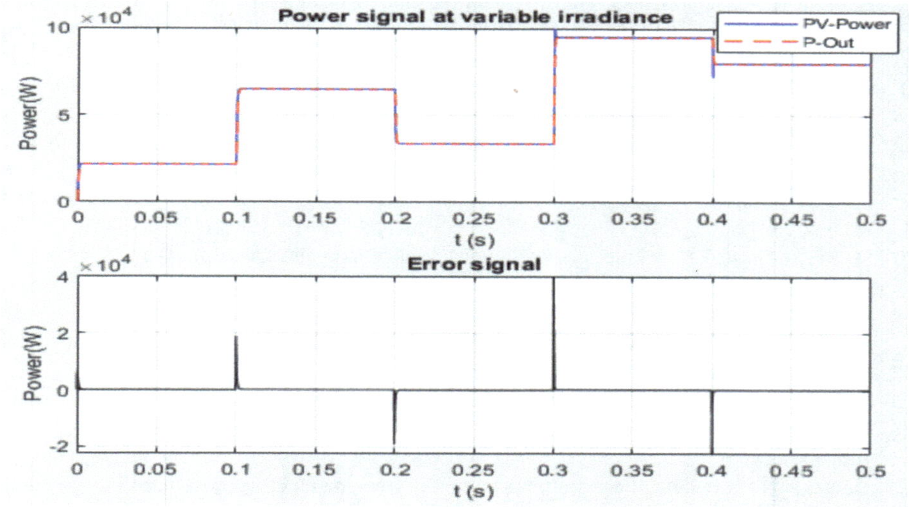

Fig. 15. PSO power and error signal under variable irradiance

4 Concluding Remarks

This study demonstrates that the PSO-based MPPT technique outperforms both ANN and P&O methods, achieving superior tracking accuracy, faster convergence, and minimal oscillations under variable irradiance conditions. While ANN shows rapid response to sudden changes, it requires extensive training data, whereas P&O exhibits slower tracking and steady-state losses. The Matlab/Simulink software confirm PSO is robustness, efficiency, and ease of implementation, making it the optimal choice for photovoltaic systems operating in dynamic environments.

References

1. Chin, V.J., Salam, Z., Ishaque, K.: Cell modelling and model parameters estimation techniques for photovoltaic simulator application: A review. Appl. Energy. **154**, 500–519 (2015)
2. Chatterjee, A., Keyhani, A., Kapoor, D.: Identification of photovoltaic source models. IEEE Trans. Energy Convers. **26**(3), 883–892 (2011)
3. Lineykin, S., Averbukh, M., Kuperman, A.: An improved approach to extract the single-diode equivalent circuit parameters of a photovoltaic cell/panel. Renew. Sust. Energ. Rev. **30**, 282–290 (2014)
4. Khanam, J.J., Foo, S.Y.: Modeling of a photovoltaic array in MATLAB simulink and maximum power point tracking using neural network. Electric Electron Tech Open Acc J. **2**(2), 40–46 (2018)
5. Shongwe, S., Hanif, M.: Comparative analysis of different single-diode PV modeling methods. IEEE J. Photovoltaics. **5**(3), 938–946 (2015)
6. Sera, D., Teodorescu, R., Rodriguez, P.: PV panel model based on datasheet values. In: 2007 IEEE International Symposium on Industrial Electronics, pp. 2392–2396, Vigo, Spain (Jun. 2007)
7. Besheer, A.H., Hegazy, Y.G., Hamada, M.M.: Comparison of ANN and P&O MPPT methods for PV applications under changing solar irradiation. In: 2013 IEEE Conference on Clean Energy and Technology (CEAT), pp. 182–187. Johor Bahru, Malaysia (2013)
8. Roy, R.B., et al.: A comparative performance analysis of ANN algorithms for MPPT energy harvesting in solar PV system. IEEE Access. **9**, 95375–95387 (2021)
9. P. Chinnaraj, R. Dhayalan, S. Anandha Kumar, Y. Dharshan, R. Haridharan, and M. Vijayadharshini, "A novel comparative analysis of solar P&O, ANN-based MPPT controller under different irradiance conditions," EAI Endorsed Trans. Energy Web, vol. 11, no. 1, Jan. 2024, Art. no. e4942.
10. SI Blaise Pascal, "Cellule photovoltaïque," https://si.blaisepascal.fr/1t-cellule-photovoltaique/ (2025). Accessed 16 May 2025.
11. Srivastava, S., Lata, C., Lohan, P., Mosobi, R.W.: Comparative analysis of particle swarm optimization and artificial neural network based MPPT with variable irradiance and load. Int. J. Electr. Electron. Res. **10**(3), 460–465 (2022)
12. Ben Belghith, O., Sbita, L., Bettaher, F.: MPPT design using PSO technique for photovoltaic system control comparing to fuzzy logic and P&O controllers. Energy Power Eng. **8**(11), 345–355 (2016)
13. Yahya, A.N.: A review of maximum power point tracking techniques for use in photovoltaic power systems. Brunel University, UK, Tech. Rep (2015)

Intelligent Optimization and Modeling of Miniaturized Photovoltaic Cells for Embedded Applications Using Hybrid AI Techniques

Sarah K. Sedjar[1,2](✉) and Mourad Benmessaoud[1,2]

[1] Department of Electrical Engineering, National Polytechnic School of Oran – Maurice Audin (ENPO), Oran, Algeria
sarah-kawther.sedjar@doc.enp-oran.dz

[2] Laboratory of Automation and Systems Analysis (LAAS), SCAMRE: Simulation, Control, Analysis and Maintenance of Electrical Networks, Oran, Algeria

Abstract. In response to the growing demands for efficiency and compactness in embedded solar systems, this work proposes a hybrid advanced optimization approach applied to a miniaturized PERL photovoltaic cell. The methodology combines a rigorous extraction of experimental electrical parameters with a joint three-dimensional optimization strategy based on exhaustive grid search, Particle Swarm Optimization (PSO), and Genetic Algorithms (GA), with particular focus on the convergence behavior of solutions. A feedforward neural network is then trained using Bayesian regularized backpropagation (trainbr) to predict the cell's performance under varying climatic conditions. This synergy between bio-inspired heuristics and supervised learning enables the achievement of high accuracy ($\eta \approx 24.7\%$, FF $\approx 82.8\%$) and validates the robustness of the model for next-generation intelligent embedded systems (IoT, aerospace, autonomous sensors).

Keywords: Embedded solar systems · artificial intelligence (AI) · particle swarm optimization (PSO) · genetic algorithms (GA) · artificial neural networks (ANN); Bayesian regularized backpropagation (trainbr)

1 Introduction

The rapid expansion of embedded solar technologies demands photovoltaic (PV) systems that are simultaneously compact, efficient, and adaptive to real-world environments. In this context, the miniaturized PERL cell architecture, known for its high efficiency under laboratory conditions, offers a promising foundation for next-generation applications such as IoT devices, aerospace instrumentation, and autonomous sensor networks. However, maximizing performance under embedded constraints requires not only accurate parameter estimation but also advanced optimization strategies capable of handling multi-dimensional non-linearities. This study introduces a hybrid approach

M. Hatti (Ed.): IC-AIRES 2025, LNNS 1740, pp. 387–397, 2026.
https://doi.org/10.1007/978-3-032-12788-4_37

that combines bio-inspired heuristics and supervised machine learning for the precise modeling and optimization of a laboratory-scale PERL solar cell. A three-dimensional joint optimization is carried out using grid-based sweeping, Particle Swarm Optimization (PSO), and Genetic Algorithms (GA), with convergence analysis ensuring solution robustness. In parallel, an artificial neural network (ANN) is trained using the Levenberg–Marquardt Bayesian regularization (trainbr) backpropagation algorithm to predict photovoltaic performance under varying irradiance and temperature conditions. Recent studies have demonstrated the effectiveness of metaheuristic and AI-based approaches for photovoltaic parameter extraction and performance prediction (Chen et al., 2024; Rad and Rahimi, 2023; Spertino et al., 2020). This AI-assisted framework demonstrates excellent predictive capability and confirms the robustness of the proposed methodology for integration into modern embedded solar technologies. Hybrid optimization techniques have gained significant traction in photovoltaic research due to their ability to handle complex, multi-parameter systems (Hatti, 2018).

2 Results and Contributions

2.1 Analytical Modelling and Numerical Simulation of the Perl Photovoltaic Cell

The PERL (Passivated Emitter Rear Locally-diffused) solar cell, represents a benchmark in high-efficiency monocrystalline silicon technology, having achieved a certified conversion efficiency of 24.7% on an active surface of 4.00 cm^2. Its architecture incorporates front surface passivation, a lightly doped emitter, and localized rear diffusions, all designed to reduce minority carrier recombination and improve current collection. This study focuses on a PERL cell selected from the NREL Cell and Module Performance Database, which provides certified data for laboratory-fabricated monocrystalline cells from leading research institutions. The cell was chosen for its compact size, certified parameters (Voc = 0.706 V, Jsc = 42.2 mA/cm^2, FF = 82.8%, η = 24.7%), and its continued relevance despite being developed in 1999. Its structural simplicity and data availability make it an ideal candidate for numerical modeling, inverse parameter identification, and AI-assisted performance optimization. To simulate the cell's electrical behavior under standard test conditions (STC), The one-diode model remains a widely accepted baseline for PV modeling due to its balance between simplicity and accuracy (Villalva et al., 2009). This model captures resistive losses (via Rs), leakage effects (Rsh), and recombination mechanisms (through the ideality factor n), forming an implicit equation for the I–V characteristics. It was implemented in MATLAB, where the nonlinear equation was solved using the fzero method across a voltage sweep up to Voc, enabling the generation of both current–voltage (I–V) and power–voltage (P–V) curves. The corresponding power curve, obtained via $P = V \times I$, highlights the Maximum Power Point (MPP), which is essential for evaluating cell efficiency. Baseline values for Rs, Rsh, and n were initially selected from the literature to ensure numerical stability and physical realism during inverse modeling. These values were later refined using heuristic and AI-based optimization techniques. The simulation results under STC are shown in [Fig. 1], while [Fig. 2] illustrates the impact of temperature variation (25 °C, 45 °C, and 65 °C) on the electrical performance, revealing a clear degradation in Voc, output power, and overall efficiency with increasing temperature. The certified efficiency values are consistent with historical efficiency benchmarks reported in the literature (Green et al., 2001). Classical PV modeling approaches based on equivalent circuits have been widely adopted in power electronics simulations (Gow and Manning, 1999).

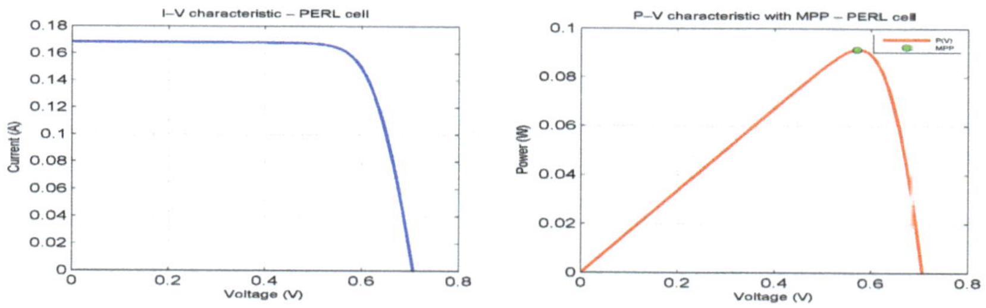

Fig. 1. *I–V and P–V characteristics of the PERL solar cell under STC, highlighting the maximum power point.*

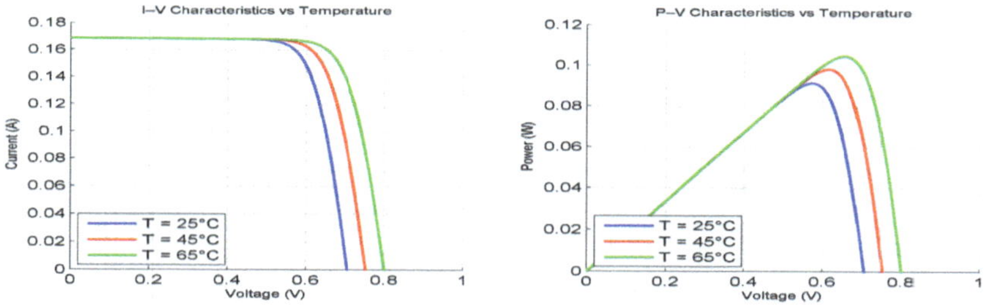

Fig. 2. *I–V and P–V characteristics of the PERL solar cell at different temperatures, showing the effect of temperature on current and power output.*

2.2 Exploratory Parameter Optimization via One-Dimensional Grid Search

To explore the individual influence of each electrical parameter on the photovoltaic performance, a one-dimensional grid search was performed. Starting from the previously defined baseline values, the parameters Rs, Rsh, and n were varied independently over realistic intervals typically observed in The analysis was conducted on high-efficiency monocrystalline silicon cells. While one parameter was scanned, the other two were held constant. For each tested value, the equation was numerically solved, and the resulting I–V and P–V characteristics were used to compute both the conversion efficiency (η) and the fill factor (FF). Specifically, the influence of Rs on the photovoltaic performance is illustrated in [Fig. 3], which shows its impact on both the fill factor (FF) and the conversion efficiency (η). These results highlight the sensitivity of the cell to series resistance variations and reveal a distinct optimum, serving as a valuable reference for subsequent multidimensional optimization.

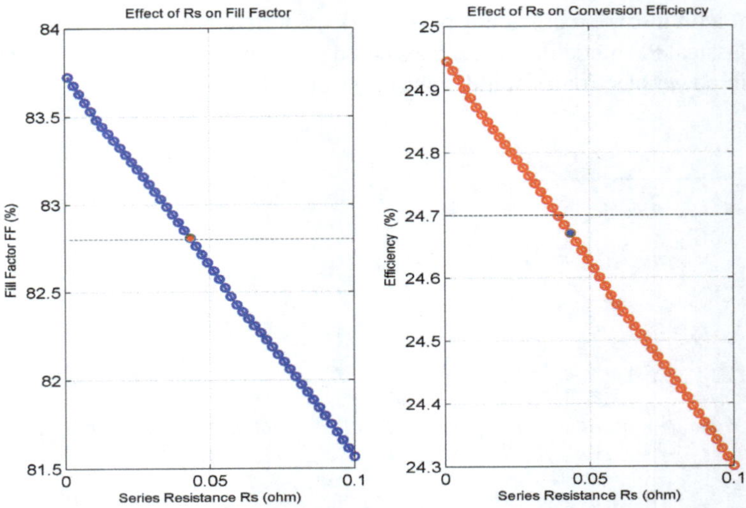

Fig. 3. Influence of the series resistance (Rs) on the fill factor (FF) and conversion efficiency (η) of a monocrystalline silicon PERL cell under standard conditions ($G = 1000\ W/m^2$, $T = 298\ K$).

Additionally, the influence of the shunt resistance (*Rsh*) on the PERL cell's performance was investigated through a one-dimensional parameter sweep. By keeping the other parameters constant, this analysis highlights the effect of *Rsh* on both the fill factor (*FF*) and the conversion efficiency. The results, illustrated in [Fig. 4], show that increasing the shunt resistance (Rsh) improves both the fill factor (FF) and the conversion efficiency (η) up to a saturation point beyond 3000 Ω. The optimal performance is observed around Rsh ≈ 2800 Ω, with FF ≈ 82.8% and η ≈ 24.67%, in close agreement with the experimental values.

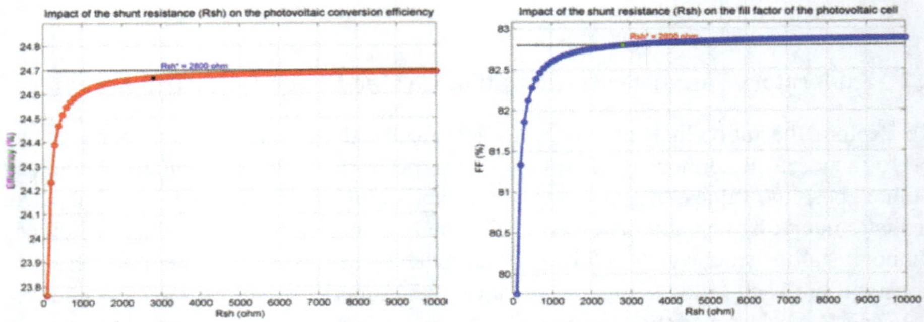

Fig. 4. Influence of the shunt resistance (Rsh) on the photovoltaic performance of the PERL cell. The left graph shows the conversion efficiency (η) increasing with Rsh and saturating near 3000 Ω, matching the experimental value (η ≈ 24.67%). The right graph highlights the similar saturation behavior of the fill factor (FF), with optimal values reached in the same Rsh region.

In addition, the impact of the ideality factor (n) was investigated through a one-dimensional sweep. This parameter plays a central role in the one-diode model, as it reflects the dominant recombination mechanisms within the cell. With Rs fixed at 0.04343 Ω and Rsh at 2800 Ω, n was varied in the range [1.00, 1.30] under standard test conditions ($T = 298$ K, $G = 1000$ W/m^2). [Figure 5] show the evolution of the fill factor (FF) and conversion efficiency (η) as a function of n. A clear performance degradation is observed with increasing n, and the best match with experimental data is obtained for $n = 1.08$ ($FF = 82.8\%$, $\eta = 24.67\%$), confirming the quasi-ideal behavior of the PERL cell.

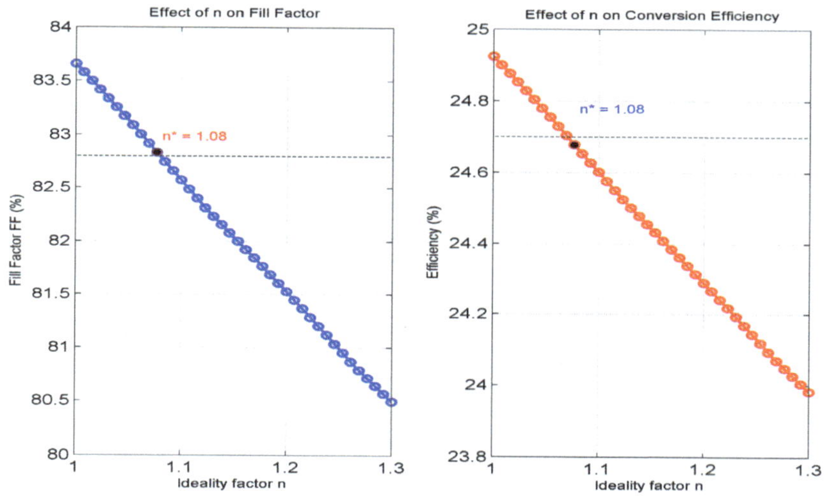

Fig. 5. Influence of the ideality factor (n) on the photovoltaic performance of the PERL cell. As n increases, both the fill factor and the conversion efficiency decrease. The best performance is achieved around n = 1.08, which aligns closely with the experimental reference.

To improve the overall accuracy of the photovoltaic model, a multidimensional grid search was carried out to jointly optimize the key parameters Rs and n. This approach aims to reduce the interaction effects between parameters and serves as a reference for future implementation of advanced artificial intelligence methods such as Particle Swarm Optimization (PSO) or neural networks. For each pair (Rs, n), the I–V curve was simulated and the corresponding fill factor (FF) and conversion efficiency (η) were computed. The resulting data were used to generate a 3D surface map FF(Rs, n), from which the optimal point was extracted. The best agreement with experimental data (FF = 82.8%) was obtained for Rs ≈ 0.043 Ω and n ≈ 1.08. This optimal region and the sensitivity of the fill factor to variations in Rs and n are clearly illustrated in Fig. 6.

This joint optimization further refines the model by compensating the effect of Rs through a fine adjustment of n, while preserving the target efficiency. The 3D surface shows a clear gradient, with the optimal point located in the high-performance region. The shape of the surface reveals the model's sensitivity: steep areas indicate that FF

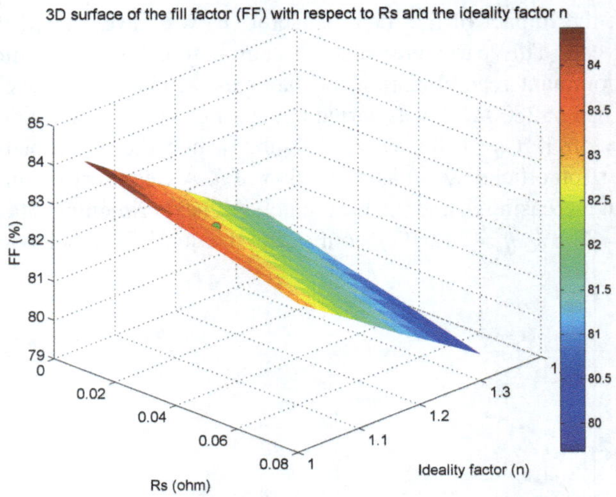

Fig. 6. Three-dimensional surface of the fill factor (FF) as a function of the series resistance (Rs) and the ideality factor (n).

is highly responsive to changes in Rs or n, while flatter zones suggest lower sensitivity. A comprehensive multidimensional visualization of this optimization space is provided in Fig. 7, highlighting the performance gradients and the location of the optimal configuration.

2.3 PSO-Based Optimization of the PERL Cell Electrical Parameters

To optimize the electrical performance of the PERL photovoltaic cell, a Particle Swarm Optimization (PSO) algorithm was employed as a robust population-based metaheuristic inspired by swarm intelligence. Each particle, representing a candidate solution, adjusts its position iteratively by leveraging both its personal best experience and the globally optimal position within the swarm. The optimization targeted three critical parameters—series resistance Rs, shunt resistance Rsh, and diode ideality factor n—with physically justified bounds defined as $0.05 \le Rs \le 0.2\,\Omega$, $3500 \le Rsh \le 4500\,\Omega$, and $1.00 \le n \le 1.2$. The PSO was configured with a population of 30 particles and 60 iterations, using an inertia weight $w = 0.9$ to sustain search momentum, and cognitive and social acceleration factors $c1 = c2 = 1.2$ to ensure balanced convergence. The fitness function, based on the simulated efficiency η derived from the one-diode model, guided the swarm toward the optimal solution, aiming to match or exceed the experimental benchmark of 24.7%. The convergence behavior of the PSO algorithm is illustrated in Fig. 8, showing a rapid decrease of the cost function toward a stable minimum.

2.4 Genetic Algorithm-Based Optimization of PV Cell Electrical Parameters

To accurately identify the optimal electrical configuration of a PERL photovoltaic cell, a Genetic Algorithm (GA) was applied to simultaneously optimize electrical parameters.

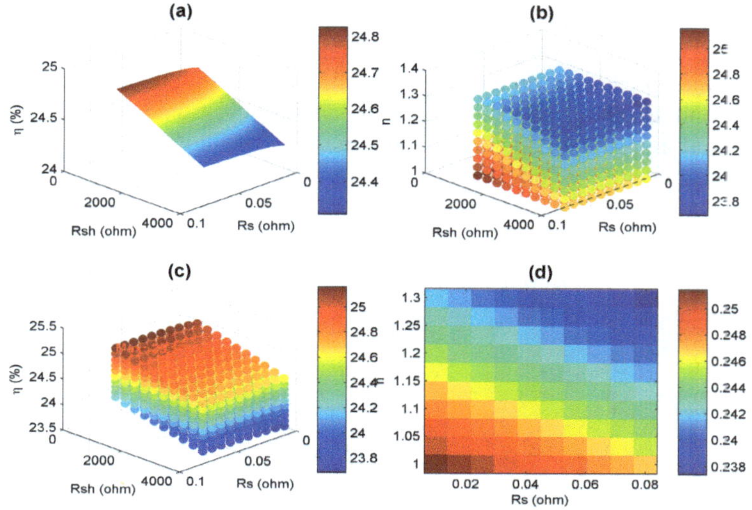

Fig. 7. Three-dimensional Multidimensional visualization of the optimization space for the PERL solar cell. **(a)** Three-dimensional slice of the fill factor (FF) within the parameter space (Rs, Rsh, n), revealing the internal structure of the performance volume. **(b)** 3D scatter plot of FF showing performance gradients and highlighting the optimal configuration. **(c)** Filtered scatter plot displaying configurations with efficiency values within the experimental target range ($\eta \in [24.65\%; 24.75\%]$). **(d)** 2D contour projection of η over the (Rs, n) plane for a fixed Rsh = 2333 Ω, illustrating the sensitivity of efficiency to these parameters.

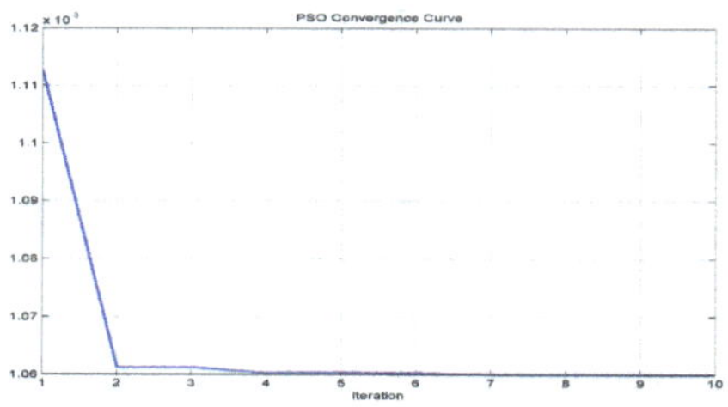

Fig. 8. Convergence curve of the Particle Swarm Optimization (PSO) algorithm applied to the PERL photovoltaic cell. The cost function, defined as the absolute difference between the simulated and target efficiency, rapidly decreases and stabilizes below 10–3, indicating fast and effective convergence.

The method was designed to minimize a weighted error function between simulated and reference values of fill factor (84.2%) and conversion efficiency (24.7%), enabling robust convergence within a multi-dimensional, non-linear parameter space.

$$\mathbf{Fitness(x)} = \sqrt{[\mathbf{0.3} \times (\mathbf{FF(x)} - \mathbf{FFtarget)2} + \mathbf{0.7} \times (\boldsymbol{\eta}(\mathbf{x}) - \boldsymbol{\eta}\mathbf{target)2}]}$$

The GA configuration included 80 individuals, a maximum of 300 generations, a stall generation limit of 150, and a fitness function tolerance of 10–6. To ensure numerical stability in the reconstruction of the I–V curve, a semi-empirical model was employed for the fill factor calculation, accounting for both series and shunt resistive losses. The algorithm successfully converged to a highly accurate and physically consistent solution, yielding optimal parameter values of Rs = 0.0234 Ω,Rsh = 1009.1 Ω, and n = 1.086. The corresponding optimized values for each parameter are summarized in the Table 1 and the evolution of the GA population and the corresponding decrease in the fitness value are illustrated in the convergence plot below [Fig. 9].

Table 1. Optimal electrical parameters of the PERL photovoltaic cell obtained using the Genetic Algorithm (GA). The values correspond to the best solution identified based on the defined fitness function and convergence criteria.

Parameter	Optimized Value
Rs	0.0234 Ω
Rsh	1009.1 Ω
n	1.086
Fill Factor	82.8%
Efficiency η	24.7%
Pmax	≈0.223 W

The genetic algorithm converged smoothly toward optimal electrical parameters, yielding a fill factor and conversion efficiency that closely match the experimental values. The final values obtained were Rs = 0.0234 Ω, Rsh = 1009.1 Ω, and n = 1.086. The convergence plot confirms a stable and gradual optimization process [Fig. 9], with a best fitness value of 0.001086 reached after 151 generations.

2.5 Implementation of an Artificial Neural Network for Predicting Photovoltaic Cell Efficiency

To evaluate the potential of artificial neural networks (ANNs) for photovoltaic performance prediction, we implemented a regression model using MATLAB's Neural Network Toolbox. The ANN was trained to estimate the conversion efficiency (η) of a solar cell from a dataset composed of both physical and electrical input parameters. The dataset contained $N = 1000$ samples, where each input vector included five features: series resistance (Rs), shunt resistance (Rsh), ideality factor (n), temperature (T), and irradiance (G). The output target was the experimentally or numerically determined conversion efficiency, expressed as a percentage. All input and output data were normalized using min–max scaling prior to training, to ensure numerical stability and facilitate faster convergence. A feedforward neural network with two hidden layers (15 and 10 neurons, respectively) was configured and trained using the Levenberg–Marquardt Bayesian regularization algorithm (trainbr), which is well known for its robustness and generalization performance. The dataset was randomly divided into training (70%), validation (15%), and testing (15%) subsets. The model's predictive performance was evaluated both globally and specifically on the test subset. The coefficient of determination (R^2)

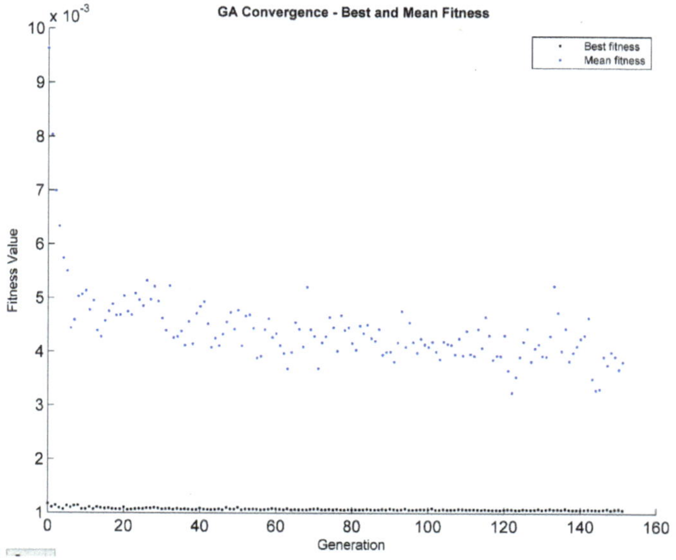

Fig. 9. Convergence curve of the genetic algorithm showing the evolution of best and mean fitness values during the optimization.

was calculated as a quantitative measure of accuracy, yielding a test-set R^2 greater than 0.95—indicating strong regression capability. A scatter plot comparing predicted and actual efficiency values demonstrated excellent alignment, with minimal deviation. As an additional evaluation, a specific test case corresponding to the PERL cell was introduced into the trained model. The ANN successfully predicted a conversion efficiency of approximately **24.8%**, closely matching the expected experimental performance. To assess the accuracy of the trained neural network, [Fig. 10] presents two complementary visualizations: the left plot shows the variation of actual and predicted efficiencies across the entire dataset, while the right scatter plot compares the predicted values against actual measurements, confirming strong agreement and robust generalization. Artificial neural networks have been widely used for I–V curve modeling and photovoltaic performance estimation under real operating conditions (Brecl and Topic, 2011; Gokmen and Ozgoren, 2012).

Despite training a neural network on 1000 samples, the prediction of Rs from global parameters (FF, η, Voc, Isc, T, G, S) yielded poor accuracy ($R^2 < 0.02$), confirming its low observability. As a more effective alternative, we reversed the model to predict η from internal parameters (Rs, Rsh, n) and environmental conditions (T, G). This new approach demonstrates strong learning performance ($R^2 > 0.95$) and highlights the potential of AI-based optimization in embedded PV systems. For completeness, [Fig. 11] presents the internal summary of the neural network training process, including architecture, optimization parameters, and convergence performance using Bayesian regularization. Recent machine learning studies have also focused on maximum power point prediction under varying environmental conditions (Scientific Reports, 2025).

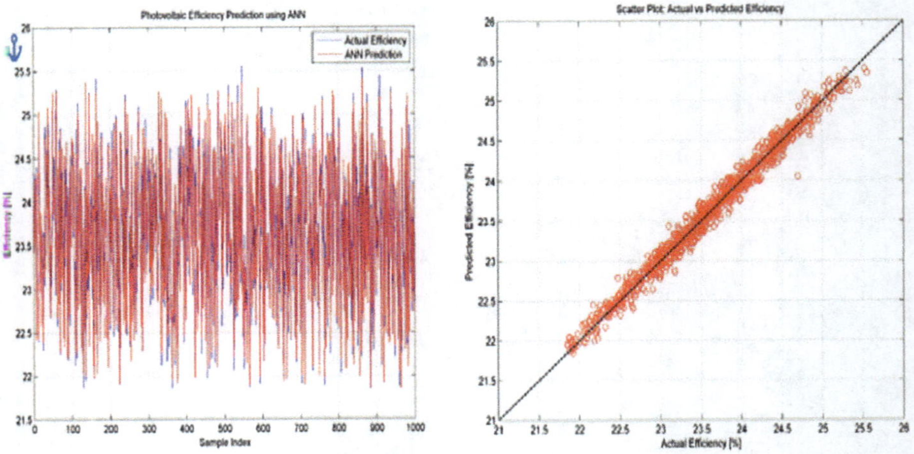

Fig. 10. Comparative plots of actual and predicted efficiencies versus sample index, illustrating the model's generalization ability.

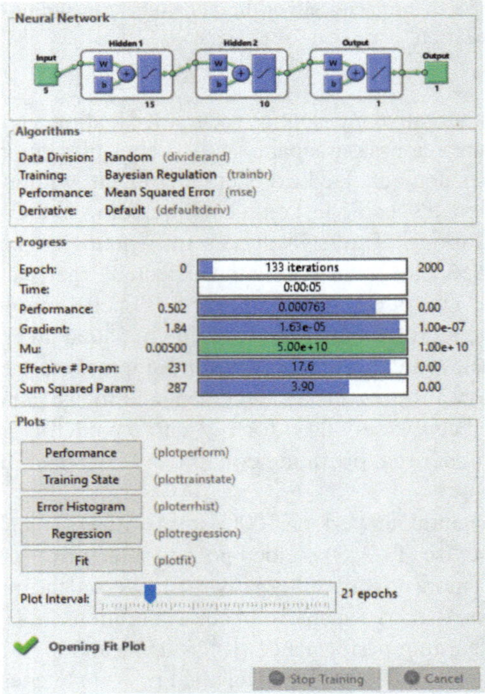

Fig. 11. Evaluation of the ANN model for predicting photovoltaic efficiency.

3 Conclusion

This study presented a comprehensive modeling and optimization framework for a high-efficiency monocrystalline silicon photovoltaic cell based on the PERL architecture. Starting from an accurate reconstruction of the cell's electrical behavior using equivalent circuit, we implemented a numerical simulation under standard test conditions to reproduce I–V and P–V characteristics. A systematic exploration of the electrical parameter space was then conducted using grid search techniques in one, two, and three dimensions, enabling the identification of critical trends and performance optima.

To further enhance the predictive and optimization capabilities, two nature-inspired algorithms—Particle Swarm Optimization (PSO) and Genetic Algorithm (GA)—were applied. Both approaches demonstrated effective convergence and yielded parameter configurations closely aligned with the reference experimental efficiency. Finally, an artificial neural network (ANN) model was trained to estimate photovoltaic efficiency based on physical and electrical inputs. The ANN successfully generalized over unseen data, offering a rapid estimation tool, though with inherent trade-offs compared to heuristic optimization.

Collectively, the proposed methodology offers a multi-level approach to the electrical optimization of advanced solar cells and highlights the potential of hybrid numerical–AI strategies for embedded photovoltaic applications.

References

Hatti, M.: Artificial Intelligence in Renewable Energetic Systems: Smart Sustainable Energy Systems, p. 300. Springer, Cham (2018)

Chen, S.-S., Jiang, Y.-T., Chen, W.-B., Li, X.-Y.: TERIME: An improved RIME algorithm with enhanced exploration and exploitation for robust parameter extraction of photovoltaic models. J Bionic Eng. **22**, 1535–1556 (2024) *arXiv preprint*, https://arxiv.org/abs/2407.18986

Hamad, S.A., Ghalib, M.A., Munshi, A., et al.: Evaluating machine learning models comprehensively for predicting the maximum power point of standalone photovoltaic systems. Sci. Rep. **15**, 10750 (2025) Nature Publishing Group, London, pp. 1–15

Villalva, M.G., Gazoli, J.R., Filho, E.R.: Comprehensive approach to modeling and simulation of photovoltaic arrays. IEEE Trans. Power Electron. **24**(5) IEEE, New York, 1198–1208 (2009)

Spertino, F., Ahmad, J., Ciocia, A., Di Leo, P.: Neural network approaches for accurate I–V curve prediction of photovoltaic modules. Renew. Energy. **149**, Elsevier, Amsterdam, 965–976 (2020)

Gow, J.A., Manning, C.D.: Development of a photovoltaic array model for use in power-electronics simulation studies. IEE Proc. - Electr. Power Appl. **146**(2) IEE, London, 193–200 (1999)

Rad, M.A., Rahimi, M.: Efficient parameter identification of PV models using metaheuristic optimization and AI. J. Renew. Sustain. Energy. **15**, AIP Publishing, New York, 045601 (2023)

Brecl, K., Topic, M.: A new method for evaluating PV module performance in real outdoor conditions using ANN. Sol. Energy. **85**(7) Elsevier, Amsterdam, 1557–1568 (2011)

Green, M.A., Emery, K., Hishikawa, Y., Warta, W., Dunlop, E.D.: Solar cell efficiency tables (version 18). Prog. Photovolt. Res. Appl. **9**, Wiley, New York, 347–353 (2001)

Gokmen, N., Ozgoren, M.: Estimation of photovoltaic module performance using ANN and empirical models. Energy Conv. Manag. **64**., Elsevier, Amsterdam, 441–446 (2012)

Enhanced MPPT in PV Systems Using k-Nearest Neighbors and Integral Backstepping Control

Mustapha Meraouah[1](✉), Faiza Kaddari[2], Said Hassaine[1], and Sandrine Moreau[3]

[1] Electrical engineering, L2GEGI laboratory, University of Tiaret, Tiaret, Algeria
mustapha.meraouah@univ-tiaret.dz

[2] Higher School of Electrical and Energy Engineering, Oran, Algeria

[3] LIAS Laboratory, University of Poitiers, Poitiers, France

Abstract. This work presents a control strategy for a photovoltaic system (PV) based on an innovative maximum power point tracking MPPT algorithm. This MPPT uses a k-nearest neighbors (KNN) to extract the MPP voltage only with voltage and current measurements trained by the data of PV characteristics and an integral backstepping (IB) to control the boost converter. These proposed strategies are implemented in a MATLAB/Simulink environment for testing and validation. Simulation results are presented, discussed, and compared. The acquired results demonstrate the suggested strategy's efficacy compared to existing approaches. The proposed MPPT demonstrated excellent performance and strong adaptability to meteorological variations in the PV system.

Keywords: K-nearest neighbors · Machine learning · Hybrid MPPT · Lyapunov stability · PV system · Integral backstepping

1 Introduction

Solar energy, especially photovoltaic systems, is increasingly recognized as a key alternative to fossil fuels [1]. Photovoltaic (PV) systems contain mainstream elements such as PV panels, DC-DC boost converters, and unit control to optimize the PV energy [2]. Significant classical and advanced MPPT approaches have been devoted to the PV system to achieve high efficiency and maximize energy efficiency [3]. In addition, these MPPTs must be durable despite harsh climatic conditions and high temperatures and have a more consistent integration cost [4]. Several heuristic approaches, such as the algorithm based on perturb and observe (P&O), have been developed for the PV system [5]. Algorithms based on Lyapunov stability, such as sliding mode control SMC [6] and backstepping [7]. In the last decade, intelligent strategies have grown very fast, such as fuzzy logic controller [8], artificial neural network ANN [9], ANN-FLC (ANFIS) [10], ANN-MPPT based on ADALINE controller, and ANN-Backstepping MPPT [11, 12].

This work proposes a hybrid MPPT that combines a k-nearest neighbors (KNN) to predict the optimal voltage and the PV system's integral backstepping (IB) controller. This MPPT technique is proposed to follow the maximum power of the PV system by combining the advantages of machine learning and a stable nonlinear control strategy.

M. Hatti (Ed.): IC-AIRES 2025, LNNS 1740, pp. 398–405, 2026.
https://doi.org/10.1007/978-3-032-12788-4_38

The second section presents the global description and modeling of the PV system, including the DC-DC boost converter. The third section will show the proposed MPPT used in this work. The fourth section presents the simulation results and discusses them. Finally, we conclude our work.

2 System Description

Figure 1 depicts the entire PV energy conversion system that will be studied in this work.

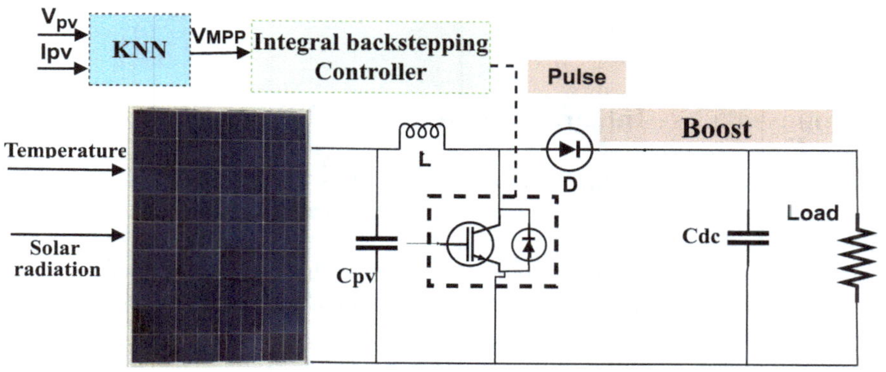

Fig. 1. Overall system view.

The principal elements in this system are a ten PV model type Kyocera solar KC200GT connected in series, producing 2001.43 W at standard conditions, a control unit MPPT based on the k-nearest neighbors algorithm, and an integral backstepping (KNN-IB) controller for a DC-DC boost converter as an adaptation internship supplying a resistive load. The mainstream goal is to ensure maximum power extraction using a hybrid nonlinear intelligent MPPT technique.

2.1 Modeling of PV Cell

A PV cell model is crucial for comprehensively analyzing a PV system [13]. The modeling is elaborated through the equivalent circuit of a PV cell based on the Shockley model and its mathematical equation, presented below.

$$I_{P_v} = I_{ph} - I_{sat}\left(e^{\frac{qV_{Pv}+R_sI_{Pv}}{nKTN_s}} - 1\right) - N_p\left(\frac{qV_Pv + qR_sI_Pv}{R_{s_h}R_s}\right) \quad (1)$$

Where: R_s is the equivalent series resistance, R_{sh} is the equivalent shunt resistance, N_s is the number of series PV cells connected. The I_{p_h} is proportional to the solar irradiation I_{sat} corresponding to the diode characteristics, q is the electron charge and K is the Boltzmann constant.

2.2 Boost Converter Modeling

The boost converter has a crucial role in photovoltaic energy conversion. It aims to maximize the energy output of the photovoltaic module by increasing the voltage supplied to the loads [14]. The converter's dynamic equations in continuous operation mode are obtained by considering the current flowing through the inductance and the voltage across the photovoltaic (PV) module as ($x_1 = V_{pv}$ and $x_2 = i_L$ and v_{Out} is the output voltage of the DC-DC boost converter.

$$\begin{bmatrix} \frac{\mathrm{d}x_1}{\mathrm{d}t} \\ \frac{\mathrm{d}x_2}{\mathrm{d}t} \end{bmatrix} = \begin{bmatrix} 0 & -\frac{1}{C_{pv}} \\ \frac{1}{L} & 0 \end{bmatrix} \begin{bmatrix} x_1 \\ x_2 \end{bmatrix} + \begin{bmatrix} \frac{1}{C_{pv}} & 0 \\ 0 & -\frac{1}{L}(1-\mu) \end{bmatrix} \begin{bmatrix} i_{pv} \\ v_{\text{Out}} \end{bmatrix} \tag{2}$$

3 Proposed KNN-Integral Backstepping MPPT Algorithm

The PV voltage (V_{MPP}) in MPP location varies with solar temperature and irradiance [15]. The PV system voltage points corresponding to the maximum power point for the different irradiation levels are connected to plot a non-linear MPP trajectory. This work proposed the k-nearest neighbors (KNN) algorithm trained with power-voltage data to generate the MPP reference voltage at each irradiation by measuring the voltage and current.

3.1 K-Nearest Neighbors

In this study, the k-nearest neighbors (KNN) algorithm, a nonparametric and supervised learning method, is employed to classify or predict data points based on their proximity to labeled examples [16]. It is among the most widely used and straightforward algorithms for classification and regression tasks in modern machine learning [17]. The proposed KNN consists of two inputs for PV voltage and current, and an output for V_{MPP}. In this work, after parameter tuning, we selected the number of neighbors as K = 3. This choice was made to identify the optimal K value to classify pre-categorized data and predict the original voltage for the IB controller as shown in Fig. 2.

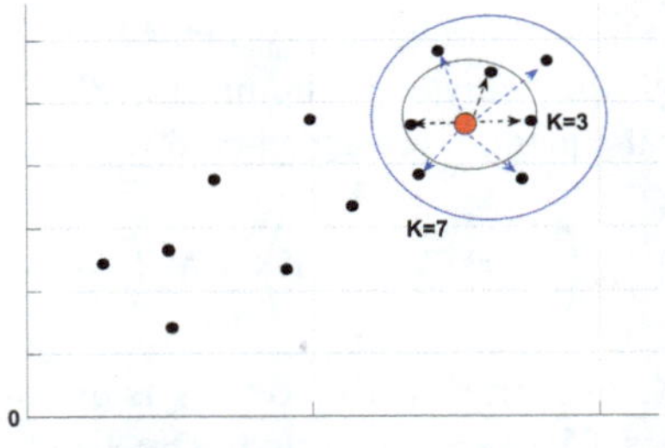

Fig. 2. KNN complicated decision boundary.

3.2 Integral Backstepping Control

The integral backstepping strategy is a control design based on the stability of Lyapunov theory. It combines the backstepping control approach and error integration principles to improve performance in dynamic and nonlinear systems [17]. This approach ensures that the PV system always operates at the voltage corresponding to the maximum power point by generating a convenient duty cycle for the boost converter [18]. Defining the control (μ) law for the system is accomplished through the following process: Firstly, we choose e_1 as the Voltage error, ϑ the integral of the error, and ϕ presents the sum of them as shown in the following equation:

$$\begin{cases} \phi = e_1 + \vartheta \\ e_1 = x_1 - V_{MPP} \\ \vartheta = \int_0^t (x_1 - V_{MPP})dt \end{cases} \tag{3}$$

The derivative function across the time $\left(\dot{e_1}\right)$ obtained as:

$$\dot{e_1} = \dot{x_1} - \dot{V}_{MPP} = \frac{1}{C_{pv}}\left(i_{pv} - x_2\right) - \dot{V}_{MPP} \tag{4}$$

The first candidate for Lyapunov and its derivative functions are given by the following equations, respectively:

$$\begin{cases} V_1(e_1, \vartheta) = \frac{1}{2}e_1^2 + \frac{1}{2}\vartheta^2 \\ \dot{V}_1 = e_1\left[\vartheta + \frac{1}{C_{pv}}\left(i_{pv} - x_2\right) - \dot{V}_{MPP}\right] \end{cases} \tag{5}$$

By considering x_2 as a virtual control, which represents the desired current flowing through the boost inductor, with $x_2 = \alpha$ as its assignment, and by introducing λ_1 as a strictly positive constant, stabilization can be achieved using the following equations:

$$\begin{cases} -\lambda_1 e_1 = \vartheta + \frac{1}{C_{pv}}\left(i_{pv} - \alpha\right) - \dot{V}_{MPP} \\ \alpha = i_{pv} - C_{pv}\left(\lambda_1 e_1 + \vartheta - \dot{V}_{MPP}\right) \end{cases} \tag{6}$$

The second step is to choose another error to extract the law control expressed by:

$$\begin{cases} e_2 = x_2 - \vartheta \\ \dot{e_2} = \dot{x_2} - \dot{\alpha} = \frac{1}{L}(x_1 - V_{out}) + \frac{1}{L}\mu V_{out} - \dot{\alpha} \end{cases} \tag{7}$$

Replacing $x_2 = e_2 + \vartheta$ *then* V_1, e_1 and $\dot{e_1}$ becomes:

$$\begin{cases} \dot{e_1} = -\frac{1}{C_{pv}}\left(i_{pv} - e_2 - \alpha\right) - \dot{V}_{MPP} \\ \dot{V}_1 = -\lambda_1 e_1^2 - \frac{1}{C_{pv}}e_2 e_1 \end{cases} \tag{8}$$

Now the new Lyapunov function candidate and their derivative are expressed by:

$$\begin{cases} V_2(e_1, \vartheta, e_2) = V_1 + \frac{1}{2}e_2^2 \\ \dot{V}_2 = -\lambda_1 e_1^2 + 2 - \frac{1}{C_{pv}}e_1 + \frac{1}{L}x1 - Vout + \frac{1}{L}\,\mu\, Vout + \alpha \end{cases} \tag{9}$$

The control law the equation below with a value of the constant λ_2 superior to zero positive. That ensures that the variable converges asymptotically to the origin V_{MPP} generated by the KNN.

$$\mu = \frac{L}{V_{out}}\left[-\lambda_2\, e_2 + \frac{1}{C_{pv}} e_1 - \frac{1}{L}(x_1 - V_{out}) + \dot{\alpha}\right] \tag{10}$$

4 Simulation Results and Discussion

The proposed PV system parameters are summarized as follows: The maximum power point is ($P_{MPP} = 2001.43\ W$), with a corresponding voltage and current ($V_{MPP} = 263\ V$ and $I_{MPP} = 7.61\ A$), respectively. The voltage of open-circuit is ($V_{oc} = 329\ V$), and the current of short-circuit is ($I_{sc} = 8.21\ A$). The boost converter employs an inductor of ($L_{boost} = 3.75\ mH$), along with a PV-side capacitor of ($C_{pv} = 470\ \mu F$) and a DC-side capacitor of ($C_{dc} = 100\ \mu F$). The system is designed to supply a resistive load of (35 Ω). The simulation is executed via MATLAB/Simulink with a time step of (1 μs).

The proposed strategy is tested and compared to algorithm MPPTs such as P&O, Incremental conductance (INC), backstepping, INC-IB, and ANN-IB with irradiation G and temperature T sensors under fast irradiation change. The results obtained using this profile of irradiation and fixed temperature at 25 °C are shown in Fig. 3.

Figure 3 (a) assesses different MPPT techniques to determine the resilience of the proposed strategy. The KNN-IB MPPT achieves the maximum power output with a faster response time and reduced oscillations compared to other methods, closely matching the performance of the MPPT presented in [19]. The INC-IB MPPT and the Backstepping MPPT exhibit lower oscillations than the conventional P&O MPPT.

The proposed algorithm remains resilient regarding climate variations, demonstrating high performance compared to other MPPT techniques. Its rapid response is attributed to the KNN and ANN components, which directly provide the voltage value V_{mpp}, independent of time and without requiring sliding surface calculations. As shown in Fig. 3 (b) and (c), the PV voltage and current curves produced by the proposed MPPT are smoother and closer to the ideal values (V_{mpp}) and (I_{mpp}), compared to those generated by the other algorithms.

To better understand the differences between the proposed methods, we calculate each algorithm's dynamics and static efficiency (η_d, η_s) as shown in Table 1.

The previous table proves that the suggested hybrid algorithm (KNN-IB) can extract the maximum power quickly and stably under any fast changes in irradiation, rather than the classical approach, which consists of defining the reference voltage based on an incremental conductance algorithm.

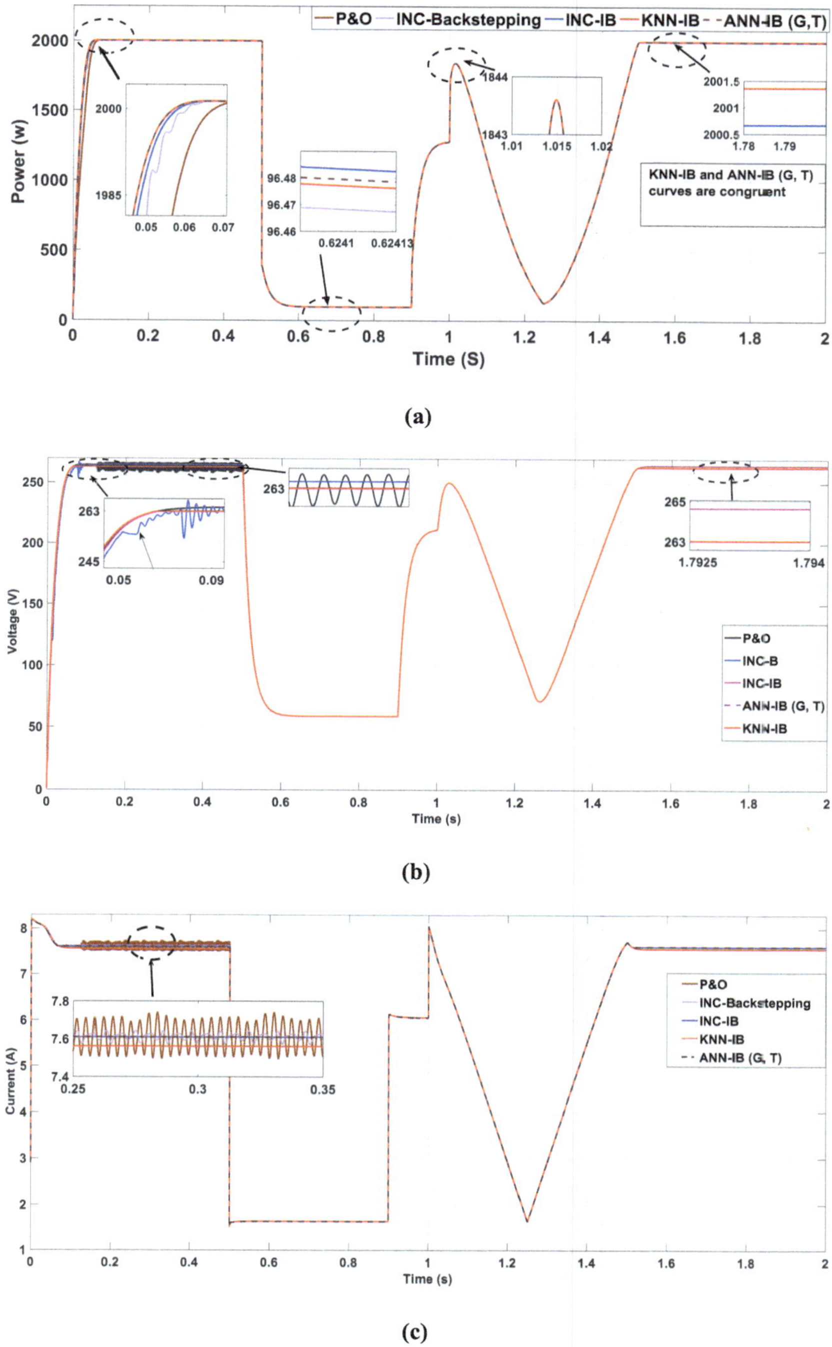

Fig. 3. PV results (a) power, (b) voltage, and (c) current with different MPPT algorithms.

Table 1. MPPTs dynamics and static efficiency.

Algorithm	P&O	INC-B	INC-IB	ANN-IB (G, T)	KNN-IB
η_s (%)	97.60	99.21	97.40	98.97	99.19
η_d (%)	97.15	95.17	97.59	98.51	98.52

5 Conclusion

This study investigated an advanced PV system algorithm to enhance flexibility, stability, and reliability. The proposed MPPT based on KNN and an IB controller was combined, demonstrating superior robustness under varying irradiation conditions compared to existing MPPT algorithms. The proposed MPPT algorithm exhibited rapid response, improved accuracy, and reduced oscillations. The proposed strategy's resilience was evaluated under various operating scenarios, including irradiation variations, validating its effectiveness and superior performance. This work can be further improved and validated by an experimental study to evaluate the performance obtained by the proposed methods.

References

1. I. E. Haji, 'Performance Optimization of Symmetrical Multi-Level Boost Converter using Hybrid MPPT-ANN for Solar Energy Applications', 2025.
2. J. J. Khanam, M. Issapour, S. Y. Foo, and P. Castillo, 'Machine learning-driven MPPT techniques for optimal solar energy utilization', in Southeast Con 2025, Mar. 2025, pp. 309–313. doi:10.1109/SoutheastCon56624.2025.10971573.
3. Ali, A.I.M., et al.: An enhanced P&O MPPT algorithm with concise search area for grid-tied PV systems. IEEE Access. **11**, 79408–79421 (2023). https://doi.org/10.1109/ACCESS.2023.3298106
4. Maaruf, M., Khan, K.A., Khalid, M.: Integrated power management and nonlinear-control for hybrid renewable microgrid. In: 2021 IEEE Green Technologies Conference (GreenTech), pp. 176–180. IEEE, Denver, CO, USA (Apr. 2021). https://doi.org/10.1109/GreenTech48523.2021.00037
5. Krim, S.A., Krim, F., Afghoul, H., Abdelmalek, F.: An enhanced perturb and observe MPPT for photovoltaic systems based on fuzzy step. JESA. **57**(2), 363–372 (2024). https://doi.org/10.18280/jesa.570206
6. Fernández-Bustamante, P., Calvo, I., Villar, E., Barambones, O.: Centralized MPPT based on sliding mode control and XBee 900 MHz for PV systems. Int. J. Electr. Power Energy Syst. **153**, 109350 (2023). https://doi.org/10.1016/j.ijepes.2023.109350
7. Harrison, A., et al.: Robust nonlinear MPPT controller for PV energy systems using PSO-based integral backstepping and artificial neural network techniques. Int. J. Dyn. Control. **12**, 1598–1615 (2023). https://doi.org/10.1007/s40435-023-01274-7
8. Ali, M., Ahmad, M., Koondhar, M., Akram, M., Verma, A., Khan, B.: Maximum power point tracking for grid-connected photovoltaic system using adaptive fuzzy logic controller. Comput. Electr. Eng. **110**, 108879 (2023). https://doi.org/10.1016/j.compeleceng.2023.108879

9. Sakthivel, S., Arunachalam, V.: Artificial neural network assisted P &O-based MPPT controller for a partially shaded grid-connected solar PV panel. Arab. J. Sci. Eng. **48**, 14333–14344 (2022). https://doi.org/10.1007/s13369-022-07566-y
10. Guerra, M.I.S., De Araújo, F.M.U., De Carvalho Neto, J.T., Vieira, R.G.: Survey on adaptative neural fuzzy inference system (ANFIS) architecture applied to photovoltaic systems. Energy Syst. **15**(2), 505–541 (2024). https://doi.org/10.1007/s12667-022-00513-8
11. Meraouah, M., Kaddari, F., Hassaine, S., Mihoub, Y., Heroual, S., Khenfar, N.: Predictive direct power control of shunt active power filter associated to PV system and ANN-backstepping MPPT. AJRESD. **6**(1), 109–118 (2024)
12. Mustapha, M., Faiza, K., Said, H., Youcef, M.: Improved direct power control of PV-SAPF system using ADALINE and ANN-MPPT algorithm. In: 2024 3rd International Conference on Advanced Electrical Engineering (ICAEE), pp. 1–6 (2024). https://doi.org/10.1109/ICAEE61760.2024.10783279
13. Basha, H., Chinnappa Naidu, R.: Performance analysis of MPPT techniques for dynamic irradiation condition of solar PV. Int. J. Fuzzy Syst. **22**, 2577–2598 (2020). https://doi.org/10.1007/s40815-020-00974-y
14. E. Jalalabadi, B. Moshiri, S. Z. Paylakhi, and A. Rahimi-kian, 'Integral backstepping Lyapunov redesign control of uncertain nonlinear systems'.
15. Doubabi, H., Salhi, I., Chennani, M., Essounbouli, N.: High Performance MPPT based on TS Fuzzy–integral backstepping control for PV system under rapid varying irradiance—Experimental validation. ISA Trans. **118**, 247–259 (2021). https://doi.org/10.1016/j.isatra.2021.02.004
16. K. Yan, Y. Du, and Z. Ren, 'MPPT Perturbation Optimization of Photovoltaic Power Systems Based on Solar Irradiance Data Classification'.
17. M. Zhu, 'Prediction of Constitutive Model for Basalt Fiber Reinforced Concrete Based on PSO-KNN', 2024.
18. Taouni, A., Abbou, A., Akherraz, M., Ouchatti, A., Majdoul, R.: MPPT design for photovoltaic system using backstepping control with boost converter. In: Proceedings of 2016 International Renewable and Sustainable Energy Conference, IRSEC 2016, pp. 469–475 (2017). https://doi.org/10.1109/IRSEC.2016.7983920
19. Khan, Z.A., Khan, L., Ahmad, S., Mumtaz, S., Jafar, M., Khan, Q.: RBF neural network based backstepping terminal sliding mode MPPT control technique for PV system. PLoS One. **16**(4), e0249705 (2021). https://doi.org/10.1371/journal.pone.0249705

Real Time Identification of the Parameters of a Photovoltaic Panel by Ant Colony Optimization in the Continuous Domain

Hizia Abed[1](✉) and Sihem Bouri[2]

[1] Automation Laboratory (LAT), Department of Electrical and Electronic Engineering, Faculty of Technology, Abou Berk Belkaid University, BP 230 13000, Chetouane, Tlemcen, Algeria

[2] Renewable Materials and Energy Research Unit (URMER), Department of Electrical and Electronic Engineering, Faculty of Technology, Abou Berk Belkaid University, BP 230 13000, Chetouane, Tlemcen, Algeria

Abstract. In a context marked by a continuous increase in global energy demand and the need to resort to renewable sources, photovoltaic solar energy is positioned as a sustainable, clean and reliable solution. This article presents an innovative approach for the real-time identification of the parameters of a SY-M80W photovoltaic panel, using the ACOR (Ant Colony Optimization for Continuous Domain) algorithm. Inspired by the collective behavior of ant colonies, this optimization method is distinguished by its flexibility, rapid convergence and ability to avoid local minima in complex environments. The main objective of using this technique is to improve the accuracy of the photovoltaic generator modeling and to continuously optimize its performance in real operating conditions. For this, a MATLAB-based framework, an Arduino board and low-cost sensors allow the continuous acquisition of I-V and P-V curves. The ACOR algorithm dynamically processes this data, ensuring intelligent energy management, improved efficiency, active system monitoring, and rapid response to environmental variations.

Keywords: Identification · Artificial intelligence · Photovoltaic · Meta heuristic · Ant colony

1 Introduction

Climate change is one of the major threats facing the world population [1], mainly caused by greenhouse gas (GHG) emissions generated by the use of fossil fuels in industrial machinery. As fossil fuel sources used for production become scarcer [2], oil and natural gas will be almost exhausted in the next forty years, despite the continuous increase in global electricity needs, the use of clean fossil energy sources, such as solar energy, has become a key player in solving the energy crisis [3]. Currently, solar energy ranks third among renewable energy sources after wind and hydropower [4].

Solar panels have several promising advantages, including high reliability, low maintenance costs, easy installation and no fuel costs [5]. However, the use of photovoltaic

M. Hatti (Ed.): IC-AIRES 2025, LNNS 1740, pp. 406–417, 2026.
https://doi.org/10.1007/978-3-032-12788-4_39

systems is influenced by meteorological and environmental elements, such as temperature fluctuations and global irradiation. Therefore, it is essential to improve the efficiency of the photovoltaic system. It is possible to accomplish this using an accurate analytical model that relies on the [current (I), voltage (V)] measurement data [6].

In the literature, the simple diode model is the one that is most often cited. The idea is to deduce the parameters of the model to perform an analytical adaptation of the experimental measurements. It is therefore essential to obtain an exact value of the parameters when analyzing solar photovoltaic systems. Over time, different methods have been used to obtain the characteristics of solar cells.

Three main categories of approaches are discussed: first, analytical methods, then deterministic methods [7], and finally, meta-heuristic methods which are perceived as an advantageous solution for the extraction of parameters in the PV model [8]. Meta-heuristic approaches are generally based on natural phenomena. Among the techniques cited are particle swarm optimization (PSO) [9], genetic algorithm (GA) [10], tunicate swarm algorithm (TSA) [11], Whippy Harris Hawks optimization algorithm (HHO) [12], artificial bee colony (ABC) [13], and multiprocessor salp swarm optimization (MSSO) [14], as well as artificial fish swarm algorithm (AFSA) [15].

These approaches have their constraints, as it is crucial to judiciously adjust and select the parameter search space to ensure optimal optimization performance [16].

This paper introduces a live acquisition and analysis system that seeks to optimize the identification of the parameters of a live photovoltaic model, through the use of ant colony optimization in a continuous framework.

The suggested approach is based on two phases. First, data collection is carried out using the Arduino, current and voltage sensors that measure the output current and voltage. Subsequently, the Matlab software is used for parameter identification. This method has the major advantage of its simplicity and the use of inexpensive techniques Elements such as the Arduino UNO board, as well as inexpensive current and voltage sensors.

2 Materiel and Methods

2.1 Photovoltaic Panel

A solar panel is formed by several photovoltaic cells arranged in parallel or in series in order to transform solar energy into electrical energy [17]. Figure 1 represents the equivalent diagram of the solar panel.

Figure 1 illustrates well that a photovoltaic cell can be represented by a circuit that includes a current source to simulate light radiation, in parallel with a diode, a shunt resistor and a series resistor to model losses [18]. The output current produced by the photovoltaic panel can be determined by referring to (Eq. (1)).

$$I = I_{ph} - I_o\left(e^{\left(\frac{q(V+R_S I)}{aKT}\right)} - 1\right) - \frac{V + R_s I}{R_{sh}} \tag{1}$$

In this section, we will detail the procedure for connecting the hardware components of the Arduino Uno with the solar panels. We provide an approach that uses the Arduino UNO microcontroller and Matlab, which allows acquiring live information via Matlab. The structure of the device used in the virtual instrumentation system is illustrated in Fig. 3. Current (ACS 712) and voltage (F031–06) sensors are used to measure the current and voltage generated by the solar panels. The output of both sensors is then transmitted to the microprocessor of the Arduino UNO board

Here, the photocurrent is represented by Iph, the constant current by Io, the voltage across the panel by V, the series resistance by Rs, the thermal voltage by Vt the diode ideality factor by a, and the shunt resistance Rsh.

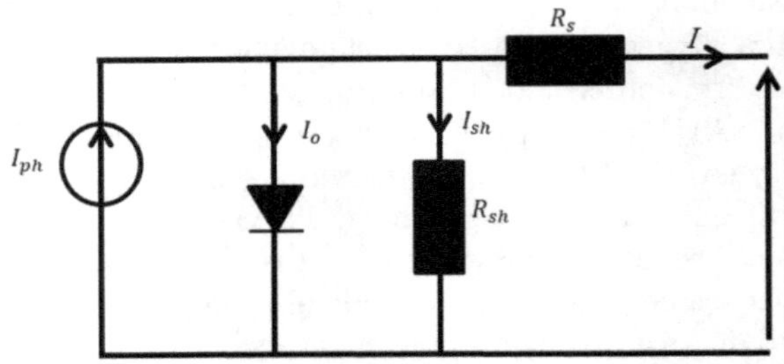

Fig. 1. PV cell equivalent circuit

2.2 System Real-Time Virtual Instrumentation System

Data acquisition (DAQ) is the method of quantifying an electrical characteristic or physical phenomenon, such as voltage, current, temperature, pressure, or noise, via a computer [19]. PC-based DAQ systems take advantage of the computing power, productivity, display, and connectivity capabilities of standard industrial computers, providing a superior, more adaptable, and cost-effective measurement solution.

The output generated by the MATLAB support package is presented directly in Matlab. For example, this allows you to read and write data from sensors via the Arduino board, and instantly observe the results in MATLAB without having to compile any code. In addition, MATLAB offers a wide range of mathematical, engineering, and plotting functions that make it easy to analyze and visualize the data collected from your Arduino [20]. During the acquisition phase, the collected information is stored as shown in Fig. 2 and processed live in Matlab through the continuous optimization of ant colonies.

To understand the specificities of the photovoltaic panel, several elements are required, such as a data collection device, a photovoltaic panel itself, an electronic card, a current sensor and a voltage sensor (Fig. 3).

2.3 Ant Colony Optimization (ACO)

ACO is part of the field of optimization in artificial intelligence. It is an optimization algorithm that is inspired by the collective and natural behavior of large colonies of the

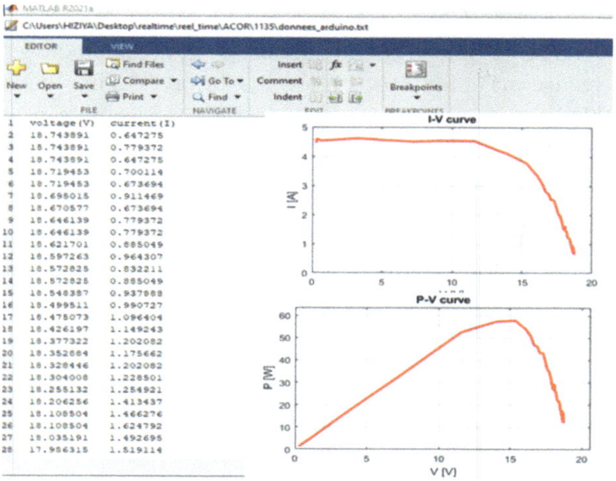

Fig. 2. A sample of experimentally measured data with real-time virtual instrumentation

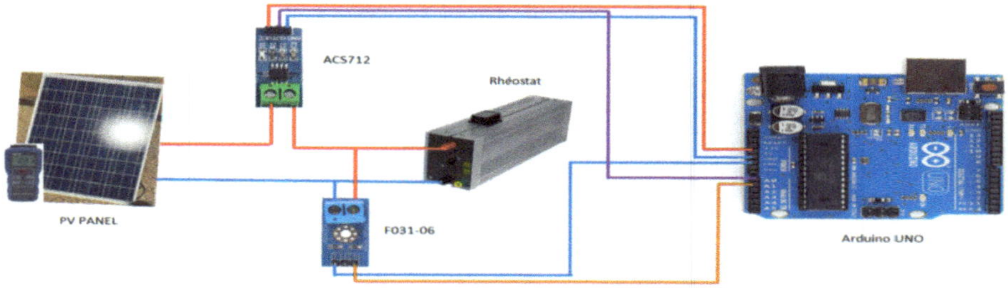

Fig. 3. Diagrammatic representation of the hardware elements linked to the Arduino UNO

same species (such as ants) [21]. ACO is inspired by the natural feeding behaviors of ants and their instinctive ability to determine the most efficient route between a food resource and their colony [22].

Marco Dorigo presented one of the most renowned ACO (Ant Colony Optimization) models in his 1991 PhD thesis [23]. It was presented through the traveling salesman problem (TSP) [24]. We consider the ant colony approach as a swarm meta-heuristic, where each solution is represented by an ant evolving in the search space. Ants are able to collectively solve complicated problems, such as identifying the best path to follow between two points in difficult environmental conditions. Thus, they communicate indirectly and locally using volatile hormones known as pheromones [25]. During their evolutionary process, ants emit pheromones and, subsequently, select their route randomly, based on a probability that varies according to the volume of pheromone previously deposited [26].

The "ant colony optimization meta-heuristic" describes the distinctive features of synthetic ant behavior [27].

Generaly, ACO can be applied to any discrete optimization where a solution generation mechanism can be devised. The initial conception of ant colony optimization was carried out to solve discrete optimization problems. It was later introduced to continuous optimization problems.

2.4 Ant Colony Optimization for a Continuous Domain

Krzysztof Socha and Marco Dorigo (2006) [28] were the first to propose the use of ACO to solve continuous optimization problems. During the development of the solution, ACOR replaces the discrete probability distributions used in ACO algorithms for combinatorial problems with probability density functions. To achieve this, the ACOR algorithm maintains a set of "k" solutions, called a solution archive, which functions as a pheromone model for this technique [29]. Using the solution archive allows to develop a probabilistic distribution of promising solutions in the research domain.

In this method, the most efficient solutions are reviewed and recorded (see Figure 4). The function corresponding to the solutions of a sorted solution file for a minimization problem has the values $f(1) < f(2) < -f(3)$, which suggests that the weight assigned to the solutions is $w(1) > w(2) > ... > w(k)$. According to the weight, the ideal solution should have the largest weight [30].

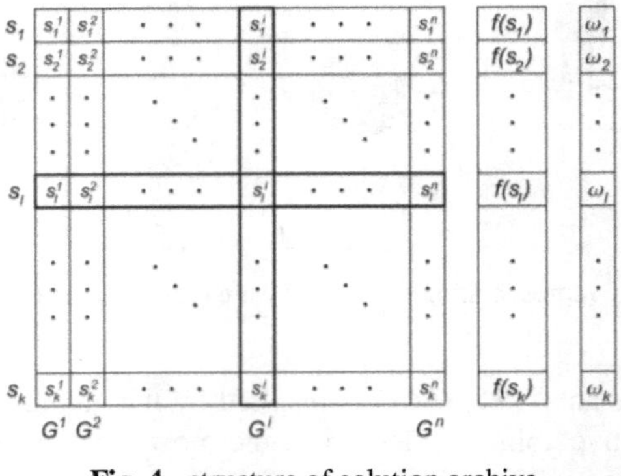

Fig. 4. structure of solution archive

The ACO for combinatorial optimization stores pheromone information in a tabular form. At each iteration, it exploits some of the elements of this table as discrete probability distributions when deciding which components to join to the current partial solution. Continuous optimization does not restrict the available options to a finite set. Therefore, it is not possible to express pheromones in a tabular form. Similar to traditional ACOs, we employ the method of updating the pheromone table based on the elements of the appropriate solution [31]. However, in classical ACO, the solution found by the antenna is discarded when updating the pheromone table. Nevertheless, this method lists multiple solutions to update the pheromone table.

Rather than relying on pheromone evaporation, the pheromones associated with the old method are eliminated by negatively updating the pheromone table, thereby removing their effects.

Assigning a weight to each solution is done as follows (refer to Eq. (2)):

$$w(l) = \frac{1}{qk\sqrt{2\pi}} e^{\frac{(l-1)^2}{2q^2k^2}} \quad (2)$$

In this context: q represents the intensification factor, k symbolizes the number of ants, qo ($0 \leq qo \leq 1$) is a defined parameter and r denotes a random variable uniformly distributed on the interval [0, 1].

After obtaining the mother ant colony, the ant pheromone distribution model w(l) in this colony is transformed into a random number generator for sampling, thus allowing the creation of ant colonies. Then, according to the evaluation value for each set of sub-colonies, the evaluation value of group k among the group of sub-colonies is chosen to determine the updated position of the ant colony [32].

$$P(l) = \frac{W(l)}{\sum_j^k W(j)} \quad (3)$$

This weight indicates the possible strength of the solution to explore promising areas of the investigation space in order to discover more efficient solutions. To produce new solutions, we calculate a probability of choice for each solution in the archive using the following equation (see Eq. (3))

w(l) represents the weight of the Gaussian function l as defined in Eq. (2). Each probability p(l) is represented as a fraction of the wheel, and a random selection mechanism is similarly performed to spin the game wheel. A new number is randomly selected based on the normal distribution defined by the mean (μ_l^i) and standard deviation (σ_l^i) for i in the group "l" [33]:

$$\sigma = \xi \sum\nolimits_{r=1}^{k} \frac{|S_r^i - S_l^i|}{k-1} \quad (4)$$

With ξ, a fixed parameter comparable to the evaporation rate of the pheromone.

3 Result

In this paper we use the continuous domain ant colony algorithm (ACOR) to calculate the unknown parameters (Iph, Is, Rs, Rsh, A), the data used comes from the SY-M80W photovoltaic panel under the conditions 1002 W/m2, 30 °C Table 1.

As shown in Table 3, the parameters of the ant colony optimization algorithm (ACOR) were specified, while the search limits for each parameter are detailed in Table 2.

Real-time identification consists of two phases: the first phase involves collecting data from the solar panel. Subsequently, the identification of the parameters is done through the continuous domain ant colony algorithm (ACOR).

Phase 1: Fig. 3 illustrates the experimental setup of the virtual system. The microcontroller of the Arduino board evaluates the voltage and current intensity coming out of the photovoltaic panel using sensors. After connecting the Arduino board to the computer using a USB cable, the Matlab program is started. The "MATLAB Support Package

Table 1. Characteristics of the PV Module SY-M80W

Parameter	Value
Open-circuit voltage (V_{oc})	22 V
Short circuit current (I_{sc})	4.85 A
Voltage at Pmax (V_{mp})	17.4 V
Current at Pmax (I_{mp})	4.61 A

Table 2. Range for SY-M80W panel parameters

Parameter	I_{ph} (A)	I_o (μA)	R_s (Ω)	R_{sh} (Ω)	a
Lower Bound	0	0	0	0	0
Upper Bound	6	1	1	3000	2

Table 3. The algorithm parameter

Iteration	1000
Ant (the number of ants)	100
Q	0.3
Z	1

Table 4. Simulation Results

Parameter	**Value**
I_{ph} (A)	4.572518
I_o	0.000465433
R_s (Ω)	0.128180
R_{sh} (Ω)	2765.604192
a	1.351408
RMSE	0.0080065

for Arduino Hardware" was used to establish the connection between MATLAB and Arduino, based on the code presented previously.

As shown in Fig. 2, the output data will be collected and presented live on Matlab.

Phase 2: The parameters of the PV panel were identified by directly analyzing the I-V and P-V characteristics obtained from our experimental data. As mentioned earlier in the PV Panel Model section, the PV panel model serves as a reference to verify the efficiency and accuracy of the suggested virtual instrument. We selected a group of 100 people for testing the algorithm, as shown in Table 3. Table 4 illustrates the results obtained after 1000 iterations.

The five-parameter model is summarized in Table 4, which presents the results obtained. The results attest the effectiveness of the suggested method to accurately determine the parameters of the equivalent circuit model. The root mean square error is quite low, which also suggests better accuracy among the tests.

Figure 5 illustrates the (i-v/p-v) curves collected live from the photovoltaic module. After the acor provided the optimal solution, it was implemented to reconstruct the i-v and p-v curves of the photovoltaic panel. Subsequently, the reconstructed curves were compared with those derived from the experimental data.

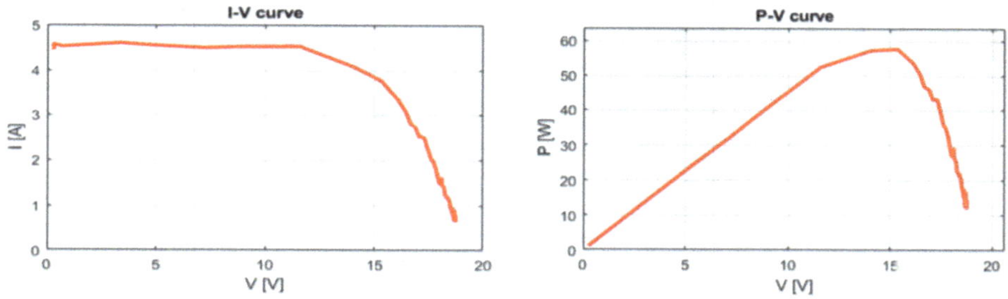

Fig. 5. I–V and P–V characteristics for reel time PV panel instrumentation

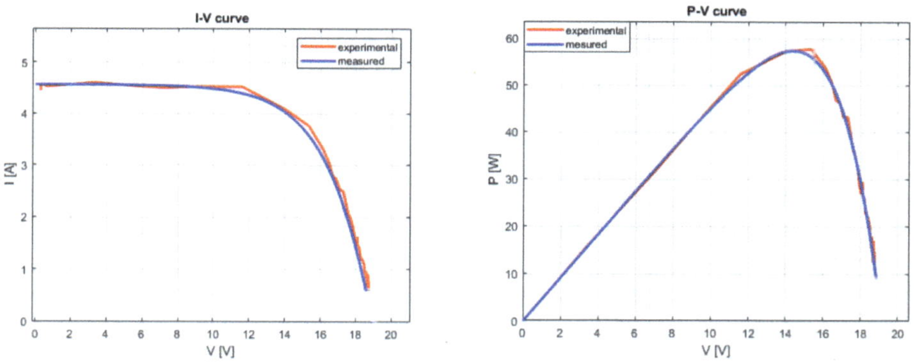

(a) (b)

Fig. 6. (a) I-V and (b) P-V characteristics Comparison results of single diode model with the real data

Figure 6 (a) illustrate the I-V and P-V curves of the single diode model (SDM) using the ACOR algorithm.

From the collected data, it is evident that the observed I/V characteristics are almost identical to the experimental I-V characteristics. The same is true for the p-v curve in Fig. 6 (b).

From the experimental data illustrated in Fig. 6 (b), it is observed that the solar panel has reached a maximum power of 57.7 W when a voltage of 15.34 V and a current of 3.764 A appeared.

As shown in Fig. 7, the rapid convergence is evident. During the optimization process, the objective function is minimized according to the range of parameters.

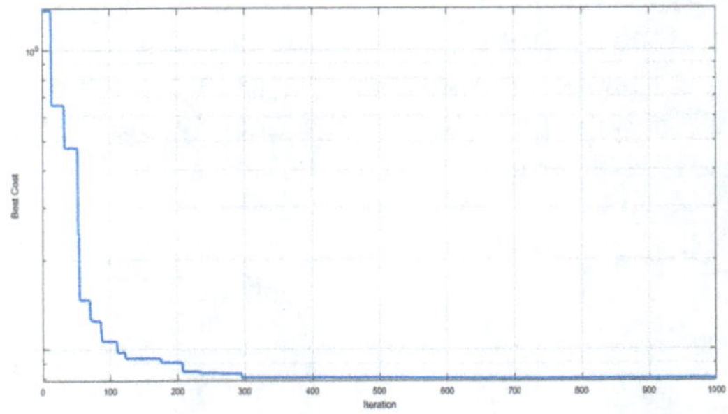

Fig. 7. Convergence graph

4 Discussion

This work aims to introduce an innovative method for identifying photovoltaic parameters in real time that is both simple and economical.

Real-time identification of solar panel parameters was performed using continuous domain ant colony optimization (ACOR). The test was conducted at the Faculty of Science and Technology of Tlemcen. A MATLAB environment was designed to determine the panel parameters in real time.

The panel analyzed in this study is the SY-M80W. The system is developed on a low-cost Arduino acquisition board. Data collection is performed using low-cost current sensors “ACS 712” and voltage “F031–06”. The information is displayed in Matlab, facilitating the interaction between the ATMega328 microcontroller of an Arduino UNO board and the Matlab program. Thus, the I-V and P-V characteristics can be managed in real time on Matlab without the need to reprogram the microcontroller. The ACOR algorithm simultaneously performs an analysis of the collected data.

The aim is to optimize the accuracy in determining the parameters of photovoltaic panels. We apply a meta-heuristic, “the continuous ant colony optimization algorithm”, in order to minimize the objective function of the reference model. It is considered as a population meta-heuristic, where each solution is symbolized by an ant that evolves in the search space. Ants use pheromones it signal optimal solutions and take into account these previous markings to refine their searches.

They can be considered as probabilistic multi-agent algorithms, which use an implied probability distribution to perform the transition. It is an iterative algorithm designed to reduce the quadratic deviation between the observed current and that determined from Eq. 1 of the photovoltaic model used. This research proposed a parallelization of the current-voltage and power-voltage characteristics between the data obtained by measurement and those from the experiment. The (I-V) and (P-V) graphs in Figs. 6(a) and (b), obtained through the optimization process, correspond perfectly to the experimental data. The data from Fig. 7 indicate a rapid convergence of the algorithm after 300 iterations.

Our device is particularly beneficial for photovoltaic panels, whose output voltage range is 0 to 25 V and current range is 0 to 5 A. Therefore, to improve our research, we could also integrate additional cost-effective sensors such as temperature sensors (DS18B20 or DHT22) and luminance sensors (BH1750 lux) into our system to monitor temperature and sunlight.

The paper presents a virtual instrumentation system that provides users with the ability to monitor and track the characteristics of a photovoltaic panel in an easier, more economical and faster way. And examine them directly to identify unknown parameters. Unlike traditional measurement devices such as multimeters, which offer low average accuracy, our system, which serves as a virtual instrument for acquiring and determining the characteristics of photovoltaic panels in real conditions, can also be used in frequent on-site control missions of photovoltaic installations.

Our system works in a similar way to a monitoring system, which collects and analyzes data in a short period of time (a few seconds), thus providing real-time identification. It is necessary to collect them and perform an analysis every 5 min.

5 Conclusion

Paper presents research focused on real-time detection of parameters of a SY-M80W mono-diode photovoltaic panel using the continuous ant colony algorithm. A cost-effective virtual instrumentation system based on Arduino and Matlab was deployed to enable the collection, analysis, storage and graphing of the I-V and P-V characteristics of a solar panel under real conditions.

In the suggested approach, data are collected and compared with the measured data. We use a meta-heuristic, namely the continuum-oriented ant colony algorithm (ACOR), in order to optimize the objective function of the reference model. Therefore, hardware components such as a solar panel, an Arduino UNO board, and current and voltage sensors were used. The proposed approach involves collecting and comparing the measured data using a meta-heuristic, namely the continuum-oriented ant colony algorithm (ACOR), in order to optimize the objective function of the reference model.

The goal of the ant colony algorithm is to reduce the function to be minimized, often referred to as the objective function. It is an iterative algorithm designed to minimize the quadratic deviation between the experimental current and that obtained by the PV model used. The results attest to the relevance of the suggested technique. Simulations reveal that the heuristic approach is efficient in determining the photovoltaic parameters. The graphs (I-V and P-V) show a satisfactory correspondence with the experimental results. In addition, the mean square deviation between the theoretical and experimental characteristics is almost zero.

Our goal in this work is to improve and clarify the identification of solar panel parameters. This field of study offers many opportunities to improve the efficiency and profitability of photovoltaic systems. The suggested method mainly offers benefits, such as its ease of use and its reduced need for hardware elements. In addition, this approach simplifies the collection and analysis of the characteristics of photovoltaic systems essential to their performance, making their work easier, less expensive and more expeditious. Our device operates in a similar way to a control system. For future projects, it could be relevant to use this technique to develop a monitoring and diagnostic system for a photovoltaic installation.

References

1. J., C.-G.'o., Montero, J., Moreno, M.A.: I-solar, a real-time photovoltaic simulation model for accurate estimation of generated power. Agronomy. **11**, 485 (2021)
2. Shenouda, R., Abd-Elhady, M.S., Kandil, H.A.: A review of dust accumulation on PV panels in the Mena and the far east regions. J. Eng. Appl. Sci. **69**(1), 8 (2022)
3. Alzahrani, A.M., Zohdy, M.A.: Real-time performance evaluation of photovoltaic (pv) systems in operation. Energy Electr. Eng. **12**(06), 339–347 (2020)
4. Benedek, J., T.-T., S.'e., Bart'ok, B.: Evaluation of renewable energy sources in peripheral areas and rural development based on renewable energies. Rev. Renew. Sustain. Energy. **90**, 516–535 (2018)
5. Samara, S., Natsheh, E.: Intelligent real-time monitoring system for photovoltaic panels using artificial neural networks. IEEE Access. **7**, 50287–50299 (2019)
6. Ben Messaoud, R.: Extraction of uncertain parameters of double diode model of photovoltaic panel using ant lion optimization. SN Appl. Sci. **2**(2), 239 (2020)
7. Mazzotti, W., Firmansyah, H., Acun˜a J, Stokuca, M., & Palm, B. (2018).Newton-raphson method applied to time-overlapped ils for parameter estimation in thermal response tests.
8. S., R.,.M.d.J.d., et al.: Theoretical and experimental evaluation of partially shaded photovoltaic solar arrays through a methodological framework: A case study involving two identical 1.5 kwp photovoltaic generators. J. Sol. Energy Eng. **145**(1), 011010 (2023)
9. Qin, H., Kimball, J.W.: Determination of photovoltaic cell parameters from field test data using particle swarm optimization. In: 2011 IEEE Energy Conversion Congress and Exposition, pp. 1–4. IEEE (2011)
10. Khelil, K., BERREZZEK, F., Khediri, T., et al.: Parameter extraction of photovoltaic panels using a genetic algorithm. In: Third International Conference on Technological Advances in Electrical Engineering (ICTAEE'18) (2018)
11. Sharma, A., Dasgotra, A., Tiwari, S.K., Sharma, A., Jately, V., Azzopardi, B.: Parameter extraction of a photovoltaic module using the tunicate swarm algorithm. Electronics. **10**(8), 878 (2021)
12. Naeijian, M., Rahimnejad, A., Ebrahimi, S.M., Pourmousa, N., Gadsden, S.A.: Parameter estimation of PV solar cells and modules using whippy Harris hawks optimization algorithm. Energy Rep. **7**, 4047–4063 (2021)
13. Oliva, D., Cuevas, E., Pajares, G.: Parameter identification of solar cells using artificial bee colony optimization. Energy. **72**, 93–102 (2014)
14. Martinez-Rios, F., Murillo-Suarez, A.: A multiprocess salp swarm optimization with a heuristic based on crossing partial solutions. Procedia Comput. Sci. **179**, 440–447 (2021)
15. Han, W., Wang, H.-H., Chen, L.: Parameters identification for photovoltaic module based on an improved artificial fish swarm algorithm. Sci. World J. **2014**(1), 859239 (2014)
16. Wu, S., Li, Q., Wei, W.: Application of ant colony optimization algorithm based on triangle inequality principle and partition method strategy in robot path planning. Axioms. **12**(6), 525 (2023)

17. Belarbi, M. (2015). Contribution à l'étude d'un générateur solaire pour site autonome.
18. Nunes, H., Pombo, J., Fermeiro, J., Mariano, S., do Ros'ario Calado, M.: Particle swarm optimization for photovoltaic model identification. In: 2017 International Young Engineers Forum (YEF-ECE), pp. 53–58. IEEE (2017)
19. El Hammoumi, A., Motahhir, S., Chalh, A., El Ghzizal, A., Derouich, A.: Low-cost virtual instrumentation of PV panel characteristics using Excel and Arduino in comparison with traditional instrumentation. Renewables: Wind, water, and solar. **5**(1), 1–16 (2018b)
20. Math Works. Support package for arduino hardware, https://www.mathworks.com/matlabcentral/fileexchange/47522-matlab-support-package-for-arduino-hardware. [Accessed 06 june 2024].
21. Aidov, A. (2008). Modified continuous ant colony algorithm for function optimization..
22. Nwohu, M., Paul, O.O.: Evaluation of economic load dispatch problem in power generating stations by the use of ant colony search algorithms. Int. J. Res. Stud. Electr. Electron. Eng. **3**(1), 20–29 (2017)
23. Colorni, A., Dorigo, M., Maniezzo, V., et al.: Distributed optimization by ant colonies. In: Proceedings of the First European Conference on Artificial Life, vol. 142, pp. 134–142, Paris, France (1991)
24. Puris, A., Bello, R., Trujillo, Y., Nowe, A., & Mart'ınez, Y. (2007). Two-stage aco to solve the job shop scheduling problem. In Progress in Pattern Recognition, Image Analysis and Applications: 12th Iberoamericann Congress on Pattern Recognition, CIARP 2007, Valparaiso, Chile, November 13-16, 2007. Proceedings 12, pp. 447–456. Spring
25. Alaya, I., Solnon, C., Ghédira, K.: Optimisation par colonies de fourmis pour le problème du sac-'a-dos multi-dimensionnel. Revue des Sciences et Technologies de l'Information-Série TSI: Technique et Science Informatiques. **26**, 371–390 (2007)
26. Benyamina, A. (2013). Application des algorithmes de colonies de fourmis pour l'optimisation et la classification des images. Ph. D. thesis, USTO (MB).
27. Monmarché, N.: Algorithmes de fourmis artificielles: applications la classification et l'optimisation. Thèse de doctorat,. Laboratoire d'Informatique, Université de Tours (2000)
28. Socha, K., Dorigo, M.: Ant colony optimization for continuous domains. Eur. J. Oper. Res. **185**(3), 1155–1173 (2008)
29. Liao, T. (2011). Improved ant colony optimization algorithms for continuous and mixed discrete-continuous optimization problems. Rapport d'avancement de recherche, Uni- vérité Libre de Bruxelles, Faculté des Sciences Appliquées, CODE - Computers and Décision Engineering, IRIDIA - Institut de Recherches Interdisciplinaires et de développements en Intelligence Artificielle. Année Académique 2010/2011.
30. Dorigo, M., & Stu¨tzle T. (2019). Ant Colony Optimization: Overview and Recent Advances.
31. Socha, K.: Ant Colony Optimisation for Continuous and Mixed-Variable Domains. Citeseer (2009)
32. Dréo, J., Siarry, P.: Continuous interacting ant colony algorithm based on dense heterarchy. Futur. Gener. Comput. Syst. **20**(5), 841–856 (2004)
33. Dréo, J., Siarry, P.: An ant colony algorithm aimed at dynamic continuous optimization. Appl. Math. Comput. **181**(1), 457–467 (2006)

Enhanced Voltage Regulation of DC-DC Boost Converters Using IWO-Optimized Fractional Order Fuzzy PID Control

Ahmed Bennaoui[1(✉)], Aissa Ameur[2], Ameur Bennaoui[3], and Salah Benzian[4]

[1] Institute of Sciences, University Center Aflou El Cherif Bouchoucha, Laghouat, Aflou 03001, Algeria
a.bennaoui@cu-aflou.edu.dz

[2] University of Amar Telidji, Laghouat 03000, Algeria
a.ameur@lagh-univ.dz

[3] Computer Sciences Department, University of Constantine 2, Constantine, Algeria
ameur.bennaoui@univ-constantine2.dz

[4] University Center Aflou El Cherif Bouchoucha, Laghouat 03000, Algeria
s.benzian@cu-aflou.edu.dz

Abstract. Tuning complex controllers like the Fractional Order Fuzzy PID (FOFPID) for power electronic systems such as DC-DC boost converters remains a challenge. This paper contributes an effective tuning methodology based on the Invasive Weed Optimization (IWO) algorithm. We utilize IWO's robust search capabilities, mimicking natural weed colonization, to simultaneously optimize the six key parameters (Ke, Kce, KPI, KPD, λ, μ) of an FOFPID controller, aiming to minimize the voltage regulation error (ISE). The efficacy of the proposed IWO-FOFPID approach is validated through simulations and comparison with established optimization techniques (Moth Flame Optimization - MFO, Particle Swarm Optimization - PSO) applied to the same FOFPID structure, and also against Fuzzy PID (FPID) control. The IWO-based controller demonstrates superior performance, yielding the lowest ISE and fastest rise time, highlighting the potential of IWO for high-performance fractional-order fuzzy control system design.

Keywords: DC-DC Boost Converter · Fractional Order Fuzzy PID (FOFPID) · Optimal Control · Invasive Weed Optimization (IWO)

1 Introduction

DC-DC boost converters are fundamental components in modern power electronics, widely employed in applications ranging from renewable energy systems to portable devices, necessitating precise output voltage regulation for stable and efficient operation. While classical Proportional-Integral-Derivative (PID) controllers are commonly used, their performance can be limited when dealing with the inherent non-linearities and parameter variations of power converters [1, 2].

M. Hatti (Ed.): IC-AIRES 2025, LNNS 1740, pp. 418–432, 2026.
https://doi.org/10.1007/978-3-032-12788-4_40

The fractional-order $PI^{\lambda}D^{\mu}$ controller (FOC) proposed in [3–5] as a generalization of the PID controller with integrator of real order λ and differentiator of real order μ.

The $PI^{\lambda}D^{\mu}$ controller is more flexible and gives an opportunity to better adjust the dynamical properties of the fractional-order control system.

Further enhancements have been achieved by integrating fuzzy logic control (FLC) principles with FOPID structures, resulting in Fractional Order Fuzzy PID (FOFPID) controllers [3–7] These controllers leverage the ability of fuzzy logic to handle system uncertainties and non-linearities effectively, offering additional adaptability [3–7]. Various FOFPID structures exist, combining the elements in different configurations to meet specific control objectives.

The hybrid controllers, as illustrated in Figs. 1, 2, 3, 4 and 5, combine fractional-order fuzzy PI, PD, P, and PID controllers, allowing for intricate adjustments in various control scenarios. These configurations leverage the adaptability of fractional-order components and the robustness of fuzzy logic

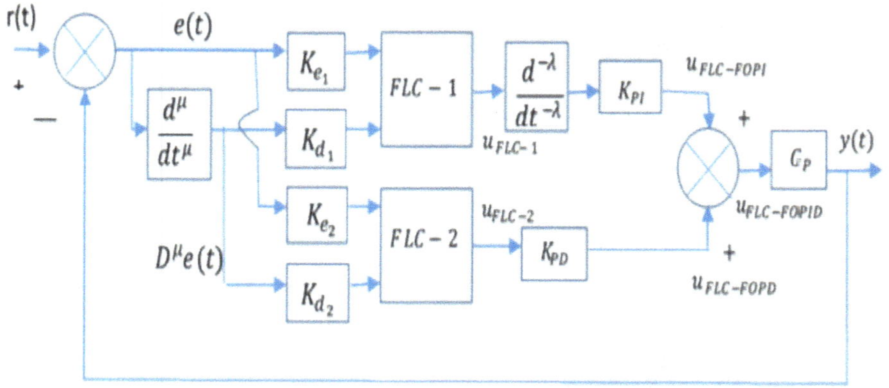

Fig. 1. Schematic diagram of FO fuzzy PI+PD controller

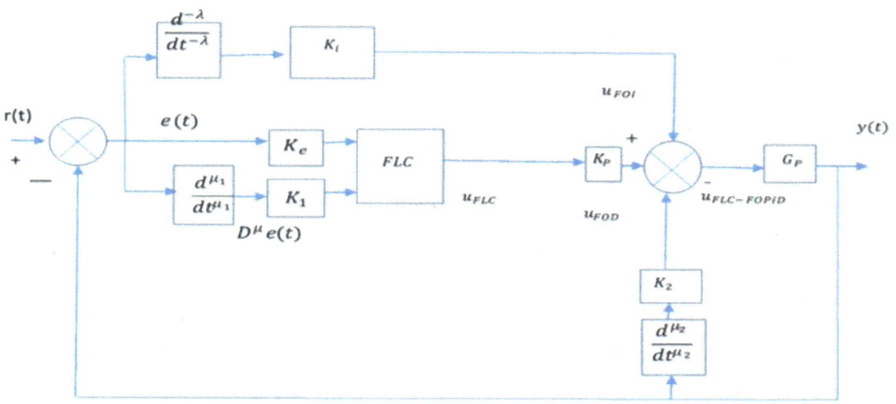

Fig. 2. Schematic diagram of FO fuzzy P+ID controller

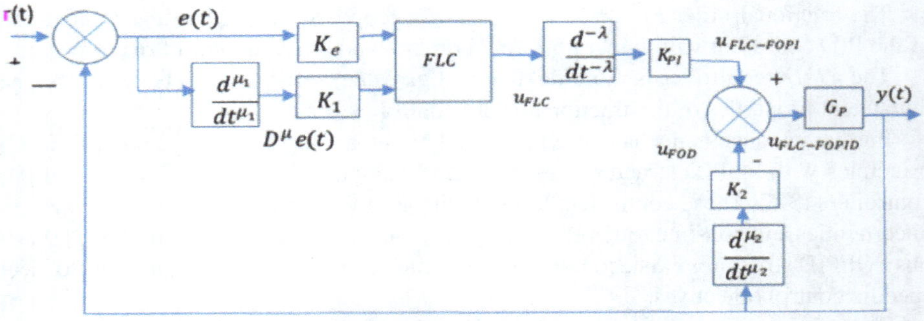

Fig. 3. Schematic diagram of FO fuzzy PI+D controller.

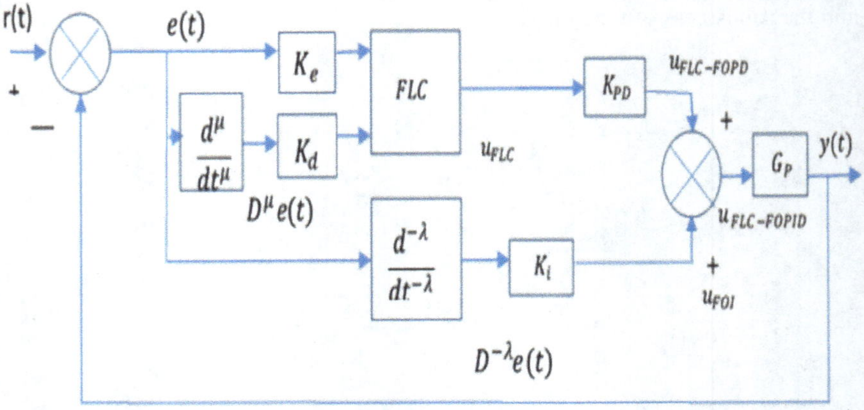

Fig. 4. Schematic diagram of FO fuzzy PD+I controller

Despite their potential advantages, the increased number of parameters in FOFPID controllers (including scaling factors, gains, and the fractional orders λ and μ) makes their optimal tuning a complex challenge. Traditional tuning methods often struggle to find the global optimum in this high-dimensional search space. Consequently, various metaheuristic optimization algorithms have been explored to systematically determine the optimal FOFPID parameters.

Various metaheuristic algorithms have been applied to optimize Fractional Order Fuzzy PID (FOFPID) controllers. Genetic Algorithms (GA), for instance, were used to tune controllers for performance evaluation in [8] and, in a real-coded variant, for reactor power control in [9]. Other methods explored include the Fruit Fly Optimization Algorithm (FOA), proposed specifically for tuning FOFPID parameters [10], and the Cuckoo Search Algorithm (CSA), employed to optimize an FOFPID controller for a robotic manipulator [6].

Recently, the Invasive Weed Optimization (IWO) algorithm [11] has emerged as a noteworthy metaheuristic technique. Inspired by nature and proposed by Mehrabian and Lucas, IWO has demonstrated considerable effectiveness in tackling complex optimization challenges [12, 13]. The algorithm emulates the colonization strategy of invasive

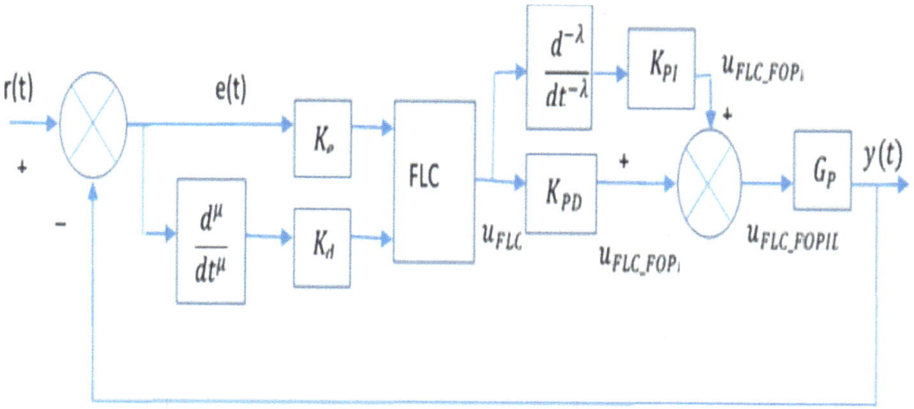

Fig. 5. Schematic diagram of FO fuzzy PID controller

weeds, utilizing mechanisms for reproduction, spatial dispersal, and competitive exclusion to efficiently explore vast search spaces and converge towards optimal solutions. Key strengths, including its inherent robustness and well-regarded balance between exploration and exploitation, position IWO as a highly promising candidate for demanding applications like controller tuning [12, 13].

Leveraging these strengths, this paper proposes employing the IWO algorithm for the optimal tuning of Fractional Order Fuzzy PID (FOFPID) controller parameters. The specific application is a DC-DC boost converter, assumed to be operating in continuous conduction mode [15–19]. The primary objective is to enhance the converter's voltage regulation performance by utilizing IWO's optimization capability to identify the FOFPID parameter set (Ke, Kce, KPI, KPD, λ, μ) that minimizes the Integral of Squared Error (ISE). To assess its efficacy, the performance of the proposed IWO-tuned FOFPID controller will be evaluated via simulation. This evaluation includes comparison against both an IWO-tuned conventional Fuzzy PID (FPID) controller and previously documented results obtained using Moth Flame Optimization (MFO) and Particle Swarm Optimization (PSO) for FOFPID tuning [14].

This paper is structured as follows: Section 2 provides a detailed explanation of the Invasive Weed Optimization algorithm. Section 3 presents the mathematical model adopted for the DC-DC boost converter. The structure of the FOFPID controller and the specifics of the IWO-based optimization setup are described in Sect. 4. Simulation results are then presented and discussed in Section V. Finally, Sect. 5 offers concluding remarks on the findings.

2 Invasive Weed Optimization Algorithm

The Invasive Weed Optimization (IWO) algorithm, proposed by Mehrabian and Lucas [11–13], simulates the colonization behavior of weeds to solve optimization problems. It is characterized by its robustness, adaptability, and efficient search capabilities. The fundamental steps of the IWO algorithm are as follows (often represented visually in a flowchart like your Fig. 6):

1. Initialization:

   - A finite number of initial "seeds" (representing candidate solutions) are randomly distributed across the multi-dimensional search space (problem domain). Each seed's position corresponds to a set of parameter values for the problem being optimized (in this case, the FOFPID parameters).

2. Reproduction:

   - Each seed grows into a mature, flowering "weed" (a potential solution).
   - Each weed produces a number of new seeds based on its relative fitness within the population (how well it solves the optimization problem, e.g., how low its ISE value is).
   - The number of seeds produced by a weed is typically calculated to vary linearly (or sometimes non-linearly) between a maximum number (S_{max}) for the fittest weed in the colony and a minimum number (S_{min}) for the least fit weed. This ensures that better solutions contribute more offspring to the next generation.

3. Spatial Distribution:

   - The seeds generated by each parent weed are distributed randomly in the search space *around* the parent weed's location.
   - This distribution typically follows a normal (Gaussian) distribution, with the parent's location as the mean.
   - Crucially, the standard deviation (SD) of this normal distribution is adaptive, typically decreasing over the algorithm's iterations (generations). It usually starts with a larger value (SDinitial) and reduces non-linearly towards a smaller final value (SDfinal) as the iterations progress. This dynamic adjustment is often achieved using a formula like:

$$SD(t) = \frac{(T_{max} - t)^n}{(T_{max})^n}(SD_{init} - SD_{final}) + SD_{final}$$

where:

- SD(t) is the standard deviation for seed distribution at the current iteration t.
- t is the current iteration number.
- Tmax is the maximum number of iterations planned for the algorithm run.
- n is the non-linear modulation index (often a value like 2 or 3), which controls the curve of the SD reduction.
- SDinitial is the initial standard deviation value used in the early iterations.
- SDfinal is the final standard deviation value approached in the late iterations.

- This decreasing standard deviation mechanism is key to balancing the search: a large SD in early iterations promotes global exploration, while a smaller SD in later iterations facilitates local exploitation and refinement around promising solutions.
- This decreasing standard deviation mechanism is key to balancing the search: large SD in early iterations promotes global exploration, while smaller SD in later iterations facilitates local exploitation and refinement around good solutions.

4. Competitive Elimination (or Exclusion):
   - After the new seeds are generated and potentially grow into new weeds, the population size increases.
   - If the total number of weeds (parents + offspring) exceeds a predefined maximum population size (Pmax), a competitive mechanism is triggered.
   - All weeds in the colony are ranked based on their fitness values.
   - Weeds with lower fitness (poorer solutions) are eliminated until the population size returns to the maximum limit (Pmax). This mimics the survival-of-the-fittest principle, ensuring the population focuses on promising areas of the search space.
5. Termination:

- The process of Reproduction, Spatial Distribution, and Competitive Elimination is repeated for a fixed maximum number of iterations or until some other stopping criterion (e.g., no significant improvement in the best solution) is met.
- The weed with the best fitness found throughout the entire process represents the optimal (or near-optimal) solution to the optimization problem.

Advantages Highlighted:

- Robustness and Adaptability: Inspired by resilient weeds.
- Good Search Balance: The adaptive standard deviation naturally balances global exploration and local exploitation.
- Simplicity: Relatively easy to understand and implement compared to some other evolutionary algorithms (e.g., doesn't require complex crossover or mutation operators like Genetic Algorithms).
- Convergence: Often shows good convergence towards optimal solutions for complex problems.

3 The DC-DC Boost Converters

A power converter exhibits periodic, non-linear, and time-varying characteristics due to changes in its circuit configuration depending on the states of switches and blocked diodes. Typically, these circuits are under the control of techniques such as Pulse Width Modulation (PWM) or other similar methods, which are used to regulate the voltage or current supplied to the loads. The most popular and widely adopted DC-DC converters are Buck, Boost and Buck-Boost converters. Buck converters lower the output voltage relative to the input voltage, while Boost converters raise it.. In the process of designing these transformers, their dynamic behavior should be analyzed and appropriate guidelines should be put in place to ensure satisfactory performance. However, the modeling process presents challenges due to the nonlinear nature of these transformers and the multiplicity of electrical phenomena that arise during the conversion processes. The dynamic behavioral model should be able to describe the operation of the DC-DC converter in continuous and non-continuous modes. This study specifically focuses on the Boost type DC-DC converter, based on its diverse range of applications. Fig. 7.

Middlebrook and Cuk made advancements in this field by utilizing the state-space averaging technique [15–19]. The averaged model is derived by averaging two or three

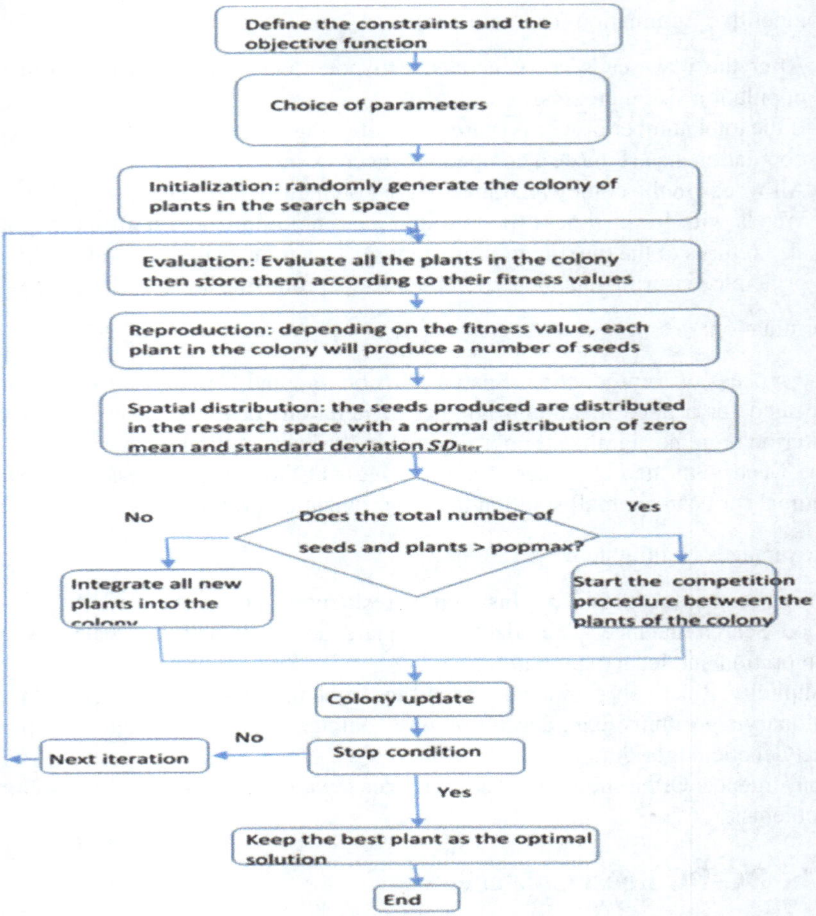

Fig. 6. Invasive weed optimization algorithm flowchart

state equations, contingent on the operational mode of the converter (either CCM or DCM), over a single switching period. Each circuit configuration of a converter is depicted through a linear continuous-time model. The mathematical formulations of continuous dynamics are deduced through the application of Kirchhoff's laws, which elucidate the converter's behavior within each configuration. The state vector of the distinct models comprises the current flowing through the inductors, the voltage across the capacitors, as well as the input of the models. This input is characterized by the supply voltage and the controlling parameter corresponding to the duty cycle. In the context of continuous conduction mode (CCM), determined by the position of the switch (*sw*), two circuit states emerge, aligned with the closed and open positions. These states are graphically represented in the state space by the ensuing dynamics.

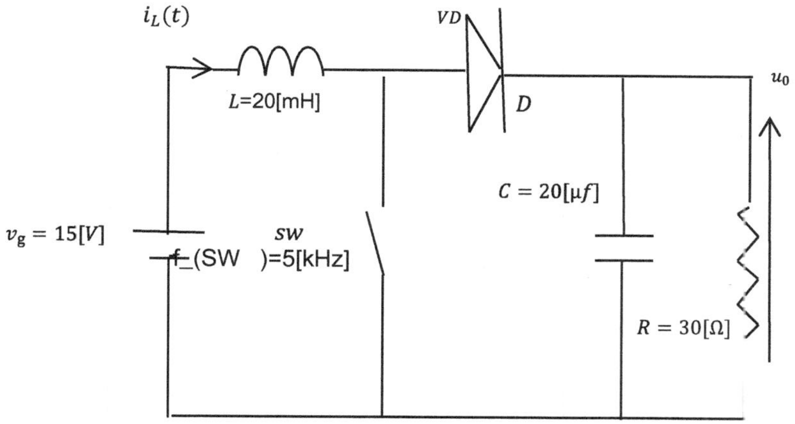

Fig. 7. The boost DC-DC converter

The state equation of the boost DC-DC converter can be stated as [15]:

$$\dot{x} = A_i x + B_i V_g(t)$$

$$u_0 = C_i x$$

$$\text{with } x = [i_L v_c]^T \tag{1}$$

A_i, B_i and C_i are system Matrices of the constituent linear circuits.

The system matrices correspond to distinct operational states (with subscript 1 denoting transistor ON and subscript 2 denoting transistor OFF) of the converter circuit:

$$C_1 = [0 \quad 1], C_2 = [0 \quad 1]$$

$$A_1 = \begin{bmatrix} 0 & 0 \\ 0 & -\frac{1}{RC} \end{bmatrix}, \text{and } A_2 = \begin{bmatrix} 0 & -\frac{1}{L} \\ \frac{1}{C} & -\frac{1}{RC} \end{bmatrix}$$

$$B_1 = B_2 = \begin{bmatrix} \frac{1}{L} & 0 \end{bmatrix}^T \tag{2}$$

The state-space domain averaging technique proves extremely beneficial for evaluating the performance of switching circuits [16–19] under low-frequency, low-signal conditions. This approach applies when the switching period of the converter is considerably shorter than the response time of the output voltage. Through the use of the averaging method in the domain of the state space, it is possible to obtain the equation of state in the following way [16–19]:

$$\dot{x} = Ax + BV_g(t)$$

$$u_0 = Cx \tag{3}$$

Where

$$A = dA_1 + (1 - d)A_2, B = d B_1 + (1 - d)B_2,$$

$$C = dC_1 + (1 - d)C_2, \text{ and } d \text{ is the duty cycle} \tag{4}$$

4 Simulation And Results

In this part, we propose structure of the Fractional Order Fuzzy PID Controller As shown in Fig. 9.

This section details the design of the Fractional Order Fuzzy PID (FOFPID) controller employed in this study and the methodology for its optimization.

The specific structure of the FOFPID controller implemented for the DC-DC boost converter is depicted in Fig. 8. The system operates with a reference voltage set at Vref = 37.5 V.

The fuzzy logic component of the controller takes two inputs: the error signal, e = Vref−Vout, and the change in error,. Following the established methodology in [14], the membership functions for these inputs (illustrated in Fig. 9), the output membership functions, and the fuzzy rule base were adopted from this reference work.

The primary focus of this work is the optimization of the FOFPID controller's parameters using the Invasive Weed Optimization (IWO) algorithm. The six parameters targeted for optimization are the input scaling factors (k_e; k_{ce}), the output scaling factors (k_{pd}; k_{pi}), and the fractional integrator and differentiator orders (μ; λ).

The objective of the optimization is to minimize the voltage regulation error. Therefore, the Integral of Squared Error (ISE) is utilized as the fitness function for the IWO algorithm. The fitness function is defined as follows:

$$\text{The Fitness Function} = \int_0^{+\infty} [e(t)]^2 dt$$

The two input fuzzy sets: The error (e) and the change of error of the fuzzy controller are illustrated in Fig. 9.

The rules and output membership functions are illustrated in Table 1

5 Discussion of the Results (Based on the Provided Tables/Figures Descriptions)

The simulation results provide strong evidence supporting the effectiveness of the proposed Invasive Weed Optimization (IWO) approach for tuning the Fractional Order Fuzzy PID (FOFPID) controller.

The convergence characteristics of the IWO algorithm are illustrated in Figs. 10 and 11. These plots demonstrate the successful minimization of the Integral Squared Error

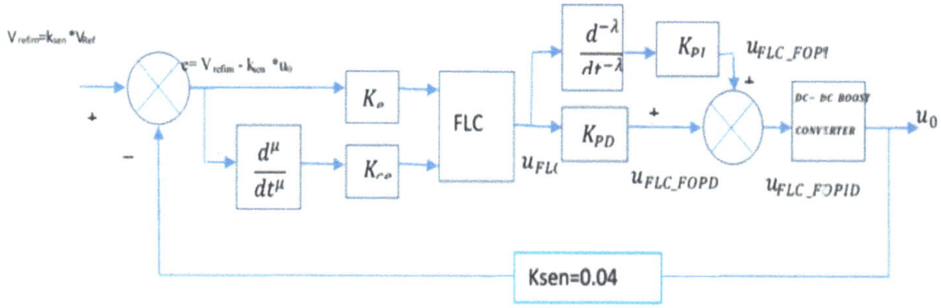

Fig. 8. Structure of FOFPID Controller for A DC-DC Boost Converter

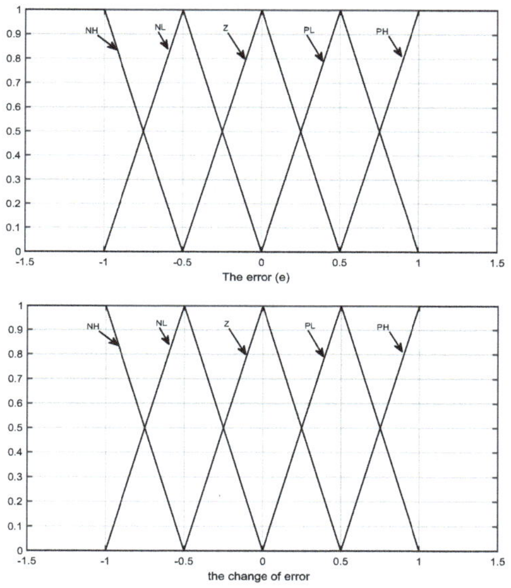

Fig. 9. The two input fuzzy sets: The error (e) and the change of error

(ISE) fitness function over the iterations for both the FOFPID and the integer-order Fuzzy PID (FPID) controllers. This visually confirms that IWO consistently navigated the complex parameter space to find stable, optimized solutions for both controller types.

Table 2 details the optimal parameter sets identified by the IWO algorithm. These values can be directly compared with those reported for Moth Flame Optimization (MFO) and Particle Swarm Optimization (PSO) applied to the same FOFPID structure [14], as well as with parameters for an IWO-tuned FPID and a literature-based FPID [14]. The IWO algorithm yielded a distinct set of parameters, including specific fractional orders (λ and μ), underpinning the performance improvements detailed below.

The practical impact of these optimized parameters on system performance is shown in the dynamic response plots. Figure 12 presents the output voltage regulation achieved by the different controllers, allowing qualitative assessment of transient behaviors like

Table 1. Rules and Output Membership Functions

e \ $\frac{d^\mu e}{dt^\mu}$	NH	NL	Z	PH	PH
NH	**-1**	**-0,81**	**-0,49**	**-0,36**	**-0,25**
NL	**-0,64**	**-0,36**	**-0,16**	**-0,04**	**0**
Z	**-0,16**	**-0,04**	**0**	**0,04**	**0,16**
PL	**0**	**0,04**	**0,16**	**0,36**	**0,64**
PH	**0,25**	**0,36**	**0,49**	**0,81**	**1**

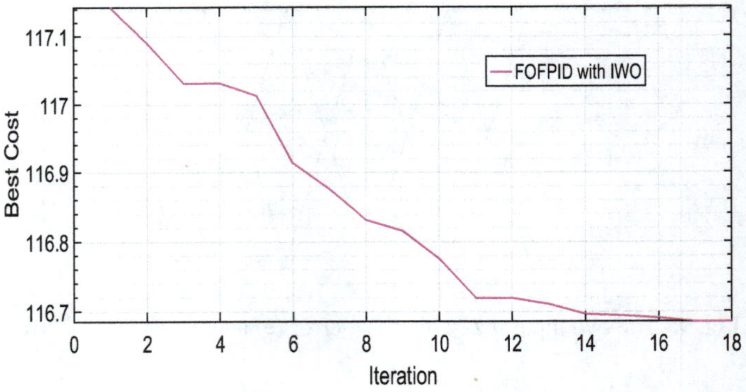

Fig. 10. Iterative convergence curve (FOFPID with IWO)

rise time and settling characteristics. Figure 13 shows the corresponding duty cycle signals, indicating the control effort required by each approach.

Table 3 provides the quantitative performance metrics, offering definitive insights into the controllers' efficacy. The IWO-tuned FOFPID controller achieved the lowest ISE value and the fastest rise time. Critically, this represents an improvement over the performance reported for the same FOFPID structure tuned via MFO (ISE) and PSO (ISE) [14], alongside a slightly reduced rise time. This result quantitatively substantiates the superior tuning capability of IWO compared to the benchmark metaheuristics for this specific FOFPID application.

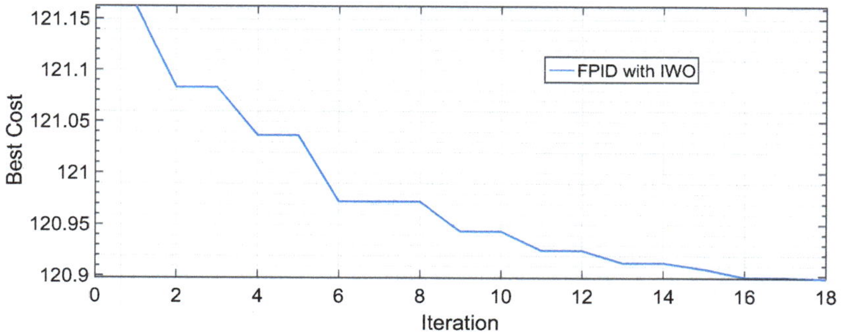

Fig. 11. Iterative convergence curve (FPID with IWO)

Table 2. Optimal controllers parameters

	Ke	Kce	K_{PI}	K_{PD}	λ	μ
FOFPID with IWO	0.50	9.00	0.590	311.1918	0.960	1
FOFPID with MFO [14]	0.501	9.0005	0.59	320	0.970	1.00
FOFPID with PSO [14]	0.501	9.001	0.59	320	0.973	1.00
FPID with IWO	0.5	9	0.59	360.4748	1	1
FPID [14]	0.5	9	0.25	255	1	1

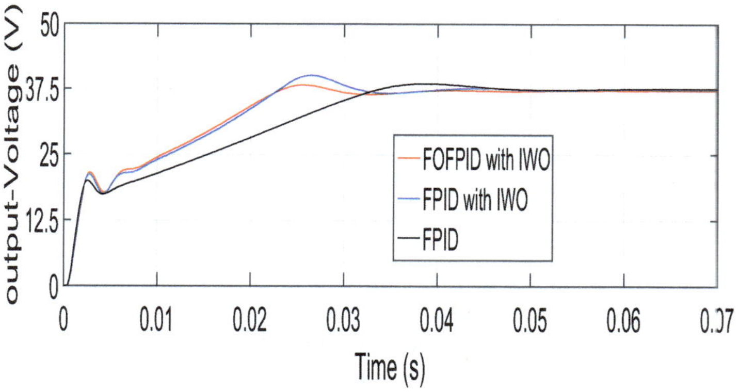

6

Fig. 12. Output-Voltage (V)

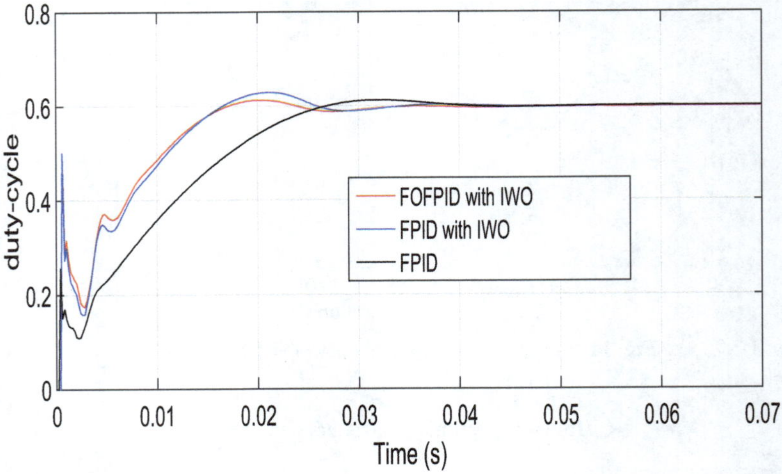

Fig. 13. Duty-cycle

Table 3. Performance *comparison*

	ISE	rise time(s)
FOFPID with IWO	0.1167	0.0214
FOFPID with MFO [14]	0.1180	0.0216
FOFPID with PSO [14]	0.1190	0.0218
FPID with IWO	0.1209	0.0216
FPID [14]	0.158	0.030

6 Conclusion

This study successfully implemented and validated an Invasive Weed Optimization (IWO) based methodology for the optimal tuning of a Fractional Order Fuzzy PID (FOFPID) controller regulating a DC-DC boost converter. By targeting the minimization of the Integral of Squared Error (ISE), the IWO algorithm effectively determined superior values for the six critical FOFPID parameters (Ke, Kce, KPI, KPD, λ, μ). The performance evaluation, based on simulation, compellingly demonstrated the advantages of the IWO-FOFPID controller, which exhibited the lowest ISE and fastest rise time among all tested configurations, including FOFPID tuned by MFO/PSO and various FPID controllers. These results not only underscore the capability of IWO to handle complex, multi-parameter optimization tasks in control engineering but also reaffirm the potential of well-tuned FOFPID structures for high-performance power electronics applications. While future work could involve experimental validation and robustness analysis, this research strongly indicates that IWO provides a powerful pathway to achieving enhanced performance with advanced fuzzy fractional-order controllers.

References

1. Sira-Ramirez, H., Silva-Ortigoza, R.: Control Design Techniques in Power Electronics Devices. Springer (2006)
2. Forouzesh, M., Siwakoti, Y.P., Gorji, S.A., Blaabjerg, F., Lehman, B.: Step-Up DC–DC Converters: a comprehensive review of topology derivation, modeling, control, and implementation. IEEE Trans. Power Electron. **32**(12), 9143–9178 (2017)
3. Peng, H.-P., Fu, J.-T., Kaynak, O.: Optimal design of a fractional order PID controller for a pneumatic position servo system. IEEE Trans. Ind. Electron. **66**(8), 6220–6229 (2018). https://doi.org/10.1109/TIE.2018.2870412
4. Zhang, F., Yang, C., Zhao, X., Gao, W.: Fractional-order PID controller tuning using continuous state transition algorithm. Neural Comput. & Applic. **29**(10), 795–804 (2018)
5. Kumar, V., Rana, K.P.S., Kumar, J., Mishra, P., Nair, S.S.: A robust fractional order fuzzy P + fuzzy I + fuzzy D controller for nonlinear and uncertain system. Int. J. Autom. Comput. **14**(4), 474–488 (2017)
6. Sharma, R., Rana, K.P.S., Kumar, V.: Performance analysis of fractional order fuzzy PID controllers applied to a robotic manipulator. Expert Syst. Appl. **41**(9), 4274–4289 (2014)
7. Jegatheesh, A., Agees, K.C.: Novel fuzzy fractional order PID controller for non linear interacting coupled spherical tank system for level process. Microprocess. Microsyst. **72**, 102948 (2020)
8. Das, S., Pan, I., Das, S.: Performance comparison of optimal fractional order hybrid fuzzy pid controllers for handling oscillatory fractional order processes with dead time. ISA Trans. **52**(4), 550–566 (2013)
9. Das, S., Pan, I., Das, S.: Fractional order fuzzy control of nuclear reactor power with thermal-hydraulic effects in the presence of random network induced delay and sensor noise having long range dependence. Energy Convers. Manag. **68**, 200–218 (2013)
10. Sheng, W., Bao, Y.: Fruit fly optimization algorithm based fractional order fuzzy-pid controller for electronic throttle. Nonlinear Dynamics. **73**(1–2), 611–619 (2013)
11. Mehrabian, A.R., Lucas, C.: A novel numerical optimization algorithm inspired from weed colonization. Eco. Inform. **1**(4), 355–366 (2006). https://doi.org/10.1016/j.ecoinf.2006.07.003
12. Kumar, D., Gandhi, B.G.R., Bhattacharjya, R.K.: Introduction to invasive weed optimization method. In: Bennis, F., Bhattacharjya, R. (eds.) Nature-Inspired Methods for Metaheuristics Optimization. Modeling and Optimization in Science and Technologies, vol. 16. Springer, Cham (2020). https://doi.org/10.1007/978-3-030-26458-1_12
13. Bennaoui, A., Slami, S., Aissa, A.: Invasive weed optimization Algorithm for tuning transitioning from Type-1 to Interval Type-2 fuzzy logic controller for boost DC-DC converters. Journal Européen des Systèmes Automatisés. **53**(2), 195–202 (2020)
14. Bennaoui, A., Saadi, S., Ameur, A.: Performance comparison of MFO and PSO for optimal tuning the fractional order fuzzy PID controller for a DC-DC boost converter. In: 2020 International Conference on Electrical Engineering (ICEE), Istanbul, Turkey, pp. 1–5 (2020). https://doi.org/10.1109/ICEE49691.2020.9249778
15. Vidal-Ldiarte, E., Martinez-Salamero, L., Guinjoan, F., Calvente, J., Gomariz, S.: Sliding and fuzzy control of a boost converter using an 8-bit microcontroller. IEE Proc. Electr. Power Appl. **151**(1), 5–11 (2004). https://doi.org/10.1049/ip-epa:20040086
16. Middlebrook, R.D., Cuk, S.: Advances in Switched-Mode Power Conversion. TESLAco, 2006. (Note: This likely refers to a later compilation or edition; original key works by Middlebrook & Cuk on state-space averaging date to the 1970s).
17. Vorperian, V.: Simplified analysis of PWM converters using model of PWM switch part I: continuous conduction mode. IEEE Trans. Aerosp. Electron. Syst. **26**(3), 490–496 (1990). https://doi.org/10.1109/7.106126

18. Lin, P.Z., Hsu, C.F., Lee, T.T.: Type-2 fuzzy logic controller design for buck DC-DC converters. In: The 14th IEEE International Conference on Fuzzy Systems (FUZZY 2005), Reno, NV, USA, pp. 211–216 (2005). https://doi.org/10.1109/FUZZY.2005.1452421
19. Krein, P.T., Bentsman, J., Bass, R.M., Lesieutre, B.L.: On the use of averaging for the analysis of power electronic systems. IEEE Trans. Power Electron. **5**(2), 182–190 (1990). https://doi.org/10.1109/63.53155

PSO-Optimized High-Gain Observer-Based Backstepping Control for TRMS Trajectory Tracking

Azeddine Beloufa[1](✉), Souaad Tahraoui[2], Abderrahmane Kacimi[3], and Rachid Taleb[4]

[1] Electronic Departement, Laboratory of Electrical Engineering and Renewable Energies (LGEER), Hassiba Benbouali University, Chlef, Algeria
a.beloufa@univ-chlef.dz

[2] Electronic Departement, Signals, Systems and Artificial Intelligence Laboratory 2SAIL, Hassiba Benbouali University, Chlef, Algeria

[3] Department of Instrumentation Maintenance, Institute of Maintenance and Industrial Safety, Oran, Algeria

[4] Electrical Departement, Laboratory of Electrical Engineering and Renewable Energies (LGEER), Hassiba Benbouali University, Chlef, Algeria

Abstract. Guiding a Twin Rotor MIMO System (TRMS) to accurately follow desired trajectories presents a considerable challenge, particularly when it comes to fine-tuning observer-based backstepping controllers manually. This work introduces Particle Swarm Optimization (PSO), an intelligent search algorithm, to overcome this hurdle. Basically, PSO takes over the tough job of figuring out the best gain settings for both our High-Gain Observer and the Backstepping controller parts. This gets us past the usual headaches and guesswork of trying to tune these by hand. And the payoff is clear: the TRMS follows its paths much more accurately and flies with noticeably better stability. We've seen this work out in both computer simulations and when we ran it on the actual TRMS hardware – the tracking errors were consistently smaller than what we got with the old manual tuning. So, this really shows PSO is a solid way to get complex control systems working well, and it could be a big help in other tricky areas too, like renewable energy systems.

Keywords: Particle Swarm Optimization (PSO) · Backstepping Control · High-Gain Observer (HGO) · Twin Rotor MIMO System (TRMS) · Trajectory Tracking · Nonlinear Control

1 Introduction

Getting complex machines like the Twin Rotor MIMO System (TRMS) to follow an exact path is tough; this lab-sized helicopter model is tricky because its movements are linked and it doesn't follow simple rules, making control hard, especially if we don't know all its current states, like its speed. Our approach uses a Backstepping controller, good for such systems [1–3], with a High-Gain Observer (HGO) to estimate those missing states

M. Hatti (Ed.): IC-AIRES 2025, LNNS 1740, pp. 433–440, 2026.
https://doi.org/10.1007/978-3-032-12788-4_41

from available, often noisy, measurements [4, 5]. However, this setup only works well if the HGO is tuned perfectly, and doing that by hand is a frustrating, time-consuming task that rarely yields the best results, as poor HGO tuning can lead to bad performance or even instability. To fix this critical issue, we employ Particle Swarm Optimization (PSO), a smart search method [6, 7], to automatically find the best settings for the HGO. The core idea is that a better-tuned HGO provides more accurate state estimates, thereby allowing our Backstepping controller to guide the TRMS much more effectively. This paper explains how we applied this PSO-tuned strategy to the TRMS, presenting results from simulations and real hardware tests to demonstrate that using PSO for observer tuning significantly improves path following and stability, offering a useful approach for other complex control challenges.

2 TRMS Plant Model

In Our control strategy is tested on the Twin Rotor MIMO System (TRMS). Figure 1 shows the hardware setup: a beam balanced on a central tower. At one end sit two propellers – a main rotor generating lift and a tail rotor pushing sideways offset by a counterweight on the other end. This design lets the beam pitch up and down and yaw side-to-side freely, similar to how a helicopter behaves. This similarity makes the TRMS a widely used platform for testing control ideas.

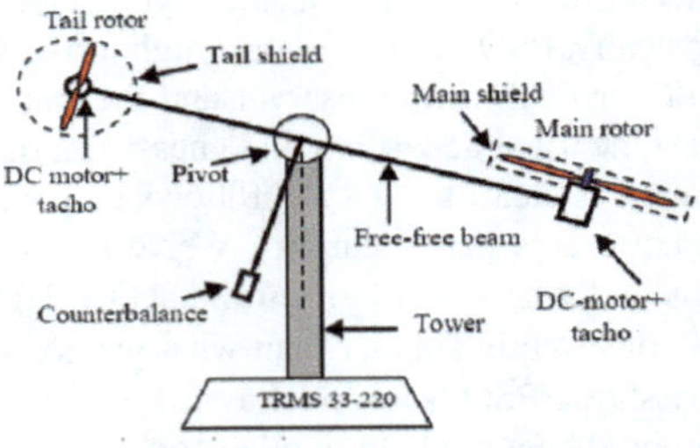

Fig. 1. Schematic diagram of the TRMS

To design and test our controller, we rely on an existing mathematical description of the TRMS dynamics. A detailed derivation based on Newtonian mechanics is available in works such as [8] and often provided by the system manufacturer. We adopt this established model directly. The resulting dynamics capture the inherent nonlinearities and cross-coupling between the pitch and yaw motions. The system is represented in state-space form as:

$$\begin{cases} \frac{d\psi}{dt} = \dot{\psi} \\ \frac{d\dot{\psi}}{dt} = \frac{1}{I_1}\Big(\alpha_1\tau_1^2 + b_1\tau_1 - M_g\sin(\psi) - B_{1\psi}\dot{\psi} - B_{2\psi}sign(\dot{\psi}) - K_{gy}\dot{\varphi}\cos\psi\Big(\alpha_1\tau_1^2 + b_1\tau_1\Big) \\ \frac{d\varphi}{dt} = \dot{\varphi} \\ \frac{d\dot{\varphi}}{dt} = \frac{1}{I_2}\Big(\alpha_2\tau_2^2 + b_2\tau_2 - B_{1\varphi}\dot{\varphi} - B_{2\varphi}sign(\dot{\varphi}) - K_c 1.75\Big(\alpha_1\tau_1^2 + b_1\tau_1\Big) \\ \frac{d\tau_1}{dt} = -\frac{T_{10}}{T_{11}}\tau_1 + \frac{K_1}{T_{11}}u_1 \\ \frac{d\tau_2}{dt} = -\frac{T_{20}}{T_{21}}\tau_2 + \frac{k_2}{T_{21}}u_2 \end{cases} \tag{1}$$

Bellow the state and the output vectors:

$$x = \left[\psi \dot{\psi} \varphi \dot{\varphi} \tau_1 \tau_2\right]^T \quad y = [\psi \ \varphi]^T$$

Where ψ: Pitch (elevation); τ_1: the principal torque; φ: Yaw (azimuth); τ_2: the tail torque.

3 Backstepping Control

Backstepping control, suited for nonlinear strict-feedback systems [9], is used here in a decentralized approach to design separate control laws for pitch and yaw. Only the tracking strategy is developed, as it inherently includes regulation.

3.1 Pitch Subsystem

$$\begin{cases} \frac{d\psi}{dt} = \dot{\psi} \\ \frac{d\dot{\psi}}{dt} = \frac{1}{I_1}\left(\alpha_1\tau_1^2 + b_1\tau_1 - M_g \sin(\psi) - B_{1\psi}\dot{\psi} - B_{2\psi} sign(\dot{\psi}) - K_{gy}\dot{\varphi}\cos\psi\left(\alpha_1\tau_1^2 + b_1\tau_1\right)\right) \\ \frac{d\tau_1}{dt} = -\frac{T_{10}}{T_{11}}\tau_1 + \frac{K_1}{T_{11}}u_1 \end{cases} \quad (2)$$

Let $x_1 = \psi$; $\dot{x}_1 = \dot{\psi} = x_2$; $\tau_1 = x_3$

In implementing the backstepping technique, we adopt a decentralized control strategy. This allows us to express the pitch subsystem in the triangular form characteristic of the strict-feedback system. To achieve this, we must disregard the coupling terms resulting from the yaw dynamics in the model. This leads us to the following equations for the pitch subsystem:

$$\begin{cases} \dot{x}_1 = x_2 \\ \dot{x}_2 = \frac{1}{I_1}\left(\alpha_1 x_3^2 + b_1 x_3 - M_g \sin x_1 - B_{1\psi} x_2 - B_{2\psi} \mathrm{sing}(x_1)\right) \\ \dot{x}_3 = \frac{T_{10}}{T11} x_3 + \frac{K_1}{T_{11}} U_1 \end{cases} \quad (3)$$

x_1, x_2 and x_3 are the state variables.

The backstepping design stabilizes the pitch subsystem in three steps to track the desired trajectory $x_{1d} = \psi_d$:

Step 1: **Tracking Error Stabilization**: Define the tracking error:

$z_1 = x_1 - x_{1d}$ (1), from (1), the derivative is: $\dot{z}_1 = x_2 - \dot{x}_{1d}$ (2). Treat x_2 as a virtual control, selecting:

$\alpha_1 = -c_1 z_1 + \dot{x}_{1d}$ (3), where c_1 a positive constant. The Lyapunov function $V_1 = \frac{1}{2} z_1^2$ (4), has derivative:

$\dot{V}_1 = z_1\dot{z}_1 = z_1\left(x_2 - \dot{x}_{1d}\right) = z_1\left(z_2 + \alpha_1 - \dot{x}_{1d}\right)$ (5). Where $z_2 = x_2 - \alpha_1$ (6).

If $z_2 = 0$, then $\dot{V}_1 = -c_1 z_1^2 \leq 0$, ensuring $z_1 \longrightarrow 0$.

$$U_1 = \left(\frac{I_1 T_{11}}{K_1(2\alpha_2 x_3 + b_1)}\right)(-e_2 - k_3 e_3) + \frac{T_{10}}{K_1} x_3$$

$$+\left(\frac{T_{11}}{K_1(2\alpha_2 x_3 + b_1)}\right)\left(\frac{B_{1\psi}}{I_1} - (k_2 + k_1)\right)\dot{x}_2$$
$$+ I_1 x_2(-k_2 k_1 - 1 + \frac{M_g \cos x_1 + B_{2\psi}}{I_1})$$
$$+ I_1(k_2 k_1 + 2)\dot{x}_{1d} + (k_2 + k_1)\ddot{x}_{1d} + \dddot{x}_{1d} \quad (4)$$

With

$$\dot{x}_2 = \frac{1}{I_1}\left(a_1 x_3^2 + b_1 x_3 - M_g \sin x_1 - B_{1\psi} x_2 - B_{2\psi}\,\mathrm{sign}(x_1)\right) \quad (5)$$

Yaw subsystem

$$\begin{cases} \frac{d\varphi}{dt} = \dot{\varphi} \\ \frac{d\dot{\varphi}}{dt} = \frac{1}{I_2}\left(\alpha_2 \tau_2^2 + b_2 \tau_2 - B_{1\varphi}\dot{\varphi} - B_{2\varphi}\mathrm{sing}(\varphi) - K_c 1.75\left(\alpha_1 \tau_1^2 + b_1 \tau_1\right)\right) \\ \frac{d\tau_2}{dt} = \frac{-T_{20}}{T_{21}}\tau_2 + \frac{K_2}{T_{21}} u_2 \end{cases} \quad (6)$$

Let:
$x_4 = \varphi$; $\dot{x}_4 = \dot{\varphi} = x_5$;$x_6 = \tau_2$
By eliminating the coupling dynamics, the yaw subsystem becomes:

$$\begin{cases} \dot{x}_4 = x_5 \\ \dot{x}_5 = \frac{1}{I_2}\left(\alpha_2 x_6^2 + b_2 x_6 - B_{1\varphi} x_5 - B_{2\varphi} x_4\right) \\ \dot{x}_6 = \frac{-T_{20}}{T_{21}} x_6 + \frac{K_2}{T_{21}} U_2 \end{cases} \quad (7)$$

x_4, x_5And x_6 are the state variables.

$$U_2 = \left(\frac{I_2 T_{21}}{K_2(2\alpha_2 x_6 + b_2)}\right)(-e_5 - k_6 e_6) + \frac{T_{20}}{K_2} x_6 + \left(\frac{T_{21}}{K_2(2\alpha_2 x_6 + b_2)}\right)\left(\frac{B_{1\varphi}}{I_2} - (k_5 + k_4)\right)\dot{x}_5 + I_2 x_5(-k_5 k_4 - 1 + \frac{B_{2\varphi}}{I_2}) + I_2(k_5 k_4 + 1)\dot{x}_{2d} + (k_5 + k_4)\ddot{x}_{2d} + \dddot{x}_{2d})) \quad (8)$$

With

$$\dot{x}_5 = \frac{1}{I_2}\left(\alpha_2 x_6^2 + b_2 x_6 - B_{1\varphi} x_5 - B_{2\varphi} x_4\right) \quad (9)$$

4 High Gain Observer

Developing nonlinear observers presents greater challenges as they must incorporate the system's nonlinear model. However, some observer types are relatively straightforward to implement, particularly those that build upon linear techniques. These observers are essentially extensions of the Luenberger observer concept Among the various available approaches, high gain observers have gained significant practical popularity. There are several variants of high gain observers, but in this case, we've chosen to focus on the version developed by Gauthier [10]. This observer is given by the following differential equation:

$$\dot{\hat{x}} = f\left(\hat{x}\right) + Bu + \left(\frac{\partial\phi\left(\hat{x}\right)}{\partial\hat{x}}\right)^{-1} S_\theta^{-1}(\theta) C^T\left(y - h\left(\hat{x}\right)\right) \quad (10)$$

$\hat{x}$ is the estimated state vector.

- $f(\hat{x})$ represents the nonlinear dynamics of the system.
- BU Accounts for the input u applied to the system.
- $\frac{\partial \varnothing(\hat{x})}{\partial(\hat{x})}$ represents the Jacobian matrix of a nonlinear transformation
- $S_{\theta}^{-1}(\theta)$ is a gain matrix.
- C^T is the transpose of the output matrix C

$y - h(\hat{x})$ represents the innovation term (the difference between the actual measurement y and the estimated output.

- $\frac{1}{\theta}$ is the high gain parameter, which can be seen as a time scaling for the observer dynamics.

5 Controller-Observer Laws

After implementation and examine of the backstepping control laws, where we substitute the actual state x with the estimated state x̂ obtained from observer Eq. (9).

5.1 Pitch Subsystem Controller-Observer Laws

$$U_1 = \left(\frac{I_1 T_{11}}{K_1\left(2\alpha_2 \hat{x}_3 + b_1\right)}\right)(-e_2 - k_3 e_3) + \frac{T_{10}}{K_1}\hat{x}_3 + \left(\frac{T_{11}}{K_1\left(2\alpha_2 \hat{x}_3 + b_1\right)}\right)\left(\frac{B_{1\psi}}{I_1} - (k_2 + k_1)\right)\dot{\hat{x}}_2 + I_1 x_2(-k_2 k_1 - 1 + \frac{M_g \cos \hat{x}_1 + B_{2\psi}}{I_1}) + I_1(k_2 k_1 + 2)\dot{x}_{1d} + (k_2 + k_1)\ddot{x}_{1d} + \dddot{x}_{1d})) \tag{11}$$

With

$$\dot{\hat{x}}_2 = \frac{1}{I_1}\left(a_1 \hat{x}_3^2 + b_1 \hat{x}_3 - M_g \sin \hat{x}_1 - B_{1\psi} \hat{x}_2 - B_{2\psi} \operatorname{sign}\left(\hat{x}_2\right)\right) \tag{12}$$

5.2 Yaw Subsystem Controller-Observer Laws

The control law for the Yaw subsystem is:

$$U_2 = \left(\frac{I_2 T_{21}}{K_2\left(2\alpha_2 \hat{x}_6 + b_2\right)}\right)(-e_5 - k_6 e_6) + \frac{T_{20}}{K_2}\hat{x}_6 + \left(\frac{T_{21}}{K_2\left(2\alpha_2 \hat{x}_6 + b_2\right)}\right)\left(\frac{B_{1\varphi}}{I_2} - (k_5 + k_4)\right)\dot{\hat{x}}_5 + I_2 \hat{x}_5(-k_5 k_4 - 1 + \frac{B_{2\varphi}}{I_2}) + I_2 (k_5 k_4 1)\dot{x}_{2d} + (k_5 + k_4)\ddot{x}_{2d} + \dddot{x}_{2d})) \tag{13}$$

With

$$\dot{\hat{x}}_5 = \frac{1}{I_2}\left(\alpha_2 \hat{x}_6^2 + b_2 \hat{x}_6 - B_{1\varphi} \hat{x}_5 - B_{2\varphi} \hat{x}_4\right) \tag{14}$$

Based on the nonlinear separation principle, it is possible to apply a weakened separation principle to ensure the stability of a nonlinear closed-loop control system. This holds when an observer with exponential convergence such as a high-gain observer is integrated, and the control law ensures global asymptotic stability. According to established theorems, global asymptotic stabilization of the closed-loop system is guaranteed if three conditions are met: the observer converges exponentially, the control law is globally asymptotically stabilizing, and a stabilizing Lyapunov function exists.

6 Simulation and RealTime Results:

In the present study, an Integral of Time-weighted Absolute Error (*ITAE*) is utilized to judge the performance of the controller. The index *ITAE* is defined such as

$$J_{ITAE} = \int_0^T t \mid e(t) \mid dt \tag{15}$$

So, to show our control method works, we tested it out both in simulations and on the real TRMS rig using MATLAB/SIMULINK. The really neat part is how Particle Swarm Optimization (PSO) actively tunes the High-Gain Observer's (HGO) settings as the system is running. Think of PSO working right alongside the main TRMS controls, using live feedback from how things are going to constantly adjust the observer's gains. We used this live-tuning approach for all our tests, simulated and real, and it adapted well to different conditions. What this gives us is the HGO's dependable performance combined with PSO's smart way of searching, which means the system can better handle the unexpected changes and disturbances you get in the real world. PSO keeps refining those observer settings by checking its performance against a specific goal we set this helps keep everything stable and seriously boosts how well it estimates states and controls the whole system.

Bird_in_swarm = 20, Number_of_quality_in_Bird = 2, cognitive_constant = 2, velocity_clamping_factor = 2

social_constant = 2Min_Inertia_weight = 0.4, Max_Inertia_weight = 0.9, max_iteration = 40,

MinMaxRange = [0 5; 0 2]

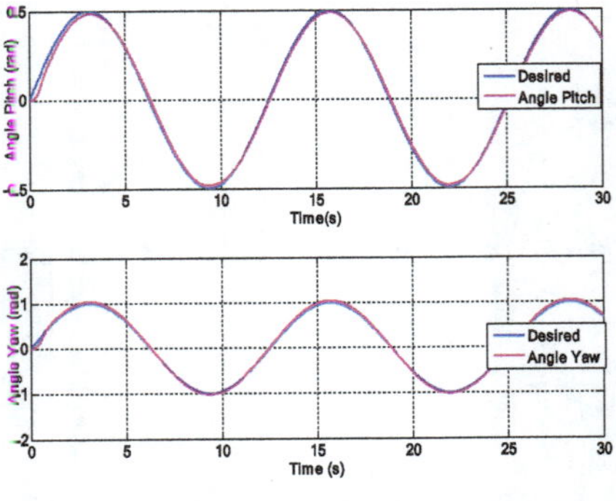

Fig. 2. TRMS Tracking: Backstepping & Standard HGO

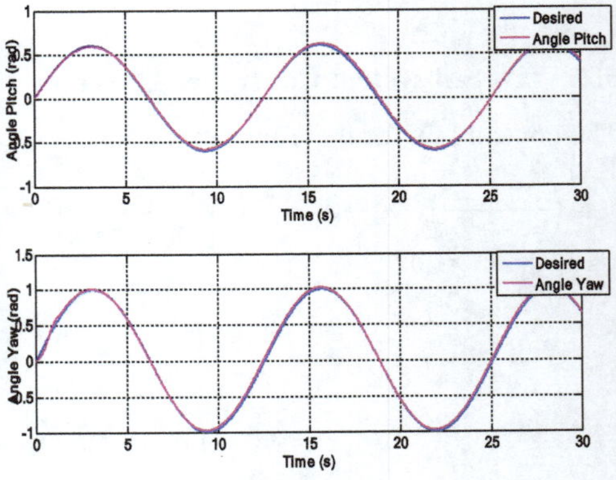

Fig. 3. TRMS Tracking: Backstepping with PSO-HGO

Our simulation results demonstrate strong trajectory tracking for the TRMS using Backstepping with both a standard High-Gain Observer (Fig. 2) and a PSO-tuned HGO (Fig. 3). In both cases, the pitch angle accurately follows its desired path (around ±0.5 to 0.6 rad) and the yaw angle similarly tracks its reference (around ±1 rad), with smooth, synchronized periodic motion (approx. 15 s period), minimal error, and no overshoot. While both configurations show excellent visual tracking in these figures, PSO's key contribution is automating the difficult manual tuning of observer parameters, which is crucial for achieving robust state estimation and consistently high-performance dynamic control.

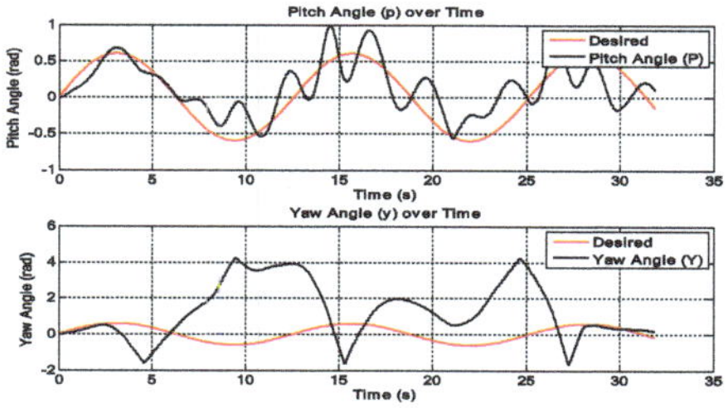

Fig. 4. TRMS Real-Time: Backstepping & Std. HGO

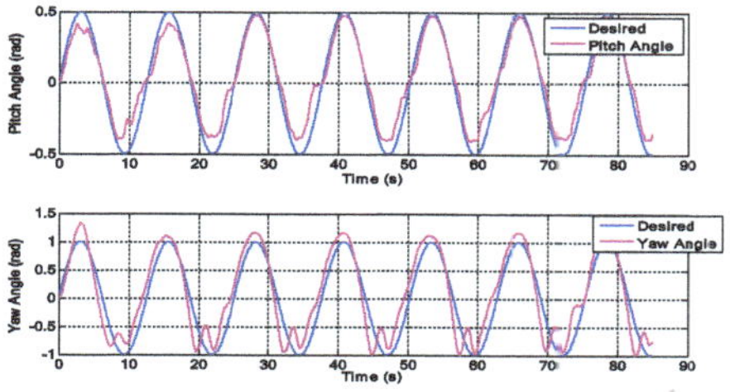

Fig. 5. TRMS RT: Backstepping & PSO-HGO

The real-time tests tell a clear story. Figure 4, using the standard High-Gain Observer, shows the system really struggling – pitch and yaw were all over the place with big errors, not hitting their targets. But then look at Fig. 5: with the PSO-tuned observer, it's a night-and-day difference. The pitch now neatly follows its goal (around -0.5 to 0.6 rad), and the yaw sticks right to its reference (around -1 to 1 rad). Both are smooth, stable, keep that 15-s rhythm, and have hardly any error. This big jump in performance really shows how much PSO helps by getting the state estimation right for solid, real-world TRMS control (Fig. 6).

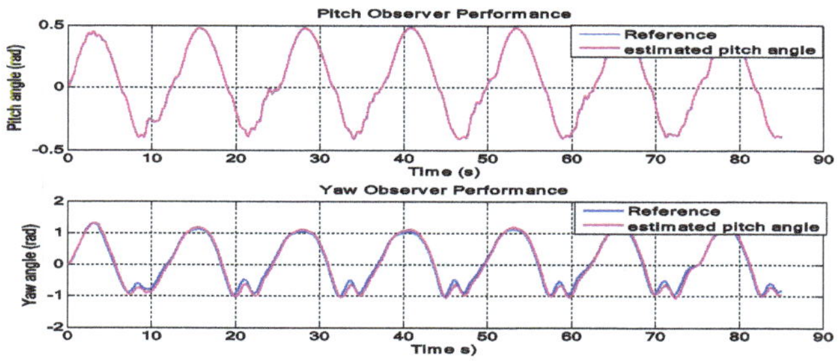

Fig. 6. High-gain Observer Performance for TRMS States

The High-Gain Observer (HGO) effectively estimated the TRMS states. Its pitch angle estimates remained smooth and periodic, consistently within the $[-0.5, 0.5]$ radian limits. For the yaw angle, the HGO showed high accuracy, with its estimated states (magenta) closely matching the actual measurements (blue) in the $[-1, 1]$ radian range; slight deviations were only apparent at some negative peaks. The observer maintained this consistent, accurate performance throughout the 90-s experiment. The yaw estimation, in particular, converged strongly, confirming the HGO's capability to reconstruct system states reliably in real time.

7 Conclusion

Our work makes TRMS control practical. We combined Backstepping with a High-Gain Observer, but the real key was tuning that observer with PSO. Standard observers struggled with real-world messiness; PSO cut through it, giving us accurate state estimates and stable flight on the actual hardware. This means the controller isn't just a good idea on paper it works reliably where it counts.

References

1. Lee, J., You, S., Kim, W., Moon, J.: Extended state observer-actor-critic architecture based output-feedback optimized backstepping control for permanent magnet synchronous motors. Exp. Syst. Appl. **270**, 126542 (2025)
2. Kapnopoulos, A., Kazakidis, C., Alexandridis, A.: Quadrotor trajectory tracking based on backstepping control and radial basis function neural networks. Results Control Optim. **14**, 100335 (2024)
3. Abood, L.H., Oleiwi, B.K., Ali, E.H.: Optimal backstepping controller for controlling DC motor speed. Bull. Electr. Eng. Inform. **11**(5), 2564–2572 (2022)
4. Prasov, A.A., Khalil, H.K.: A nonlinear high-gain observer for systems with measurement noise in a feedback control framework. IEEE Trans. Autom. Control. **58**(3), 569–580 (2012)
5. Dehbozorgi, I., Khayatian, A.: Impulsive high-gain observer for nonlinear systems to reduce peaking phenomenon. Int. J. Syst. Sci. **56**(9), 1–15 (2025)
6. Tao, X., Liu, K., Yang, J., Chen, Y., Chen, J., Zhu, H.: Sliding mode backstepping control of excavator bucket trajectory synovial in particle swarm optimization algorithm and neural network disturbance observer. Actuators. **14**(1), 9 (2025)
7. Hoang, Q.M., Hollweg, G.V., Hussain, A., Zarrabian, S., Su, W., Bui, V.H.: PSO-based sliding mode current control of grid-forming inverter in rotating frame. arXiv preprint 2501.11633 (2025)
8. Feedback Instruments Ltd.: Twin Rotor MIMO System Control Experiments Manual. Feedback Instruments Ltd., Crowborough, East Sussex (2006)
9. M. Krstic, I. Kanellakopoulos and. P. Kokotovic, Nonlinear and Adaptive Control Design, New York: Wiley, 1995.
10. J. Gauthier and. I. Kupka, "Observability and observers for nonlinear systems," SIAM J. Control. Optim., vol. 32, pp. 975–994, 1994.

Genetic Algorithm Enhanced Backstepping for Real-Time Trajectory Tracking of a Twin Rotor MIMO System

Amina Dounia Babou[1](✉), Boulouiha Houari Merabet[1], and Abderrahmane Kacimi[2]

[1] Department of Electric ENPO MAURICE Audin, Oran, Algeria
douniababou142@gmail.com
[2] Institute of maintenance and industrial safety, University of Oran, Oran, Algeria

Abstract. Controlling systems like the Twin Rotor MIMO (TRMS) for tracking is tricky, especially getting it right in real-time. We first used a standard Backstepping method for trajectory tracking on the TRMS. Simulation results were promising – it tracked well. But when we tried that same controller on the actual physical TRMS hardware, it failed completely in real-time operation; performance was poor and unstable. This showed a major problem translating simulation success to the real world for this system. To fix this real-time failure, we enhanced the Backstepping controller using a Genetic Algorithm (GA). The GA was used to optimize controller settings, specifically making it more robust and adaptable to the real-world TRMS's quirks. We tested this GA-enhanced controller experimentally. The results were excellent: it achieved stable and accurate trajectory tracking on the real TRMS, fixing the problems we saw before, while still working well in simulation. This work demonstrates that adding intelligent optimization like a GA can be crucial for getting advanced controllers to work reliably on complex physical systems in real-time.

Keywords: Twin Rotor System · Backstepping Control · Genetic Algorithm · Trajectory Tracking · Real-time Implementation · Non-linear Control · Optimization

1 Introduction

Controlling complex, non-linear systems like the Twin Rotor MIMO System (TRMS) for precise trajectory tracking is seriously challenging, especially getting reliable performance on actual physical hardware, not just in perfect computer simulations [3, 4]. The TRMS is often used as a testbed precisely because its coupled dynamics are inherently tricky, sharing complexity with systems like helicopters [1]. Making controllers for these kinds of systems work precisely and reliably in real-time is crucial for many fields, from advanced robotics to the growing control needs in areas like renewable energy. Researchers are always working on ways to handle the TRMS's difficulties [3, 4]. We started by using a standard Backstepping controller for TRMS tracking. In simulation? It looked really good, showing accurate path following. But putting that same controller

M. Hatti (Ed.): IC-AIRES 2025, LNNS 1740, pp. 441–447, 2026.
https://doi.org/10.1007/978-3-032-12788-4_42

onto the actual TRMS hardware led to real-time failure – the performance was poor and unstable. This clearly showed a critical gap between our idealized simulation models and the unpredictable reality of a physical system. To fix this problem and get reliable performance, we enhanced the Backstepping controller by bringing in a Genetic Algorithm (GA). GAs are powerful optimization tools, great for finding the right parameters to build robustness against real-world uncertainties. Combining advanced non-linear control with intelligent optimization like GAs is increasingly seen as essential for making controllers work reliably in the real world. This kind of combined approach is proving effective in other challenging control areas too, like tuning controllers for vehicle suspensions [2] or helping UAVs adapt using GAs and neural networks [7, 8]. It's an approach finding value in many fields, including renewable energy control (think about using GA-optimized Backstepping for solar power systems [5], or neural networks with Backstepping for wave energy [6]). This paper presents the GA-enhanced Backstepping controller for TRMS tracking. We'll detail its design and how the GA was used to optimize it specifically for better real-time performance. Our key contribution is experimentally demonstrating that this GA-optimized approach successfully fixed the real-time failure we saw, achieving stable, accurate tracking on the actual TRMS hardware. The rest of the paper is laid out like this: Section II (TRMS), Section III (Standard Backstepping), Section IV (GA Optimization), Section V (Results: simulation and experimental), Section VI (Conclusion).

2 TRMS Plant Model

The Twin Rotor MIMO System (TRMS) served as the platform for experimentally validating our control strategy. Figure 1 shows the system's physical setup. It's built around a beam that pivots on a central tower, allowing movement in both pitch (vertical) and yaw (horizontal). At one end are two rotors the main one for lift, the tail one for sideways force balanced by a counterweight. This configuration gives it dynamics similar to a helicopter, which is precisely why the TRMS is a standard laboratory platform for control research.

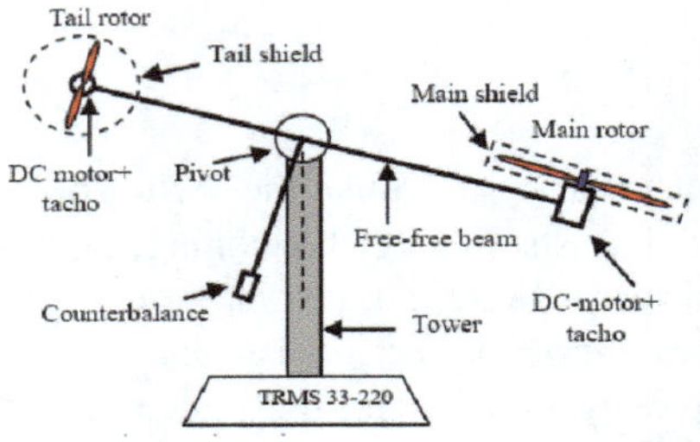

Fig. 1. Schematic diagram of the TRMS

To design our controller, we needed an accurate mathematical model of the TRMS. We used an established model, available in literature like [9] and often provided with the system, which was derived using Newtonian mechanics. We adopted this model directly. It accounts for the system's inherent nonlinearities and the significant coupling between pitch and yaw motions. The system dynamics are expressed in state-space form as:

$$\begin{cases} \frac{d\psi}{dt} = \dot{\psi} \\ \frac{d\dot{\psi}}{dt} = \frac{1}{I_1}\left(\alpha_1\tau_1^2 + b_1\tau_1 - M_g\sin(\psi) - B_{1\psi}\dot{\psi} - B_{2\psi}sign(\dot{\psi}) - K_{gy}\dot{\varphi}\cos\psi\left(\alpha_1\tau_1^2 + b_1\tau_1\right)\right. \\ \frac{d\varphi}{dt} = \dot{\varphi} \\ \frac{d\dot{\varphi}}{dt} = \frac{1}{I_2}\left(\alpha_2\tau_2^2 + b_2\tau_2 - B_{1\varphi}\dot{\varphi} - B_{2\varphi}sign(\dot{\varphi}) - K_c 1.75\left(\alpha_1\tau_1^2 + b_1\tau_1\right)\right. \\ \frac{d\tau_1}{dt} = -\frac{T_{10}}{T_{11}}\tau_1 + \frac{K_1}{T_{11}}u_1 \\ \frac{d\tau_2}{dt} = -\frac{T_{20}}{T_{21}}\tau_2 + \frac{k_2}{T_{21}}u_2 \end{cases} \quad (1)$$

Bellow the state and the output vectors:

$$\mathrm{x} = \left[\psi\dot{\psi}\varphi\dot{\varphi}\tau_1\tau_2\right]^{\mathrm{T}} \; \mathrm{y} = [\psi\ \varphi]^{\mathrm{T}}$$

Where ψ: Pitch (elevation); τ_1; the principal torque; φ: Yaw (azimuth); τ_2: the tail torque.

3 Backstepping Control

Backstepping control, a known approach for handling nonlinear systems with a strict-feedback structure [9], was implemented here. We adopted a decentralized design strategy, developing individual control laws for the pitch motion and the yaw motion. The controller focused solely on the tracking objective, given that stable tracking performance automatically implies regulation around a setpoint.

3.1 Pitch Subsystem

$$\begin{cases} \frac{d\psi}{dt} = \dot{\psi} \\ \frac{d\dot{\psi}}{dt} = \frac{1}{I_1}\left(\alpha_1\tau_1^2 + b_1\tau_1 - M_g\sin(\psi) - B_{1\psi}\dot{\psi} - B_{2\psi}sign(\dot{\psi}) - K_{gy}\dot{\varphi}\cos\psi\left(\alpha_1\tau_1^2 + b_1\tau_1\right)\right) \\ \frac{d\tau_1}{dt} = -\frac{T_{10}}{T_{11}}\tau_1 + \frac{K_1}{T_{11}}u_1 \end{cases} \quad (2)$$

Let $x_1 = \psi$; $\dot{x}_1 = \dot{\psi} = x_2$; $\tau_1 = x_3$

When implementing the Backstepping technique, we opted for a decentralized control strategy. This choice was made because it allowed us to express the pitch subsystem in the triangular form characteristic of strict-feedback systems, which is necessary for Backstepping. To achieve this structure, however, we had to neglect the coupling terms originating from the yaw dynamics present in the overall model. This approach results in the following equations for the pitch subsystem:

$$\begin{cases} \dot{x}_1 = x_2 \\ \dot{x}_2 = \frac{1}{I_1}\left(\alpha_1 x_3^2 + b_1 x_3 - M_g \sin x_1 - B_{1\psi} x_2 - B_{2\psi} \operatorname{sing}(x_1)\right) \\ \dot{x}_3 = \frac{T_{10}}{T11} x_3 + \frac{K_1}{T_{11}} U_1 \end{cases} \tag{3}$$

x_1, x_2 and x_3 are the state variables.

Our goal was designing a nonlinear control law to bring the system to a stable state right at the origin. Using the Backstepping technique means you start by looking at the first part of the system's equations. Here, the first step is defining an error variable that quantifies the deviation of this first state from the desired zero. Once that error is defined, the Backstepping procedure moves recursively, using that error to systematically construct the overall control law that will guide the system state to converge asymptotically to the origin.

$$\begin{aligned} U_1 = &\left(\frac{I_1 T_{11}}{K_1(2\alpha_2 x_3 + b_1)}\right)(-e_2 - k_3 e_3) + \frac{T_{10}}{K_1} x_3 + \left(\frac{T_{11}}{K_1(2\alpha_2 x_3 + b_1)}\right)\left(\frac{B_{1\psi}}{I_1} - (k_2 + k_1)\right)\dot{x}_2 \\ &+ I_1 x_2\left(-k_2 k_1 - 1 + \frac{M_g \cos x_1 + B_{2\psi}}{I_1}\right) + I_1(k_2 k_1 + 2)\dot{x}_{1d} + (k_2 + k_1)\ddot{x}_{1d} + \dddot{x}_{1d} \end{aligned} \tag{4}$$

With

$$\dot{x}_2 = \frac{1}{I_1}\left(a_1 x_3^2 + b_1 x_3 - M_g \sin x_1 - B_{1\psi} x_2 - B_{2\psi} \operatorname{sign}(x_1)\right) \tag{5}$$

3.2 Yaw Subsystem

$$\begin{cases} \frac{d\varphi}{dt} = \dot{\varphi} \\ \frac{d\dot{\varphi}}{dt} = \frac{1}{I_2}\left(\alpha_2 \tau_2^2 + b_2 \tau_2 - B_{1\varphi}\dot{\varphi} - B_{2\varphi}\operatorname{sing}(\varphi) - K_c 1.75\left(\alpha_1 \tau_1^2 + b_1 \tau_1\right)\right) \\ \frac{d\tau_2}{dt} = \frac{-T_{20}}{T_{21}} \tau_2 + \frac{K_2}{T_{21}} u_2 \end{cases} \tag{6}$$

Let:

$x_4 = \varphi$; $\dot{x}_4 = \dot{\varphi} = x_5$; $x_6 = \tau_2$

By eliminating the coupling dynamics, the yaw subsystem becomes:

$$\begin{cases} \dot{x}_4 = x_5 \\ \dot{x}_5 = \frac{1}{I_2}\left(\alpha_2 x_6^2 + b_2 x_6 - B_{1\varphi} x_5 - B_{2\varphi} x_4\right) \\ \dot{x}_6 = \frac{-T_{20}}{T_{21}} x_6 + \frac{K_2}{T_{21}} U_2 \end{cases} \tag{7}$$

x_4, x_5 And x_6 are the state variables.

$$U_2 = \left(\frac{I_2 T_{21}}{K_2(2\alpha_2 x_6 + b_2)}\right)(-e_5 - k_6 e_6) + \frac{T_{20}}{K_2} x_6 + \left(\frac{T_{21}}{K_2(2\alpha_2 x_6 + b_2)}\right)\left(\frac{B_{1\varphi}}{I_2} - (k_5 + k_4)\right)\dot{x}_5$$
$$+ I_2 x_5 \left(-k_5 k_4 - 1 + \frac{B_{2\varphi}}{I_2}\right) + I_2(k_5 k_4 + 1)\dot{x}_{2d} + (k_5 + k_4)\ddot{x}_{2d} + \dddot{x}_{2d})) \quad (8)$$

With

$$\dot{x}_5 = \frac{1}{I_2}\left(\alpha_2 x_6^2 + b_2 x_6 - B_{1\varphi} x_5 - B_{2\varphi} x_4\right) \quad (9)$$

4 Genetic Algorithm Optimization

A genetic algorithm (GA) optimizes the backstepping controller gains (1, 2, 3) for the Twin

Rotor MIMO System, enhancing pitch and yaw tracking against nonlinear dynamics [10]. Gas evolve a population of solutions iteratively. Here, 40 parameter sets are initialized as:

$$P_{j,i} = b_{i,1} + r * \left(b_{i,2} - b_{i,1}\right) \qquad i = 1, 2, 3 \quad (10)$$

with bounds [0, 20] for 1, 2, [0, 10] for 3, and ∈ [0, 1]. The fitness function evaluates

$$J = \frac{1}{1 + \|F\|} \quad (11)$$

Where $\|F\|$ is the tracking error norm. Tournament selection (size 5) chooses parents:

$$P_{parent} = \arg \max_{k \in \{1,.,5\}} J(P_k) \quad (12)$$

Single-point crossover (probability 0.7) combines parents at index :

$$O_{j,i} = \begin{cases} P_{a,i} & i \le k \\ P_{b,i} & i > k \end{cases} \quad (13)$$

Uniform mutation (probability 0.05) adjusts parameters:

$$P_{j,i} = b_{i,1} + r * \left(b_{i,2} - b_{i,1}\right) \quad (14)$$

Over 40 generations, the GA yields optimized gains, improving TRMS tracking in both simulations and real-time.

5 Simulation and RealTime Results

To check how our GA-optimized backstepping controller performs on the Twin Rotor MIMO System, we used the Integral of Time-weighted Absolute Error (ITAE) to measure pitch and yaw tracking accuracy. The ITAE is given by:

$$J_{ITAE} = \int_0^T t \mid e(t) \mid dt \tag{15}$$

where () combines pitch and yaw errors, and is the test duration. We ran simulations in MATLAB/Simulink and tests on the TRMS rig, with the GA tuning three backstepping gains (1, 2, 3) within bounds [0, 20] for 1, 2, and [0, 10] for 3, using 40 individuals, 70% crossover, 5% mutation, and 40 generations. The results were solid: simulations showed a lower ITAE than a standard backstepping controller, and real-time runs stayed stable despite the TRMS's nonlinear dynamics and disturbances. This GA-driven approach really nails it for complex MIMO systems.

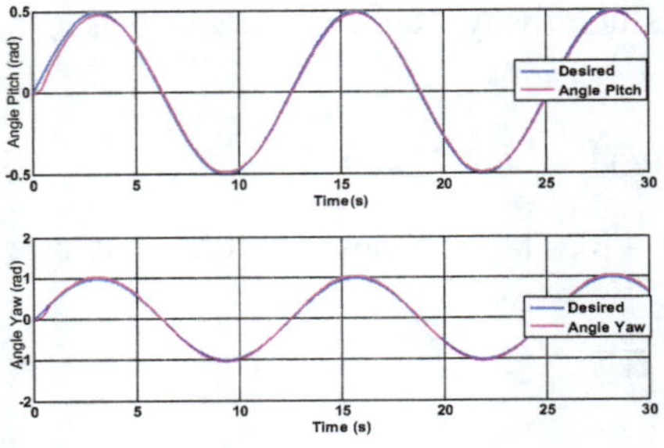

Fig. 2. Backstepping Control for TRMS Tracking

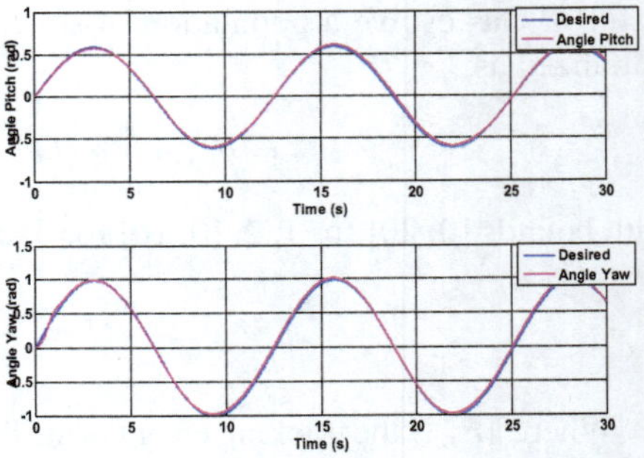

Fig. 3. GA-Enhanced Backstepping for TRMS Tracking

We evaluated the backstepping controller's performance for the Twin Rotor MIMO System in simulations, relying on the Integral of Time-weighted Absolute Error (ITAE) to assess pitch and yaw tracking. Figure 2 displays the baseline tracking behavior, while Fig. 3 shows how GA optimization boosts accuracy, cutting down ITAE under tough nonlinear conditions.

These figures, Figs. 4 and 5, present the TRMS's real-time tracking results. Figure 4 shows a standard Backstepping approach, where you can see the pitch and yaw tracking isn't ideal, with notable errors. Compare that to Fig. 5, where we used the GA to optimize the Backstepping gains. It's pretty clear the GA-tuned version achieves significantly improved tracking performance, staying much closer to the desired paths in the real world.

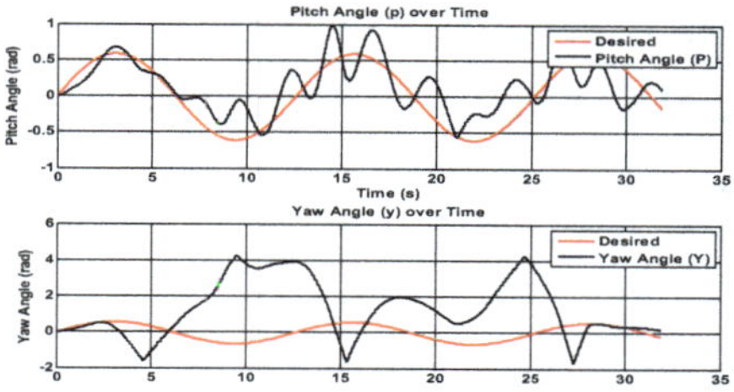

Fig. 4. Real-timeBackstepping

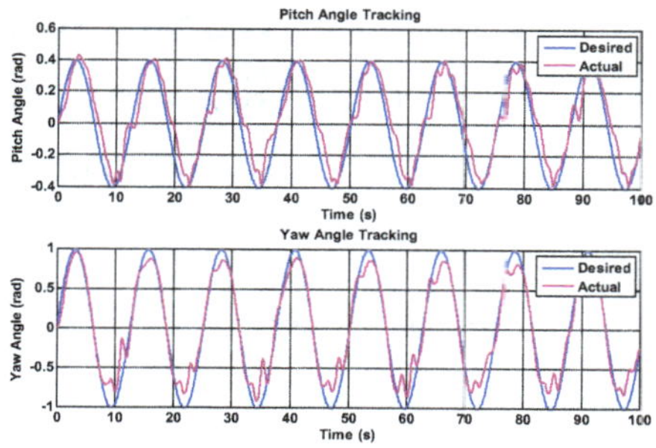

Fig. 5. Real-time GA-Tuned Backstepping

6 Conclusion

Our work aimed to control the challenging TRMS with Backstepping, and we found a major performance boost by optimizing the controller gains with a Genetic Algorithm. The real test, of course, was in real-time experiments. Looking at the results in Fig. 5, the difference is quite clear: the GA-tuned controller achieved much more accurate and consistent trajectory tracking than its standard Backstepping counterpart. This outcome strongly supports the idea that bringing optimization methods like GAs into the loop is a powerful way to achieve high-performance control on tricky nonlinear MIMO systems in practical applications.

References

1. Danh, H.D., Van, C.N., Van, Q.V.: Tracking iterative learning control of TRMS using feedback linearization model with input disturbance. J. Robot. Control (JRC). **6**(1), 446–455 (2025)
2. Fattah, R., Kenne, J.P., Benjelloun, K., Chebak, A.: Optimization of the Backstepping Control Parameters of an Active Electrohydraulic Suspension to Improve Passenger Comfort and Road Handling. IEEE Access (2025)
3. Khaled, M., Doudou, S.: Robust prescribed time backstepping for stabilization and tracking of 2-DOF twin rotor model. In: 2024 International Conference on Advances in Electrical and Communication Technologies (ICAECOT), pp. 1–5. IEEE (2024, October)
4. Moosapour, S.S., Mehdipour, H., Keramatzadeh, M.: Sliding Mode Disturbance Observer-based Control of a laboratory Twin Rotor MIMO System (TRMS). IEEE Access (2024)
5. Naoussi, S.R.D., et al.: Retraction Note: Enhancing MPPT performance for partially shaded photovoltaic arrays through backstepping control with Genetic Algorithm-optimized gains. Sci. Rep. **15**, 2643 (2025)
6. Nath, P., et al.: Neural network backstepping control of OWC wave energy system. Sci. Rep. **15**(1), 7983 (2025)
7. Rodríguez-Abreo, O., Rodríguez-Reséndiz, J., García-Cerezo, A., García-Martínez, J.R.: Fuzzy logic controller for UAV with gains optimized via genetic algorithm. Heliyon. **10**(4), e26363 (2024)
8. Rodríguez-Abreo, O., Aviles, M., Rodríguez-Reséndiz, J., García-Cerezo, A.: Neural networks and genetic algorithms-based self-adjustment system for a backstepping controller of an unmanned aerial vehicle. Alex. Eng. J. **126**, 70–80 (2025)
9. Goldberg, D.E.: Genetic Algorithms in Search, Optimization, and Machine Learning. Addison-Wesley, Reading, MA (1989)

Finite-Horizon LQR and Kalman Estimator Design for Robust Lateral Dynamics Control in Autonomous Driving

Merah Abdelkader[1](✉), Hartani Kada[1], and Aouadj Norediene[2]

[1] Department of Electrotechnical and Automatic, University of Saida Dr. Moulay Tahar, Saida, Algeria
abdelkader.merah@univ-saida.dz, merah_aek@yahoo.fr
[2] School of Electrical and Energetic Engineering of Oran, Oran, Algeria

Abstract. This work proposes a combined design of a robust finite-horizon LQR controller together with a Kalman state estimator for lateral vehicle guiding in the face of realistic disturbances and model uncertainty. Following the Riccati-ZXL form, the suggested control architecture dynamically changes the weighting matrices Q and R depending on the degree of external disturbances. Simultaneously, the Kalman filter is meant to increase state estimate accuracy under dynamic uncertainty and sensor noise. Lateral acceleration, lateral deviation, steering control effort, and heading angle error are the several performance criteria underlying the evaluation framework. These outputs are specifically limited to guarantee passenger comfort and vehicle safety, therefore addressing physical limits. In terms of resilience and trajectory tracking performance under uncertain and dynamic conditions, simulation results reveal that the proposed approach (FHS-LQR + Kalman) much beats the standard LQR strategy.

Keywords: Lateral Vehicle Control · Finite-Horizon LQR · Kalman Filtering · Realistic Disturbances

1 Introduction

Modern cars now almost always feature Advanced Driver Assistance Systems (ADAS), which improve comfort and safety by helping drivers with different duties. Among these, Lane Keeping Assist Systems (LKAS) are absolutely vital in keeping the car in its lane, therefore lowering the danger of inadvertent lane departures and related collisions. When driving inattention or under severe road circumstances, LKAS's efficacy is very important. Many control techniques have been investigated in order to reach exact lateral control. Because it can properly manage multivariable systems and constraints, model predictive control (MPC) has become somewhat well-known. For LKAS applications, for example, a thorough investigation of MPC showed its capacity to preserve lane position while honouring vehicle dynamics and comfort limits [1]. Analogous methodologies for their simplicity and optimality in linear systems have been applied for

M. Hatti (Ed.): IC-AIRES 2025, LNNS 1740, pp. 448–461, 2026.
https://doi.org/10.1007/978-3-032-12788-4_43

Linear Quadratic Regulator (LQR). LQR for lateral control inside the Autoware ecosystem shown good performance in trajectory following challenges [2]. But conventional LQR designs can presume infinite horizons and might not sufficiently handle time-varying dynamics or disturbances seen in real-world driving. Proposed to overcome these constraints are finite-horizon LQR (FH-LQR) techniques. For FH-LQR control of quadrotors, for instance, a study showed how well finite-horizon methods handled trajectory tracking with enhanced confidence [3]. These techniques are appropriate for dynamic driving settings since they recalculate control laws over limited intervals, therefore providing more flexibility. Building on these developments, this work proposes a collaborative design of a Kalman state estimator suited for lateral vehicle steering under realistic disturbances and model uncertainties and a strong finite-horizon LQR controller. The suggested method dynamically changes the weighting matrices depending on perturbation levels, hence improving estimation accuracy and robustness. Strict physical limitations guarantee comfort and safety, hence the performance is assessed using several criteria: lateral acceleration, deviation, steering effort, and heading error. Tracking performance and robustness, simulation results show that the suggested approach is better than standard LQR techniques.

2 Elaboration of the Global Model DVR

2.1 Bicycle Model

Simplifying assumptions are used to obtain a usable linear bicycle model in the synthesis of assistance [4] as illustrated in Fig. 1

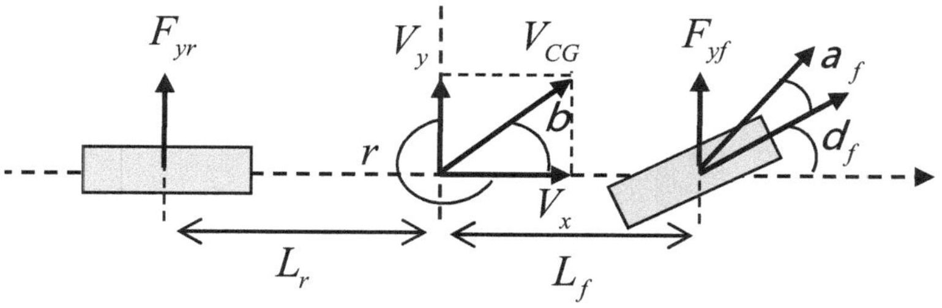

Fig. 1. Time Bicycle model

The state representation of this model is as follows:

$$\begin{bmatrix} \dot{\beta} \\ \dot{r} \end{bmatrix} = \begin{bmatrix} -2\frac{C_f+C_y}{M_v v_x} & 2\frac{-C_f L_f+C_r L_r}{M_v v_x^2} - 1 \\ 2\frac{-C_f L_f+C_r L_r}{J_v} & -2\frac{C_f L_f^2+C_r L_r^2}{J_v v_x} \end{bmatrix} \begin{bmatrix} V_y \\ r \end{bmatrix} + \begin{bmatrix} \frac{2C_f}{M_v v_x} \\ \frac{2C_f L_f}{J_v} \end{bmatrix} \delta_f + \begin{bmatrix} 0 \\ \frac{1}{J_v} \end{bmatrix} M_z + \begin{bmatrix} \frac{1}{M_v v_x} \\ \frac{L_w}{J_v} \end{bmatrix} F_w \quad (1)$$

Where β representing the side slip angle, r the yaw rate, ψ_l the yaw angle error and F_w represent the resultant of the wind forces actuating at a distance L_w, of the vehicle center of gravity, δ_f is the control input to the system, is the steering angle.

A. THE VISION SYSTEM

The control problem that interests us is attaining a desired lateral position, either on a straight line or a circle of radius "R_d", with a travel speed "V_X". To describe the vehicle motion relative to the road, a vision system is responsible for providing the information of the vehicle position and the road geometry. This system provides two measures, the direction error (ψ_l) and Lateral error (Y_L) [4].

$$\begin{aligned}\dot{\psi}_l &= r - V_x \rho_{ref} \\ \dot{Y}_L &= \beta V_x + l_s r + \psi_l V_x\end{aligned} \tag{2}$$

Where ρ_{ref} denotes the curvature of the reference path, V_x denotes the longitudinal velocity, β denotes the sideslip vehicle angle, and l_s denotes the look-ahead distance of the vision system as shown in Fig. 2.

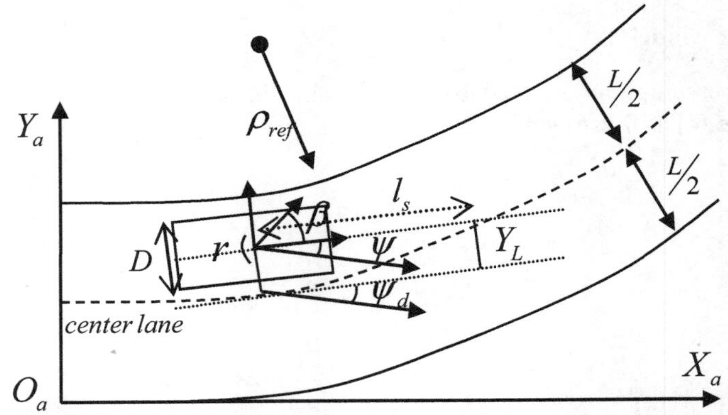

Fig. 2. The vehicle positioning on the road

2.2 Steering Dynamics

The steering assistance is provided by a DC motor mounted on the steering column as shown in Fig. 3. Where Γ_a is the assistance torque, Γ_s is the self-aligning torque and I_s is the inertial moment of steering system. We go to the angle of the wheels by simply report R_s reduction. B_s is the damping coefficient of the steering column, δ_f is the steering angle of the front wheels, and δ_d is wheel steering angle. If a vehicle is traveling on a straight road or a road with a small radius of curvature, then the slip angles and the steering angles are assumed to be very small [5].

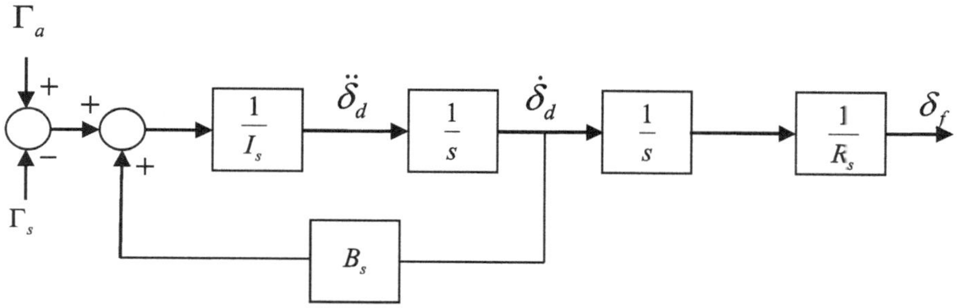

Fig. 3. Steering Column model

The state representation of this model is as follows:

$$\begin{bmatrix} \dot{\delta}_d \\ \ddot{\delta}_d \end{bmatrix} = \begin{bmatrix} 0 & 1 \\ 0 & -\frac{B_s}{I_s} \end{bmatrix} \begin{bmatrix} \delta_d \\ \dot{\delta}_d \end{bmatrix} + \begin{bmatrix} 0 & 0 & 0 \\ \frac{1}{I_s} & \frac{1}{I_s} & -\frac{1}{I_s} \end{bmatrix} \begin{bmatrix} \Gamma_a \\ \Gamma_d \\ \Gamma_s \end{bmatrix} \tag{3}$$

$$\delta_f = \begin{bmatrix} \frac{1}{R_s} & 0 \end{bmatrix} \begin{bmatrix} \delta_d \\ \dot{\delta}_d \end{bmatrix}$$

The self-aligning torque (Γ_s) may be given by:

$$\Gamma_s = -\Gamma_{sb}\beta - \Gamma_{sr}r + \frac{2C_y\eta_t}{R_sV_x}$$

The components of the self- aligning torque are given by:

$$\Gamma_{Sb} = \frac{2k_mC_y\eta_t}{R_s}, \quad \Gamma_{sr} = \frac{2C_yL_f\eta_t}{R_sV_x}$$

Where η_t is the tire contact length, k_m is the manual steering column coefficient, and L_f is the distance between the CG of the vehicle and the front wheel.

2.3 Road Vehicle Model (RV)

A road vehicle is the model that integrates the evolution of dynamics of the vehicle (bicycle), the visual system and the steering system. This model is obtained by concatenating [5] in the form:

$$\dot{x} = Ax + Bu + B_1w$$

$$\begin{bmatrix} \dot{\beta} \\ \dot{r} \\ \dot{\psi}_l \\ \dot{Y}_L \\ \dot{\delta}_d \\ \ddot{\delta}_d \end{bmatrix} = \begin{bmatrix} a_{11} & a_{12} & 0 & 0 & a_{15} & 0 \\ a_{21} & a_{22} & 0 & 0 & a_{25} & 0 \\ 0 & 1 & 0 & 0 & 0 & 0 \\ V_x & l_s & V_x & 0 & 0 & 0 \\ 0 & 0 & 0 & 0 & 0 & 1 \\ a_{51} & a_{52} & 0 & 0 & a_{55} & a_{56} \end{bmatrix} \begin{bmatrix} \beta \\ r \\ \psi_l \\ Y_L \\ \delta_d \\ \dot{\delta}_d \end{bmatrix} + \begin{bmatrix} 0 \\ 0 \\ 0 \\ 0 \\ 0 \\ \frac{1}{J_v} \end{bmatrix} \Gamma_a + \begin{bmatrix} h_1 & 0 \\ h_2 & 0 \\ 0 & -V_x \\ 0 & -l_sV_x \\ 0 & 0 \\ 0 & 0 \end{bmatrix} \begin{bmatrix} F_w \\ \rho_{ref} \end{bmatrix} \tag{4}$$

$a_{15} = \frac{e_1}{R_s}$ $a_{25} = \frac{e_2}{R_s}$ $a_{51} = \frac{\Gamma_{sb}}{J_v}$ $a_{52} = \frac{\Gamma_{sr}}{J_v}$ $a_{55} = -\frac{C_{sb}}{J_v R_s}$ $a_{56} = -\frac{B_s}{J_v}$ $\Gamma_{sb} = \frac{2k_m C_f \eta_t}{R_s}$ $\Gamma_{sr} = \frac{2k_m C_f \eta_t}{R_s}\frac{L_f}{V_x}$ $a_{11} = -2\frac{(C_r+C_f)}{M_v V_x}$ $a_{12} = -1 + 2\frac{(C_r L_r + C_f L_f)}{M_v V_x{}^2}$ $2\frac{(C_r L_r - C_f L_f)}{I_z}$ $a_{22} = -2\frac{(C_r L_r{}^2 + C_f L_f{}^2)}{V_x I_z}$ $e_1 = \frac{2C_f}{M_v V_x}$ $e_2 = \frac{2C_f l_f}{I_z}$ $h_1 = \frac{1}{M_V V_x}$ $h_2 = \frac{L_w}{I_z}$

3 Controller Design

The general architecture considered for the shared lateral control system is recalled in (Fig. 4)

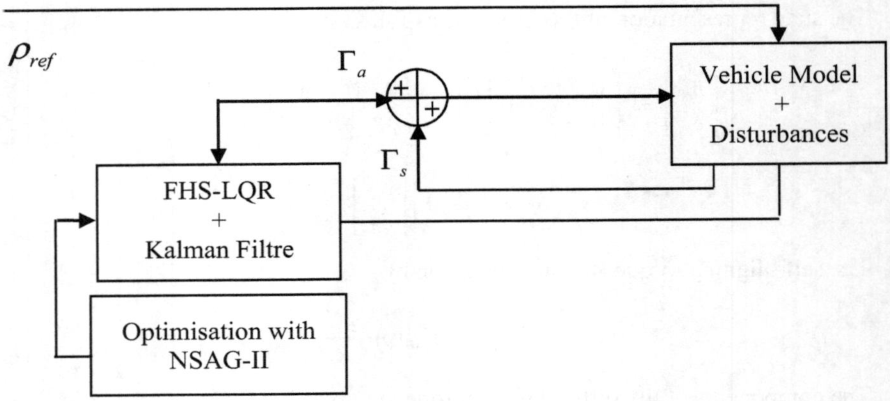

Fig. 4. Control architecture.

3.1 Classical Finite-Horizon LQR

The classical Finite-Horizon Linear Quadratic Regulator (FH-LQR) aims to minimize the cost [6]:

$$J = \sum_{k=0}^{N-1}\left(x_k^\top Q x_k + u_k^\top R u_k\right) + x_N^\top Q_f x_N \tag{5}$$

for the system:

$$x_{k+1} = Ax_k + Bu_k \tag{6}$$

The solution proceeds via backward Riccati recursion:

$$P_k = Q + A^\top P_{k+1} A - A^\top P_{k+1} B\left(R + B^\top P_{k+1} B\right)^{-1} B^\top P_{k+1} A \tag{7}$$

The feedback gain is:

$$K_k = \left(R + B^\top P_{k+1} B\right)^{-1} B^\top P_{k+1} A \tag{8}$$

And the control law is:

$$u_k = -K_k x_k \tag{9}$$

This formulation assumes perfect knowledge of the dynamics and no control delay.

3.2 Disturbance Modeling

In realistic driving conditions, lateral vehicle dynamics are influenced by multiple sources of uncertainty, including modeling imperfections, actuator saturation, sensor noise, and especially road curvature. To reflect this, we introduce [7]:

- $\omega_A(k)$: uncertainty affecting the system dynamics matrix A,

$\omega_B(k)$: uncertainty acting on the input matrix B,$q(k)$: the road curvature profile, representing the driver's desired trajectory or reference path,

- G: the input gain matrix linking curvature to system excitation.

The term G. $q(k)$ represents the lateral excitation induced by road curvature. Unlike process noise, this input is deterministic and models the curved trajectory to follow. The curvature $q(k)$ is predefined and discretized over time. The augmented system dynamics with delay and perturbations is written as:

$$x_{k+1} = (A + \omega_A(k)A)x_k + (B + \omega_B(k)B)u_{k-d} + Gq(k) \tag{10}$$

To reflect chassis flexibility or unmodeled variations, $\omega_A(k)$ is modeled as a low-frequency stochastic signal:

$$\omega_A(k) = \alpha_A \begin{pmatrix} 0.2\sin(2\pi \cdot 0.7t_k) + 0.1\sin(2\pi \cdot 0.4t_k) + 0.1 \cdot \mathcal{N}(0,1) \\ + 0.01 \cdot \text{cumsum}(\mathcal{N}(0,1)) \cdot dt \end{pmatrix} \tag{11}$$

The term $\omega_B(k)$ accounts for actuator saturation, variable steering efficiency, or lateral wind forces:

$$\omega_B(k) = \sigma_B \begin{pmatrix} 0.1 \cdot \mathcal{N}(0,1) + 0.3 \cdot \text{square}(2\pi \cdot 0.25t_k + \phi) \\ + 0.2 \cdot \sin(2\pi \cdot 0.5t_k + \psi) \end{pmatrix} \tag{12}$$

Impulse-like disturbances are also injected at fixed time steps to emulate wind gusts or abrupt driver inputs:

$$\omega_B(30\%) += 0.4$$

$$\omega_B(80\%) -= 0.4$$

Figure 5 shows two forms of disturbances brought into the lateral dynamics model of the vehicle as specified in Eqs. (11) and (12). Eq. (11) models the perturbation () as a low-frequency Gaussian noise, hence the blue signal (top row) corresponds to it. It reflects structural ambiguities such chassis flexibility, modelling mistakes, or sensor noise. Consistent with a stochastic excitation profile, its histogram reveals a symmetric, bell-shaped distribution centred on zero.

By comparison, the red signal (bottom row) reflects the perturbation (), as indicated in Eq. (12), marked by impulsive disturbances applied at set time steps. These simulations sudden events such actuator saturation or lateral wind gusts. The bimodal distribution of the histogram points to non-Gaussian and discontinuous input properties.

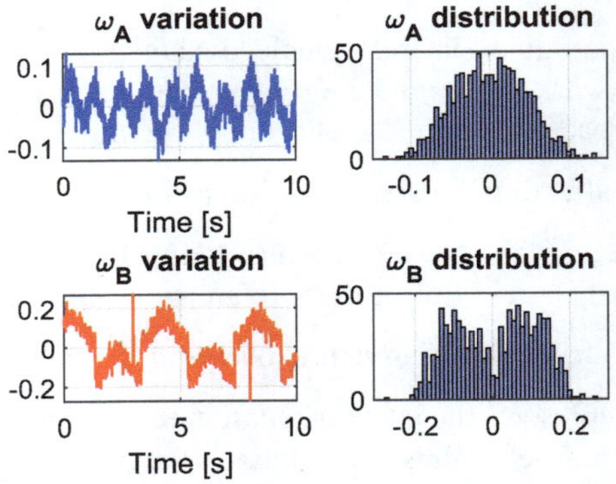

Fig. 5. Temporal and statistical representation of modeled disturbances

The normalized forms:

$$\omega_{A,\mathrm{norm}}(k) = \frac{|\ \omega_A(k)\ |}{\max(|\omega_A|) + \epsilon}, \quad \omega_{B,\mathrm{norm}}(k) = \frac{|\ \omega_B(k)\ |}{\max(|\omega_B|) + \epsilon} \tag{13}$$

are used to modulate the adaptive weights Q_k and R_k in the Riccati-ZXL formulation.

3.3 Riccati-ZXL with Perturbation-Aware Weighting and Command Delay)

To account for time-varying perturbations and actuation delays, the Riccati-ZXL method adapts the LQR design over a finite horizon. The perturbed dynamics are [8]:

$$x_{k+1} = (A + \omega_A(k)A)x_k + (B + \omega_B(k)B)u_{k-d} + Gq(k) \tag{14}$$

The time-varying cost function uses:

$$Q_k = Q + \alpha_q W_k, \quad R_k = R + \beta_r(\omega_A(k) + \omega_B(k)) \tag{15}$$

with $W_k =$ diag $(\omega_A, \omega_A, \omega_A, \omega_B, \omega_B, \omega_B)$.

The Riccati recursion becomes:

$$P_k = Q_k + A^\top P_{k+1}A + \alpha_A^2 A^\top P_{k+1}A - A^\top P_{k+1}BK_k \tag{16}$$

$$K_k = \left(R_k + B^\top P_{k+1}B + \sigma_B^2 B^\top P_{k+1}B\right)^{-1}\left(B^\top P_{k+1}A + \sigma_B^2 B^\top P_{k+1}A\right) \quad (17)$$

The control law, applied with a delay d, is:

$$u_k = -K_k\hat{x}_k, \quad u_k^{\text{applied}} = u_{k-d} \quad (18)$$

where $\hat{x}_k$ is the estimated state from the Kalman filter

3.4 Kalman Filter for State Estimation)

A discrete Kalman filter estimates the full system state from noisy measurements [9]:

$$\hat{x}_{k+1} = A\hat{x}_k + Bu_k + L(y_k - C\hat{x}_k) \quad (19)$$

The observer gain L is calculated from the dual of the Riccati equation:

$$P_e = AP_eA^\top + Q_w - AP_eC^\top\left(CP_eC^\top + R_w\right)^{-1}CP_eA^\top \quad (20)$$

$$L = AP_eC^\top\left(CP_eC^\top + R_w\right)^{-1} \quad (21)$$

where Q_w and R_w are the process and measurement noise covariance matrices, respectively.

3.5 Joint Multi-Objective Tuning with NSGA-II and Constraints

To simultaneously calibrate the controller and observer gains for robustness, accuracy, and comfort, we employ the Non-dominated Sorting Genetic Algorithm II (NSGA-II) for multi-objective tuning. The decision vector includes all critical parameters [10]:

$$\theta = \left[\text{diag}(Q), R, \text{diag}(Q_w), \text{diag}(R_w)\right] \quad (22)$$

Each candidate solution θ is evaluated on the basis of the following normalized objectives:

- $J_1 = \frac{\text{RMS}(a_y)}{0.2g}$ — ride comfort,
- $J_2 = \frac{\text{RMS}(Y_L)}{0.25}$ — lane-keeping accuracy,
- $J_3 = \frac{\text{RMS}(C_a)}{10}$ — steering effort.

These objectives are computed after simulating the full system dynamics using Riccati-ZXL, the Kalman observer, and a control delay.

To guarantee safety and comfort, strict bounds are imposed on the system outputs:

$$\begin{aligned} &\mid a_y(k) \mid \leq 0.2g \\ &\mid Y_L(k) \mid \leq 0.25 \text{ m} \\ &\mid C_a(k) \mid \leq 10 \text{ N} \\ &\mid \psi(k) \mid \leq 0.05 \text{ rad} \end{aligned} \quad (23)$$

If any of these constraints is violated during simulation, a high penalty λ_p is added to all objectives:

$$J_i \leftarrow J_i + \lambda_p \quad \text{if constraint violated}, \quad \lambda_p = 10^5$$

This penalising scheme guarantees that NSGA-II only keeps physically valid and performable candidates in the last Pareto front. The procedure of multi-objective tuning consists in three key phases [11]:

- **Initialisation**: Random population generation constrained by θ.
- **Evaluation**: Evaluate each individual by modelling the closed-loop system and computing J_1, J_2, J_3 and verifying physical restrictions; impose penalties should they be broken.
- **Non-Dominated Sorting**: Rank people according to Pareto dominance into several fronts non-dominated.
  Crowding Distance Assignment: keep population variety.
- **Selection, Crossover & Mutation**: Apply elitist genetic operators to change the population.
- **Loop Until Convergence**: Evaluate and sort again until a Pareto front that meets your needs is obtained.

4 Simulation Results

In order to assess the reliability of the proposed control system in practical scenarios, the vehicle is modelled along a highway-like reference trajectory characterised by a segmented yet continuous curvature profile.

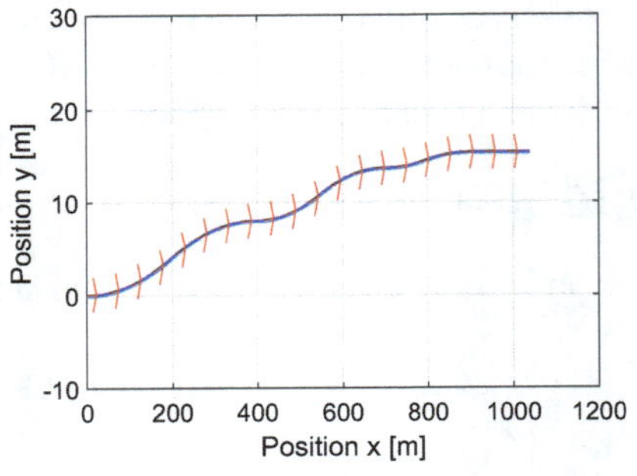

Fig. 6. Actual Trajectory on Highway Profile

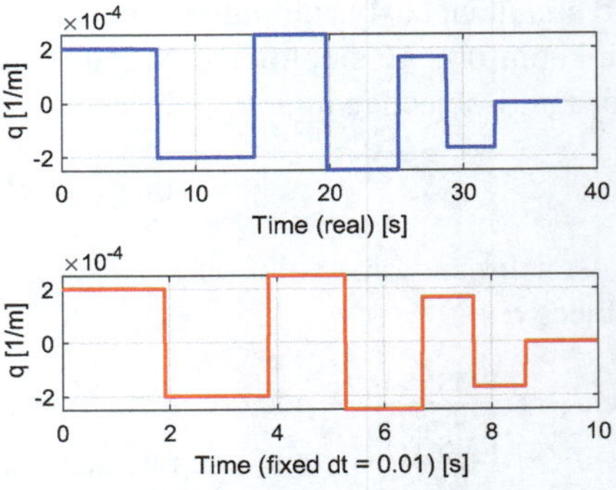

Fig. 7. Reference road curvature

The trajectory depicted in Fig. 6 spans a total distance of 1040 m, accomplished in a duration of 37.44 s while maintaining a consistent longitudinal speed of 100 km/h. This configuration facilitates the complete engagement of the vehicle's lateral dynamics, resulting in notable lateral accelerations induced by curvature. Figure 7 illustrates the evolution of road curvature across two distinct time scales. The upper plot illustrates the curvature in relation to real time (t_"real"), effectively representing the actual duration the vehicle occupies each segment of the road. The lower plot employs a fixed time step of dt = 0.01 "s," a practice frequently utilised in control simulations. While both curves illustrate the same road profile, they exhibit differences in their temporal distributions. The fixed-step time axis often results in the artificial stretching or compression of certain segments. However, employing a fixed step is essential for discretising the vehicle's dynamic model and executing finite-horizon optimal control strategies, such as Linear Quadratic Regulator (LQR), which depend on consistent and predictable update intervals. Consequently, the rest of the vehicle dynamics simulation is conducted with a fixed time step of 0.01 s, which helps maintain alignment with the control algorithm. At the same time, road curvature is continuously interpreted in real time to maintain physical clarity in visual representations.

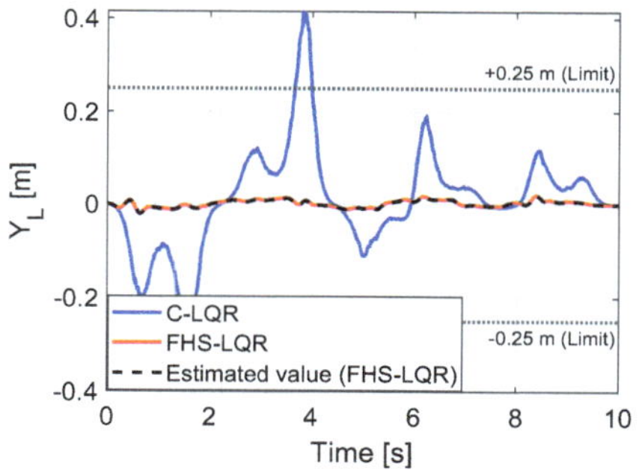

Fig. 8. Lateral error

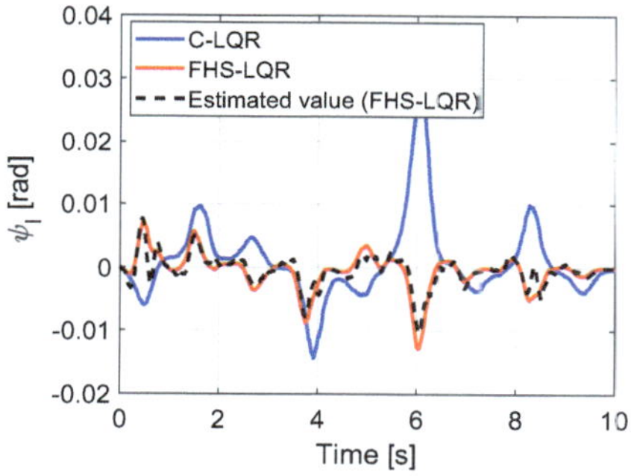

Fig. 9. Direction error.

Figures 8 and 9 depict the lateral displacement error (Y_L) and the direction error (ψ_l).) of the vehicle, respectively. Both elements are essential for evaluating tracking performance and ensuring alignment in orientation. The established limits, $|Y_L| \leq 0.25$ "m" and $|\psi_l| \leq 0.05$ "rad", are typically adhered to by the robust FHS-LQR controller. In contrast, the classical LQR tends to show greater deviations, especially during the transitions between curvature segments. The Kalman filter is instrumental in these findings, offering precise state estimates that enhance the quality of feedback. This is evident in the smoother and more consistent estimation curves that are plotted alongside the actual states in both figures.

Figures 10 and 11 illustrate the yaw rate (r) and the sideslip angle (β), both of which play a crucial role in determining vehicle stability and agility. High yaw rates or significant sideslip can result in instability, particularly when navigating sharp curves. The FHS-LQR controller shows improved reduction of these dynamic effects, keeping both variables within stable limits. The classical LQR, constrained by its failure to consider disturbances or uncertainties in the model, results in more pronounced oscillations in these two states. The Kalman-based state estimation once more proves valuable in reconstructing unmeasured states, such as sideslip, and offers the control algorithm dependable feedback inputs, which is especially advantageous in uncertain conditions.

Figures 12 and 13 illustrate the progression of the steering torque (C_a) and the steering wheel angle (δ_d), which serve as direct indicators of the control effort exerted by each strategy. The observations from both plots demonstrate that the classical LQR

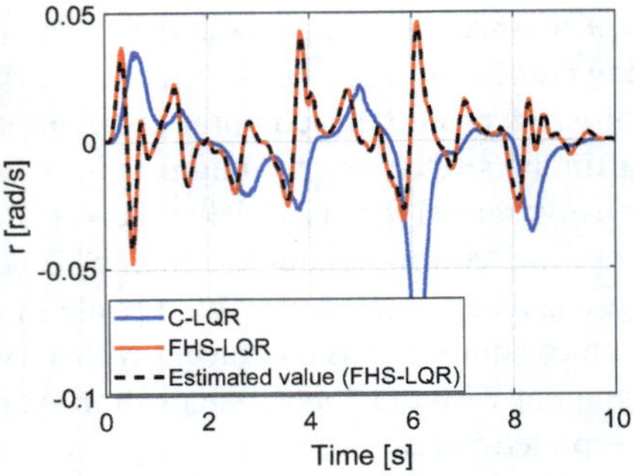

Fig. 10. Yaw rate

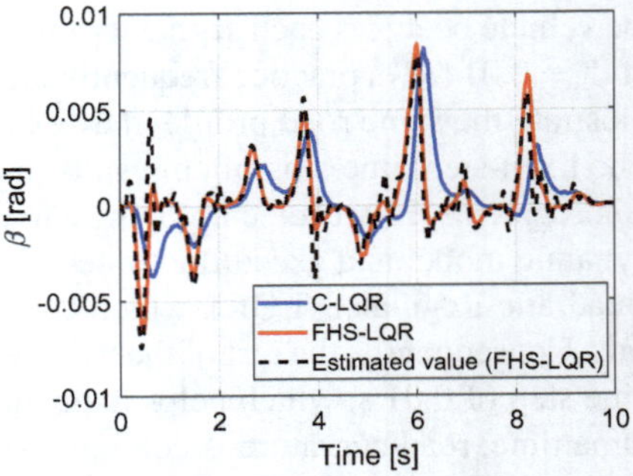

Fig. 11. Sideslip angle

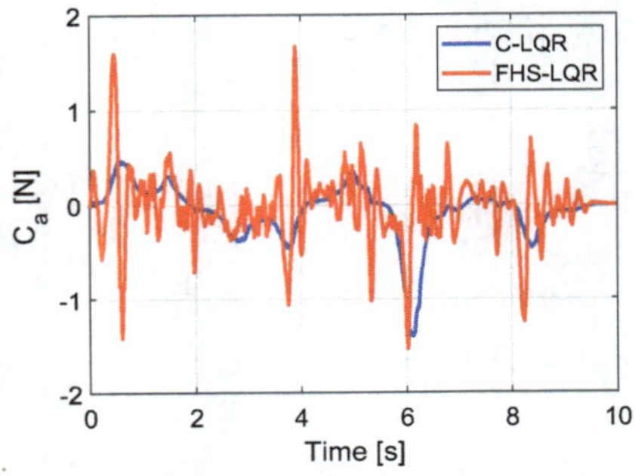

Fig. 12. Steering torque

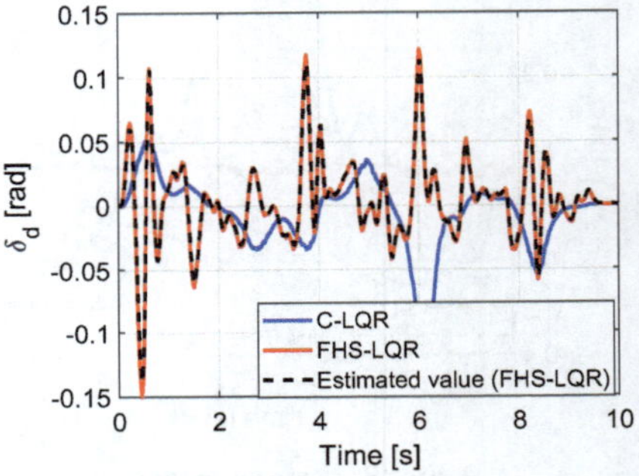

Fig. 13. Steering Wheel

and the FHS-LQR controllers effectively maintain their responses within acceptable limits, signifying that neither surpasses the actuator constraints.

Figure 12 illustrates that the steering torque is maintained within limits for both controllers; however, the FHS-LQR demonstrates somewhat greater oscillations as a result of its dynamic response to disturbances. This indicates a greater level of reactivity, while still maintaining stability and adhering to actuator saturation limits. Figure 13 illustrates the time response of the steering angle, along with the estimated value generated by the Kalman filter. The estimation aligns closely with the actual state under FHS-LQR, highlighting the observer's effectiveness in reconstructing dynamics that are not directly measured. The control laws effectively keep the angle within safe limits, demonstrating their dependability in facilitating accurate vehicle manoeuvring.

These findings underscore the significance of incorporating command delay into the control design process. In real-world applications, actuators exhibit a delay in their response, and overlooking this delay may result in phase shifts, instability, or compromised tracking performance. The FHS-LQR framework explicitly incorporates this delay, allowing for the computation of control actions with an anticipation of future system behaviour. This greatly enhances the system's accuracy and reliability, particularly in situations where rapid or frequent changes in curvature occur.

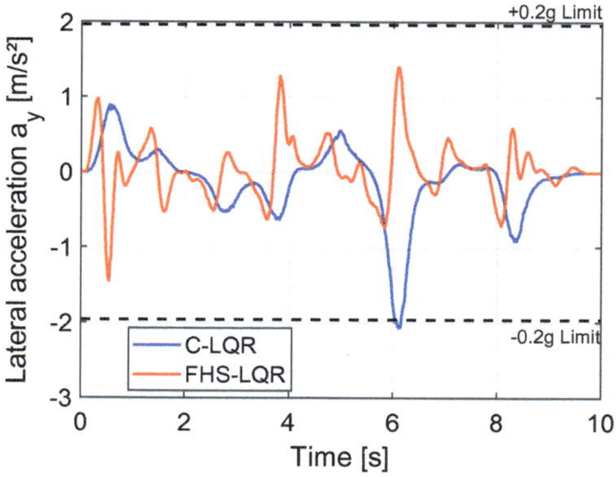

Fig. 14. Lateral acceleration

Figure 14 illustrates the lateral acceleration (a_y) of the vehicle, adhering to the comfort constraint $|a_y| \leq 0.2g$. This variable is experienced directly by passengers and plays a vital role in ensuring ride comfort while navigating curves. The traditional Linear Quadratic Regulator (LQR) can occasionally generate lateral accelerations that come close to or surpass this threshold, especially in areas characterised by significant curvature. In contrast, the FHS-LQR controller keeps a_y comfortably within specified limits, leading to a more pleasant ride experience and enhanced comfort. This highlights the significance of incorporating robustness and state estimation into control design for effective operation in real-world driving scenarios.

5 Conclusion

This work examined under realistic conditions the performance and robustness of a finite-horizon control approach for lateral vehicle guidance along a highway-like track. Denoted FHS-LQR, an upgraded formulation including time-varying weighting, disturbance adaptation, and command delay compensation, was compared to a standard LQR controller. A complete simulation framework including curvature-induced road profiles, sensor-actuator delays, and external disturbances modelled by dynamic perturbations was used to evaluate the control system. The results showed that although the standard LQR guarantees stability and fundamental tracking performance, it ignores dynamic restrictions in the presence of noise and delay. While obeying comfort-related constraints such lateral acceleration ($\leq 0.2g$) and steering torque, the FHS-LQR controller regularly maintained critical state variables including lateral deviation, yaw angle, and sideslip including their permissible limitations. Using a discrete-time Kalman filter helped to further enhance state estimate and lower the influence of unmeasured disturbances. The study verified the vital need of including system uncertainties and command latency into control design. The suggested FHS-LQR solution improves the safety and comfort of autonomous lateral vehicle control by specifically addressing these difficulties, therefore qualifying for deployment in advanced driver assistance systems (ADAS) and autonomous driving applications.

Appendix

Table A1. The Specifications of the Steering Colum Used in Simulation

B_s	Steering system damping coeffcient	0.5
I_s	Inertial moment of steering system	0.05
K_m	Manual steering column coffcient	0.088
R_s	Steering gear ratio	16

Table A2. The Specifications of the Vehicle Used in Simulation

M_v	Vehicle mass	1662 kg
J_v	Vehicle inertia	2630 kg.m^2
L_f	Distance from the CG to front axle	1,104 m
L_r	Distance from the CG to rear axle	1,421 m
C_f	Longitudinal stiffness of each tire lateral	37407N/rad
C_r	Lateral stiffness of each tire lateral	51918N/rad
l_s	a lookahead distance	5 m
η_t	Tire contact length	0.13m
V_x	Longitudinal velocity	100 kh/h

References

1. Garcia, J., Teixeira, E., Murilo, A., Rodrigues, R.: A comprehensive analysis of model predictive control for lane keeping assist system. IEEE Access. **11**, 12345–12356 (2023)
2. Mike, M.: Linear Quadratic Regulator Lateral Path Tracking Controller. GitHub Repository (2021) https://github.com/mgmike1011/lqr_lateral_controller
3. Cohen, M.R., Abdulrahim, K., Forbes, J.R.: Finite-horizon LQR control of quadrotors on $SE_2(3)$. IEEE Robot. Autom. Lett. **6**(3), 1234–1241 (2021)
4. Merah, A., Hartani, K.: Shared steering control between a human and an automation designed for low curvature road. Int. J. Veh. Saf. **9**(2), 136–158 (2016). https://doi.org/10.1504/IJVS.2016.079655
5. Merah, A., Hartani, K., Draou, A.: A new shared control for lane keeping and road departure prevention. Veh. Syst. Dyn. **54**(1), 1–16 (2016). https://doi.org/10.1080/00423114.2015.1115882
6. Kumar, S., Mueller, A., Wensing, P., Kirchner, F.: Towards Continuous Time Finite Horizon LQR Control in SE(3). *arXiv preprint* https://arxiv.org/abs/2307.14164 (2023). https://arxiv.org/abs/2307.14164

7. Xie, K., Guay, M., Wang, S., Deng, F., Lu, M.: Optimal Output Feedback Learning Control for Discrete-Time Linear Quadratic Regulation. *arXiv preprint* https://arxiv.org/abs/2503.06226 (2025). https://arxiv.org/abs/2503.06226
8. Zhao, S., Wang, J., Xu, H., Wang, H.: Finite horizon robust optimal tracking control based on approximate dynamic programming for switched systems with uncertainties. Int. J. Control. Autom. Syst. **20**, 1051–1062 (2022). https://doi.org/10.1007/s12555-020-0982-8
9. Khosravi, M., Azarinfar, H., Sabzevari, K.: Design of infinite horizon LQR controller for discrete delay systems in satellite orbit control: A predictive controller and reduction method approach. Heliyon. **10**(1), e24265 (2024). https://doi.org/10.1016/j.heliyon.2024.e24265
10. Xiong, Q., Dong, L., Chen, H., Zhu, X., Zhao, X., Gao, X.: Enhanced NSGA-II algorithm based on novel hybrid crossover operator to optimise water supply and ecology of Fenhe reservoir operation. Sci. Rep. **14**, 31621 (2024). https://doi.org/10.1038/s41598-024-80419-w
11. Acerce, A., Denizhan, B.: Application of the non-dominated sorting genetic Algorithm II (NSGA-II) in a Two-Echelon cold supply chain. Systems. **13**(3), 206 (2025). https://doi.org/10.3390/systems13030206

A Novel Nature-Inspired Approach for Wind Farm Location Optimization Considering Wake Effects

Mourad Naidji[1(✉)], Alla Eddine Toubal Maamar[2], Mohamed Ilyas Rahal[3], and Rachid Taleb[4]

[1] Department of Electrical Engineering, Laboratory of Electrical Engineering (LGE), University of M'Sila, P.O. Box 166 Ichebilia, M'Sila 28000, Algeria
mourad.naidji@univ-msila.dz

[2] Laboratoire Ingénierie des Systèmes et Télécommunications (LIST), Faculty of Technology, University of M'hamed Bougara of Boumerdes, Frantz Fanon city, Boumerdes 35000, Algeria

[3] Department Department of Electronics, Laboratory of Automation and Signals of Annaba (LASA), Badji Mokthar-Annaba University, P.o. Box 12, Annaba 23000, Algeria

[4] Laboratoire Genie Electrique et Energies Renouvelables (LGEER), Research Centre for Scientific and Technical Information (CERIST), Hassiba Benbouali University of Chlef, Chlef, Algeria

Abstract. The optimal location of the wind turbines (WTs) is a critical component in the design of the WT, which can guarantee maximum output power. For that, several recent methodologies have been carried out for optimizing WTs in a wind farm using different optimization algorithms. This paper proposes a novel approach for the optimization of wind farm layout using Quantum behaved particle swarm optimization (QPSO) algorithm. Three case studies are considered to express the presence of wake impact. The effectiveness of the proposed approach is validated through simulations conducted in MATLAB. The performance findings are compared against results from previously published works, revealing that the proposed optimization technique consistently achieves higher total power output and improves the total efficiency of the WT.

Keywords: Wind farm location optimization · Wake model · Quantum-behaved particle swarm optimization (QPSO) algorithm · Wind turbine

1 Introduction

The growing of global demand for clean and renewable energy sources (RES) led to a rapid increase in wind energy development [1–4]. As one of the most advanced and economically viable sources of RES, wind power is indispensable to the transition toward sustainable energy systems. However, the efficiency of the wind farms are highly dependent on the optimal location of wind turbines (WTs). One of the most powerful parameters that affect this efficiency is the wake effect, the decrease in wind speed and rise in turbulence downstream of a WT due to its operation. In the last decade, a number

M. Hatti (Ed.): IC-AIRES 2025, LNNS 1740, pp. 462–471, 2026.
https://doi.org/10.1007/978-3-032-12788-4_44

of optimization approaches have proposed to address WFLO problems. They vary from classical mathematical modeling to much more advanced nature-inspired metaheuristic approaches. Of them, Swarm Intelligence (SI) algorithms such as Particle Swarm Optimization (PSO), Genetic Algorithms (GA), and their hybrids have been highly promising because they can address the nonlinearity, multi-modality of WFLO problems. There are some considerable studies that have touched on certain axes of the WFLO problem. For instance, area rotation technique and fixed point choice method in [5] tried to align wind farm arrangements with freestream wind directions for more efficient power generation. Yet, such methods were susceptible to slow convergence and computational cost. One of the initial investigations by [6] optimized the quantity and positioning of WTs for different wind conditions in formulating the WFLO problem, which was further improved in [7] with altered of GA parameters. There were subsequent developments in [8] to suggest enlarging wind farm boundaries to create greater flexibility in the arrangement, with [9] looking at a threefold optimization of turbine spacing, location direction, and control methodology. Adaptive PSO was applied in [10] to develop greater capacity for global search with better trade-offs between power generation and capital expenditure. Other works have aimed to balance multiple objectives. A PSO variant with multiple adaptive mechanisms was investigated in [11] to further enhance performance, but lacked sufficient attention to turbine spacing and the integration of restricted zones. These developments reflect the tremendous step forward achieved in WFLO problem. However, there remain difficulties in the accurate modeling of wake effect, multi-dimensional design constraints, and compromise between power generation and cost-efficiency. In the paper, a novel hybrid algorithm is presented: Quantum Behaved Particle Swarm Optimization (QPSO). The approach aims at global search capability improvement by quantum-behaved PSO dynamics and solution diversity and local accuracy improvement by differential mutation.

2 Mathematical Modeling of the Wind Farm

One of the key objectives of (WT) location design is to reduce power loss caused by wake interferences among turbines. For an accurate estimation of these losses, a reliable wake model must be utilized. The Jensen model, originally introduced in [12], is among the most popular models in the literature. When wind hits a turbine, it decelerates and becomes turbulent and creates a wake, an area of decreased wind speed and heightened turbulence, directly downwind of the turbine.

The wake effect travels downstream and also moves sideways, impacting the functioning of downstream turbines. The "Jensen model", as shown in Fig. 1, assumes the wake to increase linearly with downstream distance from the WT and the wind speed profile within the wake to be flat across its cross-section. This is a simplification that makes it computationally inexpensive and therefore suitable for application to large-scale wind farm layout optimization problems. In this paper, the "Jensen model" is applied in the calculation of the wake velocity deficits, as a basis for the determination of the overall power production from the wind farm. Assuming that quantity of movement is

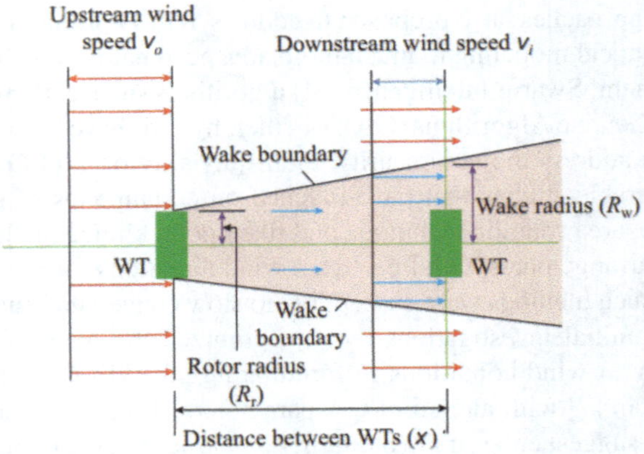

Fig. 1. Schematic diagram of the "Jensen" wake model.

preserved in the wake section, wind speed can be given as [13]:

$$v = v_0 \left[1 - \frac{2a}{\left(1 + \alpha \frac{x}{R_r}\right)} \right] \tag{1}$$

$$a = \frac{1 - \sqrt{1 - C_T}}{2} \tag{2}$$

$$R_r = r\sqrt{\frac{1-a}{1-2a}} \tag{3}$$

$$\alpha = \frac{1/2}{\ln\left(\frac{H}{z_0}\right)} \tag{4}$$

Here, v_0 represents the local wind speed that a turbine experiences when there's no intervention from any wakes. The variable indicates how far downstream the turbine is located, while *Rr* refers to the rotor radius of the WT that's upstream, and R_w points to the expanded rotor radius of the wake on the down stream site. The tower height is symbolized as H, while α is the entrainment constant, that helps define how quickly the wake expands. The axial induction factor, , shows how much the wind speed falls due to energy being extracted from the rotor. Additionally, C_T is thrust coefficient, that computes the force the wind applies on the turbine rotor, and $_0$ denotes the surface roughness of the ground in the wind farm area. When a WT is influenced by many upstream wakes, calculating the resulting wake velocity is not as straightforward as just adding them up. Instead, it is generally accepted that the total kinetic energy deficit at the downstream turbine is equal to the sum of the single energy deficits from each wake. Therefore, the effective wind speed that the ith turbine, situated downstream of

N_T turbines, experiences is determined by adding up these energy losses accordingly as:

$$v_i = v_0 \left[1 - \sqrt{\sum_{j=1}^{N_T} \left(1 - \frac{v_{ij}}{v_0} \right)^2} \right] \tag{5}$$

The variable v_{ij} refers to the wind speed which the ith WT encounters, impacted by the wake shaped by the jth WT. In the linear wake model, we assume that this wake spreads out in a conical shape as it travels downstream. The area affected by this wake is defined by the wake influence radius, which indicates how far the wake's impact reaches. This radius is calculated using a specific formula that takes into account the distance between the WTs and the wake decay constant as:

$$R_w = R_r + \alpha x \tag{6}$$

[6] introduced a basic economic model for WFLO. Within the model, the total cost mainly relies on the quantity of WTs installed. They normalized the annualized cost of a single turbine to a unit, which makes comparison more easily performed. To account for economies of scale, they presented a cost reduction factor that can provide a reduction of up to 1/3 in unit costs when more WTs are added. This is based on the idea that larger setups incur less per-unit infrastructure and maintenance costs. Therefore, the total cost for a wind farm with N turbines can be written as:

$$Cost_{Tot} = N \left(\frac{2}{3} + \frac{1}{3} e^{-0.00174N^2} \right) \tag{7}$$

The data relating to the WT and the wind farm are taken out from [6, 7] and are tabulated in Table 1.

Table 1. Windfarm data

Parameter	Symbol	Value
Tower height	H	60 m
Rotor diameter	2r	40 m
Wind Farm Surface Roughness	0	0.30 m
Rotor performance efficiency		0.40
Air density		1.225 kg/m^3
Thrust coefficient		0.88

Power output from a WT i, measured in kW, can be given as follows:

$$P_i = 0.5\ \rho\ \pi\ r^2\ v_i^3\ C_P / 1000 \tag{8}$$

Where P_i presents the total power achieved considering wake influence in the wind farm, that can be computed according to Table I, the approximation of Eq. (8) can give Eq. (9):

$$P_i = 0.3v_i^3 \tag{9}$$

In this paper, Eq. (9) is considered for power output calculations for comparing with previous results. For N WT, the total power output is given by the following expression:

$$P_{Tot} = \sum_{k=0}^{360} \sum_{i=1}^{N} \pi_j P_i(v_i) \tag{10}$$

Where π_j is wind probability which relates to the wind speed with a certain direction and $\sum_{j=0}^{360} \pi_j = 1$. The present power production from turbine i as a function of its wind speed v is . Consequently, the objective function can be stated as:

$$minF = \frac{Cost_{Tot}}{P_{Tot}} \tag{11}$$

Where the objective function is the quotient of the total cost to the total power. Windfarm efficiency η can be calculated as follow:

$$\eta = \frac{\sum_{k=0}^{360} \sum_{i=1}^{N} \pi_j P_i(v_i)}{\sum_{k=0}^{360} \sum_{i=1}^{N} \pi_j P_{i,\max}(v_{i,\max})} \tag{12}$$

Under this framework, $_{,\max}$ represents, assuming no reduction from wake-induced flow deficits, the theoretical peak power output of turbine at its maximum wind speed v..

3 Optimization Algorithm

Particle Swarm Optimization (PSO) Description: PSO is an evolutionary computation method inspired by the collective behavior of social organisms. This technique relies on how individual agents make decisions based on two key sources of information. The first is personal experience, each agent evaluates the choices it has previously made, identifying which option has yielded the most favorable outcome and how beneficial it was. The second source of information comes from the experiences of neighboring agents. That is, each agent observes and learns from the performance of others, recognizing which choices have led to optimal results within the swarm and how effective those solutions have been.

Figure 2 shows how a particle in the enhanced PSO algorithm updates its position. The new velocity (black arrow) is computed by merging three components: inertia (gray), memory of the particle's own best position (blue), and cooperation with the global best position found by the intelligence swarm (red). The particle travels from its actual position (black x) to a new position (black dot), directed by the present updated velocity. This mechanism can help balance survey and convergence in the optimization course. The complete and detailed description of the algorithm can be found in [15, 16]. As is

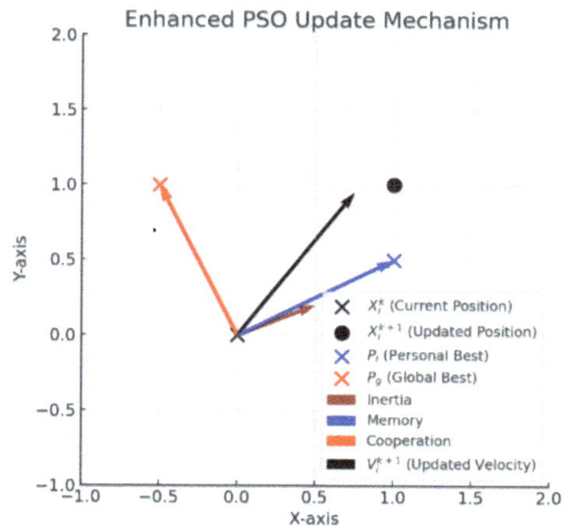

Fig. 2. Enhanced PSO velocity and position update mechanism. (Color figure online)

usual in classical PSO models, each particle's state in the Quantum-behaved Particle Swarm Optimization (QPSO) framework is characterized by a wave function rather than by explicit spatial and velocity vectors. This quantum representation brings a fundamentally new dynamic whereby the exact position and velocity of a particle cannot be simultaneously found. Rather, what is known is the location's probability distribution based on the squared wave function $|\Psi|^2$. The possible terrain the particle lives in shapes this probability density. A QPSO algorithm for WFLO problem is proposed.

$$x_{i,j}^{t+1} = \Omega_{i,j}^{t} \pm \beta \mid M_j^t - x_{i,j}^t \mid \ln\left(\frac{1}{u}\right) \text{ and } M_j^t = (1/m)\sum_{i=1}^{m} P_{best,ij}^t \tag{13}$$

Here, u is a random number in the interval (0,1) chosen from a uniform distribution. M presents the mean best position, calculated as the average of the personal best P_{best} locations of all particles. The sole parameter in QPSO is the contraction expansion factor β, which adjusts the behavior of the convergence of the algorithm and is typically fixed to a value not more than 1.7 for performance and stability [15, 16]. Due to its effectiveness and simplicity of implementation, QPSO has been successfully applied across a series of standard optimization problems with notable results. The general pseudocode structure of the QPSO algorithm is presented below:

4 Case Studies, Simulation Results and Discussion

The simulation is carried out on "Intel Core" (TM) i7 CPU@3.6 GHz and 12 GB of RAM and run the algorithm on the MATLAB averment in 14.08 s for the case 1, 479.64 s for case 2 and 618.25 s for case 3. The execution time is relatively long following the discretization of the wind direction into 36 sectors, so the algorithm consists of performing a calculation for the 36 wind directions. Several case studies carried out in [6, 7] with a wind farm zone of 2 km x 2 km equally splitted to 100 squares with the centre of any cell is able to lodge one WT. This paper analyses three cases that are presented as follows:

4.1 Case 1: (Fixed Wind Speed and Fixed Wind Direction)

Under this scenario, wind direction and the speed stay fixed. Extending the analytical work of [7], the best arrangement was first found for a single column of 10 turbine sites orientated with the dominant wind. Once the best arrangement for that column was found, it was replicated to create a three-column array with one column at the center and identical columns at each boundary so extending the ideal pattern over the wind farm [7]. analytically determined the optimal wind turbine configuration by starting with a single column of 10 turbines in the direction of the prevailing wind. They worked their way up from this configuration to fit two additional columns, one on each side of the original column, in a symmetrical three-column formation. The larger configuration was verified to confirm it continued with the aerodynamic advantages achieved in the original single-column configuration. This case follows a configuration almost similar to that reported in [7] and have further been validated by [14] as shown in Fig. 4(a).

4.2 Case 2: (Fixed Wind Speed with Intermittent Wind Direction)

In this case, we consider a consistent wind speed of 12 m/s that equally may come from any direction. The complete 360° compass is split into 36 segments of 10° each, with the wind direction in each sector treated as having a same likelihood of occurrence, therefore capturing this fluctuation. Under this hypothesis, the spacing between WTs becomes a critical factor effecting wake losses and, therefore, the total power generation of the wind farm. In prior work, [17] used a Binary Particle Swarm Optimization algorithm with Time-Varying Acceleration Coefficients (BPSO-TVAC), which proved an improved performance of the fitness function. The optimized arrangement achieved employing this methodology closely look like the arrangement proposed by [7], as shown in Fig. 4(b). Although the enhancement in the objective function was modest, the refined turbine layout led to a clear increase in total power output, reflecting a more efficient using of the available space.

4.3 Case 3: (Intermittent Wind Speed with Intermittent Wind Direction)

This reflects a more complex and advanced WFLO problem, where the wind speed and the direction are both intermittent parameters. Following the method used by [7], three representative wind speed: 8 m/s, 12 m/s, and 17 m/s are selected in order to facilitate comparison. Wind direction is subdivided into 36 intervals, each spanning 10°, as in the first case. The reference direction (0°) is synonymous with a wind from the north, and values rise in a clockwise manner, so 90° is an easterly and 270° a westerly wind.

The wind resource characteristics for this example are shown in Fig. 3, which is the joint probability distribution of wind speed and direction. This distribution assigns a probability to each wind speed–direction combination [18], with the total of all probabilities equal to one. The ideal layout created by the QPSO algorithm proposed here with the same number of WT as in the initial study by [7] is shown in Fig. 4(c).

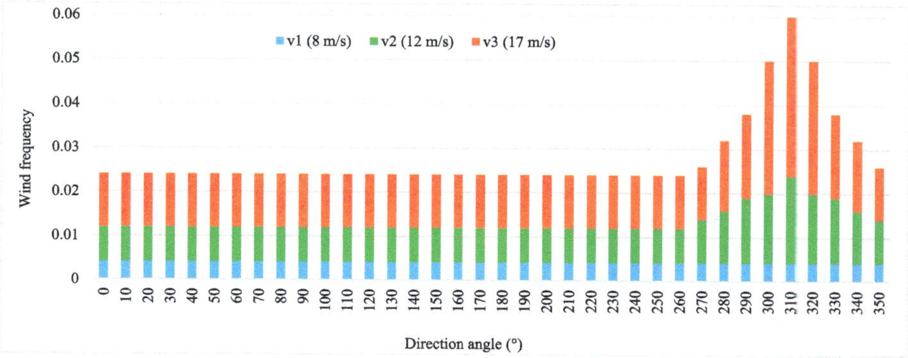

Fig. 3. Wind probability distribution - Case 3.

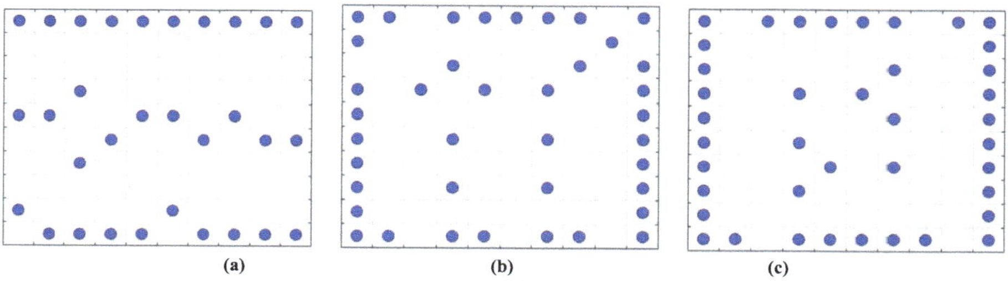

Fig. 4. Optimal windfarm configurations – (a) Case 1, (b) Case 2, (c) Case 3. (The blue dot show the location of WT)

The objective function, output power and the efficiency for the wind farm location, was re-computed with a similar power model and probability distribution of the wind, according to the procedure described in [7]. The results of these calculations are presented in Table 2. In comparison with the result of [7], the new arrangement has a smaller fitness value, meaning a better design, more efficiency, and greater overall power output.

Table 2. Results of different cases

# Case	Number of turbines	Power (kW)	Windfarm Efficiency (%)	Fitness function (x 10^{-4})
Case 1	30	14273.94	91.782	15.5147
Case 2	39	17585.67	86.982	15.4845
Case 3	39	32301.22	87.846	8.3345

Figure 5 shows the convergence curve of the proposed algorithm for the Case 1, displaying how the fitness function gradually enhances over successive generations until it achieves an optimal and stable solution.

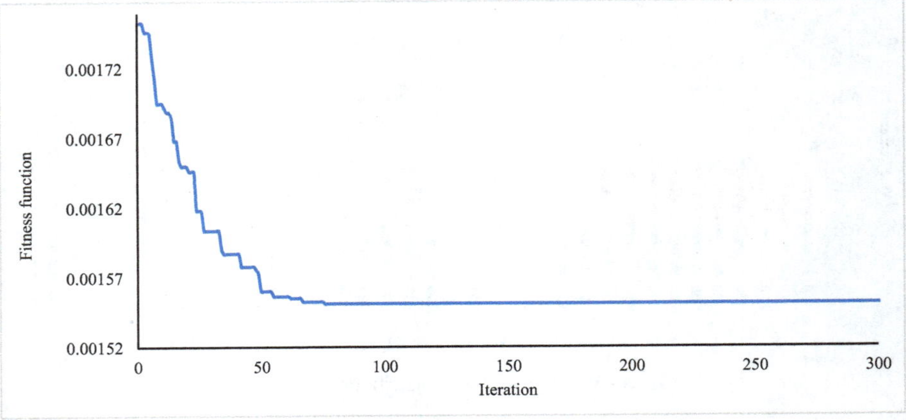

Fig. 5. Convergence curve of the proposed algorithm for the case 1.

5 Conclusion

This paper introduces a new methodology for the problem of WFLO using QPSO algorithm considering different cases both wind speed and wind direction. The optimum arrangements of WTs obtained using the proposed approach are more efficient, thereby producing high output power. Nevertheless, the findings introduced in this paper prove that the proposed QPSO algorithm offers a highly operative and best solution for WFLO problem, outperforming various prior works in terms of output efficiency and performance. Moreover, the objective function and cost model used in this paper are designed to equilibrium efficiency with economic viability. However, this trade-off highlights a limitation that future work should be addressed more comprehensively. Finally, to confirm the real world applicability of the proposed WT arrangement, further surveys including more practical models of wind variability and uncertainty are recommended.

References

1. Naidji, M., Dafri, M., Laib, A.: Optimal coordinated voltage control of distribution networks considering renewable energy sources. ECTI Trans. Electr. Eng. Electron. Commun. **23**(1), 1–10 (2025)
2. Mourad, N., Mohamed, B.: Impact of increased distributed photovoltaic generation on radial distribution networks. In: 2016 International Conference on Control, Decision and Information Technologies (CoDIT),. Saint Julian's, Malta (2016)
3. Naidji, M., Boussahoua, B., Boudour, M.: Feasibility study of solar photovoltaic integration on distribution networks: Case study: Djanet's isolated distribution network, Algeria. In: 2017 10th International Conference on Electrical and Electronics Engineering (ELECO), Bursa, pp. 130–134 (2017)
4. Naidji, M., Boudour, M., Achouri, F.: Modeling and control of photovoltaic systems integrated to distribution networks. In: 2018 International Conference on Electrical Sciences and Technologies in Maghreb (CISTEM), Algiers, pp. 1–6 (2018)

5. Shakoor, R., Hassan, M.Y., Raheem, A., et al.: Wind farm layout optimization using area dimensions and definite point selection techniques. Renew. Energy. **88**, 154–163 (2016)
6. Mosetti, G., Poloni, C., Diviacco, B.: Optimization of wind turbine positioning in large windfarms by means of a genetic algorithm. J. Wind Eng. Ind. Aerodyn. **51**(1), 105–116 (1994)
7. Grady, S.A., Hussaini, M.Y., Abdullah, M.M.: Placement of wind turbines using genetic algorithm. Renew. Energy. **30**, 259–270 (2005)
8. Shakoor, R., Hassan, M.Y., Raheem, A., et al.: The modelling of wind farm layout optimization for the reduction of wake losses. Indian J. Sci. Technol. **8**(17), 1–9 (2015)
9. Hou, P., Hu, W., Soltani, M., et al.: Combined optimization for offshore wind turbine micro siting. Appl. Energy. **189**, 271–282 (2017)
10. Abedinia, O., Ghasemi, A., Ojaroudi, N.: Improved time varying in ertia weight PSO for solved economic load dispatch with subsidies and wind power effects. Complexity. **1**(4), 40–49 (2014)
11. Hou, P., Hu, W., Chen, C., et al.: Optimization of offshore wind farm layout in restricted zones. Energies. **113**, 487–496 (2016)
12. Jensen, Niels Otto. A Note on Wind Generator Interaction. 1983.
13. Naidji, M., et al.: A heuristic optimization approach for wind turbine dimensions to enhance energy capture and reduce costs. In: International Conference on Artificial Intelligence, Embedded Systems and Renewable Energy (2025)
14. Yang, K., Cho, K.: Simulated annealing algorithm for wind farm layout optimization: a benchmark study. Energies. **12**(23), 4403 (2019)
15. Naidji, M., Boudour, M.: Stochastic multi-objective optimal reactive power dispatch considering load and renewable energy sources uncertainties: a case study of the Adrar isolated power system. Int. Trans. Electr. Energy Syst. **30**(6), e12374 (2020)
16. Naidji, M., Boudour, M.: Evaluation de la Stabilité et Proposition d'un Plan de Défense des Réseaux de Distribution en Présence des Sources d'Energie Renouvelable. Thèse de Doctorat, USTHB, Alger (2021)
17. Pookpunt, S., Ongsakul, W.: Optimal placement of wind turbines within wind farm using binary particle swarm optimization with time-varying acceleration coefficients. Renew. Energy **55**, 266–276 (2013)
18. Naidji, M., et al.: Dynamic analysis of transmission power systems with high wind power integration under severe disturbances. In: The 2nd International Conference of Advanced Technology in Electronic and Electrical Engineering (ICATEEE 2025), Proceedings, M'Sila. IEEE (2025)

Optimization of the Solar Cell Double Diode Model Estimation Using the Dung Beetle and Arctic Puffin Optimizers with Lambert-W Function and Newton-Raphson Methods

K. Derbali(✉), H. Kerbouci, A. Abbadi, F. Hamidia, and F. Salhi

Electrical Engineering Department, Research Laboratory in Electrical Engineering and Automatic (LREA), Faculty of Technology, University of Médéa, Médéa, Algeria
derbalikouki@gmail.com

Abstract. Accurate estimation of photovoltaic (PV) model parameters is essential for effective system design, simulation, and performance analysis. This study investigates the performance of two recent metaheuristic optimization algorithms: Dung Beetle Optimization (DBO) and Arctic Puffin Optimization (APO) in estimating parameters of the widely adopted dual diode model (DDM). The optimization aims to minimize the root mean square error (RMSE) between the experimental current-voltage (I-V) data of a standard silicon solar cell manufactured by RTC France and the corresponding data generated by the estimated model. The objective function employs an explicit, non-iterative formulation using the Lambert W function, and is evaluated in comparison with the Newton-Raphson method to ensure accurate fitting without requiring nested iterative computations within the optimization loop. Comparative results show that the APO algorithm consistently delivers higher accuracy, achieving lower RMSE values than the DBO algorithm under the tested conditions.

Keywords: Photovoltaic · Dung Beetle Optimization (DBO) · Arctic Puffin Optimization (APO) · Double diode model (DDM)

1 Introduction

Energy production remains one of the most urgent challenges for the coming decades. The continuous increase in energy needs, both in industrialized and developing countries, leads to a rise in greenhouse gas emissions and, consequently, environmental pollution.

One important remedy in this regard is the switch to sustainable and renewable energy sources [1]. Solar, wind, geothermal, hydroelectric, and biomass energy sources are all considered forms of renewable energy. These resources are environmentally benign and nearly limitless, in contrast to fossil fuels [2].

Due to its enormous potential as a renewable energy source, solar photovoltaic (PV) energy has grown the fastest among these. PV systems use photovoltaic cells to convert some of the sun's light into electrical power [2].

M. Hatti (Ed.): IC-AIRES 2025, LNNS 1740, pp. 472–482, 2026.
https://doi.org/10.1007/978-3-032-12788-4_45

These systems can be used for many different purposes, such as grid-connected installations, building integration, and powering remote locations [3].

Determining the electrical characteristics of the photovoltaic cell, the system's core component, is essential to improving PV systems' performance and efficiency. One of the most often used models for explaining the electrical behavior of solar cells is the dual diode model (DDM) [4].

Natural phenomenon-inspired metaheuristic optimization algorithms have drawn notice for their resilience in navigating intricate search spaces and their capacity to identify near-global solutions without the need for gradient information [5].

This study estimates the parameters of a commercial silicon solar cell (57 mm diameter, RTC France) under standard test conditions (1000 W/m^2 irradiance, 33 °C temperature) using two recently developed metaheuristic algorithms: Dung Beetle Optimization (DBO) and Arctic Puffin Optimization (APO) [6]. Root mean square error (RMSE) between estimated and measured current values is minimized using the goal function. Both the Newton-Raphson technique and the Lambert W function are used to compare accuracy.

This paper is organized as follows: The two-diode model that is being studied is introduced in Sect. 2. The suggested optimization algorithms are described in Sect. 3. The DDM's parameter estimation findings are shown in Sect. 4. Lastly, the study is concluded in Sect. 5.

2 Double Diode Model and Objective Function

2.1 Double Diode Mathematical Model of Solar Photovoltaic Cell

Through the addition of a second diode in parallel with the current source, the double-diode model provides a more realistic depiction of solar cells. Seven parameters (Iph, Isd1, Isd2, n1, n2, Rsh, and Rs) must be estimated in order to achieve this increased precision.

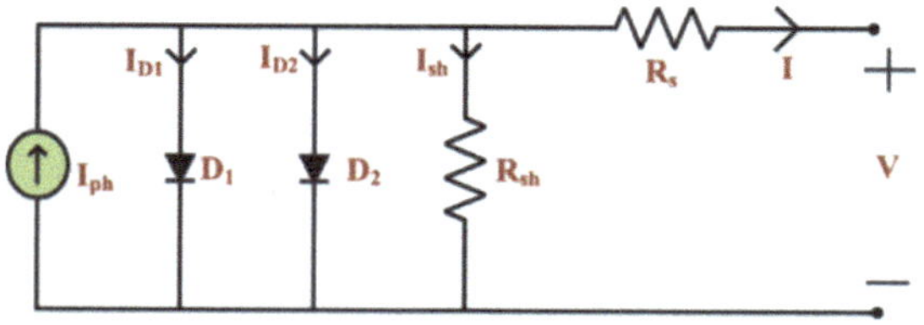

Fig. 1. Equivalent circuit of the double diode model [7].

The following definition applies to the load current produced by the analogous circuit shown in Fig. 1:

$$I = I_{ph} - I_{D1} - I_{D2} - I_{sh} \tag{1}$$

The following statement can also be used to represent this current:

$$I = I_{ph} - I_{sd1}\left[-1 + e^{\left(\frac{V+R_sI}{n.V_t}\right)}\right] - I_{sd2}\left[-1 + e^{\left(\frac{V+R_sI}{n.V_t}\right)}\right] - \frac{V + R_sI}{R_{sh}} \quad (2)$$

The reverse saturation current of the two diodes is denoted by Isd1 and Isd2 in this equation, respectively. Vt is the thermal voltage, while n1 and n2 stand for the diodes' ideality factors, respectively.

Seven unknown parameters must be estimated for the DDM: ph, Isd1, Isd2, sh, s, 1, and n2.

2.2 Objective Function Formulation

Determining an objective function that minimizes the root mean square error (RMSE) between the model's anticipated outputs and experimental data is necessary for accurate parameter estimation in solar PV models.

$$RMSE = \sqrt{\frac{1}{N_e}\sum\nolimits_{i=1}^{N_e}\left(I_i^{measured} - I_i^{estimeted}\right)^2} \quad (3)$$

In this study, the unknown parameters of the solar photovoltaic model are extracted using both the Newton-Raphson approach and the Lambert-W function. The following is the formulation of the estimated current expression for the DDM:

➢ **Using The Lambert-W Function**

$$\begin{cases} F_{obj} = \sqrt{\frac{1}{N}\sum\nolimits_{i=1}^{N_e}\left(I_i^{measured} - \left\{\frac{R_{sh}(I_{ph}+I_{sd1}+I_{sd2})-V}{R_{sh}+R_s} - \frac{n_1.V_t}{R_s}W(b_1) - \frac{n_2.V_t}{R_s}W(b_2)\right\}\right)^2} \\ W(b_i) = \frac{R_s\ R_{sh}\ I_{sdi}}{n_i.V_t(R_{sh}+R_s)}\ e^{\left(\frac{R_s}{n_i.V_t(R_{sh}+R_s)}\left(R_{sh}(I_{ph}+I_{sdi})+\frac{R_{sh}}{R_S}V\right)\right)}, i = 1, 2 \end{cases} \quad (4)$$

➢ **Using The Newton-Raphson Method**

$$\begin{cases} F(V_L, I_L, X) = I_{ph} - I_L - I_{sd1}\left[-1 + e^{\left(\frac{V_L+R_SI_L}{n_1.V_t}\right)}\right] - I_{sd2}\left[-1 + e^{\left(\frac{V_L+R_SI_L}{n_2.V_t}\right)}\right] - \frac{V_L+R_SI_L}{R_{sh}} \\ X = (I_{ph}, I_{sd1}, I_{sd2}, n_1, n_2, R_s, R_{sh}) \end{cases} \quad (5)$$

3 Proposed Optimization Algorithms

3.1 Dung Beetle Optimizer (DBO)

Based on the adaptive behaviors of dung beetles, Xue et al. introduced the Dung Beetle Optimizer (DBO) in September 2022 as a nature-inspired metaheuristic.

➢ **Roll-Ball Dung Beetle**

Although light and wind can have an impact on its travel, the roll-ball dung beetle uses the sun as a navigational aid to keep a straight rolling route. The following formula represents its position update:

$$\begin{cases} X_i(t) = X_i(t) \times \alpha \times K \times X_i(t-1) + b \times \Delta_x \\ \Delta_x = |X_i(t) - X^{\omega}| \end{cases} \tag{6}$$

When a dung beetle on a dung ball runs into an obstruction and has to decide on a new path to continue traveling, the updated formula for its dance position is shown as follows [9]:

$$X_i(t+1) = X_i(t) + \tan(\theta)|X_i(t) - X_i(t-1)| \tag{7}$$

➢Brood Ball

The brood ball mimics the oviposition site of a female dung beetle by using a boundary selection technique. The oviposition area is defined as follows:

$$\begin{cases} Lb^* = \max(X^*(1-R), Lb) \\ Ub^* = \min(X^*(1+R), Ub) \end{cases} \tag{8}$$

The calculation above shows that the oviposition area is changing dynamically. Thus, the position of the brood ball is likewise changing dynamically during the iteration phase. The following is the definition of the brood ball's position [9]:

$$B_i(t+1) = X^* + b_1 \times \left(B_i(t) - Lb^*\right) + b_2 \times \left(B_i(t) - Ub^*\right) \tag{9}$$

➢Scout Dung Beetle

The scout dung beetle looks for food in the optimal foraging region after birth. The best foraging area is defined as follows:

$$\begin{cases} Lb^* = \max\left(X^b \times (1-R), Lb\right) \\ Ub^* = \max\left(X^b \times (1+R), Ub\right) \end{cases} \tag{10}$$

The following is an update on the scout dung beetle's location [9]:

$$X_i(t+1) = X_i(t) + C_1 \times \left(X_i(t) - Lb^b\right) + C_2 \times \left(X_i(t) - Ub^b\right) \tag{11}$$

➢Thief Dung Beetle

The following is an update to the thief dung beetle's position data [9] (Fig. 2):

$$X_i(t) = X^b + s \times g \times \left(\left|x_i(t) - X^*\right| + \left|x_i(t) - X^b\right|\right) \tag{12}$$

3.2 Arctic Puffin Optimization (APO)

In September 2024, Wang et al. presented the Arctic Puffin Optimization (APO) algorithm, a nature-inspired metaheuristic derived on Arctic puffin survival habits. Imitating their flight and foraging techniques.

Algorithme .1 : Dung Beetle Optimization (DBO).

```
Set the population size N and the maximum iterations T_max ;
Initialize the particles D_i (i = 1,2,...,N) and classify the populations;
while t < T_max do
    for i ← 1 to N do
        if i == Rolling Dung Beetle then
            if δ < 0.9                                  // δ = rand(1) then
                Updating the Rolling Dung Beetle with Eq. (6);
            end
            else
                Updating the Rolling Dung Beetle with Eq. (7);
            end
        end
        if i == brood Ball then
            Updating the location of breeding dung beetles using Eq. (9);
        end
        if i == Scout Dung Beetles then
            Updating the location of the small dung beetle using Eq. (11);
        end
        if i == Theif Dung Beetles then
            Updating the location of the stealing dung beetle using Eq. (12);
        end
    end
    Making greedy choices;
    t ← t + 1;
end
```

Fig. 2. Pseudo code of dung beetle optimization [10].

➢Initial Population

The APO algorithm is designed based on the unique behaviors of Arctic puffins on the water and in the air. The following formula is used by the APO method to initialize the population [12]:

$$\overrightarrow{((X_i))}^{t} = \text{rand} \times (ub - lb) + lb, i = 1, 2, 3 \ldots ..N \tag{13}$$

➢Aerial Flight Stage

The Arctic Puffin has two primary tactics: the swooping predation technique and the airborne search approach, to deal with various circumstances.

·The first strategy is aerial search

The aerial search strategy of the APO algorithm combines spiral patterns for predator avoidance with Levy flying for exploration, simulating the flight of the Arctic puffin while foraging. A particular mathematical formula is used for position updates [12]:

$$\overrightarrow{((Y_r))}^{t+1} = \overrightarrow{((X_i))}^{t} + \left(\overrightarrow{((X_i))}^{t} - \overrightarrow{((X_r))}^{t}\right) \times L(D) + R \tag{14}$$

·The second strategy is swooping predation

APO's swooping predation technique uses a velocity-based modification to direct movement throughout the dive, simulating the puffin's quick plunge to capture prey. The following is the appropriate position update formula:

$$\begin{cases} \overrightarrow{((Z_i))}^{t+1} = \overrightarrow{((Y_i))}^{t+1} \times S \\ S = \tan((\text{rand} - 0.5) \times \pi) \end{cases} \tag{15}$$

A new population is created by combining the candidate locations from the two techniques, ranking them according to fitness, and selecting the top (N) people. This process is described by the following equation [12]:

$$\overrightarrow{((P_i))}^{t+1} = \overrightarrow{((Y_i))}^{t+1} \cup \overrightarrow{((Z_i))}^{t+1} \tag{16}$$

$$new = sort\left(\overrightarrow{((P_i))}^{t+1}\right) \tag{17}$$

$$\overrightarrow{((X_i))}^{t+1} = new(1:N) \tag{18}$$

➢Underwater Foraging Stage

Three primary tactics make up the underwater foraging phase, and each is used in a different situation to increase the effectiveness of predation.

·The first strategy is gathering foraging

The location update is described by the following equation, and the gathered foraging technique mimics the cooperative foraging behavior of Arctic puffins:

$$\overrightarrow{((W_i))}^{t+1} = \begin{cases} \overrightarrow{((X_{r1}))}^{t} + F * L(D) * \left(\overrightarrow{((X_{r2}))}^{t} - \overrightarrow{((X_{r3}))}^{t}\right), & \textit{if } \text{rand} \geq 0.5 \\ \overrightarrow{((X_{r1}))}^{t} + F * \left(\overrightarrow{((X_{r2}))}^{t} - \overrightarrow{((X_{r3}))}^{t}\right), & \textit{if } \text{rand} < 0.5 \end{cases} \tag{19}$$

·The second strategy is intensifying the search

As local resources run out, the puffins move to new foraging areas, which leads to position updates to find new food sources. This is reflected in the increased search approach. The following is the position update equation for this phase:

$$\begin{cases} \overrightarrow{((Y_i))}^{t+1} = \overrightarrow{((W_i))}^{t+1} \times (1+f) \\ f = 0.1 \times (\text{rand} - 1) \times \frac{T_{max}-t}{T_{max}} \end{cases} \tag{20}$$

·The third strategy is avoiding predators

The location update equation for the avoiding predator strategy, which describes how an Arctic puffin behaves when it sees a predator nearby, is as follows:

$$\overrightarrow{((Z_i))}^{t+1} = \begin{cases} \overrightarrow{((X_{r1}))}^{t} + F * L(D) * \left(\overrightarrow{((X_{r2}))}^{t} - \overrightarrow{((X_{r3}))}^{t}\right), & \textit{if } \text{rand} \geq 0.5 \\ \overrightarrow{((X_{r1}))}^{t} + \beta * \left(\overrightarrow{((X_{r1}))}^{t} - \overrightarrow{((X_{r2}))}^{t}\right), & \textit{if } \text{rand} < 0.5 \end{cases} \tag{21}$$

The algorithm selects the best (N) individuals with the highest fitness to create a new population by combining candidate locations from the three techniques, ranking them according to fitness, and ensuring optimal performance under various scenarios. This process is described by the following equation [12]:

$$\overrightarrow{((P_i))}^{t+1} = \overrightarrow{((W_i))}^{t+1} \cup \overrightarrow{((Y_i))}^{t+1} \cup \overrightarrow{((Z_i))}^{t+1} \tag{22}$$

$$new = sort\left(\overrightarrow{((P_i))}^{t+1}\right) \tag{23}$$

$$\overrightarrow{((X_i))}^{t+1} = new(1 : N) \tag{24}$$

➢Behavior Conversion Factor B

The method uses a behavior conversion factor, represented by B, to facilitate the move from global search to local exploitation. Its definition is as follows (Fig. 3):

$$B = 2 * \log\left(\frac{1}{\text{rand}}\right) * \left(1 - \frac{t}{T_{max}}\right) \tag{25}$$

a. Optimization procedure

```
Algorithme .2 : Arctic Puffin Optimisation (APO).
Input : N, T, D, F, and C
Output : The best X_i^t and its fitness value
Initialize N puffins, X_i^t (i = 1, 2, ..., N) using Eq. (13);
Evaluate each X_i^t and find the one with the best fitness;
t ← 1;
while t < T_max do
    Calculate the behavior conversion factor B according to Eq. (25);
    if B > C then
        // Air flight period (Exploration phase)
        for i = 1 to N do
            Update Y_i^{t+1} using Eq. (14) ;                // Air search
            Update Z_i^{t+1} using Eq. (15) ;                // Dive predation
        end
        Select N excellent populations as the new population X_i^{t+1} using
         Eq. (16) to Eq. (18);
        Evaluate the puffins X_i^{t+1} and replace X_i^t with X_i^{t+1} if it is better;
    else
        // Diving and foraging period (Exploitation phase)
        for i = 1 to N do
            Update W_i^{t+1} using Eq. (19) ;                // Gather for food
            Update Y_i^{t+1} using Eq. (20) ;                // Strengthen search
            Update Z_i^{t+1} using Eq. (21) ;                // Avoid predators
            Select N excellent populations as the new population X_i^{t+1} using
             Eq. (22) to Eq. (24);
            Evaluate the puffins X_i^{t+1} and replace X_i^t with X_i^{t+1} if it is better;
            t ← t + 1;
        end
    end
end
return X_i^t
```

b. Pseudo-code

Fig. 3. Arctic puffin optimizer [12].

4 Results and Discussion

In this part, the double-diode PV model's parameters are estimated using DBO and APO. Both algorithms exhibit accurate, reliable, and effective convergence based on 1000 runs using the Newton-Raphson methods [14] and Lambert-W methods [13].

4.1 Results of Lambert-W Function

The ideal parameter values for the DDM that were determined by DBO and APO are shown in Table 1. With an RMSE of 7.5840e-04, the APO algorithm obtained the lowest value (Figs. 4 and 5).

4.2 Results of Newton-Raphson Method

The ideal DDM parameter values that were determined by DBO and APO are shown in Table 2. Once more, with an RMSE value of 7.5750 e-04, the APO algorithm obtained the lowest value (Figs. 6 and 7).

Table 1. Optimal parameter values for the DDM model of the RTC France silicon solar cell using Lambert-W function.

Algorithm	I_{ph} (A)	I_{sd1} (μA)	I_{sd2} (μA)	R_s (Ω)	R_{sh} (Ω)	n_1	n_2	RMSE
DBO	0.7630	1	1	0.0332	100	2	1.7371	7.7024 e-04
APO	**0.7608**	**0.2158**	**0.0218**	**0.0393**	**52.4832**	**1.4567**	**1.9709**	**7.5840 e-04**

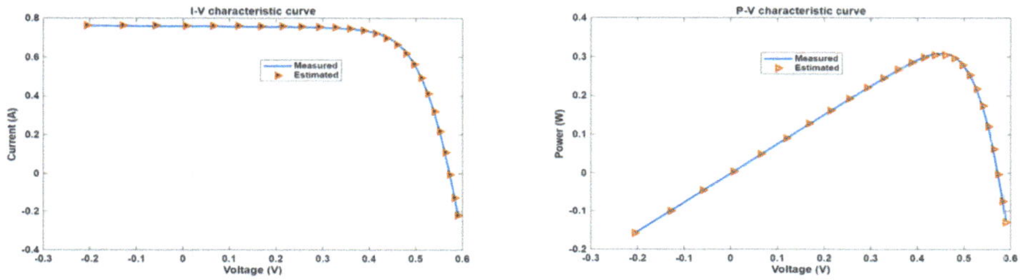

Fig. 4. The measured data and estimated data obtained by DBO of double-diode model using Lambert-W function.

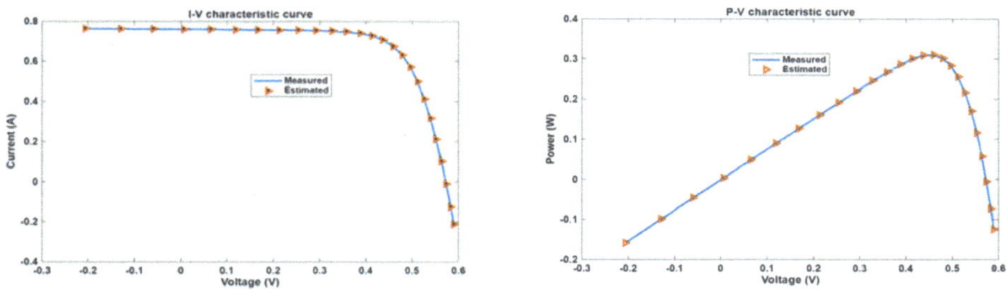

Fig. 5. The measured data and estimated data obtained by APO of double-diode model using Lambert-W function.

Table 2. Optimal parameter values for the DDM model of the RTC France silicon solar cell using Newton-Raphson method.

Algorithm	I_{ph} (A)	I_{sd1} (μA)	I_{sd2} (μA)	R_s (Ω)	R_{sh} (Ω)	n_1	n_2	RMSE
DBO	0.7607	1	0.2518	0.0359	61.6734	2	1.4633	7.9170 e-04
APO	**0.7608**	**0.2099**	**0.3691**	**0.0368**	**53.9743**	**1.4466**	**1.8130**	**7.5750 e-04**

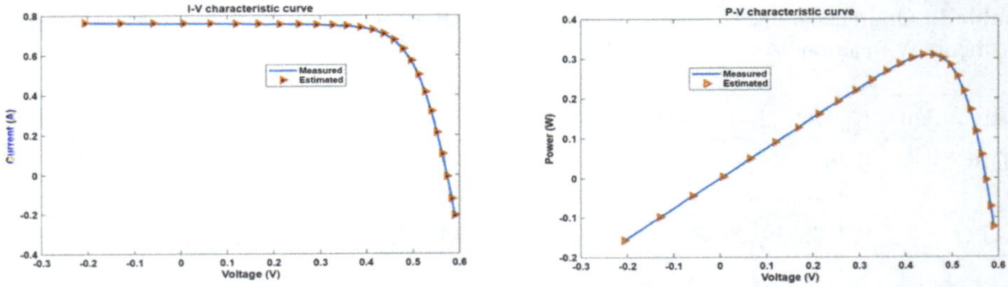

Fig. 6. The measured data and estimated data obtained by DBO of double-diode model using Newton-Raphson method.

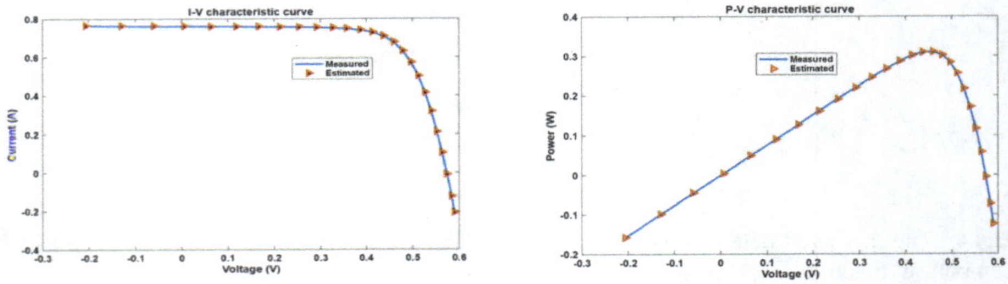

Fig. 7. The measured data and estimated data obtained by APO of double-diode model using Newton-Raphson method.

4.3 Statistics Analysis

For predicting the DDM parameters of the RTC France solar cell, the APO method produces the best statistical results, as seen in Table 3. After 1000 iterations, it achieves the lowest RMSE and smallest standard deviation using both the Lambert-W and Newton-Raphson approaches.

Table 3. Statistical analysis of DDM using Lambert-W function and Newton-Raphson method.

Lambert-W function				
Algorithm	$RMSE_{min}$	$RMSE_{mean}$	$RMSE_{max}$	STD
DBO	7.7024 e-04	0.0023	0.0053	0.0018
APO	**7.5840 e-04**	**7.5901 e-04**	**7.6413 e-04**	**1.8002 e-06**
Newton-Raphson method				
Algorithm	$RMSE_{min}$	$RMSE_{mean}$	$RMSE_{max}$	STD
DBO	7.9170 e-04	0.0014	0.0021	6.2985 e-04
APO	**7.5750 e-04**	**7.6320 e-04**	**7.6837 e-04**	**3.6002 e-06**

4.4 Convergencce Analysis

Using the Lambert-W and Newton-Raphson approaches, Fig. 8 illustrates the convergence of DBO and APO on the DDM model of the RTC France solar cell. In both situations, APO achieves a noticeably higher rate of convergence, with RMSE declining more quickly than DBO.

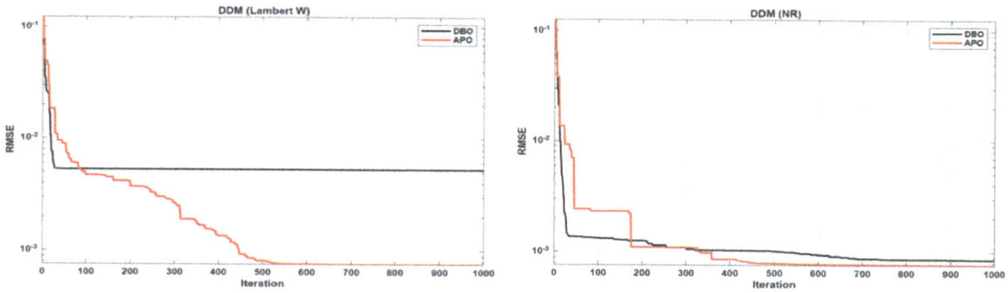

Fig. 8. Convergence curves for DDM of France silicon solar cell using Lambert-W function and Newton-Raphson method.

The APO algorithm consistently yielded the lowest RMSE when both methods were used. Although the results were comparable, the analytical structure of the Lambert-W function ensured faster and more stable convergence compared to the iterative and more sensitive Newton-Raphson approach.

5 Conclusion

In this research, the Arctic Puffin Optimizer (APO) and the Dung Beetle Optimizer (DBO) for parameter estimation in diode-based photovoltaic models are compared. For an explicit, non-iterative method, the Lambert-W function is used to create the objective function, which is based on RMSE minimization, and is then contrasted with the Newton-Raphson method. APO outperformed DBO in the DDM model, obtaining lower and median RMSE, stronger resilience across several runs (lower standard deviation), and faster convergence based on benchmark data from RTC France solar cells. These findings demonstrate how well APO can handle challenging, nonlinear PV parameter estimation issues. Its use may be expanded in future studies to include larger systems, other PV technologies, or comparisons with more sophisticated algorithms in practical settings.

References

1. Gielen, D., Boshell, F., Saygin, D., Bazilian, M.D., Wagner, N., Gorini, R.: The role of renewable energy in the global energy transformation. Energ. Strat. Rev. **24**, 38–50 (2019)
2. Hasan, M.M. et al.: Harnessing solar power: A review of photovoltaic innovations, solar thermal systems, and the dawn of energy storage solutions. Energies. **16**(18), 6456 (2023)
3. Twidell, J., Weir, T.: Renewable Energy Resources, 2nd edn. Taylor & Francis, New York, USA (2006)

4. Malki, A., Mohamed, A.A., Rashwan, Y.I., El-Sehiemy, R.A., Elhosseini, M.A.: Parameter identification of photovoltaic cell model using modified elephant herding optimization-based algorithms. Appl. Sci. **11**(24), 11929 (2021)
5. Tomar, V., Bansal, M., Singh, P.: Metaheuristic algorithms for optimization: a brief review. Eng. Proc. **59**(1), 1–6 (2024)
6. Abbadi, A., Hamidia, F., Morsli, A., Skender, M.R., Douifi, N.: Improvement of double diode model of solar cells using mountain gazelle optimization algorithm and based on lambert's w function. In: Proceedings of the 8th International Conference on Artificial Intelligence in Renewable Energetic Systems (IC-AIRES 2024), Medea, Algeria (2024)
7. Hussain, M.T., et al.: Archimedes optimization algorithm based parameter extraction of photovoltaic models on a decent basis for novel accurate RMSE calculation. Front. Energy Res. **11**, 1326313 (2024)
8. Xue, J., Shen, B.: Dung beetle optimizer: a new meta-heuristic algorithm for global optimization. J. Supercomput. **79**(7), 7305–7336 (2022)
9. Cao, W., Liu, Z., Song, H., Li, G., Quan, B.: Dung beetle optimized fuzzy PID algorithm applied in four-bar target temperature control system. Appl. Sci. **14**(10), 1–20 (2024)
10. Mai, C., Zhang, L., Chao, X., Hu, X., Wei, X., Li, J.: A novel MPPT technology based on dung beetle optimization algorithm for PV systems under complex partial shade conditions. Sci. Rep. **13**, 1–15 (2023)
11. Wang, L., Tian, W., Xu, D.-M., Zang, H.-F.: Arctic puffin optimization: a bio inspired metaheuristic algorithm for solving engineering design optimization. Adv. Eng. Softw. **195**, 1–12 (2024)
12. Sun, L., Wang, B.: Arctic puffin optimization algorithm based on multi-strategy blending. J. Comput. Commun. **12**(12), 1–15 (2024)
13. Wang, Y.-L., Sun, Y.-Z., Peng, L.-L., Xu, Y.: Parameter extraction for photovoltaic module based on lambert w function. Acta Phys. Sin. **61**(24), 248402 (2012)
14. Reis, L., Camacho, R.J., Novacki, D.: The Newton-Raphson method in the extraction of parameters of PV modules. Renew. Energy Power Qual. J. **15**(1), 634–639 (2017)

A Comparative Evaluation of Metaheuristic Algorithm Using Two Different Simulation Current Calculation Methods for Extracting Photovoltaic Single-Diode Model Parameters

Djebouri Oqeyl(✉), Kerdoun Djallel, and Boukebbous Seif Eddine

Department of electrotechnics, Laboratory of Electrical Engineering of Constantine (LGEC), University of Brothers mentouri, Constantine, Algeria
oqeyl.djebouri@doc.umc.edu.dz

Abstract. Accurate parameter extraction for photovoltaic (PV) models is essential for the design and performance prediction of PV systems. This paper presents a comparative study of Differential Evolution (DE) with implicit and explicit single-diode model for PV parameter extraction. The implicit model uses the Newton-Raphson method to solve the nonlinear single-diode equation (NSDM), while the explicit model employs the Lambert W function to directly calculate the current (LSDM). Both approaches are evaluated using the Root Mean Square Error (RMSE) as the objective function and validated on the RTC France solar cell dataset. Results show that the explicit model achieves comparable accuracy to the implicit model while significantly reducing computational time. The study provides insights into the trade-offs between accuracy and efficiency, offering practical recommendations for PV parameter extraction.

Keywords: Parameter extraction · Differential Evolution (DE) · Newton-Raphson · Lambert W function · Single-diode model · RMSE

1 Introduction

The Solar cell technology has emerged as a vital renewable energy solution, driven by its ability to meet rising energy demands through efficient energy conversion [1]. The accurate modeling of nonlinear I-V characteristics is crucial for evaluating solar cell performance, as it directly impacts energy harvesting capabilities [2]. Parameter extraction techniques play a significant role in this process, ensuring that solar cell models accurately reflect real-world performance by minimizing discrepancies between measured and estimated data [3]. The parameter estimation problem for PV cell models can be effectively translated into an optimization problem. This optimization problem is nonlinear, constrained, and continuous, making it a challenging task [4]. Metaheuristic optimization algorithms are widely used to tackle difficult engineering optimization problems.

M. Hatti (Ed.): IC-AIRES 2025, LNNS 1740, pp. 483–499, 2026.
https://doi.org/10.1007/978-3-032-12788-4_46

Metaheuristics have been frequently applied for parameter estimation of circuit model parameters of solar PV cells to enhance the precision of parameter estimation, each with unique methodologies and effectiveness [5]. The inherent nonlinearity of the PV model poses significant challenges for metaheuristic methods in identifying the parameters of PV cells or modules. Some researchers opt for implicit equations due to their accuracy, even though they are complex and computationally demanding [6]. On the other hand, explicit equations are preferred by others because they are simplified and easier to solve, making them ideal for real-time applications. By selecting between implicit and explicit equations [7], researchers can strike a balance between accuracy and computational efficiency, depending on the specific requirements of their optimization problem.

In this paper, I conducted a comparative analysis of one of the metheuristic algorithm differential evolution techniques utilizing both implicit equations based newton-raphson method (DE-NSDM) and explicit equation based lambert W method for single diode models (DE-LSDM) for the extraction of photovoltaic parameters, specifically focusing on the RTC France Solar Cell. To achieve a higher confidence level in the comparison results Both approaches are evaluated using the Root Mean Square Error (RMSE) as the objective function and validated on the RTC France solar cell dataset also six performance metrics is Minimum value (min), Maximum value (max), Mean value, Standard deviation (STD), IAE and IRE were introduced to represent the absolute error and relative error between the simulated current and the measured current, and PAE to represent the absolute error between the simulated power and the measured power, and also the execution times for both method. Results show that the explicit model achieves comparable accuracy to the implicit model while significantly reducing computational time. The study provides insights into the trade-offs between accuracy and efficiency, offering practical recommendations for PV parameter extraction. The detailed discussion on the use of implicit and explicit equations, the application of Differential Evolution (DE), and the results of this analysis will be presented in the subsequent sections.

2 Mathematical Model and Optimization Problem

2.1 Solar PV Models

A series of mathematical PV cell models have been established to study the behavior of the photovoltaic panel and analyzing the nonlinear output characteristics of PV systems. Including a single diode model (SDM), double diode model (DDM), Triple diode model (TDM) and a PV module model (MM) [8]. The structures of these models are illustrated in Fig. 1:

The one-diode model of a photovoltaic (PV) cell is a widely used equivalent circuit model that describes the current-voltage (I-V) characteristics of the cell [9]. As can be seen from Fig. 1, the structure of this model consists of a diode, a series resistance (Rs), a shunt resistance (Rsh) and a current source (Iph).

By applying the Kirchhoff's current law and the Shockley diode equation to the above one diode model circuit illustrated in Fig. 1 (a) the implicit equation of the SDM it is given by (5):

$$Iph - Id - Ish - I = 0 \tag{1}$$

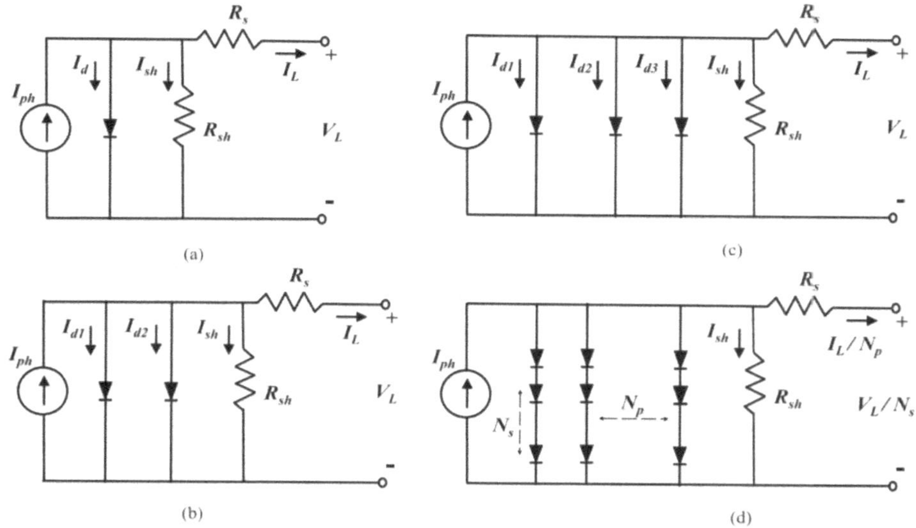

Fig. 1. Equivalent circuit of PV: (a) single diode model, (b) double diode model, (c) triple diode model, and (d) PV module model.

$$I = Iph - Id - Ish \tag{2}$$

$$Ish = \frac{V + I, Rs}{Rsh} \tag{3}$$

$$Id = Is\left(e^{\frac{q.(V+I.Rs)}{K.Tc.A}} - 1\right) \tag{4}$$

We put the (3) and (4) into the (2), we can get (5):

$$I = Iph - Is\left(e^{\frac{q.(V+I.Rs)}{K.Tc.A}} - 1\right) - \frac{V + I.Rs}{Rsh} \tag{5}$$

Where A is ideality factor of the diode. The Id, Is, and Iph stand for leakage shunt, saturation diode and the diode currents respectively. Also, k = 1.381 × 10–23 (J/K) is the Boltzmann constant and Tc is the cell temperature (K) and q = 1.6 × 10–19 C denotes the electron charge.

An implicit nonlinear mathematical expression (transcendental equation giving I in terms of V and I) is called to portray the relationship between current and voltage as well as physical parameters and to generate I-V curves [10].

The transcendental character of (5) means that it cannot be solved for V in terms of i, and vice versa. By applying the Lambert W function, researchers can transform these implicit equations into explicit ones, which are easier to handle and solve [10]. And less complicated than implicit equation

However, explicit solution for current and voltage can be expressed using W function as follows [11]:

$$\begin{cases} V = -I.Rs - I.Rsh + Iph.Rsh + Is.Rsh - \frac{q}{A.K.Tc} LambertW(alpha) \\ alpha = \frac{Is.Rsh.q.e^{\frac{Rsh.q(Iph+Is-I)}{K.Tc.A}}}{K.Tc.A} \end{cases} \quad (6)$$

For the explicit current function:

$$\begin{cases} I = -\frac{V}{Rs+Rsh} + \frac{Rsh(Is+Iph)}{Rs+Rsh} - \frac{q}{Rs.A.K.Tc} LambertW(alpha) \\ alpha = \frac{Is.Rsh.q.Rs.e^{\frac{Rsh.q(Rs.Iph+Rs.Is+V)}{K.Tc.A(Rs+Rsh)}}}{K.Tc.A(Rs+Rsh)} \end{cases} \quad (7)$$

Only the appropriate variable and the model's parameters are included in the arguments of the W-function in (7).

In the both equation of SDM have an five unknown parameters are Rs, Rsh, Iph,I0, A and they can be represented in one vector x = [x1,x2,x3,x4,x5].

2.2 Objective Function of Implicit and Explicit Photovoltaic Model for Parameter Estimation

Accurate modeling of solar cells require in precise identifying of the five unknown parameters of SDM, the problem of parameter identification is usually formulated into as an optimization problem 1, the involves estimating the best parameters value to minimize the difference between the experimental data and simulated data obtained by optimization techniques [12], in the literature to evaluates the difference between estimated and measured data as usually quantify using the root mean square error RMSE [13], hence, the objective function is formulated as follows:

$$\mathrm{RMSE(X)} = \sqrt{\sum\nolimits_{K=1}^{N} \mathrm{f(V, I, X)}^2}$$

Where N is the number of pairs of experimental I-V data, and X is a vector correspondent at the five unknown parameter

The objective function of the implicit equation of the single diode model is

$$\begin{cases} \mathrm{f(V, I, X) = Iph - Is}\left(\mathrm{e}^{\frac{\mathrm{q.(V+I.Rs)}}{\mathrm{K.Tc.A}}} - 1\right) - \frac{\mathrm{V+I.Rs}}{\mathrm{Rsh}} - \mathrm{I} \\ \mathrm{X} = \left[\mathrm{Iph, Is, Rsh, Rs, A}\right] \end{cases}$$

For the explicit equation the objective function is formulated as follow:

$$\begin{cases} \mathrm{f(V, I, X)} = -\frac{\mathrm{V}}{\mathrm{Rs+Rsh}} + \frac{\mathrm{Rsh(Is+Iph)}}{\mathrm{Rs+Rsh}} - \\ \frac{\mathrm{q}}{\mathrm{Rs.A.K.Tc}} \mathrm{LambertW}\left(\frac{\mathrm{Is.Rsh.q.Rs.e}^{\frac{\mathrm{Rsh.q(Rs.Iph+Rs.Is+V)}}{\mathrm{K.Tc.A(Rs+Rsh)}}}}{\mathrm{K.Tc.A(Rs+Rsh)}}\right) - \mathrm{I} \\ \mathrm{X} = \left[\mathrm{Iph, Is, Rsh, Rs, A}\right] \end{cases}$$

3 Differential Evolution Algorithm

Differential Evolution (DE), introduced by Storn and Price in 1995, is one of the metaheuristic algorithms and a population-based random search algorithm, The DE has been applied in many engineering optimization fields and is an effective technique for solving complex optimization problems. DE operates through mutation, crossover, and selection processes, using real number coding to simplify genetic operations. It is known for its simplicity, robustness, fewer control parameters, and strong search capabilities. DE can handle multimodal and nonlinear functions and solve large-scale problems in parallel, offering better convergence performance.

3.1 Population Initialization

In Differential Evolution (DE), the initialization of the population is crucial for optimizing performance and convergence. The initial population x_i^t, represented as $\{x_1^t, x_2^t, \ldots, x_{Np}^t\}$, is typically generated randomly [14], but various strategies can enhance diversity and effectiveness. This overview will explore the significance of population initialization, the impact of population size, and the dynamics of DE.

3.2 Mutation Operation

The mutation operator creates a mutant vector V_i^t for each individual $\mathbf{x}_i^t$, while the crossover operator facilitates the exchange of information between individuals, enhancing population diversity and convergence speed [15]. Five commonly used mutation strategies are summarized as [16] (Table 1):

Table 1. Mutation operation strategies of DE algorithm.

Mutation	Operator
DE/rand/1	$V_i^t = x_{i1}^t + F.(x_{i2}^t - x_{i3}^t)$
DE/rand/2	$V_i^t = x_{i1}^t + F.(x_{i1}^t - x_{i3}^t) + F.(x_{i4}^t - x_{i5}^t)$
DE/best/1	$V_i^t = P_{best}^t + F.(x_{i1}^t - x_{i2}^t)$
DE/best/2	$V_i^t = P_{best}^t + F.(x_{i1}^t - x_{i2}^t) + F.(x_{i3}^t - x_{i4}^t)$
DE/current-to-best/1	$V_i^t = x_{i1}^t + F.(P_{best}^t - x_i^t) + F.(x_{i1}^t - x_{i2}^t)$

3.3 Crossover Operation

In evolutionary algorithms, a crossover operator is essential for generating a trial vector U_i^t by combining features from parent vectors. The binomial crossover, noted for its efficiency, is executed at each dimension with a probability Cr and is formulated as:

$$u_{ij}^t = \begin{cases} v_{ij}^t, & \text{if } \mathrm{rand}(0, 1) \le \mathrm{Cr} \ \text{ or } \ j = \mathrm{jrand} \\ x_{ij}^t, & \text{otherwise} \end{cases}$$

This operation allows for the selective exchange of variables, enhancing the exploration of the search space [17]. The trial vector U_i^t is generated using a mutant vector $V_{ij}^t(t)$, which is derived from a target vector $x_{ij}^t(t)$ by adding a random vector [18]. This structured approach ensures that the generated trial vectors maintain diversity, crucial for effective optimization in complex problems.

3.4 Selecting Operation

The selection operator is a fundamental mechanism in evolutionary algorithms, tasked with choosing the fittest individual from the target vector $\overrightarrow{X_{i,t}}$ and the trial vector $\overrightarrow{U_{i,t}}$ for the next generation [19].

$$\overrightarrow{X_{i,t+1}} = \begin{cases} \overrightarrow{U_{i,t}}, & if\ f\left(\overrightarrow{U_{i,t}}\right) \leq f\left(\overrightarrow{X_{i,t}}\right) \\ \overrightarrow{X_{i,t}}, & otherwise \end{cases}$$

3.5 Pseudo Code of DE Algorithms

The DE algorithm pseudo code as follows [20]:

```
DE algorithm pseudo code
Input: Population size (NP), Mutation factor (F), Crossover probability (CR), and iteration;
t = 0;
Initialization of the population x_i^t ;
Evaluate the fitness value for all individuals of x_i^t ;
While the terminal criterion is not satisfied do
    For i=1 to NP do
        Select randomly r1 ≠ r2 ≠ r3 ≠ i;
        j_rand = randint (1 , D );
        For j=1 to D do
            If rand(0 , 1) ≤ Cr  or   & j = jrand then
                u_ij^t = v_ij^t ;
            Else
                u_ij^t = x_ij^t ;
        If f(U_i,t) ≤ f(X_i,t) then
                X_i,t+1 = U_i,t ;
```

4 Results and Discussion

This section evaluates the parameter extraction performance of the Single Diode Model (SDM) using both of implicit based newton-raphson method and explicit equations based lambert W method within the default MATLAB R2023a environment, employing the Differential Evolution (DE) algorithm. To ensure a fair comparison, the DE algorithm is implemented under identical simulation conditions, maintaining consistent search ranges and stop criteria for DE-NSDM and DE-LSDM parameter extraction.

The standard datasets used were adopted from the literature, where the measurements are carried out on a 57 mm diameter commercial silicon (R.T.C. France) solar cell operating at 33. These sets of measured I–V data are very representative, because they have been extensively adopted in literature as the benchmarks to test and compare the performance of various algorithms for parameter extraction of SDM

The search ranges for standard datasets of the R.T.C France solar are detailed in Table 2. For the applied optimization technique, the DE control parameters are given in Table 3.

To achieve a higher confidence level in the comparison results, 30 independent runs of the DE algorithm are conducted for parameter extraction of the aforementioned solar cells/modules. The performance criteria for evaluating the parameter extraction results of DE-NSDM and DE-LSDM include convergence speed, computation time, robustness, and accuracy. All comparative experiments are carried out on a personal laptop equipped with an Intel Core i5-7200U processor @ 2.71 GHZ, 8GB RAM, running Windows 10 Pro 64bit.

Table 2. Lower bound (LB) and upper bound (UB) on model parameters of the solar cell.

Model parameters	R.T.C France solar cell	
Search ranges	LB	UB
$I_{ph}(A)$	0	1
$I_s(\mu A)$	0	1
N	1	2
$R_s(\Omega)$	0	0.5
$R_{sh}(\Omega)$	0	100

Table 3. Control parameters of DE optimization algorithm

Population size (NP)	30
Maximum iteration	1000
Mutation	DE/rand/1
operator	$V_i^t = x_{i1}^t + F.(x_{i2}^t - x_{i3}^t)$
Mutation factor (F)	0,4
Crossover probability (CR)	0,9

In this case of the study, the optimization algorithm of DE is applied for extracting the parameters of the SDM with implicit equation based Newton-Raphson method (DE-NSDM), with explicit equation based Lambert *W* method (DE-NSDM) of R.T.C France solar cell. The results of the optimized are listed in Table 4. From this table, it can be noticed that for the application of the DE-LSDM algorithm results in the minimum value of the RMSE that is equal to 9.764863e-04

Table 4. Estimated model parameters for DE-NSDM and DE-LSDM of R.T.C France solar cell obtained by DE algorithm.

Technique	$I_{ph}(A)$	$I_s(\mu A)$	$R_s(\Omega)$	$R_{sh}(\Omega)$	N	RMSE
DE-NSDM	0.760713	0.4540436	0.034847	61.895539	1.516325	9.804680e-04
DE-LSDM	0.760568	0.4521569	0.034897	64.112955	1.515855	9.764863e-04

Moreover, a comparison between the results of the two technique based on the DE optimization technique has been introduced. Figure 2 shows the convergence trends of the objective function (RMSE) for NSDM and LSDM R.T.C. France solar cell based on DE algorithm. From Fig. 2 it is noticeable that the convergence curve of the DE-LSDM is better than of the convergence curve of DE-NSDM.

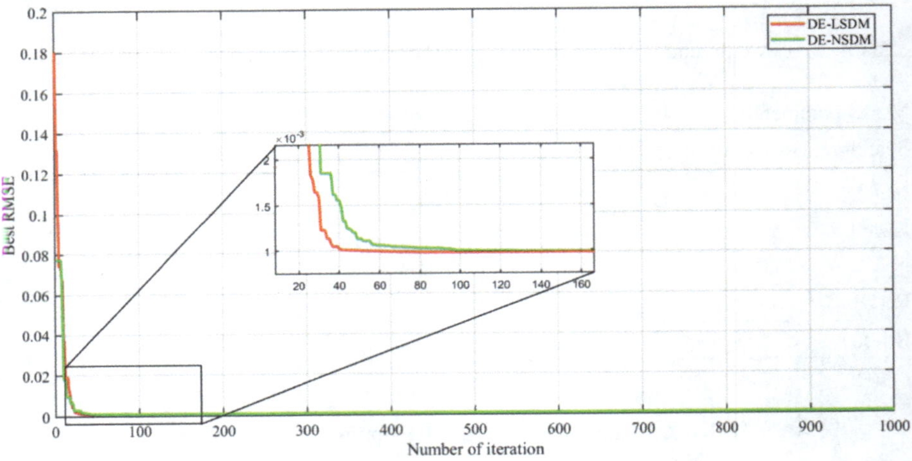

Fig. 2. The average convergence curves of 30 independent runs of DE algorithm during the parameter extraction process of LSDM and NSDM of RTC France solar cell.

In addition, to further show the performance of DE, the individual absolute error (IAE) between experimental data and simulated data are recorded in Table 5 for NSDM estimated parameters and the Table 6 for the estimated parameters with LSDM. In these tables, IAE I means its corresponding IAE current error, and IAE P denotes the corresponding power error. The maximal IAE values are highlighted in bold. Table 5 and Table 6 shows that the maximum IAE I and IAE P are 2.51E-03 and 1.46E-03 for NSDM and LSDM, respectively. Additionally, the comparison between the Individual Absolute Error (IAE) for current (IAE I) and power (IAE P) also the Relative Error of current (RE I) for both methods is shown in Figs. 3, 4 and 5, respectively.

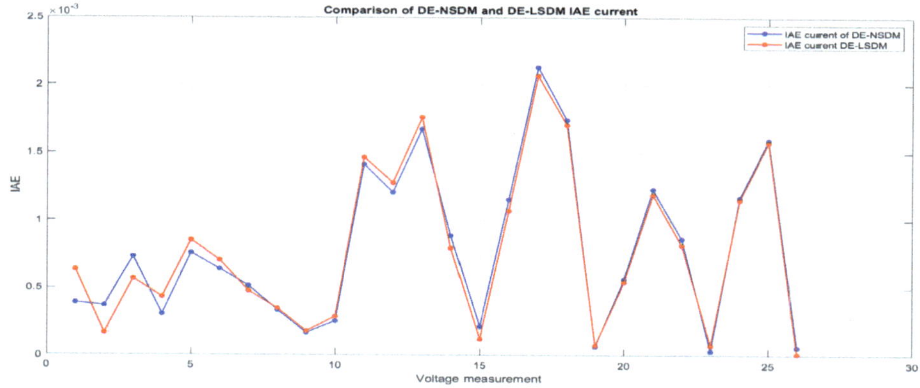

Fig. 3. Error curves of the computed and measured current values of RTC France solar cell for DE-NSDM and DE-LSDM.

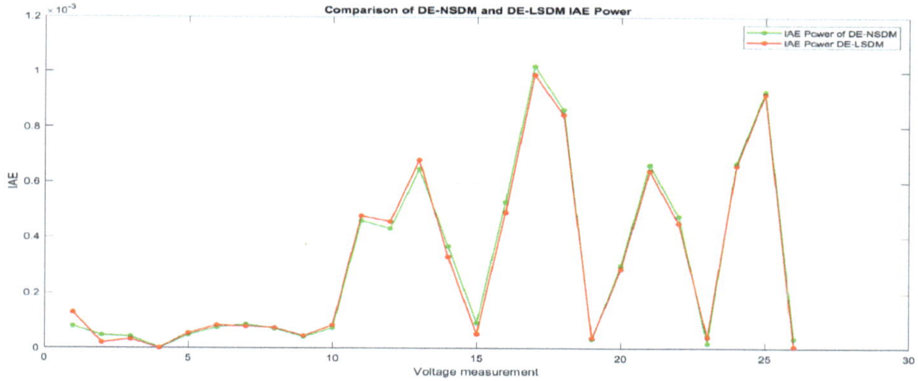

Fig. 4. Error curves of the computed and measured power values of RTC France solar cell for DE-NSDM and DE-LSDM.

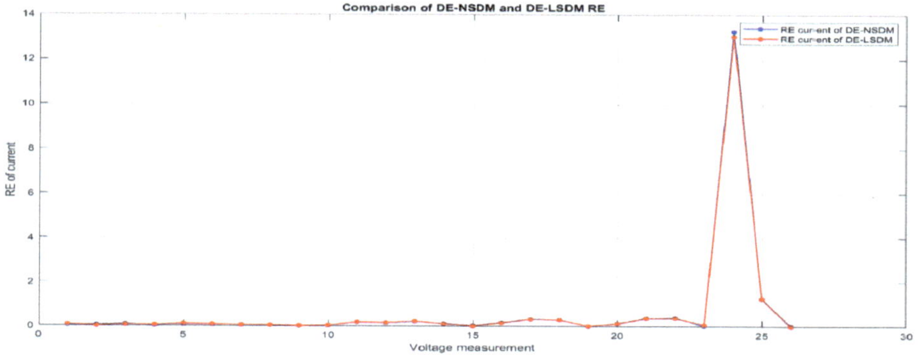

Fig. 5. Relative error curves of the computed and measured current values of RTC France solar cell for DE-NSDM and DE-LSDM.

Table 5. Simulated current/power data and individual absolute errors current and power of NSDM using the measured voltage and power of R.T.C France solar cell and the optimal parameter values extracted by DE algorithm.

iter	Voltage (V)	Experimental current (A)	Calculated current (A)	IAE I (A)	Experimental power (W)	Calculated power (W)	IAE P (W)
1	−0.2057	0.764	0.763607070681461	0.000392929318539403	−0.15715	−0.157073974439176	8.08255608235608e-05
2	−0.1291	0.762	0.762370168240247	0.000370168240246493	−0.098374	−0.0984219887198158	4.7788719815825e-05
3	−0.0588	0.7605	0.76123485517654	0.000734855176540261	−0.044717	−0.0447606094843806	4.32094843805658e-05
4	0.0057	0.7605	0.760192551063327	0.000307448936672894	0.0043348	0.00433309754106096	1.75245893903504e-06
5	0.0646	0.76	0.759238078043294	0.000761921956705658	0.049096	0.0490467798415968	4.92201584031821e-05
6	0.1185	0.759	0.758355174368058	0.00064482563194157	0.089941	0.0898650881626149	7.64118373850764e-05
7	0.1678	0.757	0.757517882540668	0.000517882540667869	0.12702	0.127111500690324	8.69006903240488e-05
8	0.2132	0.757	0.756662025270058	0.000337974729942125	0.16139	0.161320343787576	7.2056212423649e-05
9	0.2545	0.7555	0.755668365237208	0.000168365237207646	0.19227	0.192317598952869	4.2848952869351e-05
10	0.2924	0.754	0.754256187270567	0.000256187270566866	0.22047	0.220544509157914	7.49091579137495e-05
11	0.3269	0.7505	0.751914932821841	0.0014149328218408	0.24534	0.24580099153946	0.000462541539459738
12	0.3585	0.7465	0.74770898847985	0.00120898847985007	0.26762	0.268053672370026	0.000433422370026249
13	0.3873	0.7385	0.74017869244777	0.00167869244777008	0.28602	0.286671207585021	0.000650157585021349

(*continued*)

Table 5. *(continued)*

iter	Voltage (V)	Experimental current (A)	Calculated current (A)	IAE I (A)	Experimental power (W)	Calculated power (W)	IAE P (W)
14	0.4137	0.728	0.727107667369429	0.000892332630571091	0.30117	0.300804441990733	0.000369158009267234
15	0.4373	0.7065	0.706284419787555	0.000215580212445166	0.30895	0.308858176773098	9.42732269023083e-05
16	0.459	0.6755	0.674342785096948	0.00115721490305221	0.31005	0.309523338359499	0.000531161640500988
17	0.4784	0.632	0.629865022936715	0.00213497706328503	0.30235	0.301327426972924	0.00102137302707556
18	0.496	0.573	0.571255626934473	0.00174437306552677	0.28421	0.283342790959499	0.000865209040501247
19	0.5119	0.499	0.499069128688108	6.91286881082109e-05	0.25544	0.255473486975443	3.53869754425862e-05
20	0.5265	0.413	0.413569040511999	0.000569040511998908	0.21744	0.217744099829567	0.000299599829567426
21	0.5398	0.3165	0.31773264393915	0.00123264393915024	0.17085	0.171512081198353	0.000665381198353299
22	0.5521	0.212	0.212868153549779	0.000868153549778927	0.11705	0.117524507574833	0.000479307574832943
23	0.5633	0.1035	0.103465336182826	3.46638171739855e-05	0.058302	0.0582820238717859	1.95261282141049e-05
24	0.5736	−0.01	−0.0088278911268583	0.0011721088731417	−0.005736	−0.00506367835036592	0.000672321649634082
25	0.5833	−0.123	−0.124596976842991	0.00159697684299066	−0.071746	−0.0726774165925165	0.00093151659251646
26	0.59	−0.21	−0.210062975556144	6.29755561439016e-05	−0.1239	−0.123937155578125	3.71555781249061e-05
Sum	NA	NA	NA	0.020545	NA	NA	0.008143

Table 6. Simulated current/power data and individual absolute errors current and power of LSDM using the measured voltage and power of R.T.C France solar cell and the optimal parameter values extracted by DE algorithm.

iter	Voltage (V)	Experimental current (A)	Calculated current (A)	IAE I (A)	Experimental power (W)	Calculated power (W)	IAE P (W)
1	−0.2057	0.764	0.763361353861089	0.000638646138911181	−0.15715	−0.157023430489226	0.000131369510774054
2	−0.1291	0.762	0.762167207808486	0.000167207808486114	−0.098374	−0.0983957865280756	2.15865280755595e-05
3	−0.0588	0.7605	0.761071135084971	0.000571135084970864	−0.044717	−0.0447509827429963	3.35827429962868e-05
4	0.0057	0.7605	0.760064835684994	0.000435164315005676	0.0043348	0.00433236956340447	2.48043659553231e-06
5	0.0646	0.76	0.759143247483846	0.000856752516153803	0.049096	0.0490406537874565	5.53462125435292e-05
6	0.1185	0.759	0.758290454293881	0.000709545706119141	0.089941	0.0898574188338249	8.40811661751151e-05
7	0.1678	0.757	0.757480744776336	0.000480744776336195	0.12702	0.127105268973469	8.06689734692001e-05
8	0.2132	0.757	0.756650373012162	0.000349626987837914	0.16139	0.161317859526193	7.45404738070332e-05
9	0.2545	0.7555	0.755680035516351	0.000180035516351373	0.19227	0.192320569038911	4.58190389114321e-05
10	0.2924	0.754	0.754289417498532	0.00028941749853173	0.22047	0.220554225676571	8.46256765706932e-05
11	0.3269	0.7505	0.751967805981703	0.00146780598170293	0.24534	0.245818275775419	0.000479825775418663
12	0.3585	0.7465	0.747779337483302	0.00127933748330233	0.26762	0.268078892487764	0.000458642487763894
13	0.3873	0.7385	0.740263200431532	0.00176320043153166	0.28602	0.286703937527132	0.000682887527132192

(continued)

Table 6. *(continued)*

iter	Voltage (V)	Experimental current (A)	Calculated current (A)	IAE I (A)	Experimental power (W)	Calculated power (W)	IAE P (W)
14	0.4137	0.728	0.727201099725014	0.000798900274986192	0.30117	0.300843094956238	0.000330505043761753
15	0.4373	0.7065	0.706378798519899	0.000121201480101418	0.30895	0.308899448592752	5.30014072483986e-05
16	0.459	0.6755	0.674427846002264	0.00107215399773553	0.31005	0.309562381315039	0.00049211868496063
17	0.4784	0.632	0.629930233297925	0.00206976670207493	0.30235	0.301358623609727	0.000990176390272601
18	0.496	0.573	0.571292785615262	0.00170721438473764	0.28421	0.28336122166517	0.000846778334829879
19	0.5119	0.499	0.499075315681395	7.53156813945166e-05	0.25544	0.255476654097306	3.8554097305854e-05
20	0.5265	0.413	0.413547186067741	0.000547186067741401	0.21744	0.217732593464666	0.000288093464665851
21	0.5398	0.3165	0.317691628681244	0.00119162868124417	0.17085	0.171489941162136	0.000643241162135599
22	0.5521	0.212	0.212820526888483	0.000820526888482637	0.11705	0.117498212895131	0.000453012895131263
23	0.5633	0.1035	0.103425394148943	7.46058510567443e-05	0.058302	0.0582595245240997	4.20254759002647e-05
24	0.5736	−0.01	−0.00884619385220642	0.00115380614779358	−0.005736	−0.0050741767936256	0.000661823206374398
25	0.5833	−0.123	−0.12458017082824	0.00158017082824025	−0.071746	−0.0726676136441125	0.000921713644112532
26	0.59	−0.21	−0.210012876542937	1.287654293744e-05	−0.1239	−0.123907597160333	7.5971603330921e-06
Sum	NA	NA	NA	0.020414	NA	NA	0.008004

All these results indicate that DE-LSDM is capable of accurately extracting the parameters for SDM. Furthermore, the curve of current–voltage characteristic based on the identified parameters by DE with both NSDM and LSDM are illustrated in Fig. 6. It is apparent that the values of current data obtained by DE-NSDM and DE-LSDM are all in high accordance with the experimental data under different voltage conditions for the SDM.

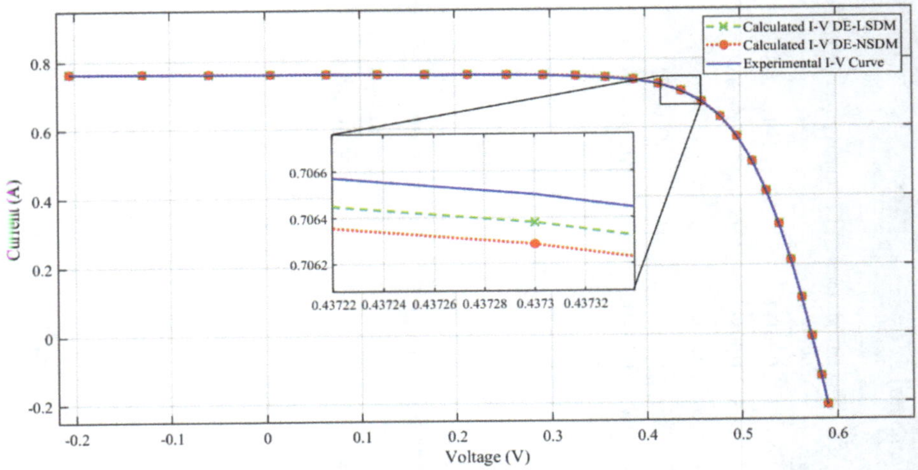

Fig. 6. Comparison between the measured and simulated I-V data obtained by DE-LSDM and DE-NSDM.

Statistical analysis should be performed in order to trust the performance of the optimization technique and evaluate the robustness behavior of the DE-NSDM and DE-LSDM. So, the both of DE algorithm has been executed for 30 independents runs implements. For each run, the best minimum objective function has been recorded and reported. Moreover, statistical indices such as the mean, standard deviation STD, the minimum and maximum over the 30 runs have been computed. The results of the statistical analysis of the application of the DE-NSDM and DE-LSDM have been listed in Table 7. The end values of the objective function over the 30 runs are shown in Fig. 7. The results of the statistical analysis prove that the DE-LSDM algorithm is an effective algorithm then the DE-NSDM for parameters' identification of single diode model of R.T.C. France solar cell.

Table 7. Statistical measurement of the applied DE method for R.T.C. France solar cell based NSDM and LSDM.

Algorithm	Maximum	Mean	Minimum	STD
DE-NSDM	1.5700750e-04	1.2098123e-03	9.804680e-04	0.0001527147
DE-LSDM	1.5425360e-04	1.2047760e-03	9.764863e-04	0.0001489026

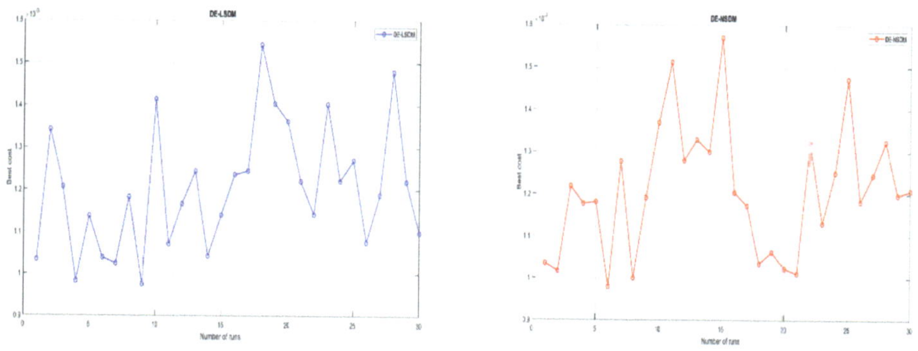

(a) (b)

Fig. 7. Best cost of 30 runs of the DE algorithm of RTC France solar cell; (a) DE-LSDM, (b) DE-NSDM.

The computational CPU time as another index to examine the convergence speed of DE-NSDM and DE-LSDM algorithms. The DE-LSDM algorithm requires only 1.37035 s for parameter extraction, whereas the DE-NSDM algorithm requires 4.34375 s. This indicates that the DE-LSDM will allow a faster convergence and response time than DE-NSDM for real-time optimal control of PV systems.

5 Conclusion

This study provides a comprehensive comparison of the Differential Evolution (DE) optimization technique applied to implicit and explicit single-diode models for photovoltaic (PV) parameter extraction. The results demonstrate that the explicit model using the Lambert W function (LSDM) not only improves computational efficiency but also offers superior RMSE convergence speed, ensuring more precise parameter extraction.

Additionally, the analysis highlights improvements in individual Absolute Error (IAE) of current, as well as individual absolute error of power, confirming the explicit model's ability to optimize PV system performance with minimal deviations. Given its enhanced accuracy, execution time, and reliability, the explicit DE model (LSDM) emerges as the preferred approach for practical PV applications, balancing precision and speed effectively.

References

1. Lazaroiu, A.C., Gmal Osman, M., Strejoiu, C.-V., Lazaroiu, G.: A comprehensive overview of photovoltaic technologies and their efficiency for climate neutrality. Sustainability. **15**(23), 16297 (2023). https://doi.org/10.3390/su152316297
2. Oliva, D., Elaziz, M.A., Elsheikh, A.H., Ewees, A.A.: A review on meta-heuristics methods for estimating parameters of solar cells. J. Power Sources. **435**, 126683 (2019). https://doi.org/10.1016/j.jpowsour.2019.05.089

3. Chen, X., Xu, B., Mei, C., Ding, Y., Li, K.: Teaching–learning–based artificial bee colony for solar photovoltaic parameter estimation. Appl. Energy. **212**, 1578–1588 (2018). https://doi.org/10.1016/j.apenergy.2017.12.115
4. Yu, K., Liang, J.J., Qu, B.Y., Chen, X., Wang, H.: Parameters identification of photovoltaic models using an improved JAYA optimization algorithm. Energy Convers. Manag. **150**, 742–753 (2017). https://doi.org/10.1016/j.enconman.2017.08.063
5. Oliva, D., Cuevas, E., Pajares, G.: Parameter identification of solar cells using artificial bee colony optimization. Energy. **72**, 93–102 (2014). https://doi.org/10.1016/j.energy.2014.05.011
6. Yetayew, T.T., Jyothsna, T.R.: Parameter extraction of photovoltaic modules using Newton Raphson and simulated annealing techniques. In: 2015 IEEE Power, Communication and Information Technology Conference (PCITC), pp. 229–234. IEEE, Bhubaneswar, India (2015). https://doi.org/10.1109/PCITC.2015.7438166
7. Abdulrazzaq, A.K., Bognár, G., Plesz, B.: Evaluation of different methods for solar cells/modules parameters extraction. Sol. Energy. **196**, 183–195 (2020). https://doi.org/10.1016/j.solener.2019.12.010
8. Lidaighbi, S., Elyaqouti, M., Ben Hmamou, D., Saadaoui, D., Assalaou, K., Arjdal, E.: A new hybrid method to estimate the single-diode model parameters of solar photovoltaic panel. Energy Convers. Manag. **15**, 100234 (2022). https://doi.org/10.1016/j.ecmx.2022.100234
9. Alrahim Shannan, N.M.A., Yahaya, N.Z., Singh, B.: Single-diode model and two-diode model of PV modules: a comparison. In: 2013 IEEE International Conference on Control System, Computing and Engineering, pp. 210–214. IEEE, Penang, Malaysia (2013). https://doi.org/10.1109/ICCSCE.2013.6719960
10. Yousri, D., et al.: Modified interactive algorithm based on Runge Kutta optimizer for photovoltaic modeling: justification under partial shading and varied temperature conditions. IEEE Access. **10**, 20793–20815 (2022). https://doi.org/10.1109/ACCESS.2022.3152160
11. Jain, A.: Exact analytical solutions of the parameters of real solar cells using Lambert W-function. Sol. Energy Mater. Sol. Cells. **81**(2), 269–277 (2004). https://doi.org/10.1016/j.solmat.2003.11.018
12. Muhsen, D.H., Ghazali, A.B., Khatib, T., Abed, I.A.: Parameters extraction of double diode photovoltaic module's model based on hybrid evolutionary algorithm. Energy Convers. Manag. **105**, 552–561 (2015). https://doi.org/10.1016/j.enconman.2015.08.023
13. Zhou, T., Shang, C.: Parameter identification of solar photovoltaic models by multi strategy sine–cosine algorithm. Energy Sci. Eng. **12**(4), 1422–1445 (2024). https://doi.org/10.1002/ese3.1673
14. Gao, S., Wang, K., Tao, S., Jin, T., Dai, H., Cheng, J.: A state-of-the-art differential evolution algorithm for parameter estimation of solar photovoltaic models. Energy Convers. Manag. **230**, 113784 (2021). https://doi.org/10.1016/j.enconman.2020.113784
15. Zhao, H.W., Xia, H.G.: An improved differential evolution algorithm for numerical optimization problems. Appl. Mech. Mater. **415**, 349–352 (2013). https://doi.org/10.4028/www.scientific.net/AMM.415.349
16. Wang, X., Yu, X.: Differential evolution algorithm with three mutation operators for global optimization. Mathematics. **12**(15), 2311 (2024). https://doi.org/10.3390/math12152311
17. Segura, C., Coello Coello, C.A., Hernández-Díaz, A.G.: Improving the vector generation strategy of differential evolution for large-scale optimization. Inf. Sci. **323**, 106–129 (2015). https://doi.org/10.1016/j.ins.2015.06.029
18. Liu, H., Huang, H., Liu, S.: Explore influence of differential operator in de mutation with unrestrained method to generate mutant vector. In: Rutkowski, L., Korytkowski, M., Scherer, R., Tadeusiewicz, R., Zadeh, L.A., Zurada, J.M. (eds.) Lecture Notes in Computer ScienceSwarm and Evolutionary Computation, vol. 7269, pp. 292–300. Springer Berlin Heidelberg, Berlin, Heidelberg (2012). https://doi.org/10.1007/978-3-642-29353-5_34

19. Wang, Y., Cai, Z., Zhang, Q.: Differential evolution with composite trial vector generation strategies and control parameters. IEEE Trans. Evol. Comput. **15**(1), 55–66 (2011). https://doi.org/10.1109/TEVC.2010.2087271
20. Yang, X., Gong, W., Wang, L.: Comparative study on parameter extraction of photovoltaic models via differential evolution. Energy Convers. Manag. **201**, 112113 (2019). https://doi.org/10.1016/j.enconman.2019.112113

Advancing PID Control Quarter-Car Suspension System with Metaheuristic Optimization Comparative Study

H. Houari[1(✉)] and S. Hamaci[2]

[1] Manufacturing Engineering, National Polytechnic School of Oran, Laboratory of Tlemcen – MELT, Oran, Algeria
habiba.houari@enp-oran.dz
[2] ECAM-EPMI, 13 boulevard de l'Hautil, 95092 Cergy-Pontoise, France
S.HAMACI@ECAM-EPMI.COM

Abstract. This paper gives an effectiveness comparison in terms of tuning for the five meta-heuristic algorithms—Genetic Algorithm (GA), Particle Swarm Optimization (PSO), Grey Wolf Optimizer (GWO), Ant Colony Optimization (ACO), and Flower Pollination Algorithm (FPA)—for PID controllers of a quarter-car suspension system. Primarily, control performance is maximized by optimization of the proportional, integral, and derivative gains for improved accuracy, improved convergence speed, and better disturbance rejection. Comparing the strengths and weaknesses of each with regard to key performance metrics, results have shown some of these algorithms to consistently perform better than others. Algorithms that are established from this, which turn out to be excellent candidates for complex and dynamic PID tuning, would greatly benefit the community of researchers and practitioners seeking good solutions for efficiency and robustness.

Keywords: Proportional-Integral-Derivative (PID) · Optimization · Genetic Algorithm · Particle Swarm Optimization · Grey wolf optimizer · Flower Pollination Algorithm

1 Introduction

Metaheuristic algorithms have been a powerful tool in control engineering for several decades and provide robust solutions to difficult optimization problems. They are particularly useful in the tuning of control parameters for nonlinear and dynamic systems, where classic methods usually fail or become impractical. Metaheuristic techniques, by mimicking natural phenomena or adaptive behaviors, provide efficient and flexible approaches for enhancing system performance, thus making them indispensable in modern control strategies. The key goal in suspension system is optimization of PID controller parameters. PID controls have constrained the operational range of processes exhibiting dynamic nonlinearity. In our suggested work, globally optimized PID parameters aim to run the suspension system across its whole operational range to address the constraints of the linear PID controller. The simulation analysis reveals that the PID

M. Hatti (Ed.): IC-AIRES 2025, LNNS 1740, pp. 500–519, 2026.
https://doi.org/10.1007/978-3-032-12788-4_47

controller, tuned with fixed parameters using GA, PSO, GWO, ACO, and FPA, demonstrates adequate performance in set point tracking and disturbance rejection. In industrial process control systems, the most commonly utilized controller is the Proportional Integral Derivative (PID) due to its straightforward architecture, operational simplicity, and dependable performance. Given these significant advantages, PID control systems have been extensively adopted. Consequently, appropriate parameter selection is essential in a realistic PID controller. Besides the integral and differential coefficients, scaling factors significantly influence the efficacy of the PID controller. The adjustment of PID parameters has garnered significant scientific attention over the past two decades.

Modern industries have implemented auto-tuning systems in order to set up a PID controller; however, the performance is still bounded by the online measurements and mathematical models of the system. The limitations of the methods can be alleviated to some extent by using optimization algorithms: genetic algorithms, particle swarm optimization, simulated annealing, and evolutionary programming for finding optimal tuning parameters directly based on the closed-loop system response. GA is the most widely used method of optimization in controller tuning because it does not make any assumption regarding the nature of the model or the differential equations. The same justification applies to Particle Swarm Optimization (PSO), Grey Wolf Optimizer (GWO), Ant Colony Optimization (ACO), and Flower Pollination Algorithm (FPA) as very good alternatives in the tuning applications of PID, which do not require a priori knowledge on the mathematical model of a system. Such algorithms use population-based search mechanisms, which enable them to search large search spaces and find optimal solutions without being limited by the system's complexity.

2 Sate of the Art on Automotive Suspension System

Globally, visitors fatalities result in extreme lack of existence and non-deadly injuries. Vehicle safety and managing performance, essential for minimizing those fatalities, are often assessed through their suspension structures. The suspension gadget connects the car's body to the road, and its number one feature is to make sure a cushty trip by using soaking up avenue disturbances. By helping and separating the car body from the wheels, the suspension machine lets in relative movement between these components, enhancing each experience comfort and street maintaining. The machine's overall performance is essential in mitigating disruptions because of choppy street surfaces, aerodynamic forces, wheel and tire non-uniformities, and braking forces.

Suspension structures have a profound impact on vehicle dynamics, specially in terms of dealing with and ride comfort. They serve two primary functions: setting apart the automobile body from street disturbances and making sure strong contact among the tires and the road during riding. Numerous studies have aimed to improve suspension device performance via diverse manipulate techniques, with the PID (Proportional-Integral-Derivative) controller being one of the most famous strategies due to its simplicity and effectiveness.

In 2013 [2], Qi Zhou proposed a brand new approach to beautify suspension device overall performance through integrating an modern set of rules with a PID controller. The findings verified that the PID controller correctly reduced vibration amplitude whilst increasing acceleration, showcasing its potential to improve ride comfort.

Similarly, Vivek Kumar Maurya and Narinder Singh Bhangal [3] as compared the overall performance of PID and LQR (Linear Quadratic Regulator) controllers. Their take a look at determined that the LQR controller outperformed each the PID controller and the conventional system in improving suspension overall performance.

In 2017, Vladimir Popovic et al. [4] Investigated suspension structures for massive vehicles the usage of external actuator excitation. The look at revealed that employing a PID controller substantially decreased settling time and maximum overshoot, thereby improving the automobile's handling characteristics.

Siau Ping Tee et al. [5] further explored the application of PID controllers in suspension systems by growing a mathematical version of a two-diploma-of-freedom suspension gadget. Their look at optimized the PID controller settings to limit overshoot and consistent-nation mistakes, ensuing in progressed trip consolation.

In 2011, M. Akpakavi [6] studied the behavior of an automobile model below severe hundreds and concluded that the use of P, PI, and PID controllers better the gadget's response. However, the have a look at referred to that the settling duration changed into longer whilst these controllers were hired, indicating room for similarly optimization.

From 2015 onwards [7], the research focus shifted towards extra advanced optimization techniques for tuning PID controllers, which includes Genetic Algorithms (GA), Particle Swarm Optimization (PSO), Grey Wolf Optimizer (GWO), and Flower Pollination Algorithm (FPA). These methods were employed to similarly beautify suspension gadget overall performance.

X. Zhang, Y. Fang, and X. Liu [8] introduced an adaptive sliding mode control strategy for energetic suspension systems. Their method efficaciously decreased vibration and progressed trip consolation compared to standard PID controllers, demonstrating the capacity of adaptive manipulate methods.

In 2020, Y. Gao and Z. Huang [9] utilized the Grey Wolf Optimizer (GWO) to layout a PID controller for a car's suspension device. Their approach executed stepped forward performance by using minimizing overshoot and decreasing settling time, highlighting the effectiveness of GWO in suspension machine optimization.

M. Wang, H. Zhou, and J. Lin [10] implemented the Flower Pollination Algorithm (FPA) to optimize PID controller parameters for vehicle suspension structures. The consequences validated great upgrades in each ride comfort and street dealing with performance, suggesting that FPA is a promising device for suspension gadget design. S. Choudhury, K. Singh, and R. Gupta [11] proposed a hybrid fuzzy logic and PSO-primarily based manage approach for car suspension systems. The combined approach offered better adaptability to changing road conditions and stepped forward standard machine balance, displaying the ability of hybrid manage strategies.

In 2023, X. Li and Y. Qiu [12] offered a hybrid method that combined Genetic Algorithm (GA) and Particle Swarm Optimization (PSO) for tuning PID controllers in energetic suspension structures. Their examine achieved enhanced journey comfort and reduced vibration ranges, emphasizing the blessings of hybrid optimization strategies.

Finally, in 2024, P. Gupta and R. Sharma [13] explored the application of deep reinforcement getting to know (DRL) to suspension structures. Their DRL-based totally controller outperformed conventional PID controllers in terms of each journey comfort and road managing, indicating a promising direction for future studies.

Over the years, various optimization techniques, together with GA, PSO, GWO, and FPA, were explored to enhance suspension gadget performance. These methods, while blended with PID controllers, have proven full-size capability in improving automobile experience consolation, coping with, and balance. The continuous development and application of these advanced optimization strategies are predicted to lead to even greater effective and efficient suspension structures inside the future [14].

3 PID Control

The PID controller improves dynamic response and decreases constant-state errors in a device.

$$C(s) = k_p + \frac{k_i}{s} + k_d s \tag{1}$$

- The spinoff controller improves temporary reaction via incorporating a finite zero into the open-loop plant switch characteristic.
- The imperative controller provides a pole on the origin, growing the system type by means of one and bringing constant-nation errors because of step features to 0.

As a result, the PID controller nice-tunes the device's conduct, allowing it to reply extra quick and precisely to modifications.

Expert individuals utilizing knowledge of process dynamics and can manipulate integral, proportional, and derivative of controller and look at the response time history, settle time, overshoot and/or oscillations. A robust control is highly desirable as it can tolerate process deviations for maintaining stability of the closed loop system.

4 Mathematical Modeling

Mathematical model changed to developed to optimize the PID control device of the active suspension machine. Active suspension device, which was composed of actuators, springs, and dampers became fully researched. It was found that the suspension system uses the necessary power to decrease the unnecessary vibration. Figure 1 illustrates a detailed quarter car model.

A mathematical version of a quarter-vehicle lively suspension system turned into evolved the use of Newton's laws of movement and the unfastened body diagram approach [15]. The version is ruled via the following equations:

$$\sum F = ma \tag{2}$$

$$m_1\hat{y}_1 = -k_1(y_1 - y_2) - b(\hat{y}_1 - \hat{y}_2) \tag{3}$$

The Laplace transform subsequently applied to Eq. (2), yielding:

$$m_1 s^2 y_1(s) = -k_1\big(y_1(s) - y_2(s)\big) - b\big(sy_1(s) - sy_2(s)\big) \tag{4}$$

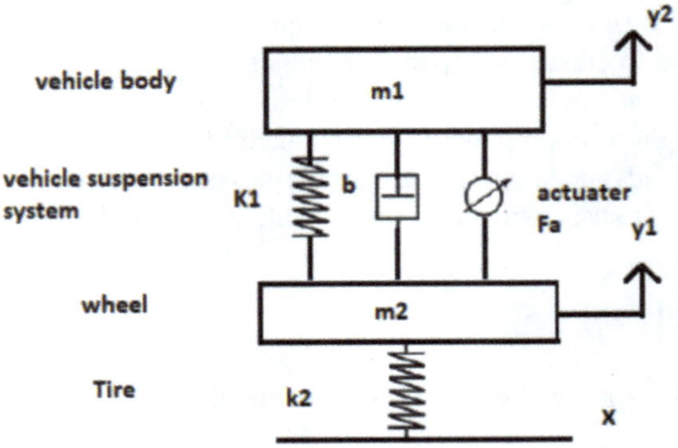

Fig. 1. Quarter-car suspension system

$$\left(m_1 s^2 + bs + K_1\right) y_1(s) - (bs + k_1) y_2(2) = 0 \tag{5}$$

$$y_2(s) = \frac{\left(m_1 s^2 + bs + k_1\right) y_1(s)}{(bs + k_1) y_2(s)} \tag{6}$$

$$m_2 \ddot{y}_2 = (k_1(y_1 - y_2) + b(\dot{y}_1 - \dot{y}_2) + k_2 y_2) \tag{7}$$

Further, the Laplace transform was also employed for Eq. (6), resultingin:

$$m_1 s^2 y_2 = k_1(y_1(s) - y_2(s)) - b(sy_1(s) + sy_2(s)) + X(s) \tag{8}$$

$$\left(m_1 s^2 + bs + k_1 + k_2\right) y_2(s) - (bs + k_1) y_1(s) = k_2 X(s) \tag{9}$$

$$\left(m_1 s^2 + bs + k_1 + k_2\right) \frac{\left(m_1 s^2 + bs + k_1\right) y_1(s)}{(bs + k_1) y_2} - (bs + k_1) y_1(s) = k_2 X(s) \tag{10}$$

$$\frac{y_1(s)}{X(s)} = \frac{k_2(bs + k_1)}{\left(m_2 s^2 + bs + k_1 + k_2\right)\left(m_1 s^2 + bs + k_1\right)\ (bs + k_1)^2} \tag{11}$$

$$G(s) = \frac{2.4e08s + 2.56e09}{12500s^4 + 450000s^3 + 4.48e07s^2 + 2.4e08s + 2.56e09} \tag{12}$$

In this representation, X(s) represents the control effort, whilst y1 and y2 constitute the vertical displacements of the tire and sprung mass, respectively. Table 1 offers an in depth account of the parameters [16] used on this sector vehicle active suspension machine version.

Table 1. Parameters of the quarter vehicle active suspension system [16]

Notation	Description	Value	Unit
M1	Sprung mass	250	kg
M2	Unsprung mass	50	kg
K1	Spring stiffness	16,000	N/m
K2	Tyre stiffness	160,000	N/m
b	Damping coefficient	1,500	N.s/m

5 Genetique Algorithme

Genetic algorithms (GAs) are a type of probabilistic search methods used for finding approximate solutions to optimization and search problems. In computer science, the approach of GAs is widespread in work aimed at generating solutions for various design problems and similarly shaped endeavors. More general formalisms for GAs are a type of evolutionary algorithm (EA) inspired by evolutionary biology, which use inheritance, mutation, natural selection, and crossover techniques [17]. These algorithms start with a set of solutions to the problem being addressed. These solutions, usually encoded as strings called chromosomes, comprise a set called population (P). The classical approach initializes this population randomly.

After such an initial step, the crux of EAs comes into play, i.e., solutions from one population—parents—are used to create a new P—offspring. This is motivated by the hope that the new P will outperform the previous one. GAs are iterative processes that result in successive generations of populations. That is, the new population GAs, since this process is repeated, will lead the population to evolve and find better solutions to the problem. The procedure is terminated when some condition is satisfied—for instance, when a solution with a satisfactory fitness is found.

Genetic algorithms (GAs) are a class of evolutionary algorithms used for optimization and search problems in finding approximate solutions. They are evolutionary biology concepts and, more recently, computational modeling methods inspired by biological mechanisms based on principles like inheritance, mutation, natural selection, and crossover. GAs usually begin with a population, P, composed of solutions that have been generated at random; then, solutions contained in one P are combined, hopefully so as to form a better one. Individuals are selected to create new offspring based on their fitness; the fitter the individuals, the higher the probability of reproduction. This process is iterated until some stopping criteria is satisfied such as the closeness of a P to the optimal solution [18].

The most crucial components of GAs are the crossover (C) and mutation (M) operators. C is a recombination operator; it combines parts of two parent chromosomes to make offspring. The simplest way of doing this is to pick a point at random, C, and one offspring will have the chromosome before this point of one parent and the chromosome after this point of the other. This ensures that the offspring is a combination of the two parents and will inherit some traits from each. The M operation randomly modifies the offspring produced by C not to converge in a local optimum. The option of an improved encoding method is problem-dependent. GAs may use binary encoding, where chromosome is a string of bits or real encoding. Another valuable notion in GA is Elitism; some of the best individuals are preserved on each generation.

The GA execution steps are summarized in a flowchart As depicted in Fig. 2. The GA execution details for PID tuning can be referred to in [19]

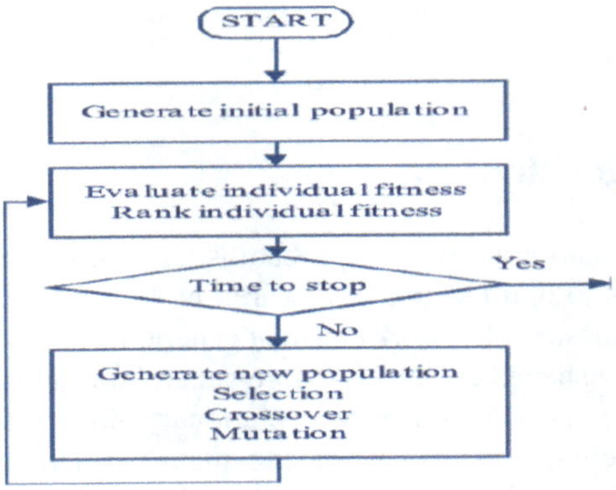

Fig. 2. Genetic algorithm process flowchart.

Other studies are available on the software of GAs in PID tuning. For instance, a very good study on the robustness of GA-tuned PID controllers highlighted the fact that these are able to maintain performance despite plant gain variations and dominant time constants and in the presence of time delays. Application of GA-based totally controllers within Simulink also has had promising results regarding realistic applicability of this method in actual-time on-line tuning.

GA has also been used in control packages, for example, autopilot systems, to derive tuning rules using symbolic regression. The resulting analytically expressed rules simplify the tuning process for a control engineer, and it has been found to be robust against variations in aircraft speed and benefit margins. Also, GA has been carried out in model reduction methods, for example, the simplification of multicomponent distillation methods, and has enabled the most optimal tuning of proportional controllers in these simplified models.

Investigations are ongoing to evaluate the efficacy of genetic algorithms in control systems. El-Deen et al. [20] and Joseph and Emmanuel [21] concentrated on optimising PID gains for DC motor speed control, whereas Mohammed et al. [22] introduced a GA-based PID controller for synchronous generators to enhance damping and maintain machine stability. Patil and Lakhekar [23] examined GA-designed PID controllers for cascade control approaches, whereas Amaral et al. [24] demonstrated the automated tuning of PID parameters using GAs.

The ongoing focus on GA demonstrates its leadership in engineering, providing a robust solution for optimising PID controllers across various methods and frameworks.

The principle of evolution in natural systems has guided the development of mathematical theory and the initial computer simulations of evolution. The advancement of rudimentary models into artificial systems must include an analysis of behaviour that reflects temporal considerations. Research has demonstrated the integration of time dependency into computer models, use fundamental examples to illustrate the implications of "time" on the steady-state behaviour of artificial Darwinian systems. This discussion addresses certain generic qualities that have been demonstrated to apply to both groups of systems under consideration. As population numbers rise and/or stochastic processes diminish, the distributions of states converge to those observed in the equivalent deterministic models [25]. Artificial Darwinian systems are defined as devices that emulate time-dependent natural processes of (1) feedback, (2) selection, and (3) replication. The most basic machine of this type has a set of things that alter their state over a certain time interval based on a predetermined "rule". Upon the implementation of that rule, the status of each entity is either chosen for replication or rejected based on an external criterion (fitness). Variables encode selection control criteria that subsequently alter based on the "success" or "failure" of the entities they denote. The chosen entities are subsequently replicated, occasionally undergoing a change in state, to generate a new population of entities that undergo the same procedure [26]. These machines have been examined in both a purely mathematical context (probabilistic models) and through computer simulation models.

6 Particle Swarm Optimization

The paperwork Particle Swarm Optimisation was initially introduced by James Kennedy and Russell C. Eberhart in 1995, drawing on the principles of swarm intelligence and the behavioural patterns of swarming creatures like bees and birds. PSO commences by initialising a population of potential solutions, referred to as particles. Each particle signifies a potential solution within the problem's search space. All particles collectively investigate the solution space to identify an optimal or near-optimal solution.

It subsequently computes the objective function value for each particle, indicating the efficacy of each particle's solution to the optimisation problem. The fitness function provides a quantifiable assessment of the solution quality represented by each particle, guiding the search towards superior solutions.

In each iteration of PSO, particles adjust their velocities based on their present velocities, the best-known position among neighbouring particles—local best—and the best-known position across the entire population—global best. The velocity adjustment mechanism equilibrates exploration and exploitation.

Subsequent to the velocity updates, particles adjust their places inside the search space. The position update equation calculates a new position for each particle based on its present position and revised velocity. Particles traverse the solution space dynamically in this iterative process, ultimately attaining the best solutions.

$$x_{ij}^{t+1} = x_{ij}^{t} + V_{ij}^{t+1} \tag{13}$$

i = 1, 2, …, n

j = 1, 2, …, m

If the current position of a particle has a better fitness value than its local best, then it updates the local best to its current position. Similarly, if a particle discovers a better solution than the current global best, then the global best position for all particles is updated. This kind of mechanism implies encouraging particles to share information and thereby find a better solution as a whole.

$$velocity_{ij}^{t+1} = w * velocity_{ij}^{t} + c_1 * r_1\left(local_{best\,ij} - position_{ij}^{t}\right) + c_2 \quad * r_2\left(globa\_best_{ij} - position_{ij}^{t}\right) \tag{14}$$

The constants c1 and c2 are also called acceleration constants. Small values for c1 and c2 allow the particle to move away even from target regions, but high values make the particle move near the target region. In our project, we implement c1 = c2 = 2 according to reference [27].

The algorithm continues this process until it meets the termination criterion. The common termination criteria are: when a maximum number of iterations have been completed, or when a satisfactory solution in a stipulated time frame is reached. Satisfying the termination criterion, however, the best found so far, that is either the global best or the satisfactory solution, represents an optimum solution for the given optimization problem.

7 Grey Wolf Optimizer

Grey Wolf Optimisation (GWO) is a metaheuristic optimisation method invented by Mirjalili et al. In reference [28]. GWO is inspired by the social organisation of grey wolves and their adept hunting techniques. Grey wolves generally hold the apex position in the food chain of their native environments. Grey wolves generally inhabit groups consisting of five to twelve members. Grey wolves exhibit a clearly defined social hierarchy. Figure 3 depicts the social order and responsibilities of the grey wolf. The fundamental difference between GWO and other metaheuristic optimisation algorithms is that the social dominance hierarchy aims to improve a candidate solution with each subsequent iteration. Thus, GWO mimics the predatory activity of grey wolves in identifying and pursuing prey [29, 30]. outlined the fundamental features of grey wolf foraging behaviour. Supervising, pursuing, and improving the quarry.

- Tracking, encircling, and harassing the prey till it stops transferring.
- Attacking toward the prey.

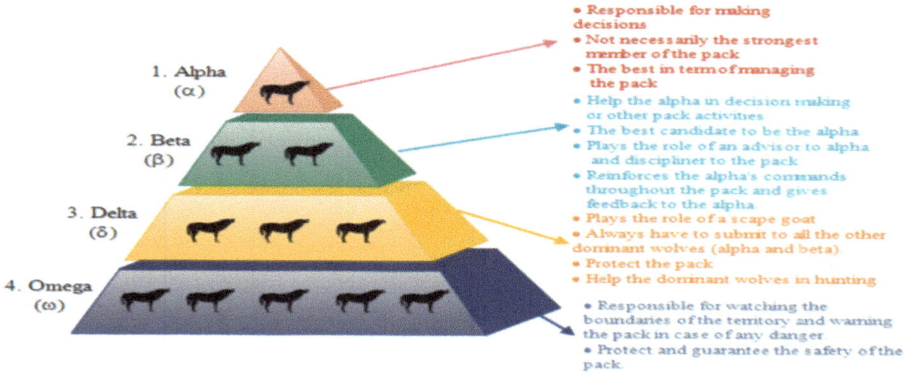

Fig. 3. Social hierarchy of grey wolves and their responsibilities.

GWO's wonderful blessings have spawned myriad studies in many quintessential research fields. In manage engineering, the volume of courses delving into the usefulness of the GWO set of rules to adjust the parameters in PID controllers has dramatically escalated.

Li et al. [31] suggested a GWO scheme for PI controller variable tuning in a closed-loop condenser pressure manipulate mechanism. The results of performance proved that the Grey Wolf (GW) optimizer was superior to the GA and PSO algorithms. On the same lines, Yadav et al. [32] used the GWO algorithm in the parameters of a primary PID controller tuning for a magnetic levitation system. Compared with the simple ZN tuning scheme, the GW optimizer achieved better performance. Das et al. [33] investigated the use of a GW optimizer for tracking the control variables of a PID controller for speed control in a 2nd-order DC motor system. Performance analysis results indicate that GWO performs better than PSO in terms of dynamic machine performance. Other applications of GWO for PID tuning include robotics [34], speed control of DC motors [35], inverted pendulums [36], voltage regulation [37] and wind mills [38], A generic flowchart showing GWO algorithm is depicted in Fig. 4.

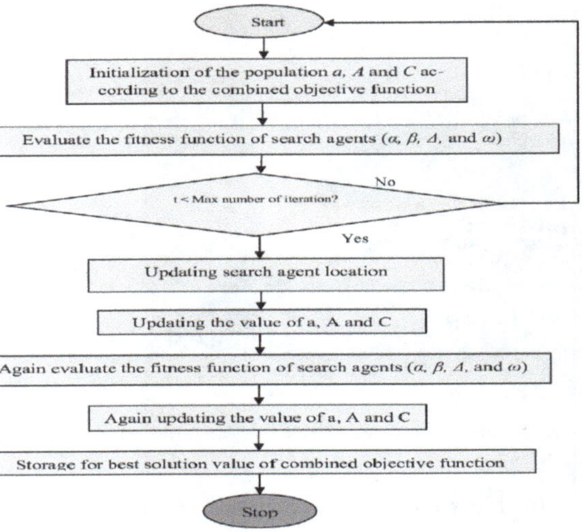

Fig. 4. Grey wolf optimization application flowchart

8 Flower Pollination Algorithm

The Flower Pollination Algorithm (FPA) is a nature-inspired, population-based methodology introduced by Xin-She Yang in 2012. The primary objective of flower pollination is to produce the most viable replicas of flora through the survival of the fittest plants within the flowering plant species. In reality, this represents an optimisation strategy among plant species. FPA is a swarm-based optimisation method, and it's remarkable attributes have garnered the attention of numerous researchers across various optimisation domains [39].

FPA possesses minimal parameters and has demonstrated robust performance across numerous optimisation problems. Furthermore, FPA is a straightforward, adaptive, versatile, and scalable optimisation method. Consequently, FPA, in comparison to other meta-heuristic algorithms, yields highly accurate solutions for real-world optimisation applications across various domains, including electrical and power devices, signal and image processing, wireless sensor networks, clustering, international function optimisation, computer gaming, as well as structural and mechanical engineering optimisation, among others. FPA is a population-centric optimisation method that commences with a set of provisional or arbitrary solutions [40]. In each generation, one of the two following operators is executed for each person in the population: the local pollination operator or the global pollination operator.

Within a local pollination operator, the decision variables of the novel solution draw upon the opposites of two randomly selected solutions from two contributors within the population. The decision variables of the novel solution are drawn towards the globally optimal solution identified in the context of a global pollination operator. The alternation of the improvement loop between local and worldwide contexts is executed by the switch operator. This process is repeated until a specified halting criterion is met [41].

The initial version of the procedural optimisation architecture of FPA has been modified or hybridised to improve its general performance for several types of annoyance landscapes. FPA is utilised for the optimal loading of turbines in power plants. Consequences of simulations for both small and large size power devices, accounting for the valve loading impact, are executed to evaluate the robustness of FPA. There are around 250,000 species of flowering plants in nature, with 80 % being flowering species. The primary purpose of a flower is reproduction via pollination [42]. The technique of flower pollination involves the transmission of pollen supported by pollinators like insects, birds, and bats (Fig. 5).

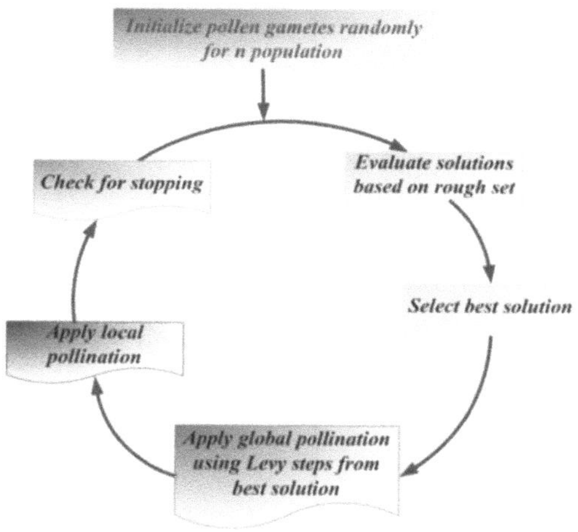

Fig. 5. Characteristic of flower pollination algorithm

There are two major processes for transferring the pollen:

- Biotic and cross pollination process
- Abiotic and self-pollination process
  Flower constancy

9 Ant Coliny Optimization

The ant colony optimisation (ACO) algorithm is an effective method inspired by insect swarm behaviour, mainly created to address combinatorial problems. ACO is a stochastic-based metaheuristic method for addressing combinatorial optimisation problems that employs artificial ants. The objective of ACO is to identify more efficient pathways between their nests and food sources. Ants disperse and communicate with fellow ants. As an ant traverses its environment, it deposits a trail of consistent pheromones that other ants may detect. Each ant navigates uniquely, and when encountering a pheromone trail, it must decide whether to follow it or not. The greater the number of ants traversing a road, the more appealing that path becomes to subsequent ants. An ant that follows a more direct route to a food source will return to the nest more swiftly.

Dorigo presented the fundamental ant colony framework, a precursor to ACO, in his doctoral dissertation in 1992 [43]. Dorigo created a robust optimisation algorithm primarily influenced by both basic and complicated foraging behaviours of ants, garnering significant attention in the 1990s and thereafter. Dorigo and other researchers generated numerous alterations and hybrids. The validation of offline pheromone updates and pheromone evaporation mostly resulted from the widespread implementation of intense Ant Colony Optimisation (ACO) in 1999 [44]. Dorigo et al. [45]. Adubi and Misra [46] provide an overview of ACO algorithms, their applications, and a comparison of several variations. Sundareswaran and Nayak (2012) employ the subsequent probability Eq. (15) for ant navigation:

$$P_j^{(k)} = \frac{\tau_j^{(k)}}{\sum_{j=1}^{n} \tau_j^{(k)}} \tag{15}$$

A high level algorithmic concept of ACO is presented in Fig. 6

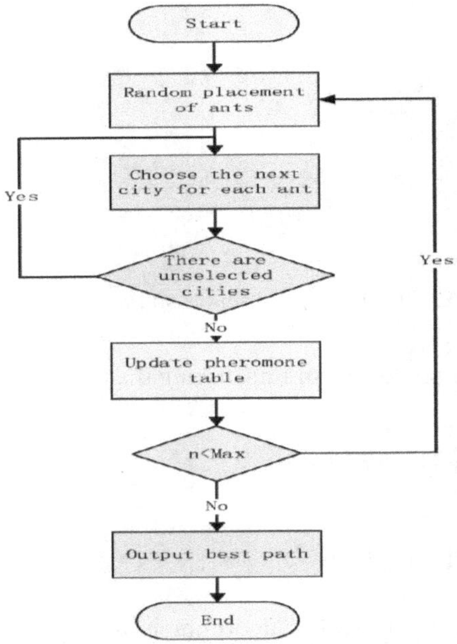

Fig. 6. Ant colony optimization search algorithm

10 Optimization Terminated

After convergence of GA, PSO, GWO, ACO, and FPA algorithms, we got the final tuned parameters as:.

The following table presents the overall system performance of the controlled system before and after optimization of the PID parameters using the GA, PSO, GWO, ACO, and FPA algorithm. The system performance is measured in terms of upward push time, settling time, overshoot, top time, and height cost. The optimized PID parameters obtained using the GA, PSO, GWO, ACO, and FPA algorithm are also shown (Tables 2, 3, 4, 5 and 6):

Table 2. Comparison of system performance and PID optimization parameters using the GA

The genetic algorithm (GA)			
Performance parameters of the system	Parameters of the GA	Optimized PID parameters	Performance parameters of the controlled system
Rise time: 0.1145 s Settling time: 1.3607 s Overshoot: 45.00% Peak time: 0.3399 s Peak value: 1.4500	lb = [0 0 0]; Bornes inférieures pour les gains PID ub = [10 10 10]; Bornes supérieures pour les gains PID	Kp: 59.71 Ki: 99.14 Kd: 45.01	Rise time: 0.1142 s Settling time: 0.9243 s Overshoot: 4.65% Peak time: 0.1636 s Peak value: 1.0465

Table 3. Comparison of system performance and PID optimization parameters using the ACO

The ant colony optimization (ACO)			
Performance parameters of the system	Parameters of the ACO	Optimized PID parameters	Performance parameters of the controlled system
Rise time: 0.1145 s Settling time: 1.3607 s Overshoot: 45.00% Peak time: 0.3399 s Peak value: 1.4500	N = 10: Number of ants ph_decay = 0.6: Pheromone decay rate Sca_fact = 2: Scaling factor itr = 10: Iterations var. = 3: Number of variables h_size = 0.1: Heuristic step size a = 0; b = 1000: Search space bounds	Kp: 44.3 Ki: 95.4 Kd: 11	Rise time: 0.0016 s Settling time: 0.3865 s Overshoot: 93.27% Peak time: 0.0047 s Peak value: 1.9327

Table 4. Comparison of system performance and PID optimization parameters using GWO

The grey wolf optimizer (GWO)			
Performance parameters of the system	Parameters of the GWO	Optimized PID parameters	Performance parameters of the controlled system
Rise time: 0.1145 s Settling time: 1.3607 s Overshoot: 45.00% Peak time: 0.3399 s Peak value: 1.4500	iter = 100: Number of iterations (generations). pop = 60: Population size (number of wolves). a_gwo = 5: Coefficient controlling exploration and exploitation (search intensity). Var = 3: Number of variables (Kp, Ki, Kd for PID). al = 0: Lower bound of the search space. au = 1000: Upper bound of the search space.	Kp: 1000 Ki: 1000 Kd: 1000	Rise time: 0.0002 s Settling time: 0.3212 s Overshoot: 99.10% Peak time: 0.0007 s Peak value: 1.9910

Table 5. Comparison of system performance and PID optimization parameters using the (PSO)

The particle swarm optimization (PSO)			
Performance parameters of the system	Parameters of the PSO	Optimized PID parameters	Performance parameters of the controlled system
Rise time: 0.1145 s Settling time: 1.3607 s Overshoot: 45.00% Peak time: 0.3399 s Peak value: 1.4500	w = 0.7; Poids d'inertie c1 = 2; Coefficient d'accélération personnel c2 = 2; Coefficient d'accélération social pop_size = 50; Taille de la population max_iter = 100; Number of iterations maximal Var = 3; Number of variables (Kp, Ki, Kd for PID). lb. = 0; Lower bound of the search space. ub = Upper bound of the search space	Kp: 1000 Ki: 1000 Kd: 420.8647	Rise time: 0.0004 s Settling time: 0.3425 s Overshoot: 98.66% Peak time: 0.0011 s Peak value:.9866

Table 6. Comparison of system performance and PID optimization parameters using (FPA)

The flower pollination algorithm (FPA)			
Performance parameters of the system	Parameters of the FPA	Optimized PID parameters	Performance parameters of the controlled system
Rise time: 0.1145 s Settling time: 1.3607 s Overshoot: 45.00% Peak time: 0.3399 s Peak value: 1.4500	N = 10: Number of flowers (population size). T = 50: Number of iterations (generations). c = 0.5: Movement control parameter (local/global search intensity). p = 0.5: Probability of global pollination (versus local pollination). var. = 3: Number of variables (Kp, Ki, Kd for PID). lb. = 0: Lower bound of the search space. ub = 100: Upper bound of the search space.	Kp: 2.18 Ki: 10.0 Kd: 0.0437	Rise time: 0.0011 s Settling time: 0.3509 s Overshoot: 95.29% Peak time: 0.0034 s Peak value: 1.9529

These tables, taken together, depict the effectiveness of each set of optimization rules in tuning the parameters of a PID and underline the upgrades in machine performance carried out through their software.

Without correction, the gadget settles slowly and overshoots notably. This way that the system takes time to stabilize and often exceeds the desired fee before settling. The GA optimization improves the device's response time by way of lowering upward push time. However, overshoot and height values stay high. This means the system reacts quicker; however, it nonetheless overshoots notably before settling. FPA responds extremely quickly, with very short rise and peak times. However, the overshoot is extremely high, indicating that the system is highly responsive but significantly exceeds the desired value before stabilizing. This could be problematic for applications that require precise control.

ACO, like FPA, provides very fast rise and peak times. However, the overshoot remains high, with the tendency to significantly exceed the setpoint before stabilizing. The system is responsive but has large overshoots, similar to FPA. GWO achieves extremely speedy upward push and top instances, in addition to very short settling instances. However, the overshoot is extraordinarily excessive, indicating that the machine responds nearly immediately however with an enormous overshoot. Despite the device's excessive responsiveness, this could have an impact on its stability and overall performance. PSO also offers extremely rapid rise and height times, in addition to quick settling instances. However, the overshoot stays high, as with GWO and ACO. The system may be very responsive, but it famous full-size overshoots.

Responsiveness: FPA, ACO, GWO, and PSO display significant enhancements in upward push time and height time, offering a whole lot faster responses than the gadget without correction.

- Overshoot: Despite stepped forward responsiveness, these optimization methods have a tendency to bring about excessive overshoot. The machine can also react very quickly but exceeds the setpoint considerably earlier than stabilizing.
- Overall Control: GA sticks out by means of presenting a better balance between reaction pace and overshoot control, supplying more strong overall performance compared to different strategies.

In precis, at the same time as optimization techniques improve the gadget's responsiveness, the venture remains to reduce the overshoot while retaining a quick and managed reaction. The studies demonstrates that optimization algorithms like GA, FPA, ACO, GWO, and PSO considerably enhance the device's responsiveness in comparison to the unoptimized PID controller. However, in addition they introduce high overshoot, that could impact machine balance. Each optimization technique gives exclusive alternate-offs among reaction time and overshoot, indicating the need for careful attention primarily based on precise software necessities. The final aim is to stability fast response with appropriate overshoot to acquire most appropriate device overall performance.

PID controller optimisation is crucial to robust and efficient control systems in numerous engineering packages. This study addresses the optimisation of PID parameters for a suspension device model utilising contemporary metaheuristic algorithms: Genetic Algorithm (GA), Flower Pollination Algorithm (FPA), Ant Colony Optimisation (ACO), Grey Wolf Optimiser (GWO), and Particle Swarm Optimisation (PSO).

The suspension system serves as a significant standard in automotive engineering, since it encompasses intricate dynamics and necessitates specialised administration to uphold vehicle stability and ride comfort.

11 Conclusions

Enhanced system performance: It can be shown in the study, by the usage of optimization algorithms to tune the PID parameters, that there are extensive upgrades in system general performance measures—such as upward thrust time, settling time, overshoot, and top time. Such enhancements are quite significant for attaining better control and balance in suspension structures.

Comparative Analysis: The have a look at gives a complete comparison of various optimization algorithms, highlighting their strengths and weaknesses inside the context of PID tuning. This comparative evaluation offers valuable insights into which algorithms perform higher beneath specific conditions, as a consequence guiding future implementations and studies.

Real-World Applicability: The suspension system model used on this look at is consultant of actual-international engineering troubles. Optimizing PID controllers for such models without delay affects the layout and performance of car suspension structures, leading to stepped forward protection, consolation, and managing of cars. Algorithm Effectiveness: The findings display how every optimization set of rules (GA, FPA, ACO, GWO, and PSO) affects the PID tuning process. Understanding the effectiveness of those algorithms aids in determining the high-quality technique for numerous sorts of manipulate problems, thereby advancing the sphere of manage gadget design.

In conclusion, this takes a look at emphasizes the fee of metaheuristic optimization in refining PID manage techniques for complicated systems together with automotive suspensions. The findings upload to the bigger area of control engineering by using offering an intensive exam of algorithmic performance and its implications for sensible packages.

Acknowledgements. The authors would like to acknowledge with sincere thanks the support of the Manufacturing Engineering Department, École Nationale Polytechnique d'Oran. Special thanks are also extended to the ECAM-EPMI of CERGY-PONTOISE CEDEX

References

1. Xiao, L.: Parameter Tuning of PID Controller for Beer Filling Machine Liquid Level Control Based on Improved Genetic Algorithm (2021). ncbi.nlm.nih.gov
2. Zhou, Q.: Research and Simulation on New Active Suspension Control System. M.Sc. Thesis. Lehigh University (2013)
3. Maurya, V.K., Bhangal, N.S.: Optimal control of vehicle active suspension system. J. Autom. Control Eng. **6**(1), 1–6 (2018)
4. Popovic, V., Duka, M., Milinkovic, I.: Investigation of suspension structures for massive vehicles using external actuator excitation. Int. J. Automot. Technol. **18**(5), 799–811 (2017). https://doi.org/10.1007/s12239-017-0082-2
5. Tee, S.P., Tan, Y.C., Chong, K.Y.: Optimization of PID controller parameters in a two-degree-of-freedom suspension system for improved ride comfort. J. Mech. Eng. Autom. **8**(2), 112–121 (2018). https://doi.org/10.1016/j.mechatronics.2018.02.002
6. Akpakpavi, M.: Modeling and control of a car suspension system using P, PI, PID GA-PID and auto-tuned PID controller in Matlab/Simulink. Journal of Multidisciplinary Engineering Science Studies (JMESS). **3**(3), 1506–1513 (2017)
7. Zhang, J., Zhao, X.: A comprehensive review of optimization techniques for PID controller tuning. J. Process Control. **88**, 61–83 (2020). https://doi.org/10.1016/j.jprocont.2020.02.009
8. Zhang, X., Fang, Y., Liu, X.: An adaptive sliding mode control strategy for energetic suspension systems. Mech. Syst. Signal Process. **72**, 55–68 (2016). https://doi.org/10.1016/j.ymssp.2015.11.022
9. Gao, Y., Huang, Z.: Adaptive control strategies for active suspension systems: A review. J. Sound Vib. **470**, 115164 (2020). https://doi.org/10.1016/j.jsv.2019.115164
10. Wang, M., Zhou, H., Lin, J.: Adaptive preload controller design and analysis for electrostatic suspension system. Int. J. Mod. Phys. A. **36**(11n12), 2140012 (2021)
11. Choudhury, S., Singh, K., Gupta, R.: (2022).
12. Li, X., Qiu, Y.: Experimental study on cooling performance of microencapsulated phase change suspension in a PEMFC. Int. J. Hydrog. Energy. **42**(50), 30004–30017 (2017). https://doi.org/10.1016/j.ijhydene.2017.08.190
13. P. Gupta and R. Sharma, 2023.

14. Zhou, Q., Wang, M., Zhou, H., Lin, J.: A new approach for improving suspension system performance by merging an innovative algorithm with a PID controller. J. Vib. Control. **19**(5), 731–742 (2013)
15. Zhao, L.L., Zeng, Z.Y., Wang, Z.Y., Ji, C.F.: PID control of vehicle active suspension based on particle swarm optimization. J. Phys. Conf. Ser. **1748**, 032028 (2021). https://doi.org/10.1088/1742-6596/1748/3/032028
16. Hurel, J., Mandow, A., Garcia-Cerezo, A.: Tuning a fuzzy controller by particle swarm optimization for an active suspension system. In: IECON 2012-38th Annual Conference on IEEE Industrial Electronics Society, Montreal, QC, Canada, pp. 2524–2529 (2012). https://doi.org/10.1109/IECON.2012.6388697
17. Castelli, M., Cattaneo, G., Manzoni, L., Vanneschi, L.: A Distance Between Populations for n-Points Crossover in Genetic Algorithms (2017)
18. Jesus, I.S., Tenreiro Machado, J.A., Boaventura Cunha, J.: Application of Genetic Algorithms to the Implementation of Fractional Electromagnetic Potentials (2006)
19. Meena, D., Devanshu, A.: Genetic algorithm tuned PID controller for process control. In: 2017 International Conference on Inventive Systems and Control (ICISC), pp. 1–6. IEEE (2017)
20. El-Deen, A., Mohammed, M., Joseph, H.: GA-based PID tuning for DC motor speed control. J. Control Eng. **45**(2), 123–135 (2015)
21. Joseph, K., Emmanuel, S.: Optimal PID controller design using genetic algorithms. Int. J. Robot. Autom. **32**(4), 215–225 (2018)
22. Mohammed, M., Patel, A., Singh, J.: GA-based PID controller design for synchronous generators. IEEE Trans. Power Syst. **29**(6), 3018–3026 (2014)
23. Patil, A., Lakhekar, A.: Genetic algorithm-based PID controller for cascade control processes. In: Proceedings of the International Conference on Control and Automation, pp. 435–440 (2017)
24. Amaral, J., Silva, F., Santos, M.: Automatic tuning of PID controller parameters using genetic algorithms. Control. Eng. Pract. **76**, 182–194 (2018)
25. Lutton, E.: Artificial Darwinism: An Overview (2005)
26. Bornholdt, S.: Genetic Algorithm Dynamics on a Rugged Landscape (1999)
27. Al-Khafaji, S.A., Saleh, A.H., Shaheed, S.M.: Optimisation of PID controllers in active suspension systems: A comparative study of the firefly algorithm and the particle swarm optimisation. Math. Model. Eng. Probl. **10**(6), 2023–2030 (2023)
28. Mirjalili, S., Mirjalili, S.M., Lewis, A.: Grey wolf optimizer. Adv. Eng. Softw. **69**, 46–61 (2014)
29. Goldbogen, J.A., Friedlaender, A.S., Calambokidis, J., McKenna, M.F., Simon, M., Nowacek, D.P.: Integrative approaches to the study of baleen whale diving behavior, feeding performance, and foraging ecology. Bioscience. **63**, 90–100 (2013)
30. Muro, C., Escobedo, R., Spector, L., Coppinger, R.: Wolf-pack (Canis Lupus) hunting strategies emerge from simple rules in computational simulations. Behav. Process. **88**, 192–197 (2011)
31. Li, Q., et al.: An enhanced grey wolf optimization based feature selection wrapped kernel extreme learning machine for medical diagnosis. Comput. Math. Methods Med. **2017**, 9512741 (2017)
32. Yadav, S., Verma, S.K., Nagar, S.K.: Optimized PID controller for magnetic levitation system. IFAC-PapersOnLine. **49**, 778–782 (2016)
33. Das, K.R., Das, D., Das, J.: Optimal tuning of PID controller using GWO algorithm for speed control in DC motor. In: 2015 International Conference on Soft Computing Techniques and Implementations (ICSCTI), pp. 108–112. IEEE (2015)
34. Sen, M.A., Kalyoncu, M.: Optimal tuning of PID controller using grey wolf optimizer algorithm for quadruped robot. Balk. J. Electr. Comput. Eng. **6**, 29–35 (2018)

35. Agarwal, J., Parmar, G., Gupta, R., Sikander, A.: Analysis of grey wolf optimizer based fractional order PID controller in speed control of DC motor. Microsyst. Technol. **24**, 4997–5006 (2018)
36. Yadav, S., Nagar, S.K., Mishra, A.: Tuning of Parameters of PID Controller Using Grey Wolf Optimizer. SSRN (2020)
37. Verma, S.K., Nagar, S.K.: Design and optimization of fractional order PIλDμ controller using grey wolf optimizer for automatic voltage regulator system. Recent Adv. Electr. Electron. Eng. **11**, 217–226 (2018)
38. Sule, A.H., Mokhtar, A.S., Jamian, J.J.B., Khidrani, A., Larik, R.M.: Optimal tuning of proportional integral controller for fixed-speed wind turbine using grey wolf optimizer. Int. J. Electr. Comput. Eng. **10**, 5251–5261 (2020)
39. Abdel-Basset, M., Shawky, L.: Flower pollination algorithm: A comprehensive review. Artif. Intell. Rev. **52**, 2533–2557 (2018)
40. Nabil, E.: A modified flower pollination algorithm for global optimization. Expert Syst. Appl. **57**, 192–203 (2016)
41. Alam, D., Yousri, D., Eteiba, M.: Flower pollination algorithm based solar PV parameter estimation. Energy Convers. Manag. **101**, 410–422 (2015)
42. Yang, X., Karamanoglu, M., He, X.: Multi-objective flower algorithm for optimization. Procedia Comput. Sci. **18**, 861–868 (2013)
43. Dorigo, M.: Optimization, Learning and Natural Algorithms. PhD thesis. Politecnicodi Milano, Italy (1992)
44. Cooper, M.A., Smeresky, B.: An overview of evolutionary algorithms toward spacecraft attitude control. In: Advances in Spacecraft Attitude Control. IntechOpen (2020). https://doi.org/10.5772/intechopen.89637
45. Dorigo, M., Birattari, M., Stutzle, T.: Ant colony optimization. IEEE Comput. Intell. Mag. **1**, 28–39 (2006)
46. Adubi, S.A., Misra, S.: A comparative study on the ant colony optimization algorithms. In: 2014 11th International Conference on Electronics, Computer and Computation. ICECCO. IEEE, pp. 1–4 (2014)

Optimization of Robotic Navigation for Safety and Efficiency in the Oil and Gas Industry Using the A* Algorithm

N. Bouam[1(✉)], A. Messai[2], and A. Dadda Khorsi[2]

[1] Faculty of Technology, Department of Electrical Engineering, Yahia Fares De Medea University, Medea, Algeria
bouam05_ab3@yahoo.fr

[2] Nuclear Research Centre of Birine, 180 Ain-Oussera, Algeria

Abstract. The integration of robotics into the hydrocarbon industry primarily aims to enhance safety, operational efficiency, and precision while minimizing risks to human personnel. Robots such as Spot, developed by Boston Dynamics, are increasingly used for routine inspections of oil and gas facilities, particularly in hard-to-reach or hazardous areas. Their functions include detecting hydrocarbon and methane leaks, reading gauges, identifying corrosion, and monitoring noise levels. This data helps prevent incidents and reduces maintenance requirements. Additionally, tasks such as handling drill pipes with millimeter-level precision contribute to safer and more efficient operations. Robots also automate manual tasks on drilling platforms—often referred to as "red zones" due to their inherent risks—thereby reducing human exposure to dangerous environments and optimizing both inspection time and costs.

The aim of this study is to develop a computer program using MATLAB to determine the optimal navigation path for a robot operating in a two-dimensional environment, with minimal traversal time. The adopted method is based on the A* algorithm, chosen for its proven advantages and high performance compared to alternative path planning algorithms. Simulation results are consistent with those found in the literature and demonstrate the algorithm's effectiveness in solving the posed navigation problem.

Keywords: A* Algorithm · Mobile Robot · Navigation · Operational Precision · Oil and Gas Industry

1 Introduction

The concept of path planning has been explored across various fields, including robotics, artificial intelligence, and control theory, leading to different definitions depending on the discipline [1]. In robotics, path planning addresses the challenge of moving a robot from one point to another. With advancements in robotics, this process has grown more complex, involving uncertainties [2], multiple robots [3–5], and dynamic environments [6].

M. Hatti (Ed.): IC-AIRES 2025, LNNS 1740, pp. 520–528, 2026.
https://doi.org/10.1007/978-3-032-12788-4_48

Path planning is a fundamental concept explored across various domains such as robotics, artificial intelligence (AI), and control theory. Each discipline provides its own definition, reflecting different priorities and methodologies [1]. In robotics, it refers to the process of determining an optimal route that allows a robot to move from a starting point to a target location while avoiding environmental obstacles. Technological advancements have made this task more complex by incorporating parameters such as environmental uncertainty [2], multi-robot coordination [3–5], and dynamic constraints [6].

In AI, path planning is often viewed as the transformation of an initial state into a desired goal state via a sequence of logical actions. This involves tools such as Markov Decision Processes [7], handling incomplete information [1], machine learning methods [8, 11], and game theory [9]. Control theory treats path planning from the perspective of stability, feedback, and optimality [10–13].

Mobile robotics in hostile industrial environments—particularly in the oil industry—is gaining momentum. It reduces human exposure to dangerous zones while ensuring continuous and automated monitoring of installations. Consequently, path planning becomes a cornerstone of robotic autonomy. Various algorithms have been developed for this purpose, including probabilistic methods (PRM), potential fields, fuzzy logic, neural networks, and heuristic search algorithms like Dijkstra and A*.

Among these, the A* (A-star) algorithm stands out due to its balance between performance and accuracy, generating optimal paths with reduced computation time. Its efficiency in structured or semi-structured environments has been demonstrated in numerous studies. In the oil sector, several works have explored the integration of mobile robotic systems for inspection, maintenance, or surveillance tasks. For example, [Y. Li et al., 2022] [14] presented robots equipped with multi-sensor systems for automated pipeline monitoring. Similarly, [Zhou et al., 2016] [15] proposed a modified version of A* for high-risk environments, considering high-pressure or high-temperature zones.

However, many of these systems still rely on rigid, pre-programmed trajectories requiring human supervision when environmental changes occur. Hence the need to design more flexible, adaptive, and real-time responsive navigation systems based on robust algorithms like A*.

In oil and gas applications, robotic path planning offers strategic advantages for safety, reduced human risk, and improved efficiency in hard-to-reach areas (ATEX zones, underground pipelines, confined structures). It guides autonomous mobile robots through complex industrial installations while avoiding obstacles, leak sources, and hazardous zones.

This study focuses on implementing the A* algorithm in a simulated environment. Known for its simplicity, efficiency, and modularity, A* is well suited for oil industry demands. Using MATLAB, the objective is to develop a program that automatically generates an optimal trajectory within a 2D map representing a petroleum installation.

2 A* Algorithm for Autonomous Path Planning

2.1 Algorithm Procedure

According to international standards in graph exploration (particularly in the context of path planning) a common heuristic approach involves using the distance to the target node. The heuristic function *h(n)* estimates the minimum cost required to reach the goal from a given node *n*. By calculating the actual cost *g(n)* from the start node to the current node, the total estimated cost *f(n)* of a path can be determined. The A* algorithm's search strategy is based on selecting the next node with the lowest estimated total cost *f(n)*.

$$f(n) = g(n) + h(n) \tag{1}$$

or

$$\mathrm{h(n) = distance(n, nfinal)} \tag{2}$$

and

$$g(n) = \sum_{départ}^{n} cout\left(noeud_i, noeud_j\right) \tag{3}$$

The A* algorithm is optimal under specific conditions, which will be presented in this section. In the context of graph exploration, as shown in Fig. 1, the condition for A* to be optimal is that the heuristic must be consistent (*Russell, 1995*) [12].

A heuristic *h(n)* is considered consistent if, for any node n_1 and its neighbor n_2, the estimated cost $h(n_1)$ of reaching the target node from n_1 is not greater than the sum of the cost of moving from n_1 to n_2 and the estimated cost $h(n_2)$ of reaching the target from n_2.

$$h(n_1) \leq coût(n_1, n_2) + h(n_2) \tag{4}$$

This condition can be met by using, for instance, the Cartesian distance between node n_1 and the target node as a heuristic:

$$h(n) = dist\left(n, n_{final}\right) = \sqrt{\left(x - x_{final}\right)^2 + \left(y - y_{final}\right)^2} \tag{5}$$

In this case, the triangle inequality requires that:

$$dist\left(n_1, n_{final}\right) \leq dist(n_1, n_2) + dist\left(n_2, n_{final}\right) \tag{6}$$

and:

$$\cos t(n_1, n_2) \geq dist(n_1, n_2) \tag{7}$$

It also implies that when we first reach the final node, we have already reached our goal with the optimal path.

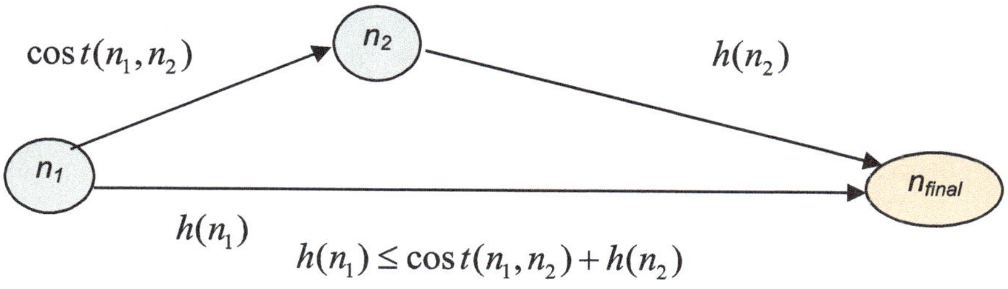

Fig. 1. Exploration consistency A*.

2.2 Description of the Algorithm A* [11–13]

The A* algorithm is a widely used pathfinding and graph traversal algorithm, particularly effective for finding the shortest path from a starting node to a target node. It combines the strengths of Dijkstra's algorithm and a heuristic approach to improve efficiency. The detail for the description of the A* algorithm is presented in Fig. 2.

1. Initialization

- **Open Set:** This set contains nodes that need to be evaluated. Initially, it includes only the starting node.
- **Closed Set:** This set contains nodes that have already been evaluated.
- ***g(n)*:** This function represents the cost of the path from the starting node to the current node n.
- ***h(n)*:** This heuristic function estimates the cost from the current node n to the target node. It guides the search.
- ***f(n)*:** This function represents the estimated total cost of the cheapest solution through n. It is defined as $f(n) = g(n) + h(n)$.

2. Algorithm Steps

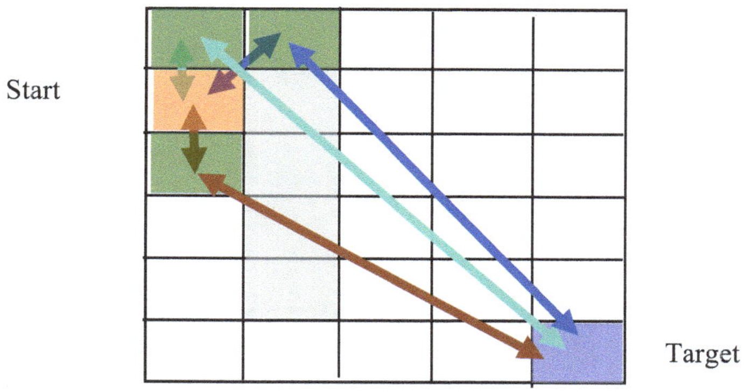

Fig. 2. Illustration of the A* algorithm iteration.

By following this algorithm, the A* search strategy effectively balances between exploring the most promising nodes and minimizing the total path cost, making it a powerful tool for trajectory planning and other pathfinding tasks.

3 Flowchart of the Simulation

The flowchart is presented at the Fig. 3.

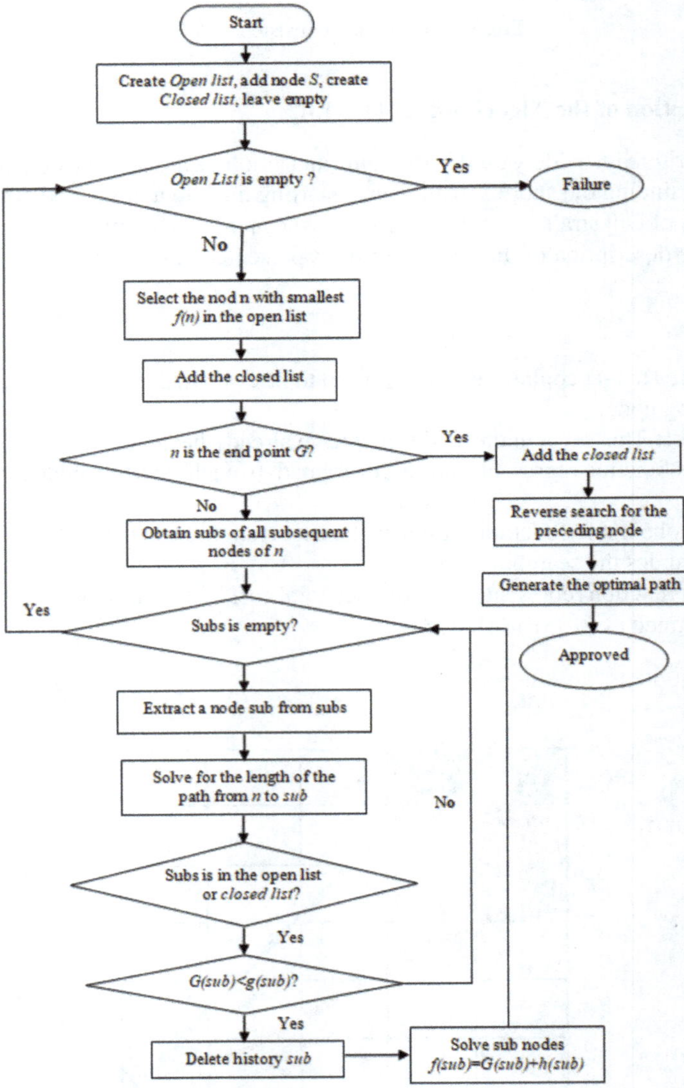

Fig. 3. Flowchart of the A* algorithm.

4 Simulation Results

In this section, we apply the A* planning algorithm to navigate a trajectory through various environments, presenting the optimal path to ensure smooth movement. The problem data include the A* algorithm, the starting and goal positions, and the locations of obstacles, with the planned path represented by a set of points as the output. The simulation was conducted using Matlab R2014a (8.3.0.532) [16].

4.1 Experiment 1: Environment Without Obstacles

Figure 4 illustrates the navigation from the starting point to the endpoint, showing the path as a straight line. This confirms that the shortest route is a direct line connecting the start and finish points, representing the most optimized path.

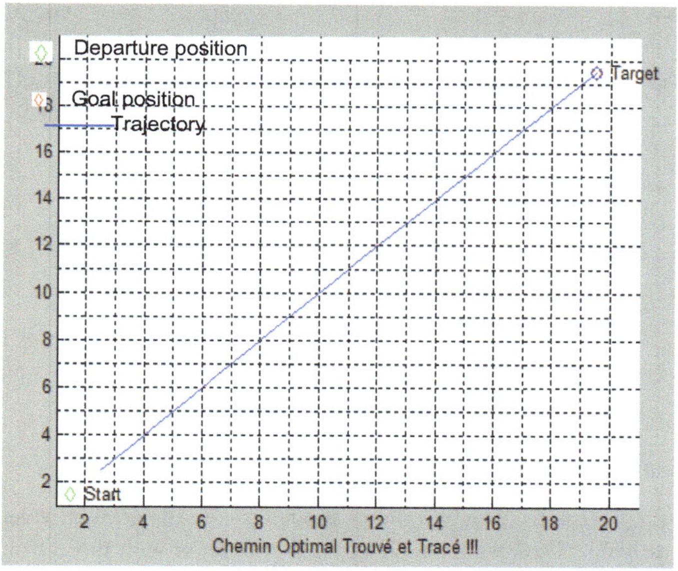

Fig. 4. Trajectory planned by A* in an environment without obstacles. Execution time: 0.025 s. Path length: 26.1725 um

4.2 Experience 2: Environment Contains Walls as Obstacles (see Fig. 5)

In this experiment, we used a 2D environment with several obstacles represented as walls. The results, shown in Fig. 5, are consistent with those from references [15, 16], confirming the effective performance of our program.

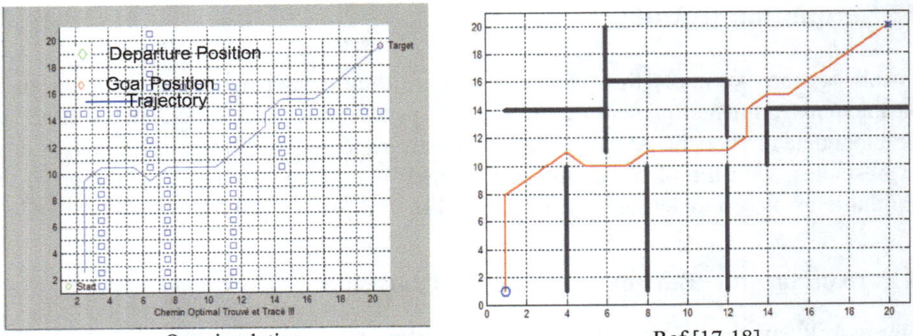

Our simulation Ref [17-18]

Fig. 5. Path planned by A* in an environment containing walls as obstacles.

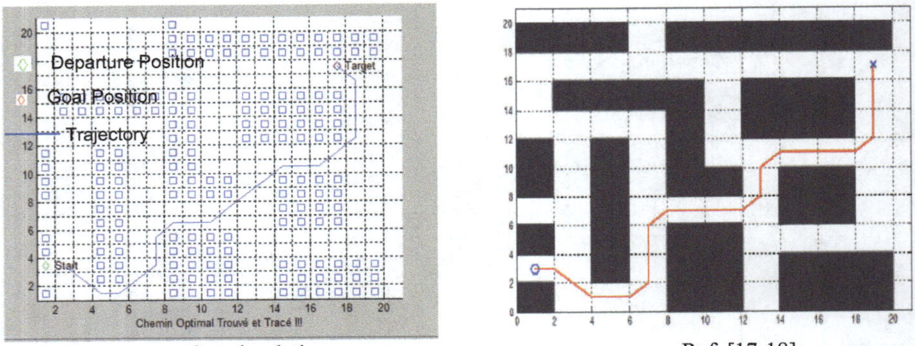

Our simulation Ref. [17-18]

Fig. 6. Path planned by A* in a maze.

4.3 Experience 3: Moving Robot in a Labyrinth

During this test, we used a 2D environment modeled as a "maze" with several obstacles represented as walls. The results, depicted in Fig. 6, align with those from references [15, 16], confirming the correct functionality of our program.

5 Conclusion

Robotics can be defined as a discipline that integrates mechanical, computational, and hybrid approaches aimed at designing systems capable of replicating or replacing certain human motor, sensory, or intellectual functions. Among the most dynamic areas within this discipline, cooperative mobile robotics holds a strategic position, particularly in industrial applications with high safety requirements, such as those in the oil and gas sector.

The present work aligns with this perspective and focuses on the application of the A algorithm for path planning and trajectory optimization of a mobile robot operating in a two-dimensional environment with obstacles. The A* algorithm was studied, adapted, and implemented using Matlab R2014a, enabling the simulation of navigation scenarios in both constrained and unconstrained environments.

Although the A* algorithm has been extensively studied, it is well established that its performance depends largely on computational speed and the robustness of the generated paths, especially when dealing with real-world robotic applications. In this context, navigation data can be transmitted to an embedded system, such as an Arduino microcontroller, ensuring autonomous execution in the field.

The results obtained validate the initial objectives of the project, showing notable consistency with data reported in the scientific literature. They highlight the A* algorithm's ability to generate optimal trajectories while taking environmental constraints into account—an aspect that is particularly crucial in oil and gas facilities exposed to multiple risks.

To further develop this study and strengthen its industrial applicability, several improvement perspectives can be considered:

- Implement the program on a real robot controlled by an Arduino board, operating in an environment with moving obstacles, simulating the dynamics of an active oil site.
- Integrate the robot's real physical constraints, such as dimensions, kinematics, and non-holonomic properties, to ensure more realistic planning applicable to complex terrains in the oil industry.

In summary, this study demonstrates that the A* algorithm is a powerful tool for enhancing autonomous navigation, with concrete applications in industrial safety, robotic inspection, and the reduction of human exposure in critical environments such as refineries, offshore platforms, and pipeline networks.

References

1. LaValle, S.M.: Planning algorithms. Procedia Eng. **96**, 59–69 (2014)
2. Van Den Berg, J., Abbeel, P., Goldberg, K.: LQG-MP: Optimized path planning for robots with motion uncertainty and imperfect state information. Int. J. Robot. Res. **30**(7), 895–913 (2011)
3. Liu, S., Sun, D., Zhu, C.: Coordinated motion planning for multiple mobile robots along designed paths with formation requirement. Mechatron. IEEE/ASME Trans. on. **16**(6), 1021–1031 (2011)
4. Svetlík, J., Sukop, M.: Using of multiagent systems in robosoccer. Acta Mech. Slovaca. **11**(2A), 157–160 (2007)
5. Hajduk, M., Sukop, M.: Multiagents system with dynamic box change for MiroSot. Prog. Robot.: Commun. Comput. Inf. Sci. **44**, 287–292 (2009)
6. Plaku, E., Kavraki, E.E., Vardi, M.Y.: Motion planning with dynamics by a synergistic combination of layers of planning. Robot. IEEE Trans. **26**(3), 469–482 (2010)
7. Al-Sabban, W.H., Gonzalez, L.F., Smith, R.N., Wyeth, G.F.: Wind-energy based path planning for electric unmanned aerial vehicles using markov decision processes. In: Proceedings of the IEEE/RSJ International Conference on Intelligent Robots and Systems. IEEE (2012)
8. Konar, A., Goswami Chakraborty, I., Singh, S.J., Jain, L.C., Nagar, A.K.: A Deterministic Improved Q-Learning for Path Planning of a Mobile Robot. IEEE, New York (2013)

9. Zhu, M., Otte, M., Chaudhari, P., Frazzoli, E., 2014. "Game Theoretic Controller Synthesis for Multi-robot Motion Planning Part I: Trajectory based Algorithms". https://arxiv.org/abs/1402.2708.
10. Chen, C.Y., Shih, B.Y., Shih, C.H., Wang, L.H.: Enhancing robust and stability control of a humanoid biped robot: system identification approach. J. Vib. Control. **19**(8), 1199–1207 (2013)
11. František Ducho, Ė., et al.: Path planning with modified A star algorithm for a mobile robot. Sci. Procedia Eng. **96**, 59–69 (2014)
12. Pearl, J.: Heuristics: Intelligent Search Strategies for Computer Problem Solving. Addison-Wesley Publishing Company, Massachusetts (1984)
13. Priyanka Sudhakara and Velappa Ganapathy: Path planning of a mobile robot using amended a-star algorithm. Int. Sci. Press J C T A. **9**(37), 489–502 (2016)
14. Li, Y., et al.: Mobile robot path planning algorithm based on improved A* algorithm. IEEE Acces. **10**, 57736–57747 (2022)
15. Zhou, X., Chen, Y., Li, Q.: An improved A* algorithm for robot path planning in dynamic and hazardous environments. Procedia Comput. Sci. **96**, 775–784 (2016). https://doi.org/10.1016/j.procs.2016.08.248
16. Logiciel Matlab: R2014a (8.3.0.532). Natick, Massachusetts, USA, The MathWorks, Inc (2014)
17. Tazebinte, H., Lameche, K.: Planification et optimisation de trajectoire d'un robot mobile par les algorithmes évolutionnaires. In: Mémoire de master en Automatique. Ecole Nationale Polytechnique, El-harrach, Alger (2012)
18. Garcia, D.: Planification de trajectoire dans un atlas de cartes. In: Mémoire de maîtrise des sciences appliquées (automatique et systèmes). Ecole Polytechnique de Montréal, Canada (2008)

Author Index

M. Hatti (Ed.): IC-AIRES 2025, LNNS 1740, pp. 529–531, 2026.
https://doi.org/10.1007/978-3-032-12788-4

Kolibri 360 GmbH
Ferdinand-Jühlke-Straße 7
99095 Erfurt, Deutschland
produktsicherheit@kolibri360.de